Fish Ecophysiology

CHAPMAN & HALL FISH AND FISHERIES SERIES

Amongst the fishes, a remarkably wide range of fascinating biological adaptations to diverse habitats has evolved. Moreover, fisheries are of considerable importance in providing human food and economic benefits. Rational exploitation and management of our global stocks of fishes must rely upon a detailed and precise insight of the interaction of fish biology with human activities.

The *Chapman & Hall Fish and Fisheries Series* aims to present authoritative and timely reviews which focus on important and specific aspects of the biology, ecology, taxonomy, physiology, behaviour, management and conservation of fish and fisheries. Each volume will cover a wide but unified field with themes in both pure and applied fish biology. Although volumes will outline and put in perspective current research frontiers, the intention is to provide a synthesis accessible and useful to both experts and non-specialists alike. Consequently, most volumes will be of interest to a broad spectrum of research workers in biology, zoology, ecology and physiology, with an additional aim of the books encompassing themes accessible to non-specialist readers, ranging from undergraduates and postgraduates to those with an interest in industrial and commercial aspects of fish and fisheries.

Applied topics will embrace synopses of fishery issues which will appeal to a wide audience of fishery scientists, aquaculturists, economists, geographers and managers in the fishing industry. The series will also contain practical guides to fishery and analysis methods and global reviews of particular types of fisheries.

Books already published and forthcoming are listed below. The Publisher and Series Editor would be glad to discuss ideas for new volumes in the series...

Available titles

1. **Ecology of Teleost Fishes**
   Robert J. Wootton

2. **Cichlid Fishes**
   Behaviour, ecology and evolution
   Edited by Miles A. Keenleyside

3. **Cyprinid Fishes**
   Systematics, biology and exploitation
   Edited by Ian J. Winfield and Joseph S. Nelson

4. **Early Life History of Fish**
   An energetics approach
   Ewa Kamler

5. **Fisheries Acoustics**
   David N. MacLennan and E. John Simmonds

6. **Fish Chemoreception**
   Edited by Toshiaki J. Hara

7. **Behaviour of Teleost Fishes**
   Second edition
   Edited by Tony J. Pitcher

8. **Genetics and Fish Breeding**
   Colin R. Purdom

9. **Fish Ecophysiology**
   J. Cliff Rankin and Frank B. Jensen

Forthcoming titles

Fish Swimming
J. Videler

Sea Bass
G. Pickett and M. Pawson

Fisheries Ecology
Second edition
Edited by T.J. Pitcher and P. Hart

Hake
Fisheries, products and markets,
J. Alheit and T.J. Pitcher

Impact of Species Change in the African Lakes
Edited by T.J. Pitcher

On the Dynamics of Exploited Fish Populations
R. Beverton and S. Holt
(Facsimile reprint)

Fish Ecophysiology

Edited by

J. Cliff Rankin and Frank B. Jensen

Institute of Biology
Odense University
Denmark

CHAPMAN & HALL

London · Glasgow · New York · Tokyo · Melbourne · Madras

Published by Chapman & Hall 2–6 Boundary Row, London SE1 8HN

Chapman & Hall, 2-6 Boundary Row, London SE1 8HN, UK

Blackie Academic & Professional, Wester Cleddens Road, Bishopbriggs, Glasgow G64 2NZ, UK

Chapman & Hall Inc., 29 West 35th Street, New York NY10001, USA

Chapman & Hall Japan, Thomson Publishing Japan, Hirakawacho Nemoto Building, 6F, 1-7-11 Hirakawa-cho, Chiyoda-ku, Tokyo 102, Japan

Chapman & Hall Australia, Thomas Nelson Australia, 102 Dodds Street, South Melbourne, Victoria 3205, Australia

Chapman & Hall India, R. Seshadri, 32 Second Main Road, CIT East, Madras 600 035, India

First edition 1993

Typeset in 10/12 pt Photina by ROM Data Corp. Ltd., Falmouth, Cornwall

Printed in Great Britain by St Edmundsbury Press, Bury St Edmunds, Suffolk

ISBN 0 412 45920 5

A catalogue record for this book is available from the British Library

Library of Congress Cataloging-in-Publication data available

∞ Printed on permanent acid-free text paper, manufactured in accordance with the proposed ANSI/NISO Z 39.48-199X and ANSI Z 39.48-1984

Contents

Contributors

Gilles Boeuf
IFREMER Centre de Brest, Station Ressources Vivantes, BP 70, 29280 Plouzane, France

C.R. Bridges
Institut für Zoologie, Lehrstuhl für Tierphysiologie, Heinrich-Heine Universität, Düsseldorf, Germany

J. Anne Brown
Department of Biological Sciences, University of Exeter, Exeter EX4 4PS, United Kingdom

Sylvie Dufour
Laboratoire de Physiologie, Museum nationale d'Histoire naturelle, Paris, France

A. Foster
Rowett Research Institute, Bucksburn, Aberdeen AB2 9SB, Scotland

Regina Fritsche
Department of Zoophysiology, University of Göteborg, Box 250 59, S-400 31 Göteborg, Sweden

Norbert Heisler
Department of Physiology, Max-Planck-Institut für experimentelle Medizin, Hermann Rein Str. 3, D-3400 Göttingen, Germany

D.F. Houlihan
Department of Zoology, University of Aberdeen, Aberdeen AB9 2TN, Scotland

Frank B. Jensen
Institute of Biology, Odense University, DK-5230 Odense M, Denmark

Malcolm Jobling
NFH/University of Tromsø, 9000 Tromsø, Norway

Ian A. Johnston
Gatty Marine Laboratory, Division of Environmental and Evolutionary Biology, School of Biological and Medical Sciences, University of St Andrews, St Andrews, Fife KY16 8LB, Scotland

Lis Olesen Larsen
August Krogh Institute, University of Copenhagen, Denmark

Pierre Laurent
CNRS, Laboratoire de Morphologie Fonctionnelle et Ultrastructurale des Adaptations, 23 Rue du Loess, B.P. 20 CR, F-67037 Strasbourg Cedex, France

Hong Lin
Department of Zoology, University of British Columbia, Vancouver, B.C., Canada

E. M. Mathers
Department of Zoology, University of Aberdeen, Aberdeen AB9 2TN, Scotland

D.G. McDonald
Department of Biology, McMaster University, Hamilton, Ontario, Canada

J.C. Montgomery
Department of Zoology, University of Auckland, Auckland, New Zealand

Mikko Nikinmaa
Department of Zoology, Helsinki University, Arkadiankatu 7, SF-00100 Helsinki, Finland

Stefan Nilsson
Department of Zoophysiology, University of Göteborg, Box 250 59, S-400 31 Göteborg, Sweden

Steve F. Perry
Department of Biology, University of Ottawa, 30 George Glinski, Ottawa, Ontario, Canada K1N 6N5

David J. Randall
Department of Zoology, University of British Columbia, Vancouver, B.C., Canada

J. Cliff Rankin
Odense Universitet, Biologisk Institut, Campusvej 55, DK-5230 Odense M, Denmark

Yoshio Takei
Laboratory of Physiology, Ocean Research Institute, University of Tokyo, Nakano, Tokyo 164, Japan

Roy E. Weber
Zoophysiology Laboratory, Aarhus University, DK-8000 Aarhus C, Denmark

R.M.G. Wells
Department of Zoology, University of Auckland, Auckland, New Zealand

C.M. Wood
Department of Biology, McMaster University, Hamilton, Ontario, Canada

Series foreword

Among the fishes, a remarkably wide range of biological adaptations to diverse habitats has evolved. As well as living in the conventional habitats of lakes, ponds, rivers, rock pools and the open sea, fish have solved the problems of life in deserts, in the deep sea, in the cold antarctic, and in warm waters of high alkalinity or of low oxygen. Along with these adaptations, we find the most impressive specialisations of morphology, physiology and behaviour. For example we can marvel at the high-speed swimming of the marlins, sailfish and warm-blooded tunas, air-breathing in catfish and lungfish, parental care in the mouth-brooding cichlids and viviparity in many sharks and toothcarps.

Moreover, fish are of considerable importance to the survival of the human species in the form of nutritious, delicious and diverse food. Rational exploitation and management of our global stocks of fishes must rely upon a detailed and precise insight of their biology.

The *Chapman & Hall Fish and Fisheries* series aims to present timely volumes reviewing important aspects of fish biology. Most volumes will be of interest to research workers in biology, zoology, ecology and physiology but an additional aim is for the books to be accessible to a wide spectrum of non-specialist readers ranging from undergraduates and postgraduates to those with an interest in industrial and commercial aspects of fish and fisheries.

Fish Ecophysiology comprises the 9th volume in the *Chapman & Hall Fish and Fisheries Series*. The editors, Cliff Rankin and Frank B. Jensen, succeeded in assembling an impressive international group of 25 authors following a recent symposium held in Denmark. This 15-chapter volume of synthesis and review is the result of their efforts.

Whole living animals must live, survive and reproduce in balance with the natural environment, so physiological machinery is not only essential for survival but has evolved to intimately reflect the challenges of ecology. Unfortunately, despite the deep ecological awareness of such founding fathers as Claude Bernard, much physiological research has traditionally been reported in an introspective way that ignores the ecological context. The interdisciplinary approach of this book provides a refreshing perspective of a field where major conceptual advances will likely be made only by studies which cut across the boundaries of traditional disciplines. Furthermore, ecophysiological insight is critically important at the present time if we are to understand and mitigate ever-increasing man-made pollution in the delicate web of ocean life which has evolved on our planet.

In *Fish Ecophysiology*, two chapters tackle the major problem of how the partitioning of metabolic energy responds to changes in the environment; we need to understand the biochemical control mechanisms for the trade-offs among energy allocated for swimming, digestion, growth and reproduction. A core of seven chapters examine environmental influences on oxygen transport, respiratory control, gill structure and function, acid base regulation, temperature change and muscle function, and regulation of water and ion balance. Two chapters examine physiological determinants of the intriguing pre-programmed life history strategies of migratory lampreys, eels and salmon. Two applied chapters address endocrine responses to pollution and how gills and the respiratory system respond to metals. The book ends with two integrated ecophysiolological case studies of antarctic and inter-tidal fishes, both groups that are challenged daily by environmental extremes.

Ecophysiology is a discipline endowed with hybrid vigour, sparkling with ideas and problems which cry out to be carried back to both its parent disciplines for detailed scrutiny. This book in the *Fish and Fisheries Series* will I trust make a major contribution to this process and serve as a valuable reference source for some years to come.

Professor Tony J. Pitcher
Editor, *Chapman & Hall Fish and Fisheries Series*
Director, Fisheries Centre, University of British Columbia, Vancouver, Canada

Preface
Fish ecophysiology: The comparative physiologist's viewpoint

The vast range of habitats inhabited by fishes varies immensely with respect to biotic and abiotic factors. Successful maintenance of populations in challenging environments requires responsive adjustments in physiology. *Fish Ecophysiology* describes how the physiology of fishes is affected by and regulated in response to environmental changes. It highlights acute problems imposed by environmental factors as well as physiological mechanisms involved in adaptation/acclimation to various environments. Such information is important to the understanding of the niche requirements of fishes, i.e. the limits of environmental factors within which animals can function and populations be sustained.

Ecophysiology forms the interface between ecology and physiology, overlapping also with other disciplines such as behaviour and morphology. Accordingly, ecophysiological problems can be approached from many different perspectives (Jørgensen, 1983). The book does not presume to provide a complete synthesis, but by presenting various studies on interactions between fish and their environments, it is intended to contribute a better understanding of ecophysiological problems and how they are studied by different disciplines. Physiologists have always been interested in the exchanges which occur between organisms and the outside world. Exchanges of respiratory gases, food, salt, water, heat, sensory information etc. are basic aspects of animal physiology, and study of these processes has been one of the major preoccupations of physiology. This has been especially true in Denmark, as exemplified by the pioneering work of August Krogh, and when we decided to hold a fish physiology conference in Odense, where the University has a chair of ecophysiology, 'Fish Ecophysiology' seemed a very appropriate theme. Additionally, we felt that the presentation and exchange of ideas across the many disciplines in which fish ecophysiologists work could be a rewarding event.

This book is not a proceedings from the conference held in Odense, 14–17 August 1991, but following the practice of previous International Symposia on Fish Physiology, we have asked some of the speakers to write chapters which will educate and inform those wishing to gain an insight into fish physiology as well as provide new research information for those who are already familiar with the field. In the interests of readability we have asked authors to limit the numbers of references cited. We hope that the book will be of interest to both physiologists and ecologists.

An important area of ecology is the study of energy flows through ecosystems. Physiologists studying metabolism are well aware of the many factors which can affect oxygen consumption, the most convenient measure of energy expenditure, and of the errors which may arise in applying data from laboratory experiments to the field situation. Chapter 1 on Fish Bioenergetics outlines the relationships between feeding and metabolism. Reliable methods of measuring growth rates in wild populations would obviously be extremely useful; this is addressed in Chapter 2. Study of the partitioning of resources between somatic growth and reproduction is essential to the understanding of life history strategies. In many fish the feeding/growth phase is temporally separated from maturation/reproduction, and study of extreme examples of this, such as in the eels and lampreys described in Chapter 3, may make understanding of resource partitioning easier.

Chapter 3 also illustrates the extent to which the endocrine system can bring about major changes not only in physiology but also in morphology and behaviour, resulting in a complete switch in mode of life. One of the best examples of this is provided by the story of salmonid smoltification, told in Chapter 4. Smoltification is considered as a 'pre-adaptation' to marine life which occurs at a certain stage in the development of juvenile fish. All the changes in physiology, biochemistry, morphology and behaviour can be brought about by injection of growth hormone. Some interdisciplinary confusion may arise from the differing usage of the word 'adaptation'. Evolutionary biologists see adaptations as properties of organisms arising gradually by natural selection, which fit them for life in particular circumstances. Physiologists study 'adaptations' defined in this restricted way, but also use the word in its ordinary English sense (defined by one dictionary as "the ability of an organism to cope successfully with its environment") on a shorter time scale and are familiar with, for example, adaptation of nerve cells within seconds to repeated stimulation or adaptation to seawater by migratory fish within days of entering it. The latter could be called acclimatization but, whilst this is a process which can lead to adaptation (in the physiological sense), in the instance described above genetically pre-programmed hormone release can ensure that salmonids are adapted to life in seawater without any acclimatization to it. Unfortunately the term in general use amongst physiologists to describe this, preadaption, has a specific and quite different meaning to evolutionary biologists. Perhaps new terminology is required.

Fish endocrinology provides examples of the enormous latent phenotypic adaptability of animals which can be triggered by hormonal stimulation. Until recently much research involved injecting mammalian hormones or using radioimmunoassays developed for higher vertebrates. In recent years much progress has been made in characterizing fish hormones and developing homologous radioimmunoassays, and the subject has progressed to the point where a peptide hormone produced by the ventricle of the heart has first been identified and sequenced in fish (Chapter 5).

The respiratory physiology of fishes is affected by numerous physical–chemical factors in the environment, including O_2 and CO_2 tensions, temperature, salinity and toxicant levels. Chapter 6 describes how the flow of oxygen from environment to cells can be perturbed at different steps and levels of organization in the gas transport system, and how various molecular, cellular and organismic mechanisms act in concert to compensate for these perturbances. One prerequisite for proper behavioural or physiological responses to be elicited in response to hypoxia is that fish are able to acquire and process information about O_2 changes. Sensing of oxygen changes and the transfer and integration of information related to ventilatory and cardiovascular control are considered in Chapter 7.

The acid–base status of fishes is strongly dependent on environmental factors, and mechanisms are needed to correct acid–base disturbances in extra- and intracellular body compartments in order not to impair metabolic processes. The characteristics and capacity of acid–base regulatory mechanisms are the subject of Chapter 8. Focus is on transepithelial transfer of acid–base equivalents and its dependence on environmental ion composition. The gill is a multifunctional organ involved in gas transfer, acid–base balance and ionic/osmotic regulation. Chapter 9 addresses the relationships between gill morphology and function, highlighting how gill morphology and ultrastructure depends upon environmental factors and how changes in morphology contribute to acid–base regulation.

Water pH is an important factor affecting both the physiology and the ecology of fishes. This has become increasingly recognized by the extensive environmental acidification of soft freshwater habitats in recent decades, which has resulted in extinction of local fish populations, altered species diversity etc. The mechanism of acid excretion across fish gills and its dependence on water pH is the subject of Chapter 10, whereas Chapter 11 describes the endocrine responses to environmental acidification and to other environmental stressors such as aluminum in acid water.

Wash out of aluminum from soil by acid rain represents just one example where aquatic ecosystems have been contaminated with metals. Anthropogenic activity has meant that other metals can also reach toxic levels in certain habitats. Aluminium and other metals exert their prime toxic action on the gills of fish. These effects are described in Chapter 12, and the question to what extent fish can acclimate to metals is discussed.

Temperature has a profound effect on all physiological processes, and fish are constrained by the temperature of the ambient water (apart from a few very active species that can keep their swimming muscles warmer). There may, however, be considerable scope for phenotypic adaptation to different water temperatures (Chapter 13). In the Southern Ocean, temperatures are uniformly low, and Antarctic fish are ideally suited for the study of how adaptations to an extreme environment have evolved during the radiation of a closely related group of animals (Chapter 14). Of course, not all special features found will be adaptations. Antarctic icefish have lost the respiratory pigments found in all other vertebrates, and it has been tempting to see this as an adaptation to life at very low temperatures by having reduced blood viscosity as a result of the absence of red blood cells. However, the reduced viscosity is much more than counterbalanced by the necessity to pump very much more blood to carry the same amount of oxygen in physical solution. More likely they survived losing expression of the haemoglobin gene because the high oxygen solubility and their low oxygen consumption meant that an adequate (but not perfect) performance could be maintained by developing the other features of their circulatory system described in the chapter.

The final chapter presents information on the ecophysiology of fish in a particularly extreme habitat, the intertidal, in which they are challenged by large diurnal variations in many of the environmental variables described in the other chapters, such as temperature, salinity, and oxygen and carbon dioxide tensions.

We are most grateful to the following for making it possible to bring authors from all over the world to Odense for the Symposium: The Danish Natural Science Research Council, The Munke Mølles Foundation, The Natural Science Faculty of Odense University and The Carlsberg Foundation.

J. Cliff Rankin
Frank B. Jensen

REFERENCE

Jørgensen, C.B. (1983) Ecological physiology: backgrounds and perspectives. *Comp. Biochem. Physiol.*, **75A**, 5–7.

Chapter one

Bioenergetics: feed intake and energy partitioning

Malcolm Jobling

1.1 INTRODUCTION

The foundation for the study of fish bioenergetics and growth was laid during the late 1940s and early 1950s (Winberg, 1956; Brown, 1957), and during the 1960s research into ecological, and fish, energetics was supported through the International Biological Programme (Gerking, 1967; Grodzinski *et al.*, 1975; Bagenal, 1978). The past three decades have seen a continued increase in research efforts devoted to the study of fish growth and bioenergetics. Interest has been generated both from fisheries management, and, not least, by the recent upsurge in the aquaculture industry. Several multiauthor books and reviews have appeared since the mid 1970s, and these give a good background into both the general principles of bioenergetics, and the methodology used (Gerking, 1978; Hoar *et al.*, 1979; Tytler and Calow, 1985; Weatherley and Gill, 1987; Schreck and Moyle, 1990; Wootton, 1990).

The basic principle of bioenergetics is relatively simple to grasp, and can be stated as follows: all energy acquired through the ingestion of food is ultimately lost as wastes in faeces or by excretion, used in metabolic processes or deposited as new body tissue (growth or energy gain). Bioenergetics is concerned with the study of changes in energy intake, energy transformations and losses in relation to time. Thus, bioenergetics not only provides a framework for the study or relationships between feeding rates and growth rates of fish subjected to different environmental conditions, but can also provide some insights into the root causes of these relationships based upon the study of the partitioning of energetic resources within the organism. The purposes of this chapter are to present a brief summary of the historical development of fish energetics, and to place the results of recent experimental work into this developmental context.

Fish Ecophysiology. Edited by J. Cliff Rankin and Frank B. Jensen. Published in 1993 by Chapman & Hall, London. ISBN 0 412 45920 5.

1.2 BASIC PRINCIPLES

The study of fish energetics involves the partitioning of ingested energy into the major physiological components of the energy budget equation. In its simplest form the equation can be represented as:

$$E(\text{In}) = E(\text{Out}) + E(\text{P}) \qquad (1.1)$$

where E(In) is the energy ingested as food, E(Out) represents energy losses and E(P) is energy retained as production or growth. The equation is usually expanded to the more general form:

$$R = F + U + M + P \qquad (1.2)$$

where R is ingested energy, F and U represent the energy losses in faeces and excretory products, respectively, M is the energy lost in metabolism and P is production (growth). Metabolism (M) can be subdivided to account for the energy losses representing different physiological processes: maintaining basic bodily functions; activity; and the digestion, absorption and processing of food. It is also usual to differentiate between energy retained as somatic (body) growth and energy channelled into the production of gametes (reproductive growth). Experimental studies are aimed at quantifying the different components of the energy budget equation to present as complete a picture as possible of the physiological transformations and pathways of energy partitioning occurring within the fish. In studies of fish energetics it is usual to express these energy transformations in terms of physiological rates, but other forms of expression have also been used.

It is important that all the components of the energy budget be expressed in the same units, with SI energy units being considered the best choice. The majority of researchers endeavour to provide information in terms of joules or kilojoules (kJ), but the caloric energy units are still quite frequently used (1 cal = 4.184 J). The energy content of a sample of food, faeces or fish tissue can be determined directly using bomb calorimetry. An alternative, approach is to undertake a chemical analysis of the samples with respect to protein, lipid and carbohydrate, and thereafter calculate the energy content using a series of conversion factors. The conversion factors most commonly employed are 24 kJ g^{-1}, 38 kJ g^{-1} and 17 kJ g^{-1} for protein, lipid and carbohydrate, respectively. Thus, in bioenergetic terms, the rates of a physiological process will be expressed in SI units of power, i.e. 1 J s^{-1} = 1 Watt.

The metabolic energy (heat) losses of fish are difficult to measure using direct methods, and it is therefore more usual to use determinations of oxygen consumption as an indirect measure of energy metabolism. Metabolic rates in terms of energy units can be estimated from rates of oxygen consumption using conversion factors. The conversion factor to be used will depend upon the type of metabolic substrate (protein, lipid or carbohydrate). For fish, which metabolize primarily protein and lipid, a conversion factor of 19.4 kJ l^{-1} O_2 (or 13.6 kJ g^{-1} O_2) has frequently been used (Brett and Groves, 1979).

In very few studies have all the components of the energy budget been measured, and it is usual to find that one or more of the major components has been estimated by difference (subtraction). Since they form a balanced equation, determinations of any four of the five main components – ingestion, faecal loss, excretion, metabolism and production – allow the fifth to be estimated by difference. All errors associated with the determinations of the measured components will, however, become a pooled error in the component estimated by difference. Further problems may arise because it is not possible to determine all the components simultaneously. For example, production is ideally studied as the change in energy content in the fish body with time, but since the energy content can only be determined after the fish has been slaughtered, it is not possible to obtain information about the energy content of a particular fish both before the start, and on completion, of an experiment. In this instance the researcher must rely on information obtained from sampling different individuals, and must therefore take care to ensure that the fish sampled before the start of the experiment are as similar as possible to those used in the experiment, with respect to genetic background, nutritional history and physiological status.

Whilst there are considerable advantages to be gained by expressing the components of the budget in terms of common energy units, it is also possible to find budgets constructed solely in terms of changes in biomass (wet or dry weight). Weight, rather than energy, gain is particularly widely used for the expression of the results of growth studies. In order for weight gain, as an expression of growth, to be a realistic reflection of energy deposition, it is essential that the relative composition of the fish tissues remains unaltered under different growth conditions. This prerequisite is rarely fulfilled, and there may be large variations in the composition of fish tissues, both with season and with age. The percentage lipid content of fish usually increases with increasing fish size, so that the energy density of large fish is often greater than that of small fish of the same species. In addition, fish that are feeding well, and growing rapidly, often deposit more lipid in their tissues than do slower-growing conspecifics. Thus, if growth is expressed solely in terms of weight gain, and there is a failure to recognize that there may be significant changes in body composition during the course of a growth study, incorrect conclusions may be drawn about the energy utilization and partitioning by the fish.

1.3 FACTORS INFLUENCING INGESTION (*R*)

When conducting laboratory studies on fish energetics and growth, the first objective will usually be to obtain information about the maximum growth rates of the fish under different environmental conditions, and thereafter conduct additional experiments in which optima can be determined. Fish will therefore often be fed according to regimes that aim to ensure that rates of ingestion are maximized, and there are a number of feeding options open to the experimenter.

Feeding to excess can be employed both when natural prey and when formulated pellets are used as food. Pelleted feeds can be provided from automatic feeders timed to distribute portions of food at short intervals. This ensures that food is available more or less continuously, and therefore gives the fish the opportunity to consume maximum rations and grow maximally under the given set of experimental conditions. The disadvantages of this method are that it is wasteful of food and, unless care is taken, the excess food may disintegrate and lead to a deterioration of water quality in the experimental tanks. Furthermore, although the fish are feeding maximally, it is impossible to determine a value for this maximum rate of feeding without applying special techniques. Talbot and Higgins (1983) described a radiographic method for the quantitative determination of the gut content of fish, and application of this method allows estimations of food intake to be made on groups of fish that have been fed in excess. The method has the advantages that fish need not be deprived of food prior to the start of a measurement period, disturbance of the fish can be kept to a minimum during feeding, and the method is non-invasive, so repeated measurements may be made on individual fish. Thus, this X-radiographic method can be a valuable tool for the study of food intake and feeding behaviour of fish exposed to a variety of conditions (Jobling *et al.*, 1990).

Satiation feeding is defined as the maximum amount of food a fish will consume, and can therefore only be achieved by having food continuously available (i.e. under conditions of excess feeding). Thus, satiation feeding as practised in laboratory growth and energetics studies usually represents a compromise, and the term has been used to describe feeding regimes employed in a wide range of experimental protocols. Most usually, satiation feeding has come to mean the maximum amount of food a fish will consume when presented with food two or three times per day. It is questionable how accurately this approximates to true satiation feeding, but it is suggested that under the experimental conditions most often employed with salmonids (temperature 10–15 °C; fish weight 10–50 g), the estimate obtained is close to maximum ingestion. Evidence for this is given by Grayton and Beamish (1977), who studied food intake and growth of small (10 g) rainbow trout, *Oncorhynchus mykiss*, fed at different frequencies at 10 °C. Feeding frequency had no significant effects on food intake and growth at frequencies above two times per day. Additional information is given by Elliott (1975), who, in a comprehensive study, examined the effects of feeding frequency, fish size and temperature on food intake in brown trout, *Salmo trutta*. There was no significant difference between intermeal times of various sizes of fish, but increasing the temperature brought about an increase in the number of meals consumed per day (meal frequency). The results suggest that two or three feedings per day should be adequate to ensure the consumption of maximum rations under normal rearing conditions. Even though the presentation of food to the fish two to three times per day should be sufficient to ensure maximum rates of ingestion, the experimenter will be faced with the problem of deciding when to present food in order to achieve maximum feeding.

Fish may not show equal propensity to feed at all times of the day, and when allowed to feed voluntarily, fish of many species display peaks and troughs of feeding activity during the course of a 24 h cycle: some species are most active at night, others during the day, and several predatory fish species feed most actively in the periods around dusk and dawn. Furthermore, the times of day at which the fish feed most avidly may also change with season. For example, during spring and autumn both rainbow trout and Arctic charr, *Salvelinus alpinus*, show peaks of feeding activity around dusk and dawn, but during winter the majority of food ingestion occurs during the hours of darkness. In summer, feeding activity is greatest during the daylight hours (Landless, 1976; Jørgensen and Jobling, 1989). By contrast, juvenile Atlantic salmon, *Salmo salar*, appear to feed almost exclusively during the hours of daylight, irrespective of season (Higgins and Talbot, 1985).

Restricted feeding: if the aim of an experiment is the investigation of a growth–rations relationship (Fig. 1.1), fish will have to be fed on a range of rations varying from zero (starvation) up to maximum. In other words, various

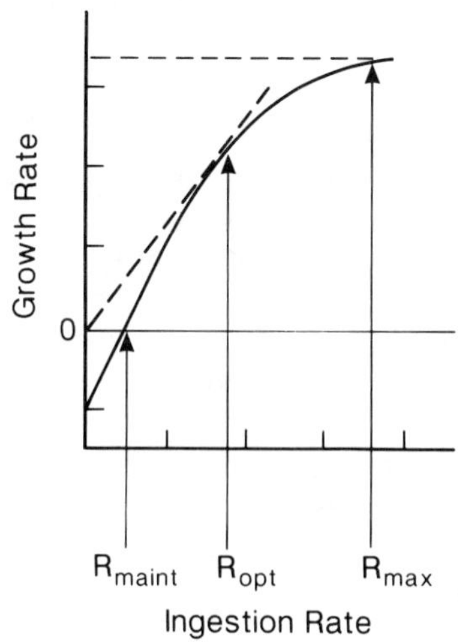

Fig. 1.1 Influence of rates of ingestion on growth rates in fish species. R_{maint} (maintenance ration) is the ingestion rate required for fish to maintain body weight; R_{opt} (optimum ration) is the ingestion rate at which gross conversion efficiency (weight gain per unit food intake) is maximized; R_{max} is maximum ration. Horizontal dashed line indicates the growth rate G_{max} to be expected at R_{max}.

restricted feeding regimes will have to be employed. This will usually entail the feeding of a fixed amount of food to the fish at some predetermined time of the day. Within medical and agricultural sciences, the effects of different feeding and rearing practices on a number of physiological variables have been investigated, but there is comparatively little information available for fish species. There is, however, evidence that both growth and fat deposition may be influenced by the time of day at which food is provided to the fish (Noeske and Spieler, 1984; Noeske-Hallin *et al.*, 1985). Thus, by arbitrarily adopting a given feeding routine (feeding once or twice per day at fixed times), the experimenter may inadvertently impose a growth restriction on the fish, whilst at the same time influencing patterns of fat deposition.

When fish are held together in groups, interactions between individuals will often lead to the establishment of social (dominance) hierarchies. Once established, a hierarchy can remain stable for an extended period of time, with the social rank of the dominant fish being reinforced by performance of threat displays or overt acts of aggression (chase and bite) against subordinate individuals. One consequence of the social hierarchy is that resources (e.g. space, feeding sites and food supplies) are not equally divided among all members of the group, and it is usually the dominant individuals that secure first access to any limited resource. Thus, when food supply is restricted, the food will not be equally apportioned, and this may lead to large differences in food consumption and growth among individual fish within the group. Acts of aggression may increase in frequency during feeding, and severe restriction of food supply will also, generally, lead to increases in agonistic interactions between the fish as they compete for the limited resource (Magnuson, 1962; Symons, 1968). When the behavioural interactions between fish have been observed, it has usually been found that the largest fish within the group are the dominants, and this has given rise to the concept of the size-related dominance hierarchy. Available evidence suggests that it is the competitively superior fish which become dominant, and, in turn, grow to being the largest individuals, either as a consequence of securing a major proportion of the food supplied or by suppression of the growth of subordinate individuals (Abbott *et al.*, 1985; Abbott and Dill, 1989; Huntingford *et al.*, 1990). Whatever the causes and effects leading to the formation of dominance–subordinate relationships, it is clear that the establishment of social hierarchies may influence the results obtained in a feeding trial, and the effects of hierarchy formation are likely to become increasingly important with decreasing food supply.

Influence of body size on ingestion

The majority of biological traits, including physiological rates, are size dependent. The relationship between a biological variable and body weight can usually be described by an equation of the form $f(W) = a\,W^b$, and b is usually < 1 so that the biological variable or function $[f(W)]$ increases allometrically

Table 1.1 Weight exponent in the relationship between body weight and food intake in a range of fish species. Information about size range and feeding methodology is also given. (From various sources: Brett, 1971; Elliott, 1975; Kirk and Howell, 1972; Niimi and Beamish, 1974; Pálsson *et al.*, 1992; Staples and Nomura, 1976; Stirling, 1977)

Species	Size range (g)	Feeding regime	Weight exponent
Oncorhynchus nerka (Sockeye salmon)	4–216	To satiation, three times per day	0.657
Oncorhynchus mykiss (Rainbow trout)	3–1300	To satiation, three times per day	0.830
Salmo trutta (Brown trout)	5–300	To satiation, four times per day	0.761
Salvelinus alpinus (arctic charr)	10–700	*Ad libitum*	0.651
Micropterus salmoides (Largemouth bass)	8–150	To satiation, two times per day	0.711
Dicentrarchus labrax (Sea bass)	2–180	To satiation, three times per day	0.727
Pleuronectes platessa (Plaice)	0.7–9.2	*Ad libitum*	0.697

with increasing body weight (W). The terms a and b are constants, with b representing the exponent relating f(W) to W (the scaling or weight exponent). The relationship between ingestion (food intake) and body weight can be described by an equation of this type, and the exponent for weight [b(1)] is almost invariably found to be less than 1 (Table 1.1). For the majority of fish species, maximum rates of ingestion have been found to scale in proportion to body weight raised to the power 0.6–0.8, and as a rule of thumb, a weight exponent [b(1)] of approximately 0.75 can be assumed.

Large fish consume more food than do small, but it is common practice to express food intake, or rates of ingestion, in relative terms, i.e. food intake per g body weight, or as % body weight consumed per day. When ingestion is expressed in these terms, i.e. $f(W) = a(1)\ W^{1-b(1)}$, it is clear that relative food intake declines with increasing body weight, and relative food intake will scale in proportion to body weight raised to approximately -0.25.

Temperature effects on ingestion

Water temperature will have a major influence on the amount of food consumed by a fish. When fish are given access to an unlimited supply of food, an increase in temperature will initially lead to increased rates of ingestion. Food consumption will then peak at some intermediate temperature before declining precipitously as the temperature approaches the upper thermal tolerance limit of the species (Fig. 1.2). For some species the relationship

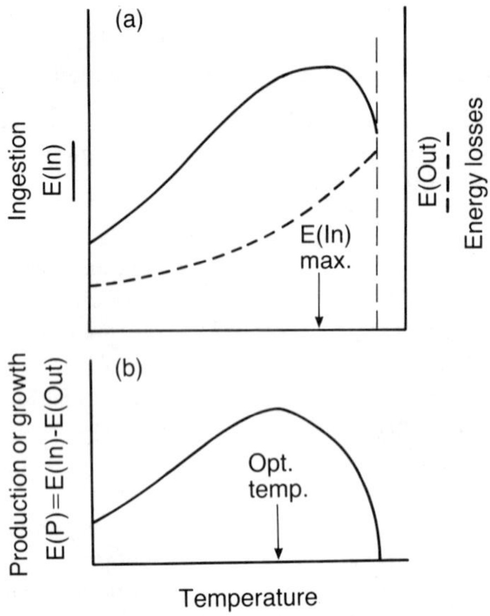

Fig. 1.2 (a) Influence of temperature on rates of ingestion and metabolism, and (b) consequences of these relationships for resources available for growth. The vertical dashed line in (a) indicates the maximum temperature tolerated by the species. Note that the optimum temperature for growth (shown in (b)) is slightly lower than the temperature at which ingestion reaches a maximum (shown in (a)).

between food consumption and temperature may be described reasonably well by a polynomial expression, but in the majority of cases an adequate description of the relationship will require the generation of a complex mathematical equation that does more to confuse than enlighten.

Other factors known to influence ingestion

Rates of ingestion may be affected by levels of dissolved oxygen, with both food intake and growth being depressed as oxygen content of the water declines (Brett, 1979). For example, the appetite of rainbow trout, *Oncorhynchus mykiss*, appears to be suppressed once oxygen saturation falls below approximately 70% (Pedersen, 1987).

Measurements of food consumption are often characterized by inter- and intra-individual variations, and large day-to-day differences in consumption have been recorded for fish of several species (Brett, 1971; Smagula and Adelman, 1982; Tackett *et al.*, 1988; Jobling *et al.*, 1989). This day-to-day variability is observed both between individuals and when food intake of specific individual fish is monitored on several occasions. In most cases the variability in food consumption will be seen as a day or two of heavy feeding followed by days on which the fish feed little or not at all.

In addition to these short-term changes in consumption, there may also be longer-term variability in rates of ingestion by fish that cannot be directly attributable to factors such as seasonal changes in water temperature. Juvenile Atlantic salmon, *Salmo salar*, for example, display seasonal differences in food intake that do not appear to be linked to changes in water temperature. In these fish, rates of food consumption may be greater at times of the year when daylength is increasing (spring) than when it is decreasing (autumn) (Higgins and Talbot, 1985). Rates of ingestion (and growth) may, however, also show long-term variations even when the fish appear to have been kept under uniform environmental conditions (constant light and temperature). For sexually mature fish, variations in food intake may be linked to different stages in the reproductive cycle, with mature fish feeding little during the spawning season. Sexually immature fish also display long-term variability in food consumption and growth, with a two- to threefold difference between peaks and troughs (Brown, 1946; Jobling, 1987), so factors other than those linked to the reproductive cycle must be important.

The potential roles of environmental (other than temperature) and endogenous factors in the long-term regulation of food consumption and growth of fishes have been relatively little studied. The investigation of the possible effects of cyclical physiological rhythms on feeding and growth is potentially rewarding, but the adoption of a chronobiological approach to the study of feeding responses and growth of fish may often be precluded by constraints on experimental facilities.

1.4 FACTORS INFLUENCING FAECAL LOSSES (*F*)

Part of the food consumed by the fish will pass through the gastrointestinal tract without being digested and absorbed. Part of the ingested food is lost as faeces. The faeces consist not only of unabsorbed food, but also mucus and cells sloughed from the walls of the gut, some digestive enzymes, some bile components and bacteria derived from the gut microflora. From this it is clear that an analysis of the faeces will not give a true picture of how much of the food consumed has remained unabsorbed during passage through the gut, but all the material that appears in the faeces can be considered to represent an energy loss for the fish. If it is assumed that mucus, dead cells, and bacteria make only a minor contribution to the total faecal material, then faecal analysis can provide a reasonably accurate estimate of how much of the nutritional content of the food has been absorbed by the fish.

Absorption efficiency (*AE*), which is also known as digestive efficiency or the digestibility of food nutrients, is a measure of the proportion of the food energy, or nutrient, content absorbed by the fish. Absorption efficiency can therefore be defined as:

$$AE = 100\,(R - F)\,/\,R \tag{1.3}$$

where R is the energy, or nutrient, content of the food and F represents faecal losses. In this definition of absorption efficiency no attempt is made to distinguish between faecal losses accruing from unabsorbed food and those derived from other sources, such as cellular breakdown products. Rarely, attempts are made to correct for the faecal components derived from non-food sources, and in such cases 'true' absorption efficiency is given by:

$$\text{'true'}\ AE = 100\ [R - (F - F')] / R \qquad (1.4)$$

where F' is the non-food component of the faeces. Corrections for the faecal components derived from cellular and bacterial sources have most usually been made when the aim of an experiment has been the determination of the digestibilities of different protein sources in foods. Protein digestibility is usually determined as the efficiency with which the nitrogen content is absorbed, and the faecal nitrogen component that is not derived from undigested and absorbed protein in the food is known as metabolic faecal nitrogen. Metabolic faecal nitrogen usually amounts to approximately 100–200 mg N per 100 g dry diet consumed. Under most circumstances, failure to take account of metabolic faecal nitrogen will lead to the estimate of absorption efficiency being 2–3% points lower than the 'true' value. When diets of low protein content are fed, however, metabolic faecal nitrogen may constitute a major part of the total nitrogen content of the faeces, and the digestibility of the protein source may be considerably under-estimated if the contribution of the metabolic faecal component is not corrected for in the calculations of absorption efficiency.

Some authors have referred to absorption efficiency as **Assimilation efficiency** (*AsE*), but the latter term is usually taken to represent the food energy or nutrients remaining after both faecal and nitrogenous losses have been accounted for. In other words, assimilation efficiency is usually defined as:

$$AsE = 100\ [R - (F + U)] / R \qquad (1.5)$$

where U is the loss via the products of nitrogenous excretion, and the other terms are as described above. Thus, care must be taken in choice of terminology if confusion is to be avoided.

A number of techniques have been developed in attempts to determine the efficiency with which fish digest and absorb their food, and these can be broadly divided into **direct and indirect methods**. The direct method requires both knowledge of the total amount of food consumed by the fish and the collection of all the faeces produced. This is usually impracticable, and most estimations of absorption efficiency have been made using indirect methods. Indirect methods involve the inclusion of an inert marker in the diet, and absorption efficiency is calculated as:

$$AE = 100 - 100\ ([XA/XB] \times [YB/YA]) \qquad (1.6)$$

where *XA* and *XB* are the concentrations of the inert marker in the food and faeces, respectively, *YA* is the nutrient concentration in the food and *YB* is the concentration of the nutrient in the faeces. Since the calculation of absorption efficiency is based upon the ratios of nutrient to marker in the food and faeces, total food consumption need not be known and there is no necessity to collect all the faeces produced. In order for the indirect method to give accurate results, a number of conditions must be fulfilled. The marker used must be inert and should not interfere with the normal processes of feeding, digestion and absorption. The food containing the marker must be fed over a sufficiently long time period to enable representative sampling of faeces to be made, i.e. samples collected should be totally free of contamination by faeces produced during consumption of previously unmarked food. Most of the markers used (chromic oxide, titanium dioxide, acid-insoluble ash) are either not normal components of the natural food of fishes or are present in such small quantity that they cannot be used for the estimation of the efficiency with which natural prey are digested and absorbed. Most studies have therefore been conducted with fish fed on artificially prepared diets to which the marker has been added. Cellulose, lignin, chlorophyll, ash and chitin are all natural products and have been suggested as being suitable markers for use in digestibility studies conducted on various animal species. All of these components may, however, be digested and absorbed to some extent and extreme caution should be exercised in the interpretation of the results obtained in experiments where these substances have been used as markers.

Collection of faeces by siphoning or netting directly from the water is cheap and simple, but there may be leaching of some components from the faeces if they are left in the water for protracted periods of time. Loss of nutrients due to leaching will obviously lead to an over-estimate of absorption efficiency, and several collection systems have been designed to ensure that the faecal contact with water is kept to a minimum. Faecal contact with the water can be excluded by collecting the faeces by anal suction, stripping or intestinal

Table 1.2 Nutrient digestibility coefficients (%) for a number of commonly used feed ingredients

Feed ingredient	Species					
	Rainbow Trout, *Oncorhynchus mykiss*		Tilapia, *Oreochromis niloticus*		Channel Catfish, *Ictalurus punctatus*	
	Protein	Energy	Protein	Energy	Protein	Energy
Fishmeal	92	91	87	80	87	85
Poultry by-products	68	71	74	59	74	67
Soya meal	96	75	91	57	77	56
Wheat middlings	92	46	76	58	84	60
Corn (maize) meal	95	39	83	76	66	59

Table 1.3 Effects of production methods and treatments on the digestibility of soya meal.

Treatment		Digestibility coefficient (%)*	
		Protein	Energy
Heat treatment			
127 °C	10 min	40	51
175 °C	10 min	70	72
204 °C	10 min	78	77
234 °C	10 min	72	70
Steam processing			
0.35 kg cm^{-2}	5 min	37	42
0.70 kg cm^{-2}	10 min	80	75
1.05 kg cm^{-2}	10 min	75	69

* Rainbow trout, *Oncorhynchus mykiss*, was used as the test species in the digestibility trials.

dissection. However, unless care is taken to collect intestinal contents from only the hindmost part of the gut, the material may not be truly representative of the faeces. Samples collected by these methods may be contaminated with urine and surface mucus, and there is also the risk of including a small proportion of incompletely digested food in the samples. All of these contaminants would lead to an underestimation of absorption efficiency.

How large a proportion of the nutrient content of the food is lost in the faeces is influenced by a wide range of factors, but food type and composition are particularly important (Jobling, 1986). For carnivorous fish consuming natural prey, absorption efficiency will usually be found to be within the range 70–95%. Plant material is digested and absorbed much less efficiently, and herbivorous fish may often absorb their food with an efficiency of 40–80%. For farmed fish that are fed artificially formulated feeds, the composition of the diet will be of overriding importance for the efficiency with which it is digested and absorbed. Dietary composition can be affected both by the choice of ingredients and by the processing to which the ingredients have been subjected, and these factors can result in large differences in digestibility, both between feed ingredients and between complete feed formulations (Tables 1.2, 1.3).

The efficiency with which food is digested and absorbed can also be influenced by feeding rate. In other words, an increase in the rate of ingestion may result in reduced absorption efficiency (Elliott, 1976; Vens-Cappell, 1984; Henken *et al.*, 1985). For example, the energy loss in the faeces of brown trout, *Salmo trutta*, amounted to 25% of the ingested energy when fish were fed maximally on *Gammarus*, but for trout given a reduced food supply the faecal

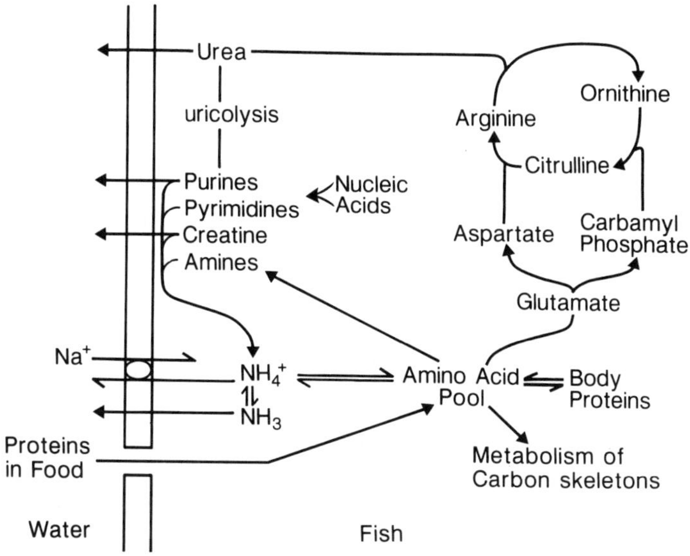

Fig. 1.3 Pathways for production and excretion of nitrogenous waste in fish. Vertical bar indicates the water–fish barrier, and the ion exchange of Na^+ for NH_4^+ is also shown. In the majority of teleosts, urea is synthesized by uricolysis since some key enzymes of the ornithine–urea cycle are absent. Urea synthesis in cartilaginous fish occurs primarily through the ornithine–urea cycle.

losses were reduced to approximately 15% of ingested energy. Similarly, when fish are fed on formulated pellet feeds, an increase in ingestion leads to an increased nutrient loss in the faeces, there being a negative linear correlation between feeding rate and absorption efficiency (Vens-Cappell, 1984; Henken *et al.*, 1985). This effect of feeding rate on absorption efficiency will rarely be of major proportions, with efficiency decreasing by about 10% as feeding rates increase from low levels to maximum rates.

1.5 PRODUCTS OF NITROGENOUS EXCRETION (*U*)

Deamination of amino acids may lead to the release of amino groups that cannot be recycled through other metabolic processes and must therefore be excreted. In fish, the major part of nitrogenous excretion occurs at the gill surface, with ammonia and ammonium ions being the main excretory products. The majority of the ammonia/ammonium is produced in the liver of the fish and is then transported in the blood system to the gills for excretion to the surrounding water. Some ammonia is also produced in the gills, kidneys and muscles. Other end products of nitrogen metabolism (e.g. urea and creatine) are produced in lesser quantity and may be excreted in the urine or through the skin or via the gills (Fig. 1.3). It is not unusual for fish to excrete over 80% of their nitrogenous wastes as ammonia/ammonium, but the proportion of nitrogen excreted as urea and other metabolic end products may reach 30–40% in some marine species. Thus, the impression is gained that marine species tend to excrete a greater proportion of their nitrogenous wastes as urea, creatine and purines than do freshwater fish species. Many teleost fish species appear to lack some of the key enzymes of the ornithine–urea cycle, and therefore cannot synthesize urea via the pathway which is the main route of ureogenesis in mammals. In these species, formation of urea appears to occur by the purine pathway, with uricolysis being the main source of urea production. Cartilaginous fish, on the other hand, have the full complement of enzymes required for production of urea through the ornithine–urea cycle, and this is the predominant route for ureogenesis in these species (Foster and Goldstein, 1969; Randall and Wright, 1987).

Many factors are known to influence rates of nitrogenous excretion by fish species, and attempts are often made to distinguish between excretion of endogenous and exogenous origin. **Endogenous nitrogen excretion** is defined as the nitrogenous excretory products resulting from the transamination and deamination of the amino acids which arise as the result of the turnover and breakdown of tissue proteins. Since the vast majority of the amino acids released as a result of protein breakdown are reutilized in synthesis (usually over 90%), rates of endogenous nitrogen excretion are generally quite low. Values of endogenous nitrogenous excretion have been reported to fall within the range 30 to 300 mg N kg^{-1} day^{-1}, depending upon fish species, size and water temperature. In practice, endogenous nitrogen excretion is often measured as the rate of excretion of a fish which has been deprived of food for a number of days. It would be more strictly correct to determine endogenous nitrogen excretion as the rate of excretion displayed by a fish that had been fed a nitrogen-free, but otherwise complete, diet. **Exogenous excretion** is considered to result from the direct deamination of amino acids ingested and absorbed from the food. The exogenous component of nitrogen excretion will therefore be influenced by factors such as feeding rate, the protein content of the food, and the amino acid composition of the diet with respect to levels of essential and non-essential amino acids.

The two factors having greatest influence on endogenous nitrogen excretion are fish size and temperature. The relationship between size and the rate of nitrogenous excretion can be described by an allometric function, and weight exponents of 0.47–0.95 have been found in experiments conducted with a range of fish species. Rates of nitrogenous excretion may increase exponentially with increasing temperature, so that a semilogarithmic transformation will linearize the data:

$$\ln U = \mathrm{k} + \mathrm{c}T \qquad (1.7)$$

where T is temperature k is a constant (ln U at T = 0 °C) and c represents the temperature coefficient. Temperature effects upon nitrogenous excretion seem to be similar to those observed for other physiological processes. For example, in a study on plaice, *Pleuronectes platessa*, the temperature coefficient for the exponential increase in nitrogenous excretion was found to be 0.061, a value similar to that found for the effects of temperature on rates of oxygen consumption (Jobling, 1981a). Results from experiments carried out on bluegill sunfish, *Lepomis macrochirus* (Savitz, 1969), suggest a temperature coefficient of 0.074 in the temperature range of 7–30 °C, but the validity of applying an exponential temperature coefficient can be questioned in this case. For the bluegill, rates of nitrogenous excretion appeared to change little between 7 and 15.6 °C, and similar instances of temperature independence of excretion rates have been reported for other species. These observations are also in accord with findings that the metabolic rates, monitored as oxygen consumption, of some fish species may be independent of temperature over part of the temperature range.

Ingestion of food will lead to an increase in the rate of nitrogenous excretion, with excretion peaking some hours after the completion of consumption of the meal. Most of the increase in nitrogenous excretion will be due to a marked rise in ammonia/ammonium excretion, with the rate of urea production and excretion being little affected by the ingestion of a meal. At its peak, the rate of ammonia excretion of a recently fed fish can be several-fold higher than that of an unfed fish. This exogenous excretion will then slowly decline until the endogenous level of excretion is reached after a period of several hours, or perhaps days. The length of time required for the rise and fall of the exogenous excretion will be determined by the size of the meal, its composition and water temperature. The composition of the food will have a marked influence on the exogenous excretion, with factors such as the protein: energy ratio and the balance of amino acids in the diet being of particular importance. Exogenous excretion may be expected to be high when the fish are fed on diets rich in protein, and also when the balance of the amino acids is unsuitable for the promotion of protein synthesis and growth (Cho and Kaushik, 1985). For a number of carnivorous fish species, 40–60% of the available dietary energy should be provided as protein if good growth is to be ensured. When fish are fed such diets, 30–50%, or in some cases more, of the nitrogen absorbed from the food may be lost in the form of exogenous excretion (Brett and Groves, 1979; Jobling, 1981a; Ramnarine *et al.*, 1987).

It is difficult to provide accurate information about the energy losses that accrue as a consequence of the excretion of nitrogenous wastes, since these will vary with both the amount and composition of excretory products. It has been suggested that a conversion factor of approximately 25 kJ per g N excreted may be realistic (Cho and Kaushik, 1985), and, based upon this, the nitrogenous excretory products may account for losses equivalent to 4–15% of ingested energy, depending upon feeding regime, food composition and the conditions under which the fish are held.

1.6 FACTORS INFLUENCING METABOLISM (*M*)

Losses due to metabolic demands usually constitute a large proportion of the energy budget of a fish. These energy costs can be partitioned into the minimal costs required for maintaining basic bodily functions [*M*(M)], those associated with activity [*M*(A)] and those related to the digestion, absorption and processing of food [*M*(F)]. Whilst this partitioning may be relatively easy to grasp from a theoretical standpoint, it is far from easy to quantify the different components experimentally.

In work with mammals, metabolism may be recorded directly as heat loss, or indirectly by the simultaneous monitoring of oxygen consumption and carbon dioxide production. The indirect method enables the calculation of a respiratory quotient, and thereby allows heat production to be estimated (Blaxter, 1989). In studies with fish, on the other hand, it is usual to measure oxygen consumption, but not carbon dioxide production, so that the estimation of heat production requires assumptions to be made about the types of respiratory substrates being metabolized. Based upon the assumption that protein and lipid are the primary respiratory substrates in fish, an oxycalorific coefficient of 19.4 kJ l^{-1} O_2 has been recommended (Brett and Groves, 1979).

Costs of maintaining bodily functions [*M*(M)]

The '**minimal metabolism**' represents a measure of the minimum rate of heat production in the absence of muscular activity, food consumption and its subsequent processing. In experimental studies, these factors have been controlled with varying degrees of success, and a number of terms have arisen to describe the types of measurements that approximate to the minimal metabolism.

In studies with human subjects, an estimate of the minimal metabolism can be obtained by measuring **basal metabolism**. There are stringent criteria that must be adhered to in the conducting of these measurements (Blaxter, 1989), but the criteria can rarely be fulfilled when conducting experiments on animals. The estimations of minimal metabolism made using domestic animals as subjects are usually referred to as the **fasting metabolic rate**, and measurements are made continuously for a period of 24 h or more. An alternative approach is to monitor heat production over a prolonged period and then to select, as the estimate of minimal metabolism, those short intervals during which heat production is at its lowest. These types of measurements may be referred to as **lowest, or least, observed metabolic rate**.

A wide range of methodologies have been used in attempts to determine the minimal metabolism of fish species, and the terminology has not been satisfactorily standardized. Terms such as 'basal', 'standard', 'maintenance' and 'resting' metabolism have been used as synonyms, but distinct definitions have also been given to each of these terms. Problems arise, however, because there is a lack of consistency in the way each of the terms has been defined by different workers, and no universally acceptable set of definitions has yet been presented.

On the basis of the foregoing discussion, it is open to question whether the term basal metabolism should be used when referring to measurements of 'minimal' metabolic rates of fish species, but there is a good case to be made for the introduction of the term fasting metabolic rate. In practice, most estimates of the minimal metabolism in studies with fish have been obtained using experimental protocols that are similar to those used for the estimation of fasting metabolic rates in domestic animals, but in the fish studies, the metabolic rate recorded has usually been termed the **resting, resting routine**, or **low routine metabolic rate**. All of these terms presuppose that measurements are being made on a quiescent fish, in a post-absorptive condition, but it is recognized that measurements include some increase over the true minimum due to the metabolic costs of low levels of spontaneous activity. Least observed metabolic rate has been reported in a number of cases, but this term is not commonly used in studies carried out on fish. Thus, on the grounds of both the promotion of standard terminology and for simplicity, it may be of value to replace the plethora of current terms with **fasting metabolic rate** and **least observed metabolic rate**. A case can, however, be made for retaining the term **standard metabolism** in fish studies, provided that the term is only used when referring to estimates of minimal metabolism obtained by specific means. It has become usual to calculate standard metabolism from data obtained on swimming fish. A relationship between swimming speed and the metabolic rate of the fish is established, and then the relationship is extrapolated to zero velocity. The estimate of the metabolic rate at zero swimming speed is defined as the standard metabolism of the fish. Whilst the validity of these extrapolations can be questioned on both statistical and physiological grounds, the method is useful for obtaining estimates of the minimal metabolism of active species which do not remain quiescent in respirometry chambers.

A prerequisite in all studies designed to measure minimal metabolism is that the animals be in a post-absorptive state, i.e. the animals have been fasted in order to ensure that the effects of previous meals, and food processing, are negligible. There have been few studies carried out on fish species to quantify the time required to reach a post-absorptive state, and a fasting period of 48–72 h has often been arbitrarily adopted prior to the conducting of measurements of minimal metabolism. Whilst a fast of this duration may be adequate to ensure a return to the post-absorptive condition in some conditions, it certainly does not have general application. The time required for the ablation of the effects of previous meals on metabolic rate is highly variable, and may be governed by factors such as meal size and composition, and, not least, by water temperature (Jobling, 1981b; p. 21). It may, however, be generally assumed that the time taken to reach the post-absorptive state is related to the retention time of food in the digestive tract. In other words, some elevation in metabolic rate due to the effects of feeding is observed for slightly longer than the time during which products of digestion are being transported across the gut mucosa into the blood stream.

Components of the minimal metabolism

Heat production is a consequence of several biochemical events, but the majority of the energy expenditure can be simply regarded as consisting of that required to meet two types of function. The first can be considered as service functions, such as expenditures by the respiratory and circulatory systems required for supplying oxygen to the various body tissues (Blaxter, 1989). The second type of function relates to cellular maintenance functions, which would include costs for ion transport and the metabolic costs associated with the synthesis and turnover of cellular constituents such as protein and lipid. Service functions may account for 35–50% of energy expenditure, and cellular maintenance functions for 40–55%, but the biochemical events contributing to both functions may not be mutually exclusive, and attempts to make strict divisions between the two types of function may be fruitless. Nevertheless, the concept remains a useful starting point for the discussion of energy expenditure related to the minimal metabolism.

The two most important processes contributing to energy demand are Na^+, K^+-ATPase activity and protein turnover (Milligan and McBride, 1985; Kelly and McBride, 1990; Hawkins, 1991; Houlihan, 1991). The enzyme **Na^+, K^+-ATPase** extrudes 3 Na^+ from the cell whilst transporting 2 K^+ into the cell. Movement of both Na and K ions occur against their respective concentration gradients at the cost of one high-energy phosphate bond. Activity of the Na^+, K^+-ATPase enzyme (also known as the '**sodium pump**') is responsible for the active transport of substrates, the maintenance of cellular homeostasis and is involved in the generation and maintenance of membrane potentials. In studies with mammals it has been estimated that the activity of the sodium pump accounts for 20–40% of the total body heat production, and a considerable proportion of the minimal metabolism is therefore related to energy expenditure associated with Na^+, K^+-ATPase activity. The exact contribution of sodium pump activity to metabolic expenditure in ectothermic vertebrates, such as fish, is not known with any degree of certainty, but it is unlikely to be negligible.

A second major component of energy expenditure in cells is that associated with protein turnover. Protein turnover involves both the synthesis and degradation of proteins, and there may be metabolic costs associated with both processes (Hawkins, 1991; Houlihan, 1991). The energetic cost of protein synthesis is assumed to be 5 mol ATP per mol peptide bond formed. The transport of an amino acid across the plasma membrane is assumed to require 1 mol ATP and 4 mol ATP are ascribed to the synthetic process. Assuming that 1 mol ATP equates to approximately 85 kJ, the energetic costs of protein synthesis amount to 4–5 kJ g^{-1} (Webster, 1981; Lobley, 1986; Kelly and McBride, 1990). Based upon this series of assumptions it has been estimated that protein synthesis contributes 12–25% to the whole-body heat production in mammals. Protein degradation may occur both through processes that do not require utilization of ATP and through those that are dependent upon the expenditure of ATP. The energetic costs of protein degradation have, however, not been studied in detail. The costs of protein turnover represent a considerable energetic drain for the animal, and may be considered to account for approximately 20–25% of ATP utilization (Hawkins, 1991).

Thus, the costs of maintaining the sodium pump, and those related to protein turnover, are considerable, but there are large differences between the various body tissues in their energy expenditures and relative contributions to the total body metabolism. For example, the tissues of the gastrointestinal tract and liver may make up only 10% of the whole-body mass, but account for 40% of the total ATP utilization. On the other hand, skeletal muscle, which makes up 50% of the body mass, contributes only 20% or so to the total energy demand. These differences are related to differences in sodium pump activity and protein turnover between tissues, with the gut and liver being considered 'active' tissues, whereas rates of protein turnover in skeletal muscle are low.

A third important process contributing to energy expenditure is substrate cycling (Blaxter, 1989; Kelly and McBride, 1990). Enzymatically catalysed biochemical pathways involved in metabolite oxidation may contain steps in which the reaction is not at equilibrium and cannot, therefore, be reversed by merely changing the concentrations of substrates and products. In cases of this sort, reversal of the reaction will require the employment of alternative enzymes and the expenditure of energy by input of ATP. When both the forward and reverse reactions occur simultaneously, with no net change of either substrate or product, substrate cycling is said to occur. There are a number of such cycles, and, since they appear to accomplish nothing more than the expenditure of energy, they have often been called 'futile cycles'. This is, however, a misnomer since their existence enables control to be maintained over the reactions of the cycle, and can lead to increased sensitivity of enzymes in response to changes in substrate concentrations. These cycles may not contribute much to the minimal metabolism, but are quantitatively much more important when there are large, short-term changes in nutrient supply, such as occurs shortly after feeding.

Other energy-consuming processes contributing to the minimal metabolism include those related to the turnover of nucleic acids and those required for urea biosynthesis, but these additional energetic costs are of minor proportions when compared with those of protein synthesis and the sodium pump.

Effects of body size

Large fish generally consume more oxygen than small fish, but on a unit-weight basis, small fish will consume more oxygen than larger conspecifics.

This allometric relationship can be described as:

$$M(\mathrm{M}) = \mathrm{a}(2)\ W^{\mathrm{b}(2)} \tag{1.8}$$

where M(M) is the rate of oxygen consumption (ml O_2 per fish per hour) (energy expenditure), W is body weight (in g), and a(2) and b(2) are constants. There are several hundred studies in the literature reporting the effects of body size on rates of oxygen consumption in fish, and the majority of the estimates of the weight exponent [b(2)] lie within the range 0.65–0.9. Winberg (1956) concluded that a weight exponent of 0.81 was suitable for use with fish species, but more recent analysis suggests that a value of 0.86 may be more appropriate (Brett and Groves, 1979). In the interests of accuracy, however, it is not advisable to apply a general value for the weight exponent uncritically. The relationship between body weight and oxygen consumption should be investigated for the particular species under study using as wide a range of fish sizes as possible, and a separate estimate of the exponent should be obtained. The reported values of the exponent are sufficiently disparate to imply that interspecific differences may exist, and there is also evidence to suggest that there may be intraspecific differences depending upon measurement conditions, such as water temperature and season during which the experiments are carried out (Xiaojun and Ruyung, 1990; Johnston *et al.*, 1991). It has been reported that the weight exponent declined from 0.94 to 0.75 as test temperature increased from 10 to 30 °C (Xiaojun and Ruyung, 1990); this was attributed to the differential effects of temperature on the various components of the minimal metabolism. This is an attractive hypothesis, since it may be suspected that the relative contributions of the sodium pump and protein turnover to the total metabolism differ between fish of different sizes, and there is no reason to believe that changes in temperature will affect both processes equally. In the absence of more convincing experimental evidence, the hypothesis must, however, be considered speculative.

Temperature effects on the minimal metabolism

When measurements of minimal metabolism are made on fish it is important to ensure that the fish are acclimated to the temperature at which the measurements are to be made. If fish are transferred from one temperature to another, several days, or even weeks, may be required before the fish become acclimated to the new conditions. Thus, the results of metabolic measurements made immediately after transfer ('acute' response) will usually be found to differ from those obtained from experiments carried out on acclimated fish (Cossins and Bowler, 1987; Evans, 1990).

It is usually found that an increase in temperature leads to an increase in the rate of oxygen consumption and, in some cases, a simple exponential (semilogarithmic) transformation comes close to linearizing the data:

$$\ln M(\mathrm{M}) = \mathrm{k} + \mathrm{c}\ T \tag{1.9}$$

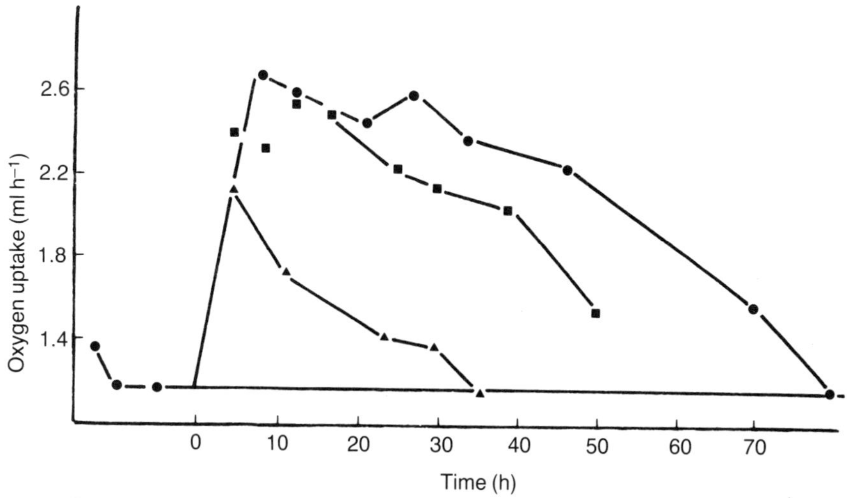

Fig. 1.4 Effects of feeding meals of different sizes on rates of post-prandial oxygen consumption in the plaice, *Pleuronectes platessa*. Example shows the time course of changes in oxygen consumption of an individual fish following the feeding of meals of different sizes. Fish were fed at time 0. Triangles, 0.25 ml (1.41 kJ) meal of white fish paste; Squares, 0.5 ml meal; circles, 1.0 ml meal. Data from Jobling and Davies (1980).

where T is temperature, k is a constant (ln M(M) when T = 0 °C) and c represents the temperature coefficient. The temperature coefficient is usually found to lie within the range 0.05–0.10, which means that metabolic rates increase 1.65 to 2.7-fold for every 10 °C rise in temperature.

The generalization that there is an exponential increase in metabolic rate with increasing temperature is an oversimplification and there are many cases in which exceptions have been reported. One way in which the pattern may deviate from the exponential is when metabolic rate becomes independent of temperature, usually within the mid-part of the temperature range (Cossins and Bowler, 1987).

Effects of feeding on metabolic rate [*M*(F)]

Following the ingestion of food there is an increase in the metabolic rate. The post-prandial increase in metabolic rate is referred to by several names, and the phenomenon may appear in the literature under the guise of the **specific dynamic effect** or **action** (SDA), **calorigenic** or **thermic effect** of feeding, **heat increment** (HI) or **dietary induced thermogenesis** (DIT). All of these terms are, presumably, being used to describe similar series of biochemical and physiological processes occurring within different animal groups.

In fish, the effect has been studied by examining the post-prandial increase in the rate of oxygen consumption. The rate of oxygen consumption increases steeply shortly after the ingestion of a meal, and has usually been found to peak at a level two to three times the pre-fed level within a few hours, before gradually declining to the pre-feeding level (Fig. 1.4; Jobling, 1981b). The duration over which the effect can be observed is affected by a number of factors including meal size and composition, with an increase in meal size leading to a prolongation of the time the metabolic rate remains elevated above the pre-feeding level. Regular provision of food to the fish can therefore lead to rates of metabolism being maintained well above those recorded prior to feeding. For fish fed a single meal, the duration of the effect of feeding on metabolic rate can be markedly affected by temperature. For example, when plaice, *Pleuronectes platessa*, were fed standard-sized meals under different temperature conditions, rates of oxygen consumption remained elevated above pre-fed levels for approximately 51, 35 and 26 h at 10, 15 and 20 °C, respectively (Jobling and Davies, 1980). There appear to be clear links between the duration of the effect and the rate of food passage through the gut, with some elevation in oxygen consumption being observed for a period several hours longer than the time required for digestion and absorption of nutrients from the food.

The post-prandial increase in metabolic rate results from the energy requirements for the digestion, absorption and storage of nutrients, for the deamination of amino acids and synthesis of excretory products and for increased turnover and deposition of tissue components, but the relative contributions of each of these processes are exceedingly difficult to evaluate with any degree of accuracy.

Following the ingestion of a meal there is a marked increase in the motor activity of the gastrointestinal tract, but results from studies in which inert bulk has been introduced into the gut suggest that the contribution of the mechanical component to the post-prandial increase in metabolic rate is small. Processing of the food in the gut also involves the digestion and absorption of nutrients, both of which may involve some energy expenditure. Energy will be required for the synthesis and secretion of digestive enzymes, and for the active transport of the products of digestion across the gut mucosa, but the energetic costs involved remain largely unknown. Nevertheless, the increase in oxygen consumption following the ingestion of a meal is probably best regarded as being a post-absorptive phenomenon, since intravenous injection of nutrients is known to give rise to a marked increase in metabolic rate (Brown and Cameron, 1991a).

Early studies with mammalian species suggested that there was a close link between the amount of protein fed and the post-prandial increase in metabolic rate, and this gave rise to the suggestion that the effect was associated with some aspect of protein metabolism. Recent work with fish has also provided evidence that infusion of amino acids directly into the bloodstream leads to a marked increase in metabolic rate (Brown and Cameron, 1991a). Since rates of nitrogenous excretion have also often been observed to increase markedly following feeding, it has been hypothesized that the post-prandial increase in metabolism was the result of the energetic costs linked with amino acid deamination and the synthesis of the nitrogenous products of excretion. The validity of this hypothesis is in question because there may not always be close links between post-prandial metabolic rates and rates of nitrogenous excretion in ureotelic and uricotelic animals (i.e. those whose main nitrogenous waste is urea and uric acid, respectively). Thus, when poor-quality proteins (zein, gelatin) are fed, there may be a marked increase in the excretion of urea or uric acid without there being any marked effect on metabolic rate, but the feeding of good-quality protein (casein, fish) may be followed by an increase in metabolic rate without a concomitant large rise in the rate of nitrogenous excretion.

Tissue constituents are in a dynamic state and there is a continuous turnover of body proteins. Both the rate of protein synthesis and the rate of protein breakdown may increase as food supply increases, and as a consequence of this there is a greater rate of protein turnover. Clearly, if the rate of protein turnover is increased by feeding, then an additional energy cost is incurred, and this must contribute to the energy expenditure associated with the ingestion of food. Infusion of essential amino acids into the blood has been shown to lead to an increase in both rates of protein synthesis and metabolic rates, and a cause-and-effect relationship has been suggested (Brown and Cameron, 1991a,b). If ingestion of food results in an increase in protein turnover, this would imply that there is an increase in energy expenditure required to maintain the status quo (i.e. metabolic costs increase without there being any net gain of body protein), but when animals are fed they also grow, and the nutrient energy supplied in the food is deposited as protein and lipid. The rate of retention of protein in the body tissues is a net quantity representing the difference between the rate of synthesis and the rate of protein degradation, and it is easy to see that the deposition of 1 g protein as tissue growth will require considerably more than 1 g protein to be synthesized. Results from research with mammals suggest that protein deposition efficiency is often within the range 30–60%, and the few results available for fish species suggest that 2–3 g protein may be required to be synthesized per g protein deposited as tissue growth (Reeds and Harris, 1981; Mulvaney *et al.*, 1985; Houlihan and Laurent, 1987; Houlihan *et al.*, 1988; Houlihan, 1991). Tissue growth is, however, more than just the deposition of protein, and the energetic costs of lipid synthesis must also be considered. Direct deposition of dietary lipids is energetically inexpensive, but the costs of *de novo* synthesis of lipids are suggested to be within the range 5–12 kJ g^{-1} depending upon the source of the substrate and the biochemical synthetic pathways involved. There is a great deal of uncertainty involved in the estimates of these energetic costs, but it has been suggested that 60–80% of the increase in metabolic expenditure associated with the ingestion of food may be the direct result of increased protein and lipid synthesis (Lobley, 1986).

The synthesis and deposition of tissue proteins and lipids represents an accretion of new tissue, or body growth, but the ingestion of food will also lead to the production of intermediate storage products that ultimately are completely oxidized. Temporary storage of nutrients as glycogen or fat imposes an energetic cost because indirect oxidation, involving substrate cycling, is less efficient than would be oxidation of the nutrient directly. A large proportion of these substrate transformations occurs in the liver, and the liver shows a rapid metabolic response to large short-term changes in nutrient supply.

Results from the earliest studies carried out on the effects of feeding on metabolic rate attest to the large contribution made by the liver to the post-prandial increase in heat production, in that the post-prandial rise in metabolic rate was much reduced in hepatectomized animals. Recent evidence suggests that there are marked increases in rates of protein synthesis in the liver shortly after the ingestion of a meal (Brown and Cameron, 1991b), and the response to feeding appears to occur earlier in the liver than in other tissues (Houlihan, 1991). This can be taken as providing evidence that the costs of protein synthesis, along with those of intermediate storage of nutrients and lipid synthesis, may be the major causal factors of the post-prandial increase in heat production.

Effects of nutritional history on metabolic rate

Results of a series of studies conducted on domestic animals suggest that the estimate of the minimal metabolism may be markedly affected by the previous nutritional history of the animal, possibly as a result of the influences of different feeding levels on rates of turnover of tissue components.

An estimate of the protein synthetic capacity of a cell, or tissue, can be obtained by examining ribosomal numbers (assuming that synthetic activity per ribosome is more or less constant). Since approximately 80% of the RNA content of a cell is ribosomal RNA (15% transfer RNA, 5% messenger RNA), determination of the RNA concentration in a cell will provide an estimate of ribosome numbers, and thereby give an indication of the protein synthetic potential (Reeds, 1987; Houlihan, 1991). Changes in the nutritional status of an animal lead to changes in tissue RNA concentrations, and these changes may take place over periods of several hours, days or weeks. In animals that are deprived of food there will be a gradual decline in tissue RNA concentration (= ribosomal numbers = protein synthetic capacity), whereas the re-feeding of a previously starved animal leads to a gradual increase in both RNA concentration and protein synthetic capacity.

In practice, the synthetic activity of ribosomes is not constant, but can vary from minute to minute. These changes can exert a short-term control over rates of protein synthesis, whilst the changes in ribosomal numbers represent a long-term adaptation of synthetic capacity to changes in nutritional status (Beitz, 1985; Reeds, 1987; Millward, 1989). Increased food availability not only leads to an increase in synthetic capacity, but may also lead to an increase in rates of breakdown. In other words, rates of turnover, and energetic costs, may increase without there necessarily being any net gain in terms of accretion of new tissue. Consequently, the energy expenditure required for the maintenance of the status quo is not constant, but varies depending upon the level of food supply and the nutritional status of the animal.

Many species of fish are capable of surviving long periods of food deprivation, and during starvation both metabolic rates and rates of protein synthesis are known to decline. The fall in protein synthesis is the result of both a reduction in ribosomal activity and a decrease in ribosomal numbers (RNA concentration). Thus, the major events that occur when fish are deprived of food resemble those reported for mammals, and the responses to re-feeding also appear to be similar in fish and mammalian species. Upon re-feeding, rates of protein synthesis increase as a result of increased ribosomal activity and due to an increase in numbers of ribosomes (Lied *et al.*, 1982, 1983; Wright and Martin, 1985). Thus, rates of tissue turnover in fish appear to be influenced by food availability, and this would be expected to have consequences for rates of energy expenditure. Some evidence for this was presented by Brown, as early as the mid 1940s. Based upon the results of growth studies with brown trout, *Salmo trutta*, she suggested that the metabolism of the fish declined as the food supply was reduced, such that fish that were fed low rations continued to grow, despite the fact that the amount of food supplied appeared to be below the maintenance ration required by better-fed fish (Brown, 1957). Thus, she hypothesized that the nutritional status of the fish was an important determinant of the level of energy expenditure, and that metabolic activities became adapted to changes in food availability. In the light of more recent work it appears that the hypothesis was correct, and it is now possible to give an acceptable explanation of these observations based on changes occurring at the cellular level.

Energy expenditure associated with swimming [*M*(A)]

Swimming is a very energy-demanding activity, and the energy expenditure of fish swimming at high speed may rise to 10–15 times that of a fish at rest (Brett, 1972; Beamish, 1978), although the factorial aerobic scope for activity is more usually found to be 4–7 (Johnston *et al.*, 1991). In other words, when fish of different species swim for prolonged periods of time they are able to maintain rates of oxygen consumption 4–7 times resting levels. Rates of oxygen consumption are usually found to increase exponentially as swimming speed increases, and an increase in swimming speed of 1 body length (BL) per second may result in the rate of oxygen consumption increasing 2- to 2.5-fold. Information about the oxygen consumption of swimming fish may be obtained from experiments conducted on fish that are voluntarily active or from those in which fish are forced to swim against a water current. Most measurements have been made using tunnel respirometers in which fish are forced to swim against currents of known speed. The advantage of this type of respirometer is that the activity level of the fish can be accurately controlled. Other types of swimming respirometer rely upon the fish being voluntarily active. These may be designed to take advantage of the optomotor response (the tendency of some fish species to hold station against a moving background), and must incorporate some form of activity counter.

To what extent measurements of oxygen consumption give a true reflection of the energetic costs of swimming activity is open to question, since anaerobic metabolism, with the production of lactate, increases as levels of swimming activity rise (Beamish, 1978; Davison, 1989; Goolish, 1991). In some species, such as flatfishes, gadoids and typical sit-and-wait predators, anaerobic respiration begins in the swimming muscle at relatively low activity levels (0.5–2.0 BL s^{-1}), whilst in salmonid species and actively foraging species higher swimming speeds (2.0–5.0 BL s^{-1}) are reached before the muscles begin to respire anaerobically and lactate begins to accumulate. There also appears to be a distinct size effect since anaerobic metabolism is induced at a lower relative swimming speed (expressed as BL s^{-1}) in large than in small fish.

Studies conducted under controlled laboratory conditions in which fish are forced to swim at a constant speed against a water current give no information about energy expended on swimming activity by fish in the wild. Attempts have been made to collect information about activity levels of fish in nature using various forms of telemetric equipment. The simplest telemetric method involves the fixing of a small radio transmitter to the fish, enabling the position and movements of the fish to be monitored at regular intervals. This method has, however, a number of shortcomings, and energy expenditure estimates based upon information obtained in this type of study are of limited accuracy. For example, in calculating the distance swum it is assumed that the fish swims in a direct line between two points and that the fish swims at constant speed. This method of calculation will result in an underestimation of the energy expenditure on activity by the fish.

Attempts have also been made to estimate activity levels by monitoring changes in a range of physiological parameters, such as ventilation frequency, heart rate and muscular contractions in the swimming muscle (Weatherley and Gill, 1987). It is, however, no simple task to use physiological measurements in the estimation of energy expenditure for activity. For example, both ventilation frequency and heart rate are influenced by a range of factors, so a change in these parameters need not necessarily imply that the swimming activity of the fish has changed. In addition, a number of physiological changes may be brought about by increased activity. The physiological response to increased activity may not only be an increase in ventilation frequency and heart rate, but there will usually also be an increase in stroke volume. In practice, the increase in cardiac output (heart rate × stroke volume) that occurs with increasing activity is more often the result of quite large changes in stroke volume than marked increases in heart rate. For example, the heart rate of the sea bass, *Dicentrarchus labrax*, appears to be relatively constant over a comparatively wide range of activity levels (Sureau and Lagardere, 1991), and the increased oxygen required by the muscles during swimming activity must be supplied by invoking physiological changes other than marked increases in heart rate.

At high swimming speed, some species of fish may switch from the usual form of gill ventilation to ram ventilation, whereupon ventilation frequency will become uncoupled from the overall activity level of the fish. Thus, there need not necessarily be close correlations between ventilation frequency, heart rate and swimming activity, so extreme caution must be exercised if this type of information is used for the estimation of energy expenditure associated with various levels of activity. Some of these difficulties can be overcome by monitoring the activity of the swimming muscles directly, but this method, too, is not completely devoid of problems. Different muscle fibre types are activated at different swimming speeds. This can lead to either changes in, or an uncoupling of, the relationship between the electrical activity recorded in the muscle tissue and swimming speeds of the fish, making accurate assessments difficult (Weatherley and Gill, 1987; Davison, 1989).

Estimates of the energy expenditure of fish in the wild have been made for a small number of species and, despite the limitations of the different methods employed, there appears to be a general consensus that relatively little time and energy is expended upon swimming activity. Under normal circumstances, approximately 10%, and seldom more than 20%, of the total energy

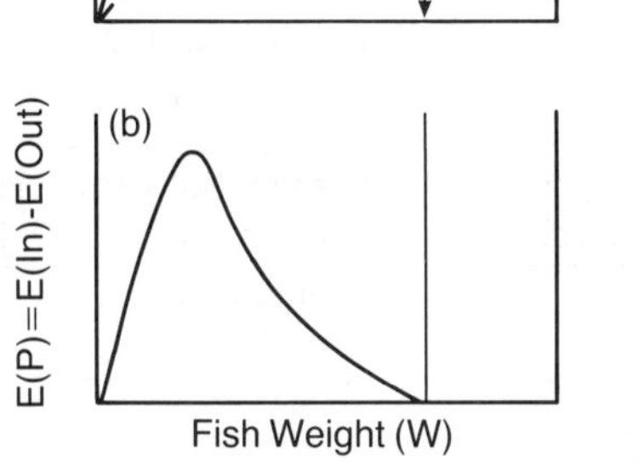

Fig. 1.5 Relationships between ingestion [*E*(In)], energy losses [*E*(Out)], production [*E*(P)] and fish weight (*W*). (a) Effects of fish size on rates of ingestion and energy losses. The point at which the lines intersect indicates the point at which *E*(In) = *E*(Out), and defines the maximum size (W_{max}) the fish can attain. (b) Influence of fish size on the resources available for production, assuming that food supplies do not limit ingestion rates.

expenditure will be used for swimming. It must be stressed, however, that the species studied (brown trout, *Salmo trutta*, rainbow trout, *Oncorhynchus mykiss*, cod, *Gadus morhua*, largemouth bass, *Micropterus salmoides*, and pike, *Esox lucius*) can be considered as being relatively sedentary, and a larger proportion of the total energy expenditure would be expected to be devoted to swimming in highly active pelagic species.

1.7 FACTORS AFFECTING GROWTH AND PRODUCTION (*P*)

A simple growth model

Ingested energy that is neither lost as faecal or excretory products, nor used for metabolism, is available for growth. Growth can take the form of either somatic or reproductive growth.

In its simplest form, the energy balance equation can be written as $E(\mathrm{P}) = E(\mathrm{In}) - E(\mathrm{Out})$, where $E(\mathrm{P})$ is energy available for growth, $E(\mathrm{In})$ is ingested energy and $E(\mathrm{Out})$ represents the energy losses due to metabolic and other processes. Both $E(\mathrm{In})$ and $E(\mathrm{Out})$ can be expressed relative to fish body weight in the form of allometric equations:

$$E(\mathrm{In}) = \mathrm{a}(1)\ W^{\mathrm{b}(1)} \tag{1.10}$$

and

$$E(\mathrm{Out}) = \mathrm{a}(3)\ W^{\mathrm{b}(3)} \tag{1.11}$$

such that

$$E(\mathrm{P}) = \mathrm{a}(1)\ W^{\mathrm{b}(1)} - \mathrm{a}(3)\ W^{\mathrm{b}(3)} \tag{1.12}$$

Figure 1.5(a) shows how $E(\mathrm{In})$ and $E(\mathrm{Out})$ are influenced by fish weight (W), assuming that $1 > \mathrm{b}(1) > \mathrm{b}(3)$. Ingestion scales in proportion to body weight raised to the power 0.75 (p. 7), and the weight exponent for the minimal metabolism is approximately 0.86 (p. 20), so the assumption made appears to be justified. The difference between $E(\mathrm{In})$ and $E(\mathrm{Out})$ gives $E(\mathrm{P})$, and the

Table 1.4 Effects of body weight W (in g) on the specific growth rate G of different salmonid species; see also Equation 1.14. (from Brett, 1979; Jobling, 1983b)

Species	Temperature	$\ln G = \ln \mathrm{a} + \mathrm{b} \ln W$	
	(°C)	Growth coefficient, a	Weight exponent, b
Oncorhynchus nerka (sockeye salmon)	15.5	5.42	-0.40
Oncorhynchus mykiss (rainbow trout)	17	6.86	-0.32
Oncorhynchus gorbuscha (pink salmon)	15	9.78	-0.45
Salmo trutta (brown trout)	13	2.79	-0.32
Salvelinus alpinus (arctic charr)	14	8.54	-0.35

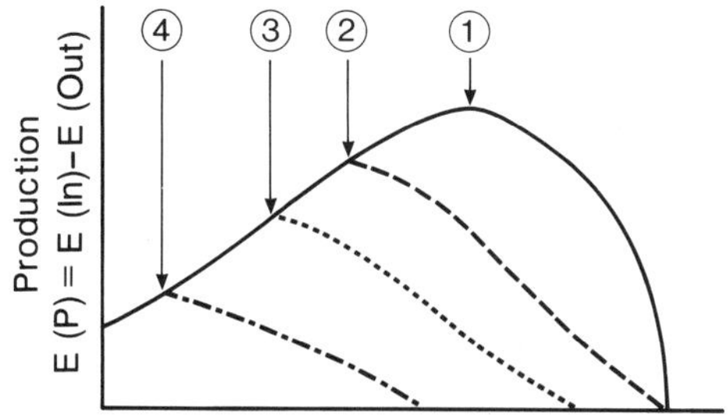

Fig. 1.6 Influence of temperature upon growth (production) of fish held under different conditions of food restriction. The numbers 1–4 indicate the temperatures at which growth rates will be greatest under the different regimes: (1) unlimited food supply; (2) slight restriction; (3) moderate restriction; (4) low level of food supply.

relationship between *E*(P) and body weight is shown in Fig. 1.5(b). Initially, *E*(P) will increase with increasing body weight, then it will peak at a given weight, then decline as body weight continues to increase. This is an expression of absolute growth, but the results of growth studies are much more frequently expressed in terms of relative growth, *E*(P)/*W*. A special case of the above occurs when reference is made to growth at a particular instant in time rather than over a protracted time interval. This is given the name the **instantaneous growth rate** (g), and multiplying by 100 gives the **specific growth rate** (G), which is the most frequently used expression of growth in experimental studies. In practice, specific growth rate (G) is calculated as:

$$G = ([\ln W(2) - \ln W(1)] \,/\, [t(2) - t(1)]) \times 100 \qquad (1.13)$$

where $W(1)$ and $W(2)$ are body weight at the start and end of the growth period, respectively, and $[t(2) - t(1)]$ is the length of the period in days. Specific growth rate will decline with increasing fish weight, and the relationship between growth rate and body size can be described by an allometric function:

$$G = a(4)\ W^{b(4)} \qquad (1.14)$$

where b(4) is the weight exponent (in this case a negative exponent) and a(4) represents the growth rate of a fish weighing 1 gram. A number of studies have been carried out in which the effects of body weight on growth have been investigated, and there is information available for several salmonid species (Brett, 1979; Jobling 1983a,b; Table 1.4). Values for the weight exponent are often close to -0.35, and are usually found to be within the range -0.3 to -0.45.

It must be stressed that in this discussion of the relationship between growth rate and body size it has been presupposed that fish are feeding at maximal rates, such that growth is not limited by food supply. It is clear that food limitation will influence growth rate; under these conditions, the body size effect on growth rate may not be observed.

In this simple modelling exercise, the effects of body weight on growth rate have been described for fish feeding maximally under a given set of environmental conditions. Factors, such as temperature, which are known to influence both ingestion and metabolism will, however, also affect growth rates. Under conditions of unlimited food supply an increase in temperature will lead to an increase in food intake, but at high temperatures there will be an abrupt decline in rates of ingestion (Fig. 1.2(a)).

Metabolic rate, on the other hand, has often been found to increase exponentially with increasing temperature (Fig. 1.2(a)). The changes in food intake and metabolic rate brought about by changes in temperature are shown in Fig. 1.2(a), and the difference between the two lines represents the resources available for growth. Figure 1.2(b) shows that growth increases with increasing temperature, peaks and then declines at high temperature. The temperature at which growth is maximized is called the optimum temperature for growth, and it should be noted that this optimal temperature is a few degrees lower than the temperature at which food intake is greatest. The term 'optimum temperature for growth' should only be used when describing the results obtained in studies in which fish have been fed maximally, since restricted feeding will have a marked influence upon the growth rate observed at a given temperature. There are large interspecific differences in temperature optima (Jobling, 1981c). The majority of salmonids, for example, have temperature optima within the range 12–17 °C, whereas many cyprinids have optima of 20 °C or higher. In addition, there may be ontogenetic changes in temperature optima for a given species, with larvae and juveniles often having a higher optimum temperature for growth than larger conspecifics. Thermal optima for growth may also be influenced by photoperiod (daylength). For example, 0-group hybrid bass (female *Morone saxatilis* × male *M. chrysops*) displayed best growth at 27.9 °C on increasing photoperiods, whereas the temperature optimal for growth was reduced to 25.7 °C when fish were held on simulated, decreasing autumnal daylengths (Woiwode and Adelman, 1991).

Under conditions of restricted feeding there will be a change in the relationship between temperature and growth rate, and as food supply becomes more and more restricted the best rates of growth will be observed at lower and lower temperatures (Fig. 1.6).

Food utilization or conversion efficiency

A discussion of production and growth processes is incomplete without some consideration being given to the efficiency with which food is utilized. Equations 1.10 to 1.12 state that $E(\mathrm{In}) = \mathrm{a}(1)\, W^{\mathrm{b}(1)}$, $E(\mathrm{Out}) = \mathrm{a}(3)\, W^{\mathrm{b}(3)}$, and $E(\mathrm{P}) = \mathrm{a}(1)\, W^{\mathrm{b}(1)} - \mathrm{a}(3)\, W^{\mathrm{b}(3)}$, where $\mathrm{b}(1) > \mathrm{b}(3)$. The proportion of $E(\mathrm{In})$ that is deposited as growth [$E(\mathrm{P})$] will therefore decrease with increasing fish weight (W), and this relationship can be described by the following function:

$$E(P)/E(In) = [a(1)\ W^{b(1)} - a(3)\ W^{b(3)}] / [a(1)\ W^{b(1)}] \quad (1.15a)$$

$$= 1 - [a(1)/a(3)]\ W^{b(3)-b(1)} \quad (1.15b)$$

This function gives an expression of the efficiency with which ingested energy is deposited as growth, and efficiency declines with increasing body size. The relationship between efficiency and fish weight can be described by an allometric function, and if the logarithm of efficiency [*E*(P)/*E*(In)] is plotted against the logarithm of body weight, a straight line, having a negative slope, is obtained. In practical studies, a slope of -0.1 to -0.2 has usually been found.

Hitherto, growth has been defined as an increase in the energy content of the fish body, and it has been assumed that an increase in body weight is synonymous with an increase in energy content. In other words, it has been assumed that the composition of fish tissue is constant and a change in weight will accurately reflect changes in the energy content of the body. The composition of the fish body is not, however, constant, and the relative proportions of protein, lipid, carbohydrate and water making up a given weight gain will vary with feeding conditions and levels of food supply. Lipid deposition tends to increase with increasing food supply, such that the bodies of fish fed maximally will generally contain proportionately more lipid than those held on restricted rations. Deposition of 1 g lipid (38 kJ) leads to a weight

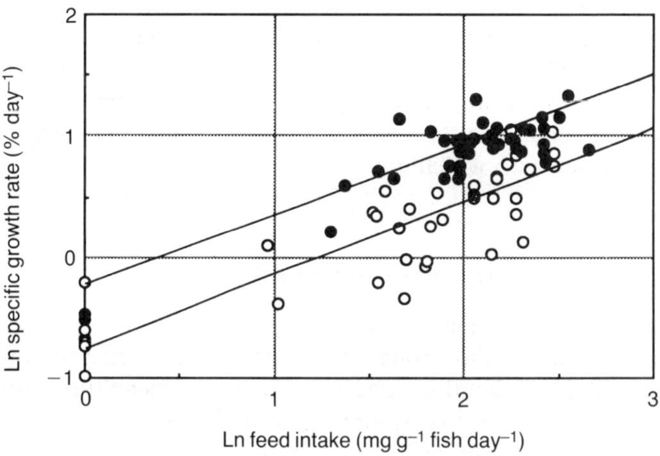

Fig. 1.7 Effects of different levels of sustained exercise (fish reared in standing water or forced to swim against a water current for a period of 9 weeks) on the relationship between rates of food intake and growth of juvenile Arctic charr, *Salvelinus alpinus*. Each point indicates an individual fish. Filled circles indicate exercised fish forced to swim against a water current (initial speed 2 BL s^{-1}). Note that exercised fish show higher growth rates, at a given level of food intake, than do the controls. Modified from Christiansen and Jobling (1990).

Table 1.5 Effects of sustained swimming on rates of protein synthesis and breakdown, and on the efficiency of protein deposition in different tissues of the rainbow trout, *Oncorhynchus mykiss* (adapted from Houlihan and Laurent, 1987)

Tissue	Protein synthesis (mg d^{-1})	Protein breakdown (mg d^{-1})	Protein gain (mg d^{-1})	Deposition efficiency* (%)
Control				
Gill	27.76	26.53	1.23	4.4
Heart	0.33	0.27	0.06	18.2
Red muscle	2.89	2.00	0.89	30.8
White muscle	27.15	5.54	21.61	79.6
TOTAL	58.13	34.34	23.79	40.9
Exercised				
Gill	31.44	29.14	2.30	7.3
Heart	0.55	0.43	0.12	21.8
Red muscle	6.81	4.65	2.16	31.7
White muscle	59.30	17.69	41.61	70.2
TOTAL	98.10	51.91	46.19	47.1

* Computed as (gain/synthesis) × 100.

increase of 1 g, whereas deposition of 1 g protein (24 kJ) also leads to the deposition of 3–4 g water. In other words, the deposition of 1 g protein will lead to a gain of 4–5 g tissue weight. Thus, changes in weight may not always give an accurate reflection of growth in terms of energy gain, and a failure to take into account the changes in body composition that may occur under different rearing conditions can result in incorrect conclusions being drawn about the efficiency with which ingested energy has been deposited as body tissue.

Sustained exercise and growth

In connection with the farming of salmonid species, attempts have been made to avoid exposing the fish to strong water currents in the rearing units. The rationale behind this is that the high levels of energy expenditure required by the fish in swimming, to hold station against the current, would lead to reduced rates of growth and/or increased food costs. Results of a number of experimental studies have not borne out these expectations, and there is an increasing body of evidence suggesting that growth rates of salmonids increase when the fish are forced to swim at moderate speeds for prolonged periods of time (Davison, 1989; Christiansen and Jobling, 1990).

There is no denying that the energy costs of swimming are high, and increased demands for oxygen during swimming have been confirmed in studies carried out on a range of fish species (Beamish, 1978; Brett and Groves, 1979). Studies on the relationship between swimming activity and metabolic

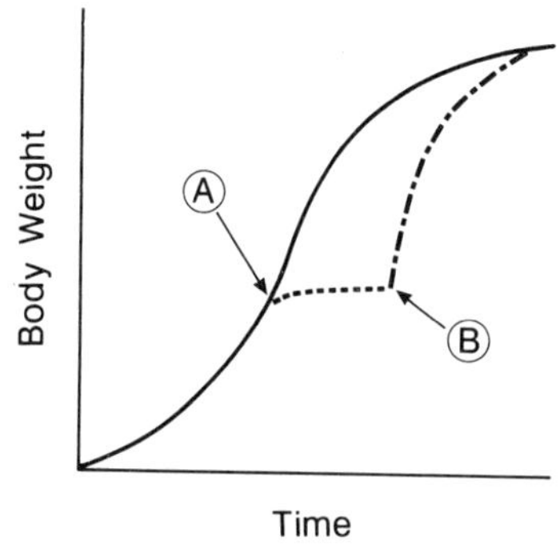

Fig. 1.8 Growth and development of normal animals (solid curve) and deviations from the normal growth curve during periods of restricted feeding and compensatory growth. A, Time point at which feeding restriction was imposed; B, return to unlimited food supply. Note that growth rate is rapid during the recovery period, and that the growth curve illustrated for restricted–re-fed animals (broken curve) shows a case of complete 'catch-up'.

rate have usually been conducted on individual fish, and when not forced to swim against a current, these fish generally show only low levels of spontaneous activity. When groups of fish have been held in respirometers, however, levels of spontaneous activity may increase and this can lead to marked, though transient, increases in metabolic rate (Brett and Groves, 1979; Umezawa *et al.*, 1983). These increases in activity levels appear to be associated with bouts of aggressive behaviour, which occur at irregular intervals. When groups of salmonid fish are exposed to water currents they orientate against the current, may form schools, and levels of aggression appear to be markedly reduced compared with those seen in groups of fish held in standing water (Christiansen and Jobling, 1990). The net result may be that the overall energy expenditures of groups of fish swimming for prolonged periods at moderate speed, and those exhibiting high levels of spontaneous activity at irregular intervals, differ very little from each other (Christiansen *et al.*, 1991).

Thus, levels of energy expenditure may differ little between groups of fish exposed to water currents and those reared in standing water, but this cannot explain why the exercised fish show more rapid rates of weight gain. The improved weight gain in the exercising fish does not, primarily, appear to be due to increased food consumption, but is more the result of these fish displaying better food conversion (Christiansen and Jobling, 1990; Fig. 1.7). In other words, fish forced to swim at moderate speeds for prolonged periods show a greater weight gain per unit food consumed than do conspecifics reared in standing water. The basis of this improved food conversion must lie in the physiological changes brought about by inducing the fish to swim for prolonged periods of time.

It is known that prolonged swimming at moderate speeds can give rise to marked changes in the swimming muscles of salmonid fishes (Davison, 1989). Since these muscles represent 50–60% of the body weight in salmonids, any changes here will have a considerable influence on the growth of the body as a whole. Prolonged exercise leads to a dramatic increase in muscle fibre size (hypertrophy) (Davison, 1989) and also induces marked increases in rates of protein synthesis, turnover, and deposition (Houlihan and Laurent, 1987). Rates of protein synthesis, breakdown and deposition are increased in all tissues during exercise, but the most marked changes occur in the swimming muscles (Houlihan and Laurent, 1987). Tissues such as gill and heart have inherently high rates of protein turnover, and protein deposition efficiency is low. In these tissues there may be only relatively small changes in rates of protein synthesis and breakdown in response to exercise. Nevertheless, protein deposition is increased during exercise, but the efficiency of deposition remains low (Table 1.5). On the other hand, protein turnover in muscle is low and deposition efficiency high. Exercise leads to marked increases in both protein synthesis and breakdown, and the net result is an increase in deposition. The efficiency of deposition of protein in the muscle may, however, decline during exercise (Table 1.5). Thus, within individual tissues, exercise leads to increased protein deposition either without any change, or with a decline, in deposition efficiency. When viewed at the whole-body level, however, a different picture emerges. Rates of protein synthesis, breakdown and deposition all increase, but, at the same time, the efficiency of protein deposition is improved by a few percentage points (Table 1.5). Thus, the disproportionately large effects of exercise on the swimming muscles may lead to both increased body protein growth and an overall improvement in protein deposition efficiency.

Compensatory or 'catch-up' growth

Growth rates of fish may be highly variable and, in many cases, growth of fish in the wild appears to be limited by food availability. When food supplies are increased following a period of starvation or restricted feeding, fish and other animals may show a growth spurt (Fig. 1.8). This type of growth response is generally referred to as **compensatory, 'catch-up'** or **recovery growth** (Wilson and Osbourn, 1960; Weatherley and Gill, 1987; Miglavs and Jobling, 1989). In some cases, recovery from starvation or food restriction has been reported as being complete (i.e. following transfer to unlimited food resources, the previously food-restricted animals achieve the same body size as animals allowed continuous access to food). In other cases recovery is only partial, in that the restricted–re-fed animals never attain the sizes of continuously fed conspecifics. There may be species differences in the responses shown to restricted–re-feeding regimes, and both the intensity and duration of the nutritional restriction also appear to influence growth during the recovery period.

The physiological basis of compensatory growth is incompletely understood, but restricted–re-fed animals may both increase food intake (become hyperphagic) and show improved food conversion efficiencies compared with continuously fed conspecifics. In some cases, however, restricted–re-fed animals have been reported as showing rapid recovery growth and improved food conversion without becoming hyperphagic. This suggests the possibility that improved conversion may be linked to differences in metabolic expenditure between restricted–re-fed and continuously fed animals.

Animals adapt to reduced levels of energy intake by reducing energy expenditure (p. 24), and restrictions in food intake generally result in lower rates of weight loss than would be expected if the metabolic rates observed prior to deprivation were maintained. These findings have been used as the basis of a hypothesis developed in an attempt to explain the changes observed during periods of compensatory growth (Wilson and Osbourn, 1960). It is suggested that in the period immediately following transfer from a restricted to a liberal feeding regime, the metabolic rate of the animal may not return abruptly to the same level as that of a continuously fed animal. In other words, re-adaptation to the fed state may require time, and low rates of energy expenditure may be maintained even though the animals are no longer food-restricted. It is proposed that the consequences of animals displaying such metabolic responses would be as follows.

1. Low metabolic expenditure occurring concurrently with either normal or high rates of food consumption would result in large amounts of food energy being available for growth. Thus, rates of growth should be rapid during periods of recovery from bouts of starvation or food restriction.
2. Low metabolic expenditure during the period of recovery would be expected to lead to a large weight gain per unit food intake, i.e. efficient food conversion.

Whilst this has been considered an attractive hypothesis, it is by no means the only one that can be invoked to explain the rapid growth and improved food conversion observed during recovery growth. According to this first hypothesis, definite predictions are made about the levels of energy expenditure to be expected during the period of recovery growth. Low rates of energy expenditure have not, however, always been reported for animals showing 'catch-up' growth, so experimental support for the hypothesis is equivocal.

A second hypothesis regarding the physiological basis of compensatory growth is related to application of the anabolism–catabolism concept of growth control. Growth is the result of a difference in the magnitude of anabolism (tissue synthesis) and catabolism (degradation). Rates of both tissue synthesis and breakdown decline with the imposition of feed restriction, and this leads to reduced rates of turnover and lower energy expenditure in food-deprived, than in continuously fed, animals. Transfer from a restricted to a liberal feeding regime is expected to result in a marked increase in synthetic activity, which, in turn, would be reflected in an increase in energy expenditure. The high growth rate, and efficient food conversion, observed during recovery is not, therefore, proposed to be the result of energy expenditure being maintained at very low levels. According to this second scenario, there is hypothesized to be an uncoupling of anabolic and catabolic processes, in that rates of synthesis and energy expenditure are suggested to increase quite rapidly, whilst rates of tissue breakdown take a longer time to return to the levels observed in continuously fed animals. Thus the rapid growth and efficient food conversion observed following re-feeding is proposed to be the result of the difference between anabolism and catabolism being greater in restricted–re-fed individuals (due to low rates of tissue breakdown) than in animals given continuous access to food. This second hypothesis appears to be more satisfactory than the first, since it enables the changes observed during recovery growth to be encompassed within the single framework of the anabolism–catabolism concept.

In both hypotheses it is proposed that the improvements in growth and food conversion observed during recovery growth are the result of restricted–re-fed animals having a greater efficiency of energy utilization than continuously fed individuals. Whether or not these changes are the result of improved energetic efficiency during the period of recovery growth is, however, open to question, since changes in growth rates (rates of weight gain) and food conversion (weight gain per unit weight of food consumed) could equally well be brought about by differences in deposition patterns and body composition of animals held on different feeding regimes (pp. 3 and 31). For example, following a period of starvation or feed restriction in Atlantic cod, *Gadus morhua*, the tissues of the swimming muscle will be the first to be repleted. Repletion involves a rapid increase in weight of the muscle, a relative increase in muscle lipid and glycogen content and a corresponding decrease in percentage muscle water. The muscle water content must drop below about 82% before the liver lipid reserves begin to show any marked increase (Black and Love, 1986). Thus, gross changes in weight gain and food conversion need not necessarily directly reflect energetic changes, and an improvement in food conversion does not of itself imply that there has been an improvement in energetic efficiency.

Growth rates increase and food conversion is improved when fish of several species are allowed access to unlimited food supplies following a period of deprivation, and several hypotheses have been proposed in an attempt to explain the phenomenon of compensatory growth. However, no single hypothesis adequately explains all the changes observed during the course of recovery from a period of food deprivation, and several different mechanisms are undoubtedly involved. These may include the development of hyperphagia, changes in energetic expenditure and efficiency of energy utilization, and changes in composition of the tissue gain compared with fish allowed continuous access to food.

Somatic (P_S) versus reproductive (P_R) growth

Compared with sperm, eggs represent a large energetic investment, and most studies of reproductive growth (P_R) have concentrated upon the examination of egg development in the female. Approximately 25% of the egg will be composed of dry matter (mostly protein and lipid), and the energy content will usually be approximately 6 kJ g^{-1}. In fish species the yolk incorporated into the oocyte during vitellogenesis is usually either the sole, or the major, source of nutrients for the developing embryo and post-hatch larva prior to the time of exogenous feeding. There may also be some reliance on nutrients from the yolk for some time after the young fish has begun to take exogenous food. Thus, the level of investment made by the female in yolk production is of critical importance for the survival of the developing young.

There is little synthesis of yolk in the ovaries and oocytes. The liver is the major site of synthesis, and vitellogenin is transported from the liver to the ovaries in the bloodstream, before being incorporated into the developing oocytes. Vitellogenesis is quite often a protracted process, requiring several days or weeks for completion, and during this period the oocyte increases markedly in size owing to the incorporation of the yolk. In salmonids, for example, the diameter of the oocyte may increase from 50 μm to approximately 5 mm during the course of vitellogenesis, and this represents a millionfold increase in volume. This increase in oocyte volume will lead to a marked increase in the size of the ovaries, and immediately prior to spawning, the ovaries of some fish species may represent 20–30% of the total body weight. This increase in size of the ovaries is usually most dramatic in species, such as salmonids and cyprinids, that have a relatively short spawning season. In species which have a longer spawning season the oocytes may develop in batches, such that spawning events may occur several times during the course of the year. In these species the ovaries may never exceed 5–10% of the total body weight, but a measurement of ovarian size will not give a true reflection of the total annual investment in reproductive growth (Wootton, 1990).

In the context of the energy budget, production can be divided into somatic (P_S) and reproductive (P_R) growth such that

$$P_S + P_R = R - (F + U + M) \qquad (1.16)$$

Provided that the other components of the budget remain unchanged during the period of reproductive growth, P_S and P_R can be considered as being directly competitive processes for finite resources. There is, however, evidence that changes in components other than somatic growth may contribute to partially offset the costs of reproduction. Levels of activity may, for example, be lower during the period prior to spawning than at other times of the year, and it is suggested that the energetic savings made are channelled towards reproductive growth (Koch and Wieser, 1983). Nevertheless, somatic and reproductive growth can be viewed as being competitors for limited resources, and the drop in somatic growth rate that occurs when fish mature is often taken as indicative of this competition. Production of gametes may be given priority over somatic growth in mature fish, such that the annual increment of body weight increase in mature individuals is often small. Thus, the investment in somatic growth may fall from approximately 25% of the annual energy budget in young, immature fish to 0–5% in mature individuals, whilst at the same time reproductive growth increases from 0% of the budget in immatures to 20% or so in mature fish (Wootton, 1990).

Several species breed only once before dying, and in these semelparous fish species the energetic investment made in gametes and spawning activities is considered to compromise survival. The majority of fish species are, however, iteroparous, with mature individuals being active for several breeding seasons. However, even amongst the iteroparous species, the costs of reproduction may be considerable. In the period immediately following spawning, the fish may be in poor condition, and several weeks or months may be required for recovery. Amongst the iteroparous species there is often a temporal alternation between the repletion of the somatic reserves and the development of the gonads. This leads to the fish showing distinct patterns of energy storage and depletion linked to cycles of reproductive growth. In this way, the fish may build up sufficient reserves to enable reproductive growth to continue during periods when food supplies are limited and rates of ingestion are insufficient to support growth.

Thus, somatic energy reserves are repleted during times of plentiful food supply, whilst the energetic demands of reproductive growth will lead to a depletion of these energy depots at a later stage during the cycle. It is, for example, often observed that the condition ($K = (\text{weight}/\text{length}^3) \times 100$) of the fish changes with season, and the condition factor (K) usually reaches its nadir in the immediate post-spawning period. At the same time, the water content of the muscle may increase and the relative proportion of other constituents decrease. In extreme cases, such as some species of flatfish, the swimming musculature may contain approximately 96% water (less than 3% protein and 1% lipid) at the end of the spawning season. It is, first and foremost, the white, anaerobic muscle that shows these changes in composition, with the red muscle composition being little affected.

In mature individuals, reproductive growth will occur to the detriment of somatic growth, but when food supplies are restricted there will also be some limitation imposed upon reproductive growth. The effects of food restriction on the reproductive growth of fish have been studied for a small number of species under laboratory conditions. Iteroparous species of fish held under conditions of very low food availability may not reproduce each year, but may take one or more years to build up the energy reserves required to support gamete development and spawning activities. When conditions are adequate to enable the fish to spawn each year, food supply may influence reproductive growth through effects upon fecundity (egg number/body weight), egg size or quality. Restricted food supply can result in reduced fecundity either by bringing about a reduction in the numbers of oocytes that begin development (reduced oocyte recruitment) or by influencing the numbers of recruited oocytes that complete development. Food restriction can also lead to females producing smaller eggs containing reduced quantities of yolk. For any given species, a limitation in food supply can lead to a simultaneous reduction in egg size and female fecundity, with severe food restriction usually leading to a reduction in the proportion of the female population that spawns. To what extent the different effects of food limitation will be observed depends both upon the degree of food restriction and upon the species under study.

1.8 CONCLUDING REMARKS

The balanced energy budget is being increasingly used as a tool in fisheries science, and it is now common practice to use some form of bioenergetics modelling both in aquaculture production planning and in fisheries management. The accuracy and reliability of such modelling will be determined both by the methods used in estimating, and by the precision of, the input terms. If the penalty for incorrect prediction is high, then increased effort should be channelled into obtaining precise estimates of the critical input parameters. Sensitivity analysis is a useful, cheap and rapid way of evaluating how uncertainty in the input parameters affects the model predictions. This type of analysis also reveals the relative importance of the different components of the energy budget, enabling identification of the input parameters that are required to be estimated with the greatest precision.

The majority of bioenergetics information has been obtained from laboratory studies conducted on young, actively growing fish, and it has been possible to construct energy budgets for the young of a number of carnivorous fish species, growing actively under conditions of plentiful food supply. In most cases, it has been found that 20–35% of the ingested energy is deposited as growth. These bioenergetics studies give a reasonably good picture of the energy partitioning in fish held under laboratory conditions. Energy expenditure of fish in the wild has rarely been quantified, and until recently it was not possible to collect such information. The recent developments within telemetry have enabled researchers to monitor activity patterns of wild fish. Correlations between the energy expenditure on activity and some physiological parameters have also been sought, and information of this type has contributed to the quantification of field metabolic rates. Much remains to be done, but further advances are expected within the near future.

A major problem within fish bioenergetics has been the accurate assessment of food ingestion. The recently developed radiographic method allows the food intake of individual fish to be monitored, and is therefore an important advance which enables in-depth studies of feeding and growth relationships to be conducted. The method does not contribute to solving the problem of the estimation of food intake by wild fish, but has important applications within the field of aquaculture. In traditional growth trials the relationships between food supply and growth are investigated, but the information that can be obtained from these types of study has severe limitations. Since data are restricted to groups, and not individuals within the group, results must be reported in terms of the gross changes occurring within the group as a whole. Thus, although differences between groups may be observed and reported, the traditional growth trial does not permit any detailed analysis of the causal factors underlying these differences. The monitoring of food intake and growth of individuals within the group permits such an analysis to be made, and this can provide valuable additional information about the relative importance of the different biotic and abiotic factors influencing growth.

Other recent developments within fish bioenergetics include the application of techniques that enable studies of anabolic and catabolic processes to be made at the tissue level, and the study of hormonal control of metabolic and growth processes. To date, such studies are comparatively few, but they have already made important contributions to our understanding of fish growth physiology.

Thus, the study of fish bioenergetics enables the investigation of problems related to fisheries management, and also has a central place within aquacultural research. This chapter has discussed the basic principles of bioenergetics and the methods used for obtaining estimates of the components of the energy budget, and attempts have been given to indicate areas requiring further research effort and development.

REFERENCES

Abbott, J.C. and Dill, L.M. (1989). The relative growth of dominant and subordinate juvenile steelhead trout (*Salmo gairdneri*) fed equal rations. *Behaviour*, **108**, 104–13.

Abbott, J.C., Dunbrack R.L. and Orr, C.D. (1985) The interaction of size and experience in dominance relationships of juvenile steelhead trout (*Salmo gairdneri*). *Behaviour*, **92**, 241–53.

Bagenal, T. (ed.) (1978) *Methods for Assessment of Fish Production in Fresh Waters: IBP Handbook 3*, Blackwell Scientific, Oxford, 366 pp.

Beamish, F.W.H. (1978) Swimming capacity, in *Fish Physiology*, Vol. VII (eds W.S. Hoar and D.J. Randall), Academic Press, London, pp. 101–87.

Beitz, D.C. (1985) Physiological and metabolic systems important to animal growth – an overview. *J. Anim. Sci.*, **61**, 1–20.

Black, D. and Love, R.M. (1986) The sequential mobilisation and restoration of energy reserves in tissues of Atlantic cod during starvation and refeeding. *J. comp. Physiol.*, **156B**, 469–79.

Blaxter, K. (1989) *Energy Metabolism in Animals and Man*, Cambridge Univ. Press, Cambridge, 336 pp.

Brett, J.R. (1971) Satiation time, appetite and maximum food intake of sockeye salmon (*Oncorhynchus nerka*). *J. Fish. Res. Bd Can.*, **28**, 409–15.

Brett, J.R. (1972) The metabolic demand for oxygen in fish, particularly salmonids, and a comparison with other vertebrates. *Respir. Physiol.*, **14**, 151–70.

Brett, J.R. (1979) Environmental factors and growth, in *Fish Physiology*, Vol. VIII (eds W.S. Hoar, D.J. Randall and J.R. Brett), Academic Press, London, pp. 599–675.

Brett, J.R. and Groves, T.D.D. (1979) Physiological energetics, in *Fish Physiology*, Vol. VIII (eds W.S. Hoar, D.J. Randall and J.R. Brett), Academic Press, London, pp. 279–352.

Brown, C.R. and Cameron, J.N. (1991a) The induction of Specific Dynamic Action in channel catfish by infusion of essential amino acids. *Physiol. Zool.*, **64**, 276–97.

Brown, C.R. and Cameron, J.N. (1991b) The relationship between Specific Dynamic Action (SDA) and protein synthesis rates in the channel catfish. *Physiol. Zool.*, **64**, 298–309.

Brown, M.E. (1946) The growth of brown trout (*Salmo trutta* Linn.). II. The growth of two-year-old trout at a constant temperature of 11.5°C. *J. exp. Biol.*, **22**, 130–44.

Brown, M.E. (ed.) (1957) *The Physiology of Fishes, I. Metabolism; II. Behavior*, Academic Press, New York.

Cho, C.Y. and Kaushik, S.J. (1985) Effects of protein intake on metabolizable and net energy values of fish diets, in *Nutrition and Feeding in Fish* (eds C.B. Cowey, A.M. Mackie and J.G. Bell), Academic Press, London, pp. 95–117.

Christiansen, J.S. and Jobling, M. (1990) The behaviour and the relationship between food intake and growth of juvenile Arctic charr, *Salvelinus alpinus* L., subjected to sustained exercise. *Can. J. Zool.*, **68**, 2185–91.

Christiansen, J.S., Jørgensen, E.H. and Jobling, M. (1991) Oxygen consumption in relation to sustained exercise and social stress in Arctic charr (*Salvelinus alpinus* L.). *J. exp. Zool.*, **260**, 149–56.

Cossins, A.R. and Bowler, K. (1987) *Temperature Biology of Animals*, Chapman and Hall, London, 340 pp.

Davison, W. (1989) Training and its effects on teleost fish. *Comp. Biochem. Physiol.*, **94A**, 1–10.

Elliott, J.M. (1975) Number of meals in a day, maximum weight of food consumed in a day and maximum rate of feeding for brown trout, *Salmo trutta* L. *Freshwat. Biol.*, **5**, 287–303.

Elliott, J.M. (1976) Energy losses in the waste products of brown trout (*Salmo trutta* L.). *J. Anim. Ecol.*, **45**, 561–80.

Evans, D.O. (1990) Metabolic thermal compensation by rainbow trout: effects on standard metabolic rate and potential usable power. *Trans. Am. Fish. Soc.*, **119**, 585–600.

Foster, R.P. and Goldstein, L. (1969) Formation of excretory products, in *Fish Physiology*, Vol. I (eds W.S. Hoar and D.J. Randall), Academic Press, London, pp. 313–50.

Gerking, S.D. (ed.) (1967) *The Biological Basis of Freshwater Fish Production*, Blackwell Scientific, Oxford, 496 pp.

Gerking, S.D. (ed.) (1978) *Ecology of Freshwater Fish Production*, Blackwell Scientific, Oxford, 520 pp.

Goolish, E.M. (1991) Anaerobic swimming metabolism of fish: sit-and-wait versus active forager. *Physiol. Zool.*, **64**, 485–501.

Grayton, B.D. and Beamish, F.W.H. (1977) Effects of feeding frequency on food intake, growth and body composition of rainbow trout (*Salmo gairdneri*). *Aquaculture*, **11**, 159–72.

Grodzinski, W., Klekowski, R.Z. and Duncan, A. (eds) (1975) *Methods for Ecological Bioenergetics: IBP Handbook 24*, Blackwell Scientific, Oxford, 368 pp.

Hawkins, A.J.S. (1991) Protein turnover: a functional approach. *Funct. Ecol.*, **5**, 222–33.

Henken, A.M., Kleingeld, D.W. and Tijssen, P.A.T. (1985) The effect of feeding level on apparent digestibility of dietary dry matter, crude protein and gross energy in the African catfish, *Clarias gariepinus* (Burchell, 1822). *Aquaculture*, **51**, 1–11.

Higgins, P.J. and Talbot, C. (1985) Growth and feeding in juvenile Atlantic salmon (*Salmo salar* L.), in *Nutrition and Feeding in Fish* (eds C.B. Cowey, A.M. Mackie and J.G. Bell), Academic Press, London, pp. 243–63.

Hoar, W.S., Randall, D.J. and Brett, J.R. (eds) (1979) *Fish Physiology*, Vol. VIII, Academic Press, London, 786 pp.

Houlihan, D.F. (1991) Protein turnover in ectotherms and its relationships to energetics. *Adv. comp. env. Physiol.*, **7**, 1–43.

Houlihan, D.F. and Laurent, P. (1987) Effects of exercise training on the performance, growth and protein turnover of rainbow trout (*Salmo gairdneri*). *Can. J. Fish. aquat. Sci.*, **44**, 1614–21.

Houlihan, D.F., Hall, S.J., Gray, C. and Noble, B.S. (1988) Growth rates and protein turnover in cod, *Gadus morhua*. *Can. J. Fish. aquat. Sci.*, **45**, 951–64.

Huntingford, F.A., Metcalfe, N.B., Thorpe, J.E., Graham, W.D. and Adams, C.E. (1990) Social dominance and body size in Atlantic salmon parr, *Salmo salar* L. *J. Fish Biol.*, **36**, 877–81.

Jobling, M. (1981a) Some effects of temperature, feeding and body weight on nitrogenous excretion in young plaice, *Pleuronectes platessa* L. *J. Fish Biol.*, **17**, 325–34.

Jobling, M. (1981b) The influences of feeding on the metabolic rate of fishes: a short review. *J. Fish Biol.*, **18**, 385–400.

Jobling, M. (1981c) Temperature tolerance and the final preferendum – rapid methods for the assessment of optimum growth temperatures. *J. Fish Biol.*, **19**, 439–55.

Jobling, M. (1983a) Growth studies with fish – overcoming the problems of size variation. *J. Fish Biol.*, **22**, 153–7.

Jobling, M. (1983b) Influence of body weight and temperature on growth rates of Arctic charr, *Salvelinus alpinus* (L.). *J. Fish Biol.*, **22**, 471–5.

Jobling, M. (1986) Gastrointestinal overload – a problem with formulated feeds? *Aquaculture*, **51**, 257–63.

Jobling, M. (1987) Growth of Arctic charr (*Salvelinus alpinus* L.) under conditions of constant light and temperature. *Aquaculture*, **60**, 243–9.

Jobling, M. and Davies, P.S. (1980) Effects of feeding on the metabolic rate and the Specific Dynamic Action in plaice, *Pleuronectes platessa* L. *J. Fish Biol.*, **16**, 629–38.

Jobling, M., Baardvik, B.M. and Jørgensen, E.H. (1989). Investigation of food–growth relationships of Arctic charr, *Salvelinus alpinus* L., using radiography. *Aquaculture*, **81**, 367–72.

Jobling, M., Jørgensen, E.H. and Christiansen, J.S. (1990) Feeding behaviour and food intake of Arctic charr, *Salvelinus alpinus* L., studied by X-radiography. *Proc. Third Int. Symp. on Feeding and Nutr. in Fish*, (Eds. M. Takeda and T. Watanabe), Tokyo University of Fisheries, Tokyo, pp. 461–9.

Johnston, I.A., Clarke, A. and Ward, P. (1991) Temperature and metabolic rate in sedentary fish from the Antarctic, North Sea and Indo–West Pacific Ocean. *Mar. Biol.*, **109**, 191–5.

Jørgensen, E.H. and Jobling, M. (1989) Patterns of food intake in Arctic charr, *Salvelinus alpinus*, monitored by radiography. *Aquaculture*, **81**, 155–60.

Kelly, J.M. and McBride, B.W. (1990) The sodium pump and other mechanisms of thermogenesis in selected tissues. *Proc. Nutr. Soc.*, **49**, 185–202.

Kirk, R.G. and Howell, B.R. (1972) Growth rates and food conversion in young plaice, *Pleuronectes platessa* L. fed on artificial and natural diets. *Aquaculture* **1**, 29–34.

Koch, F. and Wieser, W. (1983) Partioning of energy in fish: can reduction of swimming activity compensate for the cost of reproduction? *J. Exp. Biol.*, **107**, 141–6.

Landless, P.J. (1976) Demand feeding behaviour of rainbow trout. *Aquaculture*, **7**, 11–25.

Lied, E., Lund, B. and von der Decken, A. (1982) Protein synthesis *in vitro* by epaxial muscle ribosomes from cod, *Gadus morhua*. *Comp. Biochem. Physiol.*, **72B**, 187–93.

Lied, E., Rosenlund, G., Lund, B. and von der Decken, A. (1983) Effect of starvation and refeeding on *in vitro* protein synthesis in white trunk muscle of Atlantic cod (*Gadus morhua*). *Comp. Biochem. Physiol.*, **76B**, 777–81.

Lobley, G.E. (1986) The physiological bases of nutrient responses; growth and fattening. *Proc. Nutr. Soc.*, **45**, 203–14.

Magnuson, J.J. (1962) An analysis of aggressive behaviour, growth and competition for food and space in medaka (*Oryzias latipes* (Pisces: Cyprinodontidae)). *Can. J. Zool.*, **40**, 313–63.

Miglavs, I. and Jobling, M. (1989) Effects of feeding regime on food consumption, growth rates and tissue nucleic acids in juvenile Arctic charr, *Salvelinus alpinus*, with particular reference to compensatory growth. *J. Fish Biol.*, **34**, 947–57.

Milligan, L.P. and McBride, B.W. (1985) Energy costs of ion pumping by animal tissues. *J. Nutr.*, **115**, 1374–82.

Millward, D.J. (1989) The nutritional regulation of muscle growth and protein turnover. *Aquaculture*, **79**, 1–28.

Mulvaney, D.R., Merkel, R.A. and Bergen, W.G. (1985) Skeletal muscle turnover in young male pigs. *J. Nutr.*, **115**, 1057–64.

Niimi, A.J. and Beamish, F.W.H. (1974) Bio-energetics and growth of large mouth bass (*Micropterus salmoides*) in relation to body weight and temperature. *Can. J. Zool.*, **52**, 447–56.

Noeske, T.A. and Spieler, R.E. (1984) Circadian feeding time affects growth of fish. *Trans. Am. Fish. Soc.*, **113**, 540–44.

Noeske-Hallin, T.A., Spieler, R.E., Parker, N.C. and Suttle, M.A. (1985) Feeding time differentially affects fattening and growth of channel catfish. *J. Nutr*, **115**, 1228–32.

Pálsson, J.Ö., Jobling, M. and Jørgensen, E.H. (1992) Temperal changes in daily food intake of Arctic charr, *Salvelinus alpinus* L., of different sizes monitored by radiography. *Aquaculture*, **106**, 51–61.

Pedersen, C.L. (1987) Energy budgets for juvenile rainbow trout at various oxygen concentrations. *Aquaculture*, **62**, 289–98.

Ramnarine, I.W., Pirie, J.M., Johnstone, A.D.F. and Smith, G.W. (1987) The influence of ration size and feeding frequency on ammonia excretion by juvenile Atlantic cod, *Gadus morhua* L. *J. Fish Biol.*, **31**, 545–59.

Randall, D.J. and Wright, P.A. (1987) Ammonia distribution and excretion in fish. *Fish Physiol. Biochem.*, **3**, 107–20.

Reeds, P.J. (1987) Metabolic control and future opportunities for growth regulation. *Anim. Prod.*, **45**, 149–69.

Reeds, P.J. and Harris, C.I. (1981) Protein turnover in animals. Man in his context, in *Nitrogen Metabolism in Man* (eds J.C. Waterlow and J.M.L. Stephen), eds.), Applied Science Publ., London, pp. 391–408.

Savitz, J. (1969) Effects of temperature and body weight on endogenous nitrogen excretion in bluegill sunfish (*Lepomis macrochirus*). *J. Fish. Res. Bd Can.*, **26**, 1813–21.

Schreck, C.B. and Moyle, P.B. (eds) (1990) *Methods for Fish Biology*, American Fisheries Society, Bethesda, MD, 694 pp.

Smagula, C.M. and Adelman, I.R. (1982) Day-to-day variations in food consumption by largemouth bass. *Trans. Am. Fish. Soc.*, **111**, 543–8.

Staples, D.J. and Nomura, M. (1976) Influence of body size and food ration on the energy budget of rainbow trout, *Salmo gairdneri* Richardson. *J. Fish. Biol.* **9**, 29–43.

Stirling, H.P. (1977) Growth, food utilization and effects of social interaction in the European bass *Dicentrachus labrax*. *Mar. Biol.*, **40**, 173–84.

Sureau, D. and Lagardere, J.-P. (1991) Coupling of heart rate and locomotor activity in sole, *Solea solea* (L.), and bass, *Dicentrarchus labrax* (L.), in their natural environment by using ultrasonic telemetry. *J. Fish Biol.*, **38**, 399–405.

Symons, P.E.K. (1968) Increase in aggression and in the strength of the social hierarchy among juvenile Atlantic salmon deprived of food. *J. Fish. Res. Bd Can.*, **25**, 2387–401.

Tackett, D.L., Carter, R.R. and Allen, K.O. (1988) Daily variation in feed consumption by channel catfish. *Progve. Fish Cult.*, **50**, 107–10.

Talbot, C. and Higgins, P.J. (1983) A radiographic method for feeding studies on fish using metallic iron powder as a marker. *J. Fish Biol.*, **23**, 211–20.

Tytler, P. and Calow, P. (eds) (1985) *Fish Energetics – New Perspectives*, Croom Helm, London, 350 pp.

Umezawa, S.-I., Adachi, S. and Taneda, K. (1983) Group effect on oxygen consumption of the ayu (*Plecoglossus altivelis*) in relation to growth stage. *Jap. J. Ichthyol.*, **30**, 261–7.

Vens-Cappell, B. (1984) The effects of extrusion and pelleting of feed for trout on the digestibility of protein, amino acids and energy, and on feed conversion. *Aquacult. Eng.*, **3**, 71–89.

Weatherley, A.H. and Gill, H.S. (1987) *The Biology of Fish Growth*, Academic Press, London, 444 pp.

Webster, A.J.F. (1981) The energetic efficiency of metabolism. *Proc. Nutr. Soc.*, **40**, 121–8.

Wilson, P.N. and Osbourn, D.F. (1960) Compensatory growth after undernutrition in mammals and birds. *Biol. Rev.*, **35**, 324–63.

Winberg, G.G. (1956) Rate of metabolism and food requirements of fish. *Fish. Res. Bd Can Transl. Ser.*, No. 194 (1960), 202 pp.

Woiwode, J.G. and Adelman, I.R. (1991) Effects of temperature, photoperiod and ration size on growth of hybrid striped bass × white bass. *Trans. Am. Fish. Soc.*, **120**, 217–29.

Wootton, R.J. (1990) *Ecology of Teleost Fishes*, Chapman and Hall, London, 404 pp.

Wright, D.A. and Martin, F.D. (1985) The effect of starvation on RNA:DNA ratios and growth of larval striped bass, *Morone saxatilis*. *J. Fish Biol.*, **27**, 479–85.

Xiaojun, X. and Ruyung, S. (1990) The bioenergetics of the southern catfish (*Silurus meridionalis* Chen). I. Resting metabolic rate as a function of body weight and temperature. *Physiol. Zool.*, **63**, 1181–95.

Chapter two

Biochemical correlates of growth rate in fish

D.F. Houlihan, E.M. Mathers and A. Foster

2.1 INTRODUCTION

There are a number of reasons for attempting to correlate fish growth rates with measurable characteristics of the animals' tissues. On one level, the recent increase in understanding of the rates of protein turnover underlying protein accretion has been accompanied by a number of studies demonstrating that rates of protein synthesis are closely correlated with the synthetic capacity of the tissue. This chapter reviews these correlates of protein synthesis at a time when we are expanding our knowledge about the ways in which different fish species in different environments manipulate the rates of protein synthesis and degradation to generate growth rates. The commercial importance of fish growth to Man's well-being provides an equally forceful impetus for studying mechanisms of growth. There is a need to expand our understanding of the mechanisms controlling the growth rate of fish, so that decisions can be made when attempting selection programmes to improve growth performance or efficiency of food conversion or when selecting genes for transgenesis. In addition, there is a need for suitable markers which will indicate the recent growth rates of larvae, juvenile and adult fish caught from natural populations.

This chapter is constructed around the thesis that protein synthesis is a major contributor to energy metabolism in fish, and it is likely that correlates of protein synthesis and oxidative metabolism will be linked together. Rapidly growing fish may be expected to have high levels of protein synthesis-supporting machinery and the oxidative capacity to fuel protein synthesis. Most of the experimental data in this chapter have been drawn from fish in the North Sea, where we have correlated protein synthesis rates with RNA and oxidative enzyme levels in laboratory experiments and have used these to estimate growth rates of wild fish.

Fish Ecophysiology. Edited by J. Cliff Rankin and Frank B. Jensen. Published in 1993 by Chapman & Hall, London. ISBN 0 412 45920 5.

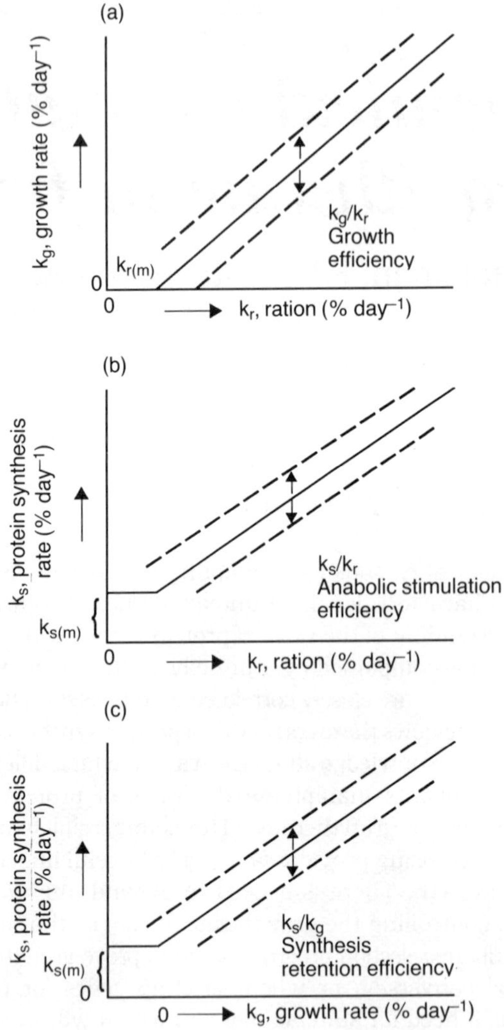

Fig. 2.1 Diagram of the likely relationships between the amount of food eaten (ration level, k_r) and growth and protein synthesis in fish. The diagram or model represents a single species at one temperature and holds for only a narrow size range. The solid lines, representing a mean response, are bracketed by dashed lines indicating the extent of deviations in stimulated (upper arrow) or stressed (lower arrow) conditions. (a) Amount of protein eaten and protein growth rate; $k_{r(m)}$ is the consumption ration for maintenance. (b) Relationship between amount of protein eaten and rates of protein synthesis; $k_{s(m)}$ is the maintenance value for protein synthesis. (c) Relationship between protein synthesis and protein growth. The model is for whole-body responses and the units are in percentage of the protein mass per day.

In this chapter, growth rates have been equated with protein growth rates, as protein is the major component of the body tissues and the energy costs of protein deposition are much higher than those of other body components (e.g. lipids and carbohydrates (Hommes, 1980)).

2.2 PROTEIN SYNTHESIS AND GROWTH

We have constructed a model to explain the way in which food consumption may regulate protein growth in fish through a stimulation of protein synthesis (Fig. 2.1). This model is both an idealized version of what might be happening in any particular species of fish and a blueprint for the way in which experiments should be constructed. Thus, the protein consumption, measured in individual fish by, for example, radiographic methods (e.g. Jobling *et al.*, 1989; Carter *et al.*, 1992a and see Jobling, Chapter 1, this volume), is combined with long-term growth studies and measurements of whole-body protein synthesis using reliable methods (Fauconneau, 1985; Houlihan, 1991). The end results are in the same units as those used for the food consumption, growth studies and protein synthesis rates (in this case percentage of the protein mass per day, frequently termed fractional rates).

In many animals, increases in food intake seem to drive increases in protein synthesis and hence growth rate (Fig. 2.1(a) and (b)). The relationships between ration and growth rate have been investigated in many studies, and there is continued discussion about whether a curvilinear relationship gives the best fit to the available data (Jobling *et al.*, 1989; McCarthy *et al.*, 1992; Jobling, Chapter 1, this volume). Calculating growth rates in wet weight units may be partly responsible for the curved relationship, because alterations occur in the nitrogen/carbon ratio of the bodies of salmonids such that at high food intake, carbon in the form of lipid increases (McCarthy, unpublished data). When the graph is constructed in terms of nitrogen ingested and nitrogen deposited, a linear relationship has been found (Birkett, 1969; Carter *et al.*, 1992a).

At a level of food consumption that maintains weight (maintenance ration, $k_{r(m)}$), protein synthesis can be described as being at a maintenance value ($k_{s(m)}$) (Fig. 2.1(b) and (c)). If the fish eat at or below this level or are starved, rates of protein synthesis do not seem to decrease further, although they might be expected to do so because of the saving in energy that would accrue. Perhaps maintenance synthesis can be taken to mean also minimal synthesis.

It is likely that protein synthesis rates above the maintenance ration will increase with the amount of protein eaten (Fig. 2.1(b)). In cod the slope of the line is one, i.e. each gram of protein eaten is correlated with the synthesis of one gram of protein (Houlihan *et al.*, 1989). Thus the consumption of protein brings about an anabolic stimulation of protein synthesis, probably through

Table 2.1 Results from mammals and fish relating growth rate (X, % day^{-1}) and protein synthesis rate (Y, % day^{-1}) using the formula $Y = a + bX$. The intercept values (a) indicate protein synthesis rates at zero growth rates and the slopes are the b values. The values are for whole bodies except for *O. mykiss* and rats, where the data come from muscle

Species	Intercept	Slope	Conditions (°C)	Reference
rats	5	2		Jepson *et al.* (1989)
Carassius auratus	1.8	2.5	25	Heba and Houlihan, unpublished data
Gadus morhua	1.42	1.6	10	Houlihan *et al.* (1988)
Oreochromis mossambicus	0.57	0.93	25	Heba and Houlihan, unpublished data
Oncorhynchus mykiss	0.21	1.3	8	McCarthy *et al.*, unpublished data

a combination of amino acid (p. 54) and hormonal stimulation (Houlihan, 1991; McMillan and Houlihan, 1992).

We know very little about the anabolic stimulation efficiency of various diets in fish. In pigs (Reeds *et al.*, 1981), rats (Laurent *et al.*, 1984), mice (Bernier *et al.*, 1987) and chickens (Muramatsu *et al.*, 1987), increases in

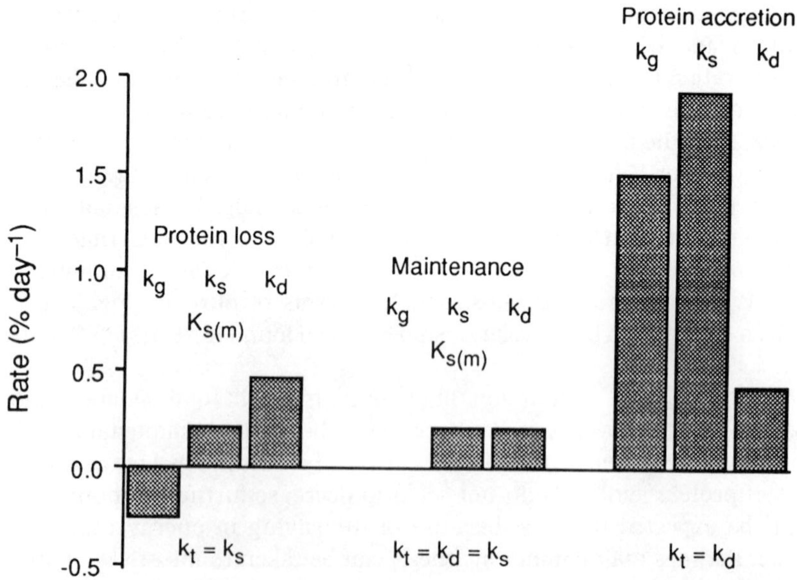

Fig. 2.2 Protein growth (k_g), synthesis (k_s) and degradation (k_d) in white muscle of rainbow trout (40 g live weight animals, 8 °C). When the animals are starving (protein loss), the rate of turnover (k_t) is equal to the rate of synthesis. When the animals are in steady state (maintenance), k_t equals synthesis or degradation, whereas in growing animals (protein accretion) the rate of turnover equals the rate of degradation ($k_t = k_d$). Unpublished data of McCarthy *et al.*

dietary protein bring about an elevation of whole-body protein synthesis, whereas restriction of energy intake results in a reduction of protein synthesis. We have little idea what effect variations in protein and energy intake have on protein turnover in fish, although ribosomal synthesis is believed to change in response to variations in protein and energy (Rosenlund and Lied, 1986).

A proportion of the synthesized proteins will be retained as growth (Fig. 2.1(c)), and the slope of the synthesis/growth line expresses the efficiency of retention of synthesized protein. Efficiency of retention of synthesized protein can also be calculated from the ratio of growth rate to synthesis rate ($[k_g/k_s] \times 100$); this ratio increases as the maintenance synthesis makes a smaller and smaller contribution to the growth synthesis. Thus there is no single value for the efficiency of retention of synthesized proteins but a range, depending upon the level of protein consumed; the slope of the synthesis/growth line may be a better descriptor of retention of efficiency than the ratio (values for various species are given in Table 2.1).

Protein degradation rates also appear to increase in fish as protein intake increases (Houlihan *et al.*, 1988), i.e. there is an increase in anabolic protein degradation, a phenomenon that is well known for mammals (Millward, 1989). Thus the relative balance between protein synthesis, growth and degradation will vary considerably in different nutritional conditions and this may give rise to some confusion in the definition of the term 'protein turnover'. In the steady state of nitrogen balance, the rate of synthesis will equal the rate of degradation, which in turn will equal the rate of turnover, k_t (Fig. 2.2). In the non-steady state of weight loss, the rate of degradation exceeds that of synthesis and $k_s = k_t$. In positive growth there will be the synthesis of newly made proteins and resynthesis of degraded molecules. Thus the turnover rate will be equal to the degradation rate (Wiesner and Zak, 1991).

2.3 PERTURBATIONS IN THE GENERAL MODEL

Individual differences within a species

It is becoming increasingly clear that there are wide differences in the food conversion efficiencies between individuals within a species. Some individuals within a group of rainbow trout, for example, will eat relatively little but grow at a high rate whereas other individuals will eat a lot but will grow relatively slowly (Fig. 2.3(a); McCarthy *et al.*, 1992). It is therefore necessary to indicate that within a species a range of food conversion efficiencies are possible (Fig. 2.1(a)). These individual differences become apparent when individual food consumption and growth rates are measured. The differences between individuals could be due to, among other things, differences in digestive absorption or differences in anabolic stimulation efficiency (Fig. 2.1(b)) and/or the

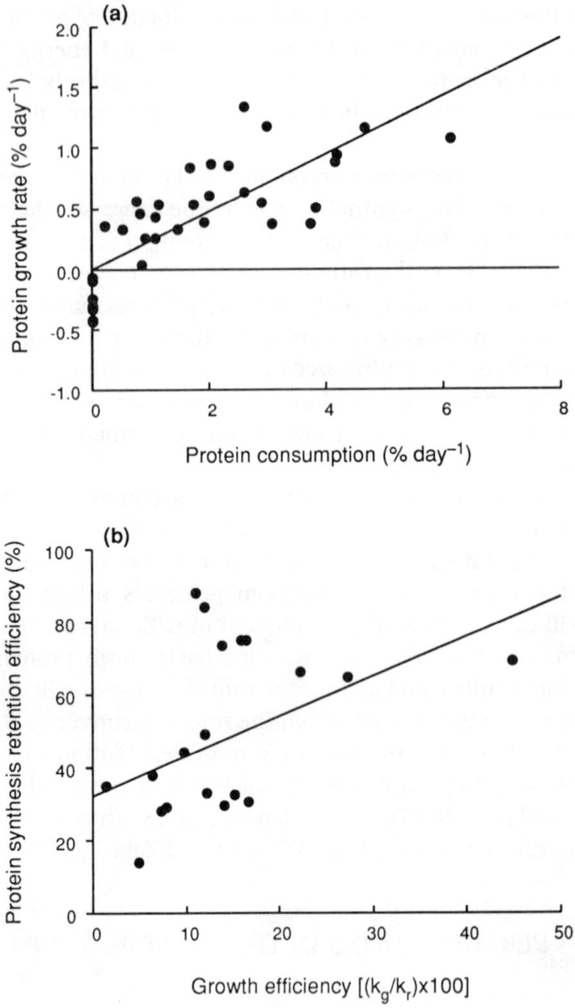

Fig. 2.3 (a) The relationship in individual rainbow trout, *Oncorhynchus mykiss* (40 g, 8 °C) between protein consumption (protein ingested as a percentage of the final protein mass per day) and white muscle protein growth (as a percentage of the protein mass per day). The regression equation is $y = 0.276x - 0.0110$, $n = 37$, $r^2 = 0.691$, $P < 0.001$. (b) the relationship between efficiency of protein utilization (protein growth/protein consumption) × 100) and efficiency of protein retention in white muscle ((k_g/k_S) × 100). The regression equation is $y = 1.01x + 32.16$, $n = 19$, $r^2 = 0.191$. $P < 0.05$. Taken from McCarthy *et al.*, unpublished data.

efficiency of the retention of synthesized proteins as growth (Fig. 2.1(c)). Initial results indicate that the trout with the highest food conversion efficiencies are also the fish with the highest efficiencies of retention of synthesized protein (Fig. 2.3(b)). Because of the influence that protein synthesis has on energy consumption (e.g. Reeds *et al.*, 1985; Hawkins, 1991; Houlihan, 1991), it therefore seems likely that maximization of food conversion efficiency is at least partly a question of reduction of protein turnover. It would be interesting to know whether the most efficient animals are also those with the lowest values for protein synthesis at maintenance, as has been suggested for mussels (Hawkins, 1991).

Within a species, environmental factors or changes in metabolism may significantly alter food/growth relationships. For example, for a given amount of food consumed, growth rate may be increased (increased food conversion efficiency) in 'stimulated' conditions such as slow-speed continuous swimming (Christiansen and Jobling, 1990) or after the application of growth hormone (Foster *et al.*, 1991). Protein synthesis always seems to increase in such stimulated conditions; protein degradation also increases in some tissues (i.e. increased turnover rates) (e.g. Houlihan and Laurent, 1987; Foster *et al.*, 1991).

Protein synthesis: are there real species differences?

When this question was asked in 1980 with respect to mammals the conclusion was that protein synthesis, when expressed relative to body weight to the power 0.75, is relatively constant among rats, rabbits, pigs, sheep, humans and cattle (Reeds and Lobley, 1980). In the beginning of this chapter it was emphasized that a single, species-specific value for protein synthesis is difficult to justify, except perhaps for maintenance synthesis. It is much more useful to analyse protein metabolism in terms of the relationship between protein growth and synthesis (Table 2.1). Differences were found between pigs and human infants in the 1980 study which suggest that pigs required a greater amount of protein synthesis per unit protein deposited and that a considerable amount of protein synthesis occurred at N equilibrium, i.e. maintenance. In fish there also appear to be considerable differences between species (Table 2.1); these may be due to differences in temperatures and body size (pp. 62–66). However, it seems that fish with the highest maintenance values are also those with the steepest slopes, i.e. for a given level of growth the highest protein synthesis rates (and the most inefficient retention of synthesized proteins as growth) result from both high maintenance costs and high anabolic stimulation. A similar pattern is seen for individual tissues (p. 56). Clearly selection of fish species for aquaculture based on rates of protein synthesis would not necessarily yield the fastest-growing fish. However, selection on the basis of high retention efficiency is likely to yield individuals with high food conversion efficiencies.

Stressful environments

Physiological and biochemical responses of fish to stressful environments have recently been reviewed (Adams, 1990; Holwerda and Opperhuizen, 1991) but little is known about stress and alterations in food/protein synthesis relationships and/or growth/synthesis relationships. A number of scenarios are possible: maintenance synthesis could be increased; anabolic stimulation could be depressed and there could be reduced growth for a given level of synthesis (suggested by the dashed lines in Fig. 2.1(a),(b),(c)). In a recent study of dab, *Limanda limanda*, exposed to sewage for 3 months, some of these predictions were confirmed (Houlihan *et al.*, unpublished data). Efficiencies of retention of synthesized proteins were found to be higher in control fish than in those exposed to sewage sludge, and there was evidence for the loss of body protein in the sewage-treated group, even though the animals were fed *ad libitum*. Thus long-term exposure to sewage sludge resulted in increased protein degradation rates. The effect was evident in most of the tissues, protein turnover rates in the white muscle were clearly increased in sewage-treated animals, and food consumption apparently could not be increased to compensate for the increased cost of protein turnover.

2.4 PROTEIN SYNTHESIS AND ENERGY CONSUMPTION

That there is a relationship between rates of protein synthesis and rates of oxygen consumption in animals seems beyond dispute. In the short term, the rate of protein synthesis is known to increase following a meal in fish (e.g. McMillan and Houlihan, 1988; Brown and Cameron, 1991; Lyndon *et al.*, unpublished data) and also in at least one invertebrate (Houlihan *et al.*, 1990). The post-prandial increase in oxygen consumption, often termed **specific dynamic action (SDA)**, may be due to this increase in whole-animal protein synthesis (Brown and Cameron, 1991; Lyndon *et al.*, unpublished data) accompanied by post-prandial changes in hormones and amino acids (Houlihan, 1991). However, the whole animal's response to a meal is a result of the complex changes in the individual tissues as different tissues respond to a meal with different time courses (McMillan and Houlihan, 1988, 1989, 1992, and unpublished data; Lyndon, 1990). For example, the liver protein synthesis rate in cod and trout (*Oncorhynchus mykiss*), have been found to increase very rapidly following a meal whereas other tissues are hardly responding. Thus the initial component of the SDA may be due to liver protein synthesis, with later stages being due to stimulation of muscle or intestinal protein synthesis.

In addition to these short-term correlations between protein synthesis and oxygen consumption, there is evidence from a whole range of animals that protein synthesis and oxygen consumption are closely linked (e.g. Waterlow and Millward, 1989; Houlihan, 1991).

How can the contribution that protein synthesis makes to oxygen consumption be calculated? One strategy is to calculate the energy cost to the whole animal of protein synthesis (MO_{2syn}, mmol animal^{-1} day^{-1}) from the experimentally determined rate of protein synthesis of the whole animal (k_s, the percentage of the total protein mass synthesized per day), the protein mass of the whole body (M_p) and a theoretical minimum cost of peptide bond elongation (MO_{2theo}, mmol O_2 g^{-1} protein synthesized) (e.g. Reeds *et al.*, 1985): $MO_{2syn} = (k_s/100) \times M_p \times MO_{2theo}$.

Using this method in fish it has been estimated that protein synthesis can account for 20–40% of oxygen consumption (Houlihan *et al.*, 1988; Houlihan, 1991; Lyndon *et al.*, unpublished data).

The theoretical minimal cost can be replaced by experimentally determined energy costs of protein synthesis generated from measuring oxygen consumption

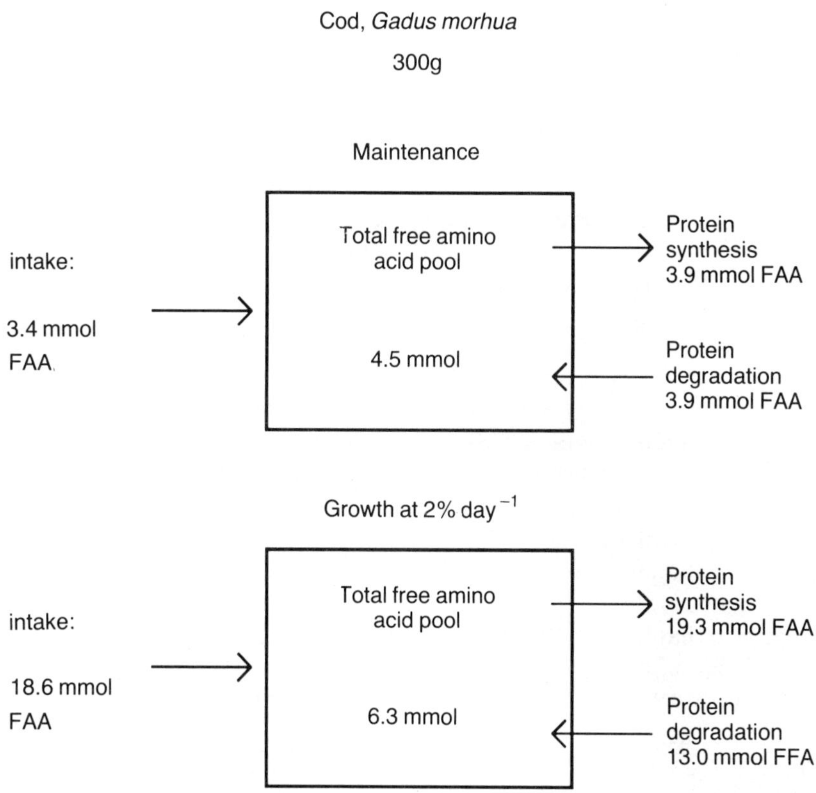

Fig. 2.4 Free amino acid pools, protein synthesis and degradation. The data are taken from cod, *Gadus morhua*, amino acid levels from Lyndon *et al.* (unpublished), protein intake, synthesis and degradation (per day) from Houlihan *et al.* (1989). The protein intake, synthesis and degradation have been converted to amino acid levels using a molecular weight for amino acids correcting for 1mole of water excluded per mole of peptide bonds formed (Fyhn, 1989).

and protein synthesis in fish tissues before and after the application of a protein synthesis inhibitor (Pannevis and Houlihan, 1992). Unfortunately these measurements have complicated the issue by suggesting that different tissues have different protein synthesis energy costs and that faster rates of synthesis may yield reduced costs (Pannevis and Houlihan, 1992).

2.5 FREE AMINO ACIDS AND PROTEIN TURNOVER

It is likely that the free amino acids resulting from digestion will play a key role in the regulation of protein synthesis rates and hence growth rates described in Fig. 2.1, although it is by no means clear how such regulation may be achieved (Waterlow *et al.*, 1978; Millward and Rivers, 1988). Thus measurement of free amino acids in tissues cannot at present be used to indicate growth rates. However, with the recent data on tissue free amino acid concentrations of cod (Lyndon *et al.*, unpublished data) it is possible to begin to integrate amino acid levels into protein synthesis and growth rates. In cod at maintenance rations in terms of protein, the quantity of amino acid derived if all the protein eaten is hydrolysed is a little less than the whole-animal total free amino acid pool (Fig. 2.4). The measured rates of protein synthesis can be converted into mmol of amino acid incorporated into protein and, as indicated above, these measurements suggest that the synthesis rate equals the absorption rate. As the animals are in steady state, the protein synthesis rate will be equalled by the protein degradation rate so that overall the nitrogen ingested in the form of amino acids will be lost as ammonia. There appears to be a good correlation between independent estimates of ammonia excretion in cod and protein turnover values (Houlihan *et al.*, 1989).

The situation is quite different in cod feeding at high rates. Unpublished data of Lyndon *et al.* indicate that there is little change in the free amino acid concentration in the tissues at various times after a meal; a small increase based upon white muscle values 18 h after a meal is indicated in Fig. 2.4. However, the intake of amino acids derived from the protein in the meal is approximately 3 × the total free amino acid pool size in the cod. It has been suggested that the accumulation of essential amino acids (EAA) in the tissues may be harmful and that EAA pools are maintained at low levels (Millward and Rivers, 1988). From these data it appears that protein synthesis in cod

Fig. 2.5 The relationships between tissue protein synthesis rates (% protein mass synthesized day^{-1}) and (a) tissue RNA concentration (μg RNA g^{-1} tissue), (b) tissue RNA:protein ratio (μg RNA mg^{-1} protein) and (c) tissue RNA:DNA ratio (mg RNA mg^{-1} DNA) for cod, *Gadus morhua*. The regression equations and significance values for (a) are 5 °C: $y = 0.002x - 3.839$, $r^2 = 0.853$, $P < 0.05$, DF = 4. 15 °C: $y = 0.003x - 3.201$, $r^2 = 0.875$, $P < 0.05$, DF = 4. The tissues were gill (G), intestine (I), ventricle (V), stomach (S) and white muscle (WM). The data are taken from Foster *et al.* (1992). (Note that the white muscle protein synthesis rates were very low but not zero.)

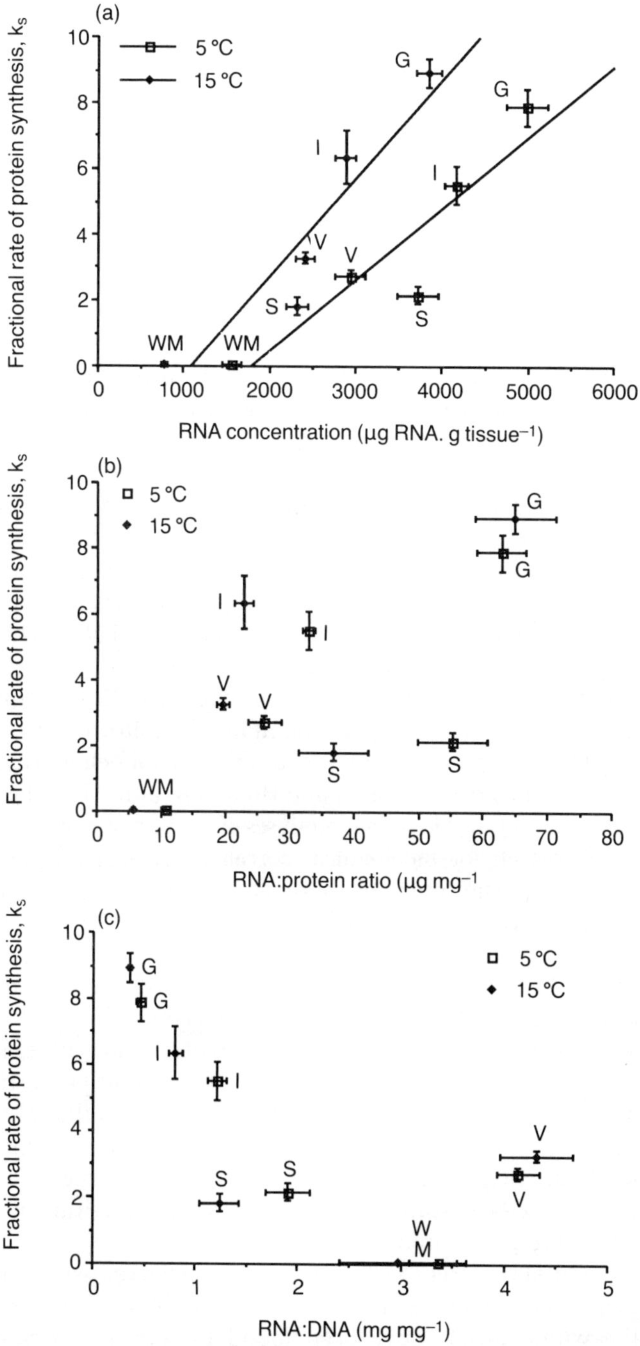
(a)
5 °C
15 °C
G
G
I
I
V
V
S
S
WM
WM
Fractional rate of protein synthesis, ks
0
2
4
6
8
10
0
1000
2000
3000
4000
5000
6000
RNA concentration (μg RNA. g tissue−1)
(b)
5 °C
15 °C
G
G
I
I
V
V
S
S
WM
Fractional rate of protein synthesis, ks
0
2
4
6
8
10
0
10
20
30
40
50
60
70
80
RNA:protein ratio (μg mg−1
(c)
5 °C
15 °C
G
G
I
I
S
S
V
V
W
M
Fractional rate of protein synthesis, ks
0
2
4
6
8
10
0
1
2
3
4
5
RNA:DNA (mg mg−1)

could act as a major factor in the removal of free amino acids which would otherwise seriously destabilize their concentrations in the tissues. This is not to deny the role of amino acid oxidation, transamination, the biosynthesis of non-essential amino acids from EAA when in surplus, and the possibility that amino acids play an osmoregulatory role in fish. From the calculations shown in Fig. 2.4 it seems highly likely that protein degradation will occur at some time after protein synthesis, and it may have an important role in the slow release of amino acids into the free pool.

2.6 GROWTH AND ITS CORRELATES IN THE TISSUES

It is an unsettled point whether all the tissues demonstrate the same stimulation in protein synthesis with ingested protein as modelled for the whole animal in Fig. 2.1. Certainly food reduction reduces protein synthesis rates in fish tissues (Fauconneau *et al.*, 1989; Houlihan, 1991; McMillan and Houlihan, 1992, and unpublished data; Carter *et al.*, 1992b). There is evidence that the different tissues in fish have different growth/protein synthesis relationships. Interestingly, just as for different species of fish, tissues with the highest maintenance values also have the highest increases in protein synthesis for a given increase in growth rate (Houlihan *et al.*, 1988). This suggests that there is a high cost in terms of protein turnover in those tissues with a high expansibility of protein synthesis; approximately 20% of this maintenance synthesis is needed simply to maintain the mitochondria, and it is notable that at times of protein synthesis stimulation, mitochondrial synthesis is preferentially turned on (McMillan and Houlihan, unpublished data).

Except for larvae it is unlikely that analyses of whole bodies will be a routine method in the search for biochemical correlates of growth. For practical reasons a blood or tissue sample is more likely to be taken, to be treated or frozen rapidly. This raises the question of which tissue is to be sampled and the relationships between individual tissue and whole animal growth. The total protein synthesis rate of the whole body results from the protein synthesis rates of the individual tissues and the total protein content of each tissue. Fish tissues clearly have very different fractional rates of protein synthesis (e.g. Fig. 2.5 (a)), yet if they are growing isometrically (in constant proportion to the whole body weight) their growth rate will be the same as that of the whole body and consequently the turnover rates will be very different between tissues. Clearly in choosing a tissue that will act as a marker of whole-fish growth rate it would be best to choose one whose synthesis rate most closely matches that of the whole animal.

The white muscle emerges as the strongest candidate tissue which could act as a sensitive marker for the growth rate of the whole fish. Protein synthesis rates in the white muscle have been found to correlate very well with whole-fish growth rates (Fauconneau *et al.*, 1990).

The partitioning of the energy demand for protein synthesis between the various tissues of the body results from the complex amalgam of the tissue-specific rates of protein synthesis, the tissue-specific energy cost of protein synthesis and the mass of the tissue. Thus the total energy cost of protein synthesis in a fish (MO_2 total syn/day^{-1} will be made up as follows:

$$MO_{2\ \mathrm{total\ syn.}} = \Sigma\ (\mathrm{Tissue}_{\mathrm{tot.p}} \times k_s/100 \times \mathrm{Tissue}_{\mathrm{syn.\ cost}}) \qquad (2.1)$$

where $\mathrm{Tissue}_{\mathrm{tot.p}}$ is the total protein content of the tissues (white muscle, intestine, liver etc.), k_s is the fractional rate of protein synthesis for that tissue (% day^{-1} divided by 100) and the tissue-specific protein synthesis rate will have its tissue-specific protein synthesis cost ($\mathrm{Tissue}_{\mathrm{syn.\ cost}}$, mmol O_2 g^{-1} protein synthesis). Such an analysis of the contribution of the different tissues to the total energy cost for synthesis has not yet been carried out, principally because of the lack of information on the protein synthesis rates of the individual tissues and uncertainty of the tissue-specific energy costs. However, the white muscle appears to be the greatest contributor to total protein synthesis (Houlihan *et al.*, 1988) (although not in sexually maturing fish, Fauconneau *et al.*, 1990) and also contributes > 60% of whole-body cytochrome oxidase activity in largemouth bass, *Micropterus salmoides* (Goolish and Adelman, 1987).

2.7 BIOCHEMICAL INDICES OF GROWTH RATE

The data thus far discussed have been drawn from laboratory studies. We now turn to methods which could be applied to estimating growth rates of fish at the time of capture. The most direct way of estimating growth rates would be to measure protein synthesis and protein degradation rates in the whole body and to calculate growth rate by difference (e.g. Fig. 2.2). The measurements could be made on an individual tissue if whole body/tissue growth rates were known. *In vivo* protein synthesis measurements have not been made on wild-caught animals, although the results, when combined with laboratory studies, may be useful in describing the state of the animals. Protein synthesis rates of isolated ribosomes correlate with growth rates (e.g. Lied *et al.*, 1982), although the efficiency of ribosomal translation changes markedly after a meal (e.g. Lyndon, 1990). Protein degradation occurs through different mechanisms (e.g. Rivett, 1989) and there is no clear method at present of estimating total protein degradation rates in whole bodies or selected tissues, although degradative enzyme activities have been measured in fish (Yamashita and Konagaya, 1990a,b).

One strategy for estimating the growth rates of wild fish populations is to compare growth correlates from laboratory studies with measurement of these correlates from wild animals. In this approach laboratory growth studies are conducted where, through manipulation of the amount of food given, a range of individual growth rates are generated for the species of interest. Ideally growth rate would be measured several times for the same individual and then samples of tissues taken for measurement of the likely growth correlate. Samples from the wild population are compared with those generated from the laboratory and likely growth rates of the wild animals are estimated. There are clearly a number of problems to be overcome with this approach. Firstly, which correlate is to be used and what is the physiological/biochemical basis for its selection? Secondly, the laboratory studies must be carried out in conditions that closely resemble those of the wild fish; as an example, the effects different temperatures have on growth physiology are described below. Body size is clearly important as growth rates are body size dependent. We will first review the most likely candidates for growth correlates.

RNA concentrations have frequently been used as measures of growth rate (Bulow, 1987; Busacker *et al.*, 1990). However, there has been little attempt in these studies to directly link RNA concentrations with protein synthesis and growth in fish tissues. As a starting point it could be hypothesized that feeding rates may determine the concentration of RNA in the tissues much as feeding rates determine protein synthesis rates (Fig. 2.1). There is evidence for this from studies with cod (Houlihan *et al.*, 1989) and saithe, *Pollachius virens* (Fig. 2.6), and from measurements of trout liver in the transition from starved to fed conditions (McMillan and Houlihan, 1992). A number of studies with a variety of ectotherms have found that RNA concentrations in the tissues are directly related to protein synthesis rates (Houlihan, 1991). Hence RNA concentrations may be directly related to growth rates, and tissue RNA concentrations are therefore prime candidates for correlates of growth rate.

A key feature of the value of using RNA concentrations in tissues is the time it takes for total RNA concentrations to change in response to a new level of food intake. Recent experiments have revealed that it takes approximately 2 weeks for RNA and cytochrome oxidase activity in the tissues to increase to stable levels in animals fed after a period of starvation (Foster, 1990). The responsiveness of tissue RNA measurements to refeeding after a period of starvation was generally: intestine >stomach >white muscle >ventricle >gill. In contrast, the responsiveness of tissue cytochrome oxidase activity to refeeding was ventricle >stomach >intestine, gill and white muscle. Because of the apparently slow relative turnover of ribosomes it takes some time before the tissues respond to a new nutritional plane. In the transition from starvation to fed conditions the RNA initially demonstrates increased efficiency of translation (Houlihan, 1991), but the extent of the increase in protein synthesis must be still dependent upon the tissue RNA concentration. Therefore the adaptive response is to increase the quantity of RNA allowing, presumably, an even greater expansion of protein synthesis if food continues to be available. The cost of maintaining RNA concentrations in the tissues may be thought of as a necessary drain needed to make maximum use of available food.

Three methods of expressing the RNA content of tissues have been used in previous studies:

1. RNA concentration per gram of tissue, as a direct measure of the ribosomal content of a tissue or whole animal (Bulow, 1987); a recent study used RNA g^{-1} dry weight (Wang and Stickle, 1988);
2. the ratio of RNA concentration to protein concentration (RNA:protein) as an index of ribosomal capacity for protein synthesis (Goldspink and Kelly, 1984);
3. the ratio of RNA concentration to DNA concentration (RNA:DNA) on the premise that the amount of DNA per cell is constant while the amount of RNA is proportional to the rate of protein synthesis (Bulow, 1987).

There is no consensus as to which is the most appropriate indicator of RNA concentration. However, there is very good evidence that RNA concentration and RNA:protein ratios correlate positively with protein synthesis rates when different tissues are compared (Fig. 2.5(a),(b)). Thus tissues such as the liver and gill have high fractional rates of protein synthesis and high levels of RNA when it is expressed relative to protein or fresh weight. The efficiency of RNA translation may be variable between tissues, and changes rapidly after a meal. It is therefore difficult to predict likely protein synthesis rates from tissue RNA concentrations, although with an increased data base this may be possible (Houlihan, 1991). Nevertheless the strong positive relationship between RNA expressed relative to protein or wet or dry weight of tissues and protein synthesis rates gives good support for the use of these ratios for estimating growth rate.

The strength of the RNA:DNA ratio has been said to lie in the belief that DNA content of cells is relatively constant while the RNA concentration varies with the rate of protein synthesis. Increasing evidence points to DNA content being variable, at least in respect to total protein. For example, in a recent study on cod, Foster (1990) observed that there was a significant positive correlation between white muscle cell size (protein/DNA ratios) and growth rate. A more serious problem with RNA:DNA ratios is that when correlations are drawn between them and protein synthesis, a negative correlation is found (Fig. 2.5(c); Foster *et al.*, unpublished data). In fact a number of publications have presented RNA:DNA ratios with similar values for muscle and for liver without commenting upon the significance of the values in relation to the likely wide differences in protein synthesis rates (Miglavs and Jobling, 1989). A negative correlation between protein synthesis and RNA:DNA ratios has been commented upon for mammalian tissues (Houlihan, 1991). This negative correlation does not necessarily diminish the value of RNA:DNA ratios, indeed we show below (p. 61) that these ratios correlate well with growth rate. However, the basis for this correlation remains problematical.

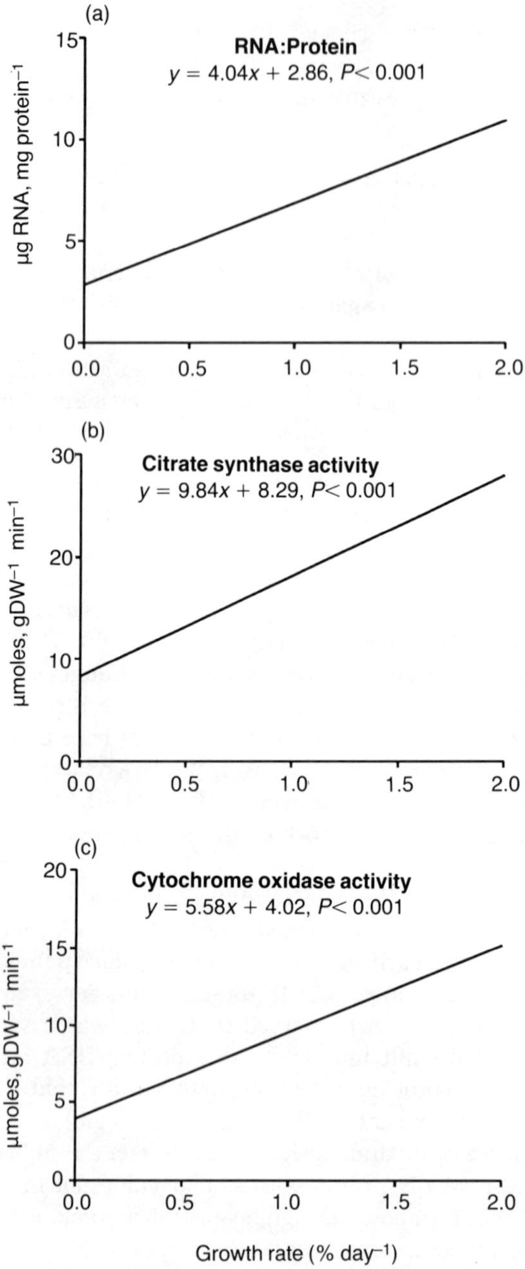

Fig. 2.6 The relationships between growth rate (% day^{-1} and (a) RNA:protein (µg RNA mg^{-1} protein), (b) citrate synthase activity (µmole g^{-1} dry weight min^{-1}) and (c) cytochrome oxidase activity (µmole g^{-1} dry weight min^{-1}) for saithe, *Pollachius virens*, white muscle. Data from Mathers (1991).

2.8 LINKAGE BETWEEN AEROBIC ENZYMES AND RNA

A variety of other correlates of growth rate have been used by various authors. Among them are scale protein synthesis rates and enzyme activities (Adelman, 1987; Bulow, 1987). The measurement of aerobic enzyme activities has been gaining acceptance as a growth rate index as cytochrome oxidase has been found to change in activity with feeding regimes (Goolish and Adelman, 1987). Cytochrome oxidase activity has been shown to be reduced during starvation (Sullivan and Somero, 1983) and increased after feeding when the postprandial increase in oxygen consumption can approach the active metabolic rate (Goolish and Adelman, 1987). It is tempting to correlate postprandial increases in aerobic enzyme levels with the postprandial increase in protein synthesis described above. However, at least one study has shown that the increased protein synthesis is involved with increased turnover and not net accretion of cytochrome oxidase (McMillan and Houlihan, 1992). Whether the observed differences in aerobic enzyme activity are the result of actual changes in activity and/or changes in enzyme concentration is unclear. Oxidative enzyme levels in white muscle may also be related to activity (Kiessling *et al.*, 1991).

In a recent study on saithe (Mathers *et al.*, 1992), correlations with growth rate were established for RNA expressed as RNA g^{-1} muscle, RNA:protein and RNA:DNA and for the activities of the enzymes citrate synthase and

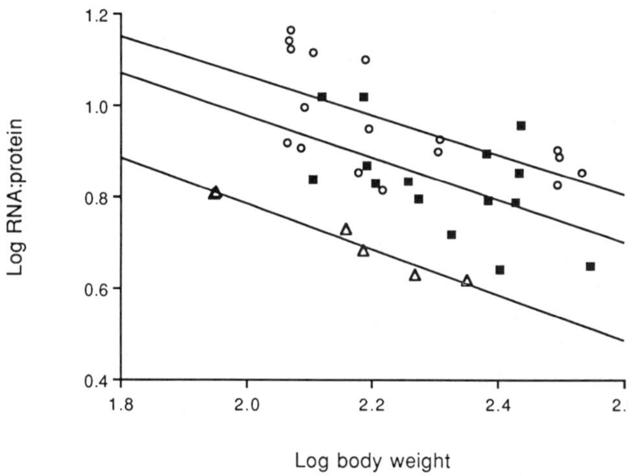

Fig. 2.7 The relationships between white muscle RNA:protein and body size for saithe, *Pollachius virens*, maintained at three different ration levels at 12 °C (fed to satiation (circles), log y = - 0.431 log x + 1.926, n = 17, r = 0.625, $P < 0.01$; fed half of satiation (squares), log y = -0.458 log x + 1.891, n = 15, r = 0.533, $P < 0.05$; starved (triangles), log y = -0.497 log x + 1.779, n = 6, r = 0.981, $P < 0.01$). Data from Mathers *et al.* (1992).

cytochrome oxidase (Fig. 2.6); all these parameters increased with increased feeding and growth rate and fell to a minimum level with starvation.

From the above discussion it would be expected that protein synthetic capacity and aerobic capacity would be linked because of the energy cost of synthesis (Goolish and Adelman, 1987). This has, in fact, been found to be the case in cod and saithe (Mathers, 1991).

Other enzymes may be valuable correlates of nutritional state/growth rate (Kiessling *et al.*, 1991). Whole-body trypsin content has been considered as an index of feeding status and growth potential of larvae because of its importance both in the activation of the pancreatic zymogens and in the breakdown of food proteins (Hjelmeland *et al.*, 1984), and several studies have shown that trypsin levels are higher in fed than in starved larvae (Ueberschär, 1988).

2.9 BODY SIZE EFFECTS ON PROTEIN GROWTH, SYNTHESIS AND RNA CONCENTRATIONS

A decrease in fish growth rate (when expressed in weight-specific terms) with increasing body size is well known (Jobling, 1983), but there are few data on accompanying protein turnover rates except for tissues of rainbow trout (Houlihan *et al.*, 1986). As body size increases, fractional rates of protein synthesis (a weight-specific unit) decline with slope values which are similar to those for weight-specific oxygen consumption (Houlihan, 1991). These results provide a further link between protein synthesis rates and oxygen consumption. RNA concentrations expressed relative to protein concentrations have also been found to decline with increasing body size in saithe white muscle (Fig. 2.7). The slope of the RNA:protein ratio did not change with level of feeding, but the elevation of the line is nutritionally sensitive. These results emphasize the need to compare animals of the same size or to make use of a scaling relationship when comparing RNA concentrations in tissues in animals of different sizes. RNA:DNA ratio is also negatively correlated with whole-body dry weight in larval cod (Buckley and Lough, 1987) and saithe (Mathers *et al.*, unpublished data).

The allometric relationships with aerobic enzyme activity appear to be both species specific and tissue specific (Goolish and Adelman, 1988), and a negative relationship between aerobic enzyme activity and body weight appears to be a general rule.

Fig. 2.8 The effects of acclimation to 5 °C (hatched columns) and 15 °C (open columns) on (a) tissue protein synthesis rates (% protein mass synthesized day^{-1}), (b) RNA concentration (μg RNA g^{-1} tissue), and (c) RNA (translational) activity (g protein synthesized g^{-1} RNA day^{-1}) in juvenile cod, *Gadus morhua*. The significance levels are $P < 0.01$, and $P < 0.05$ for ventricle and intestine, respectively. Data from Foster (1990).

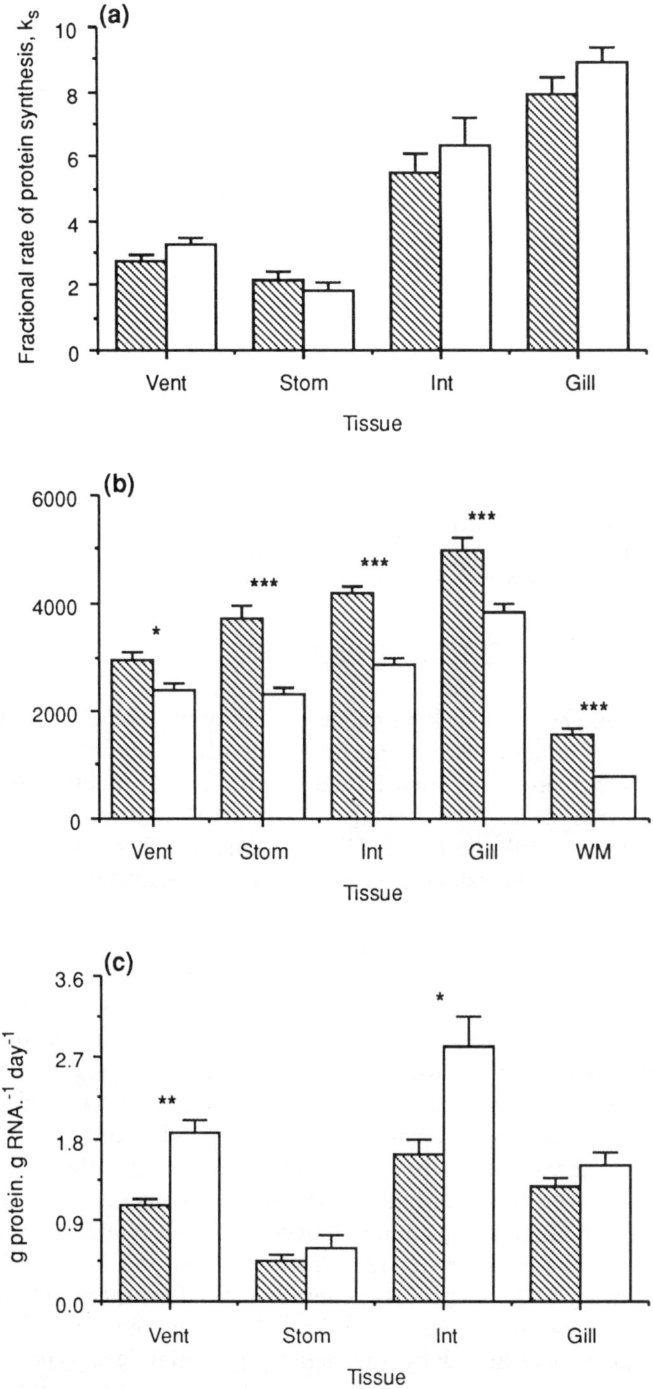
(a)
Fractional rate of protein synthesis, k_s
10
8
6
4
2
0
Vent
Stom
Int
Gill
Tissue
(b)
μg RNA. g tissue^{-1}
6000
4000
2000
0
*
***
***
***
***
Vent
Stom
Int
Gill
WM
Tissue
(c)
g protein. g RNA.$^{-1}$ day^{-1}
3.6
2.7
1.8
0.9
0.0
**
*
Vent
Stom
Int
Gill
Tissue

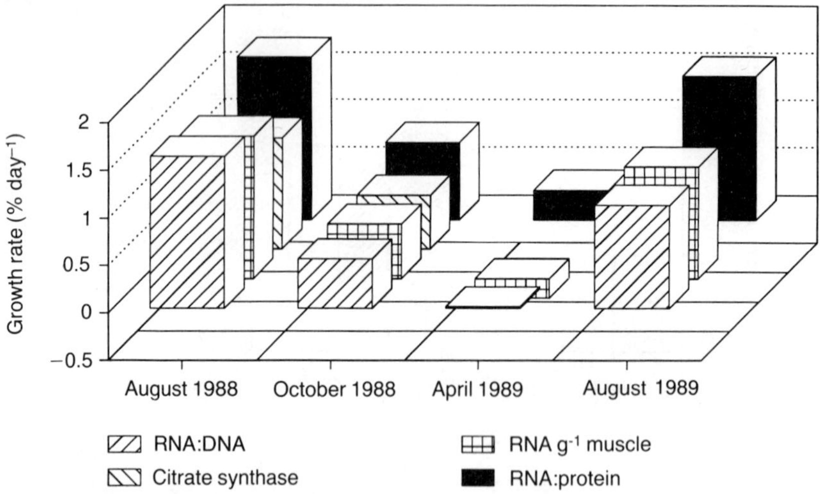

Fig. 2.9 Mean estimated growth rates of four groups of saithe, *Pollachius virens*, caught in the Beryl Oil Field between August 1988 and August 1989, from white muscle RNA g muscle^{-1}, RNA:protein, RNA:DNA and citrate synthase activity. Data from Mathers (1991).

2.10 TEMPERATURE

Temperature changes have striking effects on many physiological and biochemical processes in many ectotherms including fish. Temperature changes affect, among other processes, food intake (Elliott, 1982), maintenance requirements (Hawkins *et al.*, 1985), metabolic rates, enzymic processes, diffusion of small molecules (Sidell and Hazel, 1987), membrane functions (Cossins, 1983) and protein synthesis rates (Hazel and Prosser, 1974; Cossins and Bowler, 1987).

Generally speaking, an acute change of 10 °C in environmental temperature causes a significant change in the rate of a physiological process in a similar direction. The alteration in a rate following a 10 °C temperature change is called the Q_{10}, or temperature coefficient. Recently it has been found that acute changes in temperature result in parallel increases in rates of oxygen consumption, protein synthesis and the efficiency of ribosomal translation (k_{RNA}) in isolated trout hepatocytes (Pannevis and Houlihan, 1992).

In contrast to these responses to rapid temperature changes, the long-term exposure of ectotherms to low or high environmental temperatures may lead to a certain degree of physiological and biochemical independence from the acute effects of a change in water temperature (Hazel and Prosser, 1974; Cossins and Bowler, 1987), a process generally termed **temperature acclimation**. The limits within which such independence (or compensation) may fall are ultimately determined by the genetic potential (genotype) of an organism, and since gene expression is dependent upon the synthesis of

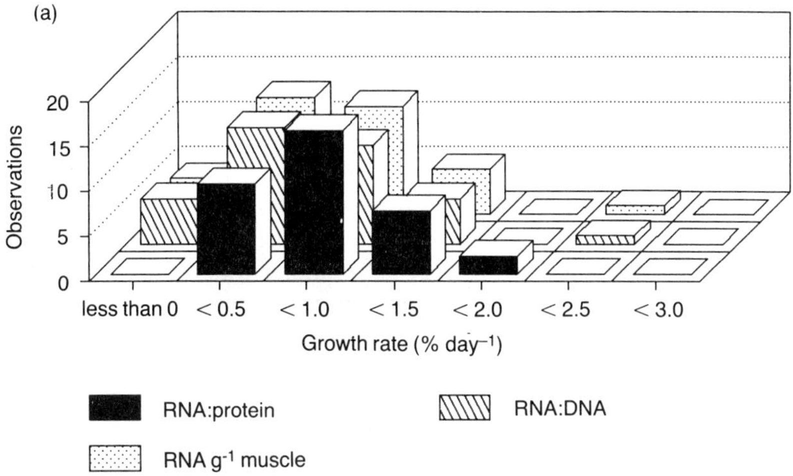

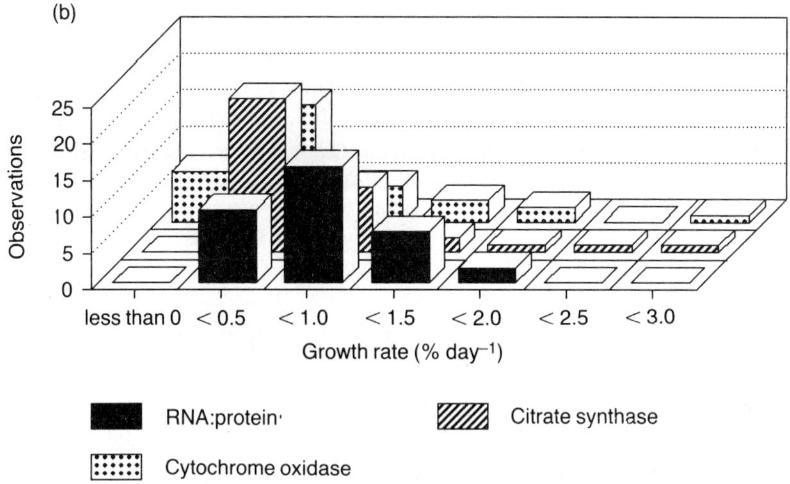

Fig. 2.10 Distribution of growth rate estimates from (a) white muscle nucleic acid concentrations and ratios, and (b) white muscle aerobic enzyme activities for saithe, *Pollachius virens*, caught in the Beryl Oil Field during October 1988. Data from Mathers (1991).

proteins, it seems likely that metabolic adjustments may be accompanied by changes in protein synthesis.

To date, the majority of publications which describe the effects of temperature acclimation on the growth and physiology of teleosts have simply 'adapted' fish to different water temperatures with feeding *ad libitum*. This experimental design is not ideal, however, since unequal ration sizes may be consumed by fish at different water temperatures. To circumvent this problem, experimental designs investigating temperature effects must use groups of fish acclimated to different temperatures and fed a similar ration size. In a recent study (Foster *et al.*, unpublished), cod were acclimated to 5 and 15 °C in individual tanks where it was possible to record the food intake of each fish. In this way, cod at 5 and 15 °C were fed a similar ration size, allowing the dissection of temperature effects from those of ration size. Figure 2.8(a) shows that there were no significant differences in tissue protein synthesis rates after at least 40 days' acclimation to each temperature. At the same time, every tissue examined from the 5 °C treatment had a significantly higher RNA concentration (μg RNA g^{-1} tissue) compared with its counterpart in the 15 °C treatment. Goolish *et al.* (1984) hypothesized that an increased tissue RNA concentration in cold-acclimated carp, *Cyprinus carpio*, compensated for a reduction in RNA (translational) activity at low water temperatures. The experiments with cod have indeed shown that tissue RNA activities (g protein synthesized g^{-1} RNA day^{-1}) are significantly decreased in cold-acclimated ventricle and intestine, though not significantly in cold-acclimated gill and stomach (Fig. 2.8(c)). Overall, these results appear to support the hypothesis of reduced RNA (translational) activity with low water temperatures (as demonstrated in the isolated hepatocytes described above) and the subsequent thermal compensation of protein synthesis by an increase in tissue RNA concentration. Cold adapted rainbow trout, *Onchorynchus mykiss*, also exhibit decreased liver and digestive tract RNA activities, although no differences were found in the RNA activity of muscle, or whole trout following acclimatization to 10 or 18 °C (Fauconneau and Arnal, 1985).

2.11 ESTIMATION OF GROWTH RATE OF FISH IN THE NORTH SEA

There have been a few studies in which growth rate correlates determined in the laboratory have been used to estimate the growth rates of wild populations (e.g. Wilder and Stanley, 1983; Buckley, 1984; Buckley and Lough, 1987), and these have been based on RNA:DNA measurements.

Mathers (1991) used multiple growth rate correlates to assess the seasonal growth rate patterns of the fish population resident around a North Sea oil platform. On four different occasions between August 1988 and August 1989 samples were taken of saithe, which was the predominant resident species.

Control groups of saithe were maintained in the laboratory with monitored feeding regimes to provide samples of known growth rates with which to compare the offshore fish. The correlates of growth determined in the laboratory were RNA g^{-1} muscle, RNA:protein, RNA:DNA, citrate synthase and cytochrome oxidase activities.

Seasonal patterns in all the parameters were observed in the fish from around the oil platform; the highest levels of RNA concentration and enzyme activity were recorded in the summer and the lowest in early spring. Significant relationships were found between RNA concentration and body weight (see above section on effects of body size). Therefore, scaling factors were used to correct all values to that of a standard size of animal before estimates of growth rates were made.

Figure 2.9 shows the mean growth rate estimates for each group of wild fish at each sampling time and from four different growth rate correlates. Comparisons of growth rate estimates within each group revealed no significant differences between the three methods of expressing RNA concentration. In most cases the estimates from aerobic enzyme activity were not significantly different from the RNA estimates.

The range and distribution of growth rate estimates for the October 1988 sample shows the agreement between the three expressions of RNA concentration (Fig. 2.10(a)) and the aerobic enzyme activity (Fig. 2.10(b)). Comparisons of the actual values for RNA concentration and aerobic enzyme activity measured for individual fish were made and the relationships were found to be significant in four out of six comparisons made. Therefore, fish with high values for RNA concentration also have high levels of aerobic enzyme activity. The use of multiple growth rate correlates may therefore be useful for assessing the overall growth rates of wild fish rather than protein synthesis alone.

ACKNOWLEDGEMENTS

This research was supported with grants from the Ministry of Agriculture and Fisheries and Food, the Natural Environment Research Council and Mobil North Sea Limited.

REFERENCES

Adams, S.M. (ed.) (1990) *Biological Indicators of Stress* (American Fisheries Society Symposium 8), American Fisheries Society, Bethesda, MD, 192 pp.

Adelman, I.R. (1987) Uptake of radioactive amino acids as indices of current growth rate of fish: a review, in *Age and Growth of Fish* (eds R.C. Summerfelt and G.E. Hall), Iowa State University Press, Ames, Iowa, pp. 65–79.

Bernier, J.F., Calvert, C.C. and Baldwin, R.L. (1987) Energetics of protein synthesis in mice with a major gene for growth. *J. Nutr.*, **117**, 2036–45.

Birkett, L. (1969) The nitrogen balance in plaice, sole and perch. *J. exp. Biol.*, **50**, 375–86.

Brown, C.R. and Cameron, J.N. (1991) The relationship between specific dynamic action (SDA) and protein synthesis rates in the channel catfish. *Physiol. Zool.*, **64**, 298–309.

Buckley, L.J. (1984) RNA–DNA ratio: an index of larval fish growth in the sea. *Mar. Biol.*, **80**, 291–8.

Buckley, L.J. and Lough, R.G. (1987) Recent growth, biochemical composition and prey field of larval haddock, *Melanogrammus aeglefinus* and Atlantic cod, *Gadus morhua* on Georges Bank. *Can. J. Fish. aquat. Sci.*, **44**, 14–25.

Bulow, F.J. (1987) RNA:DNA ratios as indicators of growth in fish: a review, in *Age and Growth of Fish* (eds R.C. Summerfelt and G.E. Hall), Iowa State University Press, Ames, Iowa, pp. 45–64.

Busacker, G.P., Adelman, I.R. and Goolish, E.M. (1990) Growth, in *Methods for Fish Biology* (eds C.B. Schreck and P.B. Moyle), American Fisheries Society, Bethesda, MD, pp. 363–87.

Carter, C.G., Houlihan, D.F., Brechin, J., McCarthy, I.D. and Davidson, I. (1992b) Protein synthesis in grass carp, *Ctenopharyngodon idella* (Val.) and its relation to diet quality, in *Fish Nutrition in Practice (Proc. IV Int. Symp. Fish Nutrition)* (eds S.J. Kaushik and P. Luquet), INRA, Paris (in press).

Carter, C.G., Houlihan, D.F. and McCarthy, I.D. (1992a) Feed utilization efficiencies of Atlantic salmon (*Salmo salar* L.) parr: effect of a single supplementary enzyme. *Comp. Biochem. Physiol.*, **101A**, 369–74.

Christiansen, J.S. and Jobling, M. (1990) The behaviour and the relationship between food intake and growth of juvenile Arctic charr, *Salvelinus alpinus* L., subjected to sustained exercise. *Can. J. Zool.*, **68**, 2185–91.

Cossins, A.R. (1983) The adaptation of membrane structure and function to changes in temperature, in *Cellular Acclimatisation To Environmental Change* (eds A.R. Cossins and P. Sheterline), Cambridge Univ. Press, Cambridge, pp. 3–32.

Cossins, A.R. and Bowler, K. (1987) *Temperature Biology of Animals*, Chapman and Hall, London, 340 pp.

Elliott, J.M. (1982) The effects of temperature and ration size on the growth and energetics of salmonids in captivity. *Comp. Biochem. Physiol.*, **73B**, 81–91.

Fauconneau, B. (1985) Protein synthesis and protein deposition in fish, in *Nutrition and Feeding in Fish* (eds. C.B. Cowey, A.M. Mackie and J.G. Bell), Academic Press, London, pp. 17–45.

Fauconneau, B. and Arnal, M. (1985) *In vivo* protein synthesis in different tissues and the whole body of rainbow trout *(Salmo gairdneri*). Influence of environmental temperature. *Comp. Biochem. Physiol.*, **82A**, 179–87.

Fauconneau, B., Breque, J. and Bielle, C. (1989) Influence of feeding on protein metabolism in Atlantic salmon (*Salmo salar*). *Aquaculture*, **79**, 29–36.

Fauconneau, B., Aguirre, P. and Blanc, J.M. (1990) Protein synthesis in different tissues of mature rainbow trout (*Salmo gairdneri* R.). Influence of triploidy. *Comp. Biochem. Physiol.*, **97C**, 345–52.

Foster, A.R. (1990) Growth and protein turnover in fish, PhD thesis, University of Aberdeen. (158 pp.)

Foster, A.R., Houlihan, D.F., Hall, S.J. and Burren, L.J. (1992). The effects of temperature acclimation on protein synthesis rates and nucleic acid content of juvenile cod (*Gadus morhua* L.). *Can. J. Zool.*, in press.

Foster, A.R., Houlihan, D.F., Gray, C., Medale, F., Fauconneau, B., Kauskik, S.J. and Le Bail, P.Y. (1991) The effects of ovine growth hormone on protein turnover in rainbow trout. *Gen. comp. Endocrinol.*, **82**, 111–20.

Fyhn, H.J. (1989) First feeding of marine fish larvae: are free amino acids the source of energy? *Aquaculture*, **80**, 111–20.

Goldspink, D.F. and Kelly, F.J. (1984) Protein turnover and growth in the whole body, liver and kidney of the rat from foetus to senility. *Biochem. J.*, **2A**, 507–16.

Goolish, E.M. and Adelman, I.R. (1987) Tissue-specific cytochrome oxidase activity in largemouth bass: the metabolic costs of feeding and growth. *Physiol. Zool.*, **60**, 454–64.

Goolish, E.M. and Adelman, I.R. (1988) Tissue-specific allometry of an aerobic respiratory enzyme in a large and small species of cyprinid (Teleostei). *Can. J. Zool.*, **66**, 2199–208.

Goolish, E.M., Barron, M.G. and Adelman, I.R. (1984) Thermoacclimatory response of nucleic acid and protein content of carp muscle tissue: influence of growth rate and relationship to glycine uptake by scales. *Can. J. Zool.*, **62**, 2164–70.

Hawkins, A.D., Soofiani, N.M. and Smith, G.W. (1985) Growth and feeding of juvenile cod, *Gadus morhua* L. *J. Cons. perm int. Explor. Mer*, **42**, 11–32.

Hawkins, A.J.S. (1991) Protein turnover: a functional appraisal. *Funct. Ecol.*, **5**, 222–33.

Hazel, J.R. and Prosser, C.L. (1974) Molecular mechanisms of temperature compensation in poikilotherms. *Physiol. Rev.*, **54**, 620–77.

Hjelmeland, K., Huse, I., Jorgensen, T., Molvik, G. and Raa, J. (1984) Trypsin and trypsinogen as indices of growth and survival potential of cod (*Gadus morhua L*) larvae. *Flodevigen rapportser*, **1**, 189–202.

Holwerda, D.A. and Opperhuizen, A. (eds) (1991) Physiological and biochemical approaches to the toxicological assessment of environmental pollution. *Comp. Biochem. Physiol.*, **100C.**

Hommes, F.A. (1980) The energy requirements for growth. A reevaluation. *Nutr. Metab.*, **24**, 110–13.

Houlihan, D.F. (1991) Protein turnover in ectotherms and its relationship to energetics, in *Advances in Comparative and Environmental Physiology*, Vol. 7 (ed. R. Gilles), Springer-Verlag, Berlin, pp. 1–43.

Houlihan, D.F. and Laurent, P. (1987) Effects of exercise training on the performance, growth, and protein turnover of rainbow trout (*Salmo qairdneri*). *Can. J. Fish. aquat. Sci.*, **44**, 1614–21.

Houlihan, D.F., Hall, S.J. and Gray, C. (1989) Effects of protein turnover in Cod. *Aquaculture*, **79**, 103–10.

Houlihan, D.F., Hall, S.J., Gray, C. and Noble, B.S. (1988) Growth rates and protein turnover in the Atlantic cod, *Gadus morhua*. *Can. J. Fish. aquat. Sci.*, **45**, 951–64.

Houlihan, D.F., McMillan, D.N. and Laurent, P. (1986) Growth rates, protein synthesis and protein degradation rates in rainbow trout: effects of body size. *Physiol. Zool.*, **59**(4), 482–93.

Houlihan, D.F., Waring, C.P., Mathers, E. and Gray, C. (1990) Protein synthesis and oxygen consumption of the shore crab, *Carcinus maenas*, following a meal. *Physiol. Zool.*, **63**, 735–56.

Jepson, M.M., Bates, P.C. and Millward, D.J. (1988) The role of insulin and thyroid hormones in the regulation of muscle growth and protein turnover in response to dietary protein in the rat. *Br. J. Nutr.*, **59**, 397–415.

Jobling, M. (1983) Growth studies with fish – overcoming the problems of size variation. *J. Fish Biol.*, **22**, 153–7.

Jobling, M., Baardvile, B.M. and Jorgensen, E.M. (1989) Investigation of food–growth relationships of Arctic charr *Salvelinus alpinus* L., using radiography. *Aquaculture*, **81**, 367–72.

Kiessling, A., Kiessling, K.H., Storebakken, T. and Asgård, T. (1991) Changes in the structure and function of the epaxial muscle of rainbow trout, *Oncorhynchus mykiss*, in relation to ration and age. II. Activity of key enzymes in energy metabolism. *Aquaculture*, **93**, 357–72.

Laurent, B.C., Moldawer L.L., Young, V.R., Bistrian, B.R. and Blackburn, G.L. (1984) Whole body leucine and muscle protein kinetics in rats fed varying protein intake. *Am. J. Physiol.*, **246**, E444–51.

Lied, E., Lund, B. and Decken, A. von der (1982) Protein synthesis *in vitro* by epaxial muscle polyribosomes from cod, *Gadus morhua. Comp. Biochem. Physiol.*, **72B**, 187–93.

Lyndon, A.R. (1990) Protein synthesis in fish in relation to oxygen consumption and free amino acids. PhD thesis, University of Aberdeen. (98 pp.)

Lyndon, A.R., Houlihan, D.F. and Hall, S.J. (1992). The effect of short term fasting and a single meal on protein synthesis and oxygen consumption in cod. *J. Comp. Physiol. B.*, **162**, 209–15.

McCarthy, I.D., Carter, C.G. and Houlihan, D.F. (1992) Individual variation in consumption in rainbow trout using radiography, in *Fish Nutrition in Practice (Proc. IV int. Symp. Fish Nutrition)* (eds S.J. Kaushik and P. Luquet), INRA, Paris (in press).

McMillan, D.N. and Houlihan, D.F. (1988) The effect of refeeding on tissue protein synthesis in rainbow trout. *Physiol. Zool.*, **61**(5), 429–41.

McMillan, D.N. and Houlihan, D.F. (1989) Short term responses of protein synthesis to re-feeding in rainbow trout. *Aquaculture*, **79**, 37–46.

McMillan, D.N. and Houlihan, D.F. (1992) Protein synthesis in trout liver is stimulated by both feeding and fasting. *Fish Physiol. Biochem.*, **10**, 23–34.

Mathers, E.M. (1991) Correlates of growth in fish. MSc thesis, University of Aberdeen. (112 pp.)

Mathers, E.M., Houlihan, D.F. and Cunningham, M.J. (1992) Nucleic acid concentrations and enzyme activities as correlates of growth rate of the saithe *Pollachius virens*: growth rate estimates of open sea-fish. *Mar. Biol.*, **112**, 363–9.

Millward, D.J. (1989) The nutritional regulation of muscle growth and protein turnover. *Aquaculture*, **79**, 1–28.

Millward, D.J. and Rivers, J.P.W. (1988) The nutritional role of indispensable amino acids and the metabolic basis for their requirements. *Eur. J. clin. Nutr.*, **42**, 367–93.

Miglavs, I. and Jobling, M. (1989) Effects of feeding regime on food consumption, growth rates and tissue nucleic acids in juvenile Arctic charr, *Salvelinus alpinus*, with particular respect to compensatory growth. *J. Fish Biol.*, **34**, 947–57.

Muramatsu, T., Kita, K., Tasaki, I. and Okumara, J. (1987) Influence of dietary protein intake on whole-body protein turnover in chicks. *Br. Poult. Sci.*, **28**, 471–82.

Pannevis, M.C. and Houlihan, D.F. (1992) The energetic cost of protein synthesis in isolated hepatocytes of rainbow trout (*Oncorhynchus mykiss*). *J. comp. Physiol.*, **B**, **162**, 393–400.

Reeds, P.J. and Lobley, G.E. (1980) Protein synthesis: are there real species differences? *Proc. Nutr. Soc.*, **39**, 43–52.

Reeds, P.J., Fuller, M.F., Cadenhead, A., Lobley, G.E. and McDonald, J.D. (1981) Effects of changes in the intakes of protein and non-protein energy on whole body protein turnover in growing pigs. *Br. J. Nutr.*, **45**, 539–46.

Reeds, P.J., Fuller, M.F. and Nicholson, B.A. (1985) Metabolic basis of energy expenditure with particular reference to protein, in *Substrate and Energy Metabolism* (eds J.S. Garrow and D. Halliday), CRC Press, London, pp. 46–57.

Rivett, A.J. (1989) High molecular mass intracellular proteases. *Biochem. J.*, **263**, 625–33.

Rosenlund, G. and Lied, E. (1986) Growth and muscle protein synthesis *in vitro* of saithe, *Pollachius virens*, and rainbow trout, *Salmo gairdneri*, in response to protein-energy intake. *Acta agric. scand.*, **36**, 195–204.

Sidell, B.D. and Hazel, J.R. (1987) Temperature affects the diffusion of small molecules through cytosol of fish muscle. *J. exp. Biol.*, **129**, 191–203.

Sullivan, K.M. and Somero, G. (1983) Size- and diet-related variations in enzymic activity and tissue composition in the sablefish, *Anoplopoma fimbria*. *Biol. Bull. mar. biol. Lab., Woods Hole*, **164**, 315–26.
Ueberschär, D.F.R. (1988) Determination of the nutritional condition of individual marine fish larvae by analyzing their proteolytic enzyme activities with a highly sensitive fluorescence technique. *Meeresforschung*, **32**, 144–54.
Wang, S.Y. and Stickle, W.B. (1988) Biochemical composition of the blue crab *Callinectes sapidus* exposed to the water-soluble fraction of crude oil. *Mar. Biol.*, **98**, 23–30.
Waterlow, J.C. and Millward, D.J. (1989) Energy cost of turnover of protein and other cellular constituents, in *Energy Transformations in Cells and Organisms* (eds W. Wieser and E. Gnaiger), Georg Thieme, Stuttgart, pp. 277–82.
Waterlow, J.C., Garlick, P.J. and Millward, D.J. (1978) *Protein Turnover in Mammalian Tissues and the Whole Body*, North Holland, Amsterdam, 804 pp.
Weisner, R.J. and Zak, R. (1991) Quantitative approaches for studying gene expression. *Am. J. Physiol.*, **260**, L179–88.
Wilder, I.B. and Stanley, J.G. (1983) RNA–DNA ratio as an index to growth in salmonid fishes in the laboratory and in streams contaminated by carbyl. *J. Fish Biol.*, **22**, 165–72.
Yamashita, M. and Konagaya, S. (1990a) High activities of cathepsins D, H and L in the white muscle of chum salmon in spawning migration. *Comp. Biochem. Physiol.*, **95B**, 149–52.
Yamashita, M. and Konagaya, S. (1990b) Purification and characterisation of cathepsin L from the white muscle of chum salmon, *Oncorhynchus keta*. *Comp. Biochem. Physiol.*, **96B**, 247–52.

Chapter three

Growth, reproduction and death in lampreys and eels

Lis Olesen Larsen and Sylvie Dufour

3.1 GENERAL INTRODUCTION

This chapter will focus on the European river lamprey, *Lampetra fluviatilis*, and on the European eel, *Anguilla anguilla*. Their life stories are dramatic, well studied, and still with challenging mysteries to be elucidated.

Both species feed, grow, and accumulate fat and protein in preparation for spawning migration. Before vitellogenesis and spermiogenesis start, they stop feeding and growing, and show atrophy of the digestive tract. Gonadal growth and energy expenditure depend on mobilization of body tissue. Both species were thought to die after spawning; this is probably true for *Lampetra fluviatilis*, but not necessarily for *Anguilla anguilla*.

Lampetra fluviatilis undergo sexual maturation under laboratory conditions, even in darkness and at constant temperature. They normally spawn in fresh water. *Anguilla anguilla* will not mature sexually under hitherto-tested laboratory conditions. They spawn in the ocean; probably they need conditions related to the deep sea. These differences with regard to sexual maturation are the reason that basic knowledge about the physiological control of reproduction in lampreys has been obtained by methods like extirpation and transplantation of endocrine organs, and in the eel by 'substitution therapy' with hormones.

Besides their dramatic life stories, attractive from the point of view of basic research, both species also have economic importance. They provide meat, highly esteemed in some places. Eels are grown in aquaculture, and much effort is directed at making a complete life cycle under aquaculture conditions. Today aquaculture must rely on catching of glass eels, elvers, and yellow eels.

Fish Ecophysiology. Edited by J. Cliff Rankin and Frank B. Jensen. Published in 1993 by Chapman & Hall, London. ISBN 0 412 45920 5.

3.2 LAMPREYS

Introduction

The European river lamprey, *Lampetra fluviatilis*, is one of the most extensively studied lamprey species. Its life story is presented schematically in Table 3.1. In order to make a reasonably coherent description of the characteristics of the life story, it is necessary to draw on knowledge obtained in other lamprey species. The review of Malmqvist (1986) can be consulted for information on variation within one species and for a general picture of other lamprey species.

Spawning takes place near stone nests in rivers in spring. Fertilization occurs outside the female body. The male has a penis-like structure, the urogenital papilla, but it is not inserted into the female during their spawning behaviour. The male attaches itself with the sucking mouth to the front end of the stone-attached female, and with the hind part of its body it 'strips' the female, from which eggs are expelled or pour out, and immediately become fertilized by sperm emitted from the male. The egg is ellipsoidal, about 1.1 mm long and yolk-laden. The female has only one ovary (and the male only one testis), and no trace of a second one, and it has no oviduct. During follicular development the egg is provided with some special mucous structures that become functional after spawning (Larsen, 1970). A tuft or cap in the animal pole of the egg provides a substrate through which the spermatozoa have to travel to penetrate the chorion, which has no micropyle (Kille, 1960). Perhaps the most vigorous spermatozoa become selected during this travel. A sticky layer in the vegetative pole of the egg allows the eggs to become attached to gravel, whirled up from the river bed during the violent movements of the couple during copulation. These eggs, attached to gravel, then become deposited in the nest.

Embryonic development starts, and hatching takes place after some weeks. The rate of development depends upon the temperature of the water.

The larvae, after some weeks of further development, drift downstream and bury in mud, where they live as suspension feeders for several years. Perhaps some or all leave their habitats during larval life and find new habitats further downstream (Malmqvist, 1983: *Lampetra planeri*). Analysis of distribution curves of body length of larvae captured by electrofishing at a given site indicated that larval life normally lasts 3–7 years. However, many complicating factors make these distribution curves difficult to interpret. One interesting experiment on length of larval life was made on *Petromyzon marinus* by Manion and co-workers (Purvis, 1980). In 1960 sexually mature lampreys were released into a river not containing any sea lampreys at that time. The larvae of the river were then examined for the following years, and even after 18 years unmetamorphosed larvae were found. Since the body length at metamorphosis is fairly constant, it can be concluded that growth rate may vary enormously. Growth rate of *P. marinus* larvae was shown to vary not only

Table 3.1 The life story of a river lamprey, *Lampetra fluviatilis*

	Phase of life											
	Fertilization	*Embryonic development*	*Hatching*	*Larval growth*	*Metamorphosis*	*Migration*	*Juvenile growth*	*Migration*	*Preparation for reproduction*	*Secondary sex characters develop*	*Nest building, spawning*	*Death after spawning*
Duration	Minutes	Weeks	Seconds?	Years	Weeks	Days–weeks	Year(s)?	Days–weeks	Months	Weeks	Days	Days (weeks)
Habitat	In stone nest in river	In stone nest		Buried in mud		Down-stream	In sea (coastal?)	Up-stream	← Attached to stones or buried →		Close to nest	Drift down-stream
Food		Yolk		Algae, bacteria and detritus from mud	No food intake	No food intake	Parasitic (fish flesh and blood)	No food intake	← No food intake →		No food intake	
Growth				Positive	Negative	Negative	Positive	Negative	← Negative →		Negative	
Reproductive system				Gonad: indeterminate→ ovary or testis				Slow vitellogenesis Sp. gon→ sp.cytes		Rapid vitellogenesis Spermiogenesis	Ovulation Spermiation	
Eyes				Eyes develop	Eyes appear	Eyes (look) very large						
External and internal factors	Genome established	External factors permissive, genome dominant	Not studied	Food and temperature important	In summer and autumn, pituitary involved	Control not known	Control not known	Control not known	Pars dist little importance	Gonadotropin and sex hormones essential; interrenals involved?	In spring, temp. important; control not known	Gonadotropin important; interrenals involved?

between individuals, but also from time to time in the same individual, even under conditions favourable for growth (Murdoch *et al.*, 1992; Docker, M.F., unpublished data). To what extent different growth rates depend on internal and external factors is not known.

The lamprey larva is called 'ammocoetes', once believed to be a species different from the lamprey, which had then only been described after metamorphosis, during its migratory phases and on the spawning grounds. The larva has no visible eyes. Its suspension feeding depends on water currents and mucus bands, leading the suspended material to the digestive tract (Moore and Mallatt, 1980).

The gonad differentiates into a testis or an ovary at some time during larval life; before then, intersex periods may be seen, and perhaps sex reversal may occur. The factors involved are not clarified. Sex ratio among adults may vary from place to place and from year to year, and from time to time during upstream migration. In the Danish Varde River, for instance, the percentage of males during upstream migration in October or November varied over 15 years from 33 to 65% (Larsen, 1980). To what extent varying sex ratios depend upon different growth rate, different mortality, different migratory behaviour or other differences between the two sexes is not known.

The length of larval life may vary, but metamorphosis takes place at a fixed season in summer and early autumn. The larva changes over some weeks into an eel-like fish, with large eyes appearing (the stage is called 'macrophthalmia') and a round sucking mouth with long and pointed horny teeth (keratinized structures). The pattern of teeth is an important taxonomic character.

The juvenile lamprey migrates downstream in late winter or spring and spends an unknown number of years in the sea feeding and growing. Hardly anything is known about the rate of growth at sea of river lampreys or sea lampreys, but in the case of the 'land-locked' sea lamprey, *P. marinus*, many data exist. They are locked to the upper Great Lakes and never migrate as far as to the sea. They feed on teleosts to which they attach themselves by their sucking mouth. They seem to prefer the same species as human beings. With their horny teeth they rasp through the scales and skin to suck out flesh and blood. Sometimes the attacked fish survive, sometimes they die. Laboratory studies showed that quite often lampreys attach to fish without feeding, for up to 18 days (Farmer, 1980). In the Great Lakes a huge, but not very successful effort has been made to eradicate the sea lamprey (the detested 'killer eel') by treating rivers with toxic substances that kill the larvae. A critique of this enterprise can be found in Maitland (1980). Based on lampreys captured by commercial fishermen, the growth curve shown in Fig. 3.1 was constructed. The starting point and end point were established on the basis of length data from lampreys captured during their migration downstream and upstream. Based on this curve it was tentatively concluded, and later repeated again and again as the truth, that the feeding stage lasts approximately 1.5

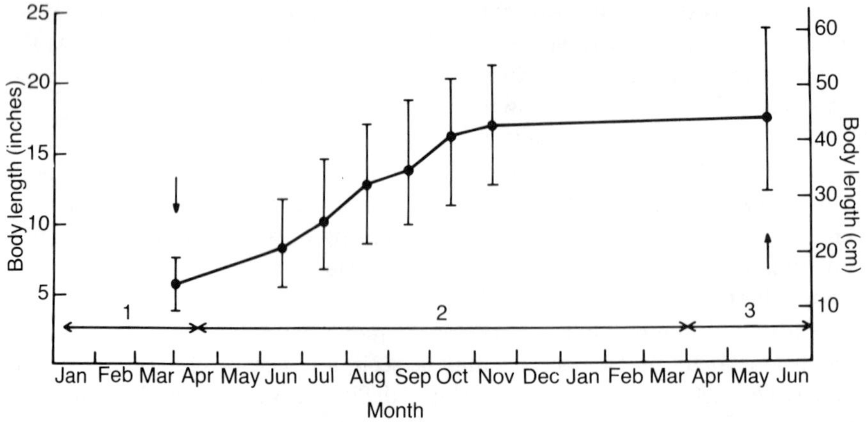

Fig. 3.1 Growth in length of the land-locked sea lamprey, *Petromyzon marinus*. After Applegate (1950). The data are based on catches of lampreys in 1947, 1948 and 1949 at various sites in lakes and rivers. The growth rate is probably the fastest possible. Phase 1: Downstream migration of newly metamorphosed lampreys. Phase 2: The feeding stage, mainly Lake Huron. Phase 3: Upstream migration. Vertical arrows: the time when most individuals migrate. Data points are mean body length in a given month; vertical lines indicate range, excluding a few extraordinary values.

years. However, this must be the shortest possible time spent in the lake, and so must represent the fastest possible growth rate. The growth rate was based on lampreys attached to fish. Lampreys not attached to fish are very rarely caught, except during their migrations or at the spawning sites. So the non-attached lampreys were not observed and may spoil the simple and nice pattern of growth represented by the (chance) catches at many different sites in the lakes. The pattern of catches shows that in the months after their downstream migration, the majority of the feeding lampreys are found in deep waters, and later they occur in shallower waters, closer to the rivers, in the period before upstream migration. The constructed (maximal) growth rate was compared with growth rates in the laboratory and was found to be realistic. The range of body length in the downstream migrants is rather low compared with the variation at upstream migration, and supports the idea that growth in the lake may show great individual variation.

At upstream migration the females are, on average, larger than males. In the Danish river mentioned above, the fresh run river lamprey males over a period of 15 years had average body lengths from 32 to 36 cm and body masses from 58 to 81 g, while the corresponding figures for females were 34–36 cm and 62–89 g. In other places in Europe, river lampreys may be considerably larger or smaller.

During the juvenile growth phase, there is some gonadal development; in the testis spermatogonia divide, and some spermatocytes begin to appear; in the ovary vitellogenesis starts.

The river lamprey migrates upstream, typically in autumn, but with a huge variation from place to place; in England, for instance, the earliest migrants are captured in July, the latest in March, and they all spawn in the spring months. This migration and change in habitat has been better studied than the downstream migration, especially with regard to osmoregulation, but the factors responsible for change of morphology, physiology and behaviour are little known. High migratory activity seems related to dark nights and high tide. Interestingly, the gadgets used for catching eels, *Anguilla anguilla*, on their downstream migration seem to be as efficient for catching river lampreys on their upstream migration, at least in Western Jutland, Denmark. There is no evidence (yet?) of any marked homing in lampreys.

At the time when upstream migration starts, the lampreys seem to stop feeding as judged from their empty intestinal tract, which undergoes atrophy. This phase of negative growth is extensively studied with regard to physiology and endocrinology and is treated on pp. 77–78. It lasts until death after spawning. All lamprey species seem to die very soon (i.e. days or at most weeks) after spawning in spring. An exception seems to be found in the Pacific lamprey, *Lampetra tridentata* (Michael, 1984). If this can be verified, it is opposite to the situation in salmonids, where the Pacific species are supposed all to die after their first spawning. The phase of reproduction and the phenomenon of natural or 'programmed' death has also been extensively studied with regard to physiology and endocrinology and is dealt with on pp. 79–85.

Table 3.1 is mainly based on knowledge about the European river lamprey. This scheme is general for other river lampreys. Brook lampreys are smaller and stay in fresh water. After metamorphosis they soon become sexually mature, spawn and die without any juvenile growth phase. Sea lampreys show the same development as the river lampreys, but they grow to a larger size and probably go further out in the sea and perhaps stay there for a longer period of time. Altogether a little more than 30 lamprey species are recognized.

Negative growth phase

During metamorphosis and downstream migration, growth is negative; the lampreys stop feeding while the oral region and the digestive tract are reorganized. The second negative growth phase starts in the sea or estuary, before upstream migration.

Knowledge about the development of adult river lampreys in their freshwater phase was initially based on morphological descriptions of lampreys caught during upstream migration and at the time of spawning, when they are more easy to catch than in the intervening period. During winter they hide, attached to stones or perhaps buried (hibernation in cold waters?). It was also known that the lack of food intake was correlated with a reduction in body mass and also in body length. The latter phenomenon is related to their lack of a backbone; the chorda allows shrinkage in all dimensions, and the

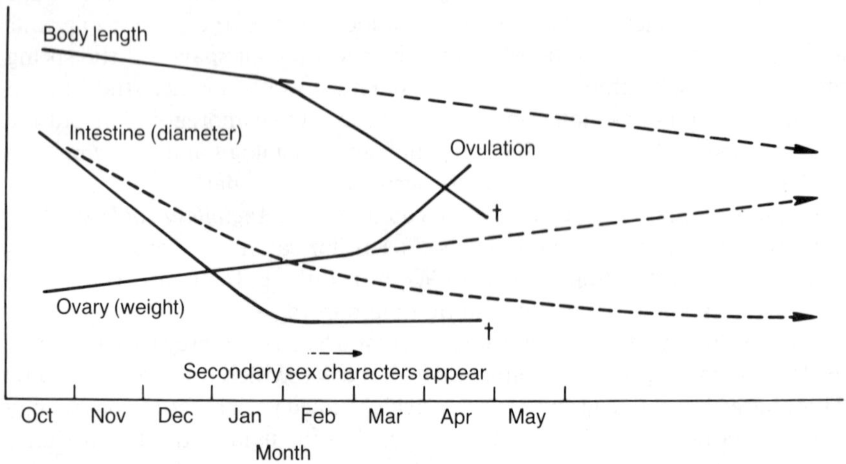

Fig. 3.2 Development of a river lamprey, *Lampetra fluviatilis*. A very schematic presentation based on laboratory studies of lampreys over about 15 years. Solid curves, intact lampreys; broken curves, pars distalis extirpated in autumn; †, natural death. All parameters were never measured in one individual or during one year; therefore no units appear on the y-axis.

lampreys may keep their proportions unchanged in spite of a marked reduction in mass (Larsen, 1962).

With lampreys kept in the laboratory from upstream migration until death after spawning, it was possible to follow length reduction as a criterion of the rate of negative growth, that is the rate of mobilization of body tissue used to cover energy requirements and used for gonadal growth. The latter is small in males, but considerable in females. The mass of the ovary, as a proportion of initial body mass at the time of capture at upstream migration in lampreys from Varde River, is about 5% and increases to about 16% before ovulation. Material for this vitellogenic growth comes mainly from the fat and protein of the body wall, which consists of skin, musculature, and connective tissue. The mobilized substances are converted in the liver to vitellogenin, at least in the later part of ovarian vitellogenesis. As mentioned, the intestine undergoes atrophy, but the amount of tissue mobilized is small compared with the body wall. The body wall represents about 68% of the initial body mass at upstream migration, and is reduced to about 51% of the initial body mass in males and to about 46% in females. The corresponding figures for the intestine are about 0.5% and 0.1%.

When the rate of change in the various tissues and organs was followed, it turned out that it was not constant. The phase called 'preparation for reproduction' (Table 3.1) is characterized by a slow reduction in body length (and a comparably slow reduction in mass of the body wall), whereas intestinal atrophy takes place rapidly (Fig. 3.2).

Sexual maturation

When secondary sex characters start to develop, the intestine has reached its minimal size; body length and mass of the body wall are now reduced at a high rate, especially in females. In this period spermiogenesis starts, and vitellogenesis becomes rapid (Fig. 3.2). The rates of all these changes are positively correlated with the temperature of the water, but the pattern is also seen at constant temperature. The river lampreys were studied in the interval from 6 to 15 °C; above 15 °C they die. Light (quality and quantity/intensity) probably plays little role; the development in the freshwater phase takes place even in near-total darkness.

Spawning and death

In nature lampreys emerge from their hiding places at the time of spawning in spring and become active in daytime. The sucking mouth is now used for a third purpose in the life story; the mature lampreys suck to small stones and move them, so that a nest is formed, close to which copulation takes place. The male is most active in nest building, but the female may assist him. After a period of mutual stimulation, the male uses his sucking mouth to attach himself to the female, as mentioned earlier. This behaviour precedes release of eggs from the female and sperm from the male. The behaviour does not occur under ordinary laboratory conditions, but when river water is let through an aquarium, and gravel and stones are present, the behaviour does occur and has been filmed and described (Hagelin, 1959). In some vertebrate species ovulation and spermiation (release of eggs from the ovary and of sperm from the testis) only take place when a mate is present or when mating has started, but in lampreys it may take place spontaneously, in isolated males and females. Ovulation of all follicles in the ovary takes place within some hours and involves release of about 40,000 eggs to the body cavity, from which they are expelled by contractions of the body and/or by the stripping movements of the male's body or the experimenter's fingers. Spermiation takes place over several days and involves gradual dissolution of the testis from its posterior end. It seems never to be totally spent, neither in the laboratory nor in nature at the time of death. In nature the female spawns its eggs over a couple of days with several short copulations.

After spawning both male and female die. In the laboratory death takes place within at most a couple of weeks from the time that stripping became possible, often with a shorter lag period in females than in males. In nature the dying lampreys are often covered by fungus, *Saprolegnia parasitica*. In order to make river lampreys survive in the laboratory, it is important to keep them free of fungal infections, which kill them. It was once believed that death after spawning was caused by fungal infections combined with the prolonged starvation period and the strenuous exercise related to upstream migration,

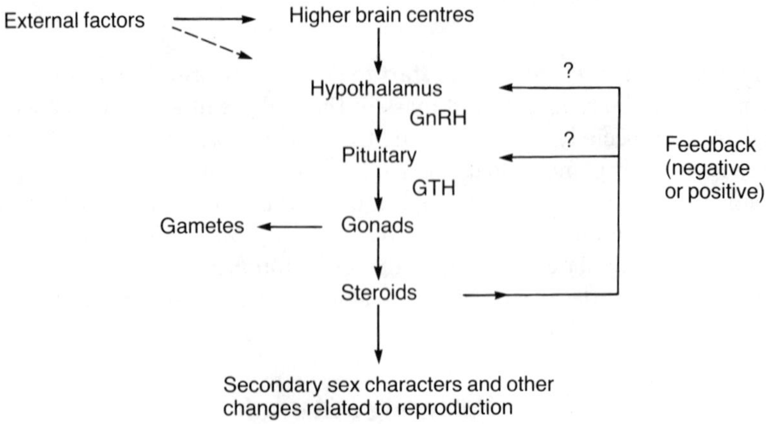

Fig. 3.3 Mammalian-based scheme for neuroendocrine control of (sea) lamprey reproduction. Sower (1990) used a similar scheme for her discussion. GnRH, peptides with gonadotropin-releasing activity; GTH, gonadotropic hormone(s). The external factors may act via sense organs or sensory cells (e.g. light and temperature), but in poikilothermic animals temperature may act on all temperature-sensitive processes in the body.

nest building and copulation. However, river lampreys in the laboratory, which are kept free of fungus, and which do not migrate and build nests or copulate, but which do ovulate and spermiate, still all die within a short period after ovulation and spermiation.

Role of hormones in development of the adult river lamprey in its freshwater phase

According to expectations based on knowledge of the neuroendocrine system of vertebrates, the scheme presented in Fig. 3.3 is a reasonable starting point for an analysis of the phase just before, during, and after sexual maturation. The endocrinological data are presented in the order they were obtained in the river lamprey, *L. fluviatilis*; they will be supplemented with data from the sea lamprey, *P. marinus*.

In order to make a sound proposal about the role of a certain hormone in an animal, you need a description of normal development, followed by interventions in the normal development by abolishment of the production, secretion or action of the hormone; in this way you can describe abnormalities. If sham controls do not reveal these abnormalities, it may be assumed that the eliminated hormone normally plays some role in the process. It may initiate and maintain or stimulate the process, or it may be permissive for other essential factors that initiate, maintain or stimulate. Further understanding of the role may be obtained by substitution therapy with the hormone, by studies of changing hormone levels in the blood and by studies of the action of the hormone at the level of cell, tissue, organ or the whole body. Modern

Table 3.2 Effects of extirpation of endocrine organs on development in adult river lampreys, *Lampetra fluviatilis**

Experimental procedure	Catabolic changes (intestine, whole body)	Vitellogenesis spermatogenesis	Secondary sex characters	Ovulation/ spermiation	Natural death
Extirp. of pars distalis, autumn (♂♀)	Slow	Slow	Absent	Absent	Postponed (months)
Extirp. of pars distalis, spring (♀)	Slow			+!	
Gonadectomy, soon after migration (♂♀)	Slow intestinal atrophy or hypertrophy		Absent		Postponed (weeks)
Gonadectomy, around New Year (♂♀)	Intestinal hypertrophy		Absent		Postponed (weeks)

* See Larsen (1980) and Fig. 3.2.

methods allow identification, purification, and synthesis of presumptive hormones to a degree never seen before. This provides a temptation to take short cuts and go into detail without laying the foundation.

Fig. 3.2 demonstrates some effects of removal of the pro- and mesoadenohypophysis of the river lamprey; these two parts of the pituitary gland are believed to correspond to the pars distalis of tetrapods and thus to be essential for growth via growth hormone and for reproduction via gonadotropin(s). This was shown to be the case, as Table 3.2 further demonstrates. The lack of pars distalis hormones does not become apparent until the time when secondary sex characters are expected to appear, except in the rate with which the intestine undergoes atrophy. The period of sexual maturation is characterized by high speed of processes, and these processes are either slowed down by removal of the pars distalis or do not occur at all. Secondary sex characters do not appear, and ovulation and spermiation do not occur, although spermatozoa and normal-sized eggs are finally obtained. The structures necessary for spermiation and ovulation do not differentiate. So secretion of a gonadotropic hormone must be postulated, although it has not yet been isolated and characterized. Rapid vitellogenesis does not occur, but the slow vitellogenesis continues, and no atrophy is seen. This is different from experience in all other investigated vertebrates with yolk-laden eggs. The reduced rate of reduction in body length indicates that the pars distalis is directly or indirectly involved in mobilization of body tissue. Substitution therapy has been successful with lamprey pituitary extract or with pregnant mare serum gonadotropin plus human chorionic gonadotropin (HCG) (Evennett and Dodd, 1963).

According to expectations, gonadectomy (Table 3.2) prevented development of secondary sex characters. Until the time when these normally start to develop, the two sexes cannot be distinguished by appearance or behaviour. The mature river lamprey male has a urogenital papilla, about 7 mm long, and dorsal 'fins' (without fin rays!) which are high; the female shows a swollen body because of the large ovary, a swelling of the anterior part of the second dorsal fin, and develops an anal fin. These features are easy to quantify. Substitution therapy with large doses of testosterone and oestradiol, as implants, induced secondary sex characters in gonadectomized lampreys, and in intact lampreys they could force development in the opposite direction; for instance testosterone-treated females showed growth of a urogenital papilla, and oestradiol-treated males an anal fin. Sex-specific gonadal steroids are thus likely to be produced by the lamprey gonad.

However, some of the processes typical of the period of sexual maturation could still take place after gonadectomy and were thus supposed not to depend on gonadal sex hormones but on the presence of the pars distalis. These were, among others, the rapid reduction in body length (although slow after some weeks in survivors) and swelling of the anal region. These two features are typically seen in stressed lampreys. Furthermore, gonadectomy around New

Table 3.3 Phenomena related to sexual maturation, but not (solely) dependent on gonadal hormones*

1. Atrophic state of intestine (gonadectomy →hypertrophy, but substitution has no effect)
2. Early secondary sex characters (♀, anal swelling; ♂, first growth of 'penis')
3. Rapid reduction in body length (in some gonadectomized individuals)
4. Biliverdin-green livers (frequent in late spring; not after extirpation of pars ditalis, sometimes after gonadectomy)
5. Natural death (sex hormones do not induce death – on the contrary)

1, 3, 4, 5: Both in males and females, but to varying degrees or with varying rates
2, 3: Elicited also by stress (e.g. toxic substances)

* See Larsen (1990), Fig. 3.2 and Table 3.2.

Year was in many cases followed by hypertrophy and histological normalization of the atrophied intestine, even in lampreys deprived of the pars distalis or treated with testosterone and oestradiol.

The idea of explaining these and other unexpected findings (Table 3.3) by considering interrenal steroids came from consideration of the prolonged survival, which was very marked in some lampreys deprived of their pars distalis; the longest survival was 13 months beyond the time when sham-operated or control lampreys had died. Also gonadectomy prolonged survival, but at most for 4 months. Death after spawning in Pacific salmon species is supposed to be correlated with hypersecretion of corticosteroids (Larsen, 1985). The interrenal cells of lampreys most probably do secrete steroids, but their identity has not yet been established, and the evidence for their role in the processes leading to natural death is circumstantial, but ample (Larsen, 1990). That steroids from the interrenal cells play an important role before and during sexual maturation is also supported by evidence from other vertebrates; the best known is the role of adrenal androgens in human beings for growth and development in puberty and for libido of women.

Gonadectomy results in cytological changes in the pituitary, but the effect does not seem to be specific for one cell type. Partial gonadectomy does not lead to compensatory hypertrophy of the remaining part of the gonad, so two essential criteria for the existence of a negative feedback relationship between gonadal sex hormones and gonadotropin are absent, and there are no indications of positive feedback either. The latter is generally seen in relation to initiation of sexual maturation in e.g. eels (see p. 94) and in mammals, and in relation to the ovulatory surge of gonadotropin(s) necessary for ovulation in mammals and birds. Probably gonadotropin does not need to be present at the time of ovulation in lampreys. Normal ovulation took place in 6 out of 19 females in which the pars distalis was extirpated shortly before expected ovulation. In these 6 females, ovulation took place between 4 and 13 days

Table 3.4 Sexual maturation in *Lampetra fluviatilis* with transplanted pars distalis *

Treatment	♂♂	♀♀
Heterotopic transplant	56% (9)	35% (6)
Orthotopic transplant	50% (5)	71% (5)
Intact controls	100% (Numerous)	100% (Numerous)
Pars distalis extirpated	0% (17)	0% (19)

* Entries are % of lampreys that developed secondary sex characteristics and spermiated or ovulated, with total number of individuals in parentheses (Larsen, 1990). Only part of the glandular tissue could be removed surgically: the rest had to be sucked out and could not be transplanted.

after the operation (Larsen, 1970), so gonadotropin is in lampreys necessary for growth and development of the gonads, but spermiation and ovulation may occur in its absence.

The question of hypothalamic control of the pars distalis was investigated by transplantation studies. The paper by Harris and Jacobsohn (1952) demonstrates exceptionally well how carefully planned transplantation experiments gave basic information about sexual differentiation, puberty, gonadal function and lactation in rats. Results of transplantation studies in river lampreys are shown in Table 3.4. They indicate that even a part of the pars distalis, when removed from the normal position below the hypothalamus and placed in a muscle in the pharynx or in the eye chamber, allowed sexual maturation to occur normally, although often with a delay. This was not totally unexpected, because in lampreys the pars distalis is separated from the brain by connective tissue through which neither portal vessels nor nerves penetrate (Gorbman, 1980). However, the demonstration of the presence in *Petromyzon marinus* of peptides similar to other vertebrate gonadotropin-releasing hormones (GnRH) has led to a series of attempts to demonstrate that such substances stimulate gonadotropin secretion. Since no method for measuring lamprey gonadotropin is available, actions *in vivo* and *in vitro* on gonadal function have been tested (Sower, 1990). The peptides have effects, but their biological role is not clarified. The evidence cited in the careful review by Hardisty and Baker (1982) indicates a role of gonadotropin-releasing hormone or other brain substances in cell differentiation.

Hormone(s) in the pars distalis are probably indirectly responsible for natural death after spawning. In females gonadotropin seems important, because females invariably die shortly after ovulation. In males, however, 'corticotropin' may be important, too. In a few males in which pars distalis function had been modified by transplantation or by partial extirpation, spermiation took place, but the males still survived for months after spermiation. Corticotropin has not been isolated and characterized. However, there is cytological and bioassay-based evidence for its existence (Hardisty and Baker, 1982), and possibly another derivative of the prohormone, pro-opiomelanocortin, may be secreted from the pars intermedia (Larsen, 1990).

That natural death is not a simple consequence of the influence of gonadotropin and gonadal sex steroids is further demonstrated by the evidence mentioned in Tables 3.2 and 3.3. Gonadectomy postponed natural death, but treatment of intact lampreys with testosterone or oestradiol, surprisingly, often caused prolonged survival (Larsen, 1974, 1987).

Mobilization of body tissue in the phase of negative growth is thus believed to some extent to depend on the postulated existence of interrenal steroids. A negative growth phase may also be caused by or depend upon lack of anabolic hormones. There is no physiological evidence of growth hormone, but prespawning *P. marinus* contain mammalian-like immunoreactive growth hormone cells in the caudal pars distalis. Such cells could be demonstrated neither in larvae nor in juvenile (growing!) lampreys (Wright, 1986). However, insulin is produced in and secreted from the pancreatic B-cells, and more is known about this hormone than about any other hormone in lampreys thanks to the work of Plisetskaya and her co-workers in Leningrad and Seattle (Plisetskaya, 1990). All aspects of its endocrinology have been investigated, including its isolation and amino acid sequence, its blood levels, its receptor, its actions in various types of muscle, liver, and brain, and its effects on metabolite levels in blood. It has to a large extent the effects known from mammals and other vertebrates. However, its biological role in blood sugar regulation is not clear, and its role in the anabolic and catabolic processes related to the freshwater phase is only a hypothesis, not so easy to test. It is, however, remarkable that its concentration in the blood of both river lampreys and sea lampreys decreases before spawning (Plisetskaya, 1985). It would be useful to know the precise stage at which the minimal levels are reached. Completion of vitellogenesis is indicated, but the relationship in time to, for instance, secondary sex characters or intestinal atrophy is not known.

As mentioned above, there is no physiological evidence of growth hormone in lampreys, and somatomedin *alias* insulin-like growth factor has not been looked for/not been found (?). However, insulin might have an effect on growth in body length. This was in fact demonstrated in 1976, as illustrated in Table

Table 3.5 Increase in body length 24 h after an injection of insulin in female river lampreys, *Lampetra fluviatilis**

Treatment	Month	Dose (IU)	Length increase (mm)	*P*-value†
Intact	Feb.	Control	0.0 ± 0.7 ($N = 6$)	n.s.
Intact	Feb.	0.03	1.1 ± 0.7 ($N = 7$)	n.s.
Intact	Feb.	0.1	2.2 ± 0.7 ($N = 6$)	0.025–0.05
Intact	Feb.	0.3	7.1 ± 0.5 ($N = 6$)	<0.001
Intact	Jan.	0.1	6.8 ± 1.0 ($N = 6$)	<0.001
Pars dist. ect.	Jan	0.1	7.0 ± 0.6 ($N = 10$)	<0.001
Intact	March	0.1	3.2 ± 1.0 ($N = 6$)	0.025–0.05
Pars dist. ect.	March	0.1	5.4 ± 0.6 ($N = 9$)	<0.001

* Data from Larsen (1976).

† For increase, as evaluated by Student's t-test; n.s., not significant.

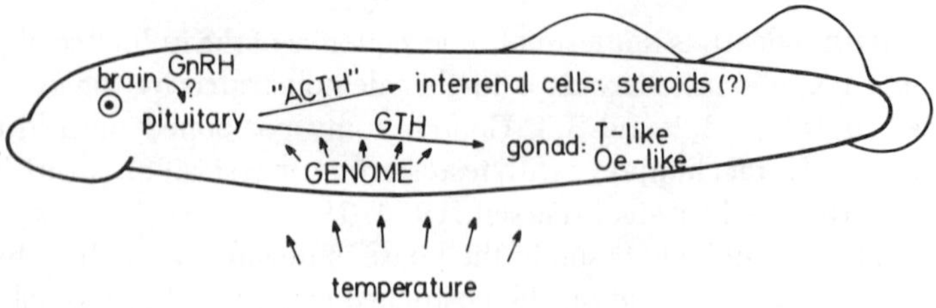

Fig. 3.4 Lamprey-based scheme for neuroendocrine control of (river) lamprey reproduction. Larsen (1990) used a similar scheme for her discussion. GnRH (Fig. 3.3) is present in the brain (of sea lampreys), but whether it affects the pituitary gland (in river lampreys) is not clarified. "ACTH", (adreno)corticotropic hormone or another derivative of the large pro-hormone, proopiomelanocortin; GTH, gonadotropic activity (chemistry unknown). Gonadal steroid hormones are similar to, but probably not identical with, testosterone and oestradiol. The interrenal cells probably secrete unidentified steroids, different from those known today. The genome starts 'programming' the life history of the lamprey after fertilization and is continuously operating in interaction with external and internal factors during the whole life story. Temperature is an important external factor in seasonality and in synchronizing development of lampreys at a given biotope. Internal factors may be hormones or metabolites. The genome is directly or indirectly responsible for the observed increase in sensitivity towards gonadotropins in the gonads and towards sex steroids in many tissues/organs at the time when development of secondary sex characters starts. Further comments: see Fig. 3.3.

3.5, but was not interpreted as growth, because the lampreys were starving and showed no increase in body mass.

Conclusion

To summarize, the essential factors supposed to be involved in lamprey reproduction are the pituitary with its secretion of gonadotropin, the ovary with its secretion of a steroid similar to oestradiol, and the testis with its secretion of a steroid similar to testosterone (Fig. 3.4). These organs and their hormones are mediators in the control which the genome basically exerts. The processes related to sexual maturation are to an unusual extent independent of environmental factors, except that temperature is important. Temperature, however, probably does not exert its effect through sense organs, but presumably through temperature-dependent processes in cells in the brain, the pituitary, the gonad and other tissues which take part in reproduction.

3.3 EELS

Introduction

The European eel, *Anguilla anguilla*, is one of the most extensively studied eel species. Its life story is schematically presented in Table 3.6. The spawning grounds of *A. anguilla* or any other eel species have never been localized.

Table 3.6 The life story of an eel, *Anguilla anguilla*

	Phases of life						
	Larval growth	Metamorphosis	Postlarval stage	Migration towards land	Juvenile growth	Migration and preparation for reproduction	Final sexual maturity, spawning and death (?)
Name	Leptocephalus		Glass eel	Elver	Yellow eel	Silver eel	
Duration	About 1 year	About 2 months	Few months	Weeks	3–30 years	0.5 year or more	
Habitat	Atlantic Ocean, drifting	Continental shelf	←	Coastal, estuarine, and freshwater	→	Oceanic	Sargasso Sea
Food	Unknown	No food intake	No food intake	Zooplankton	Invertebrates and fish	No food intake	No food intake
Growth	Positive	Negative	None or negative	Positive	Positive	Negative	Negative
Reproductive system					Gonad: indeterminate→ ovary or testis	A little yolk; spermatogonia or spermatocytes	Vitellogenesis, spermiogenesis, release of gametes
Eyes						Grow larger	Grow even larger

However, there is strong indirect evidence that *A. anguilla* spawns in the Sargasso Sea. This evidence was provided by catches of eel larvae in the Atlantic Ocean from 1904 to 1922 at various sites and times of the year by Johannes Schmidt; he studied the geographical distribution and length of the larvae. By extrapolation, the Sargasso Sea was supposed to be the spawning site; the smallest larvae found there were 5 mm long. The spawning could be assumed to start in March. A critical evaluation of the somewhat biased strategy behind the catching efforts has been published (Boëtius and Harding, 1985); the hypothesis is still the best one available, but not proven until spawning adults or eggs are found. Recently the spawning area for the Japanese eel, *A. japonica*, has been located in the same way (Tsukamoto, 1992). *A. anguilla* is the only eel species, among 15, which is known to migrate more than 6000 km from the place where juvenile growth takes place: along the coasts or in rivers and lakes of Europe.

The larva has a large mouth and big teeth, but its food is still unknown. The larvae drift with the Gulf Stream and probably also swim. At a length of about 85 mm they arrive at the continental shelf where they undergo metamorphosis and become eel-like in shape; they are still transparent and accordingly are called glass eels. During metamorphosis they stop eating and shrink in all dimensions like the starving lamprey; the larval teeth are exchanged for adult teeth, and the digestive tract is reorganized. Age determination (based on daily rings displayed by otoliths) allows estimation of the duration of the travel from the Sargasso Sea to Europe; it is about 1 year (Lecomte-Finiger and Yahyaoui, 1989); a reinterpretation of Schmidt's data showed the same duration, and not 3 years as previously believed (Boëtius and Harding, 1985).

When the glass eels arrive in coastal waters they start eating and become pigmented, and are now called elvers (6–8 cm long). Some elvers stay in coastal waters, others (especially future females?) migrate upstream, when the water temperature of cold rivers reaches 7–9 °C (Boëtius and Boëtius, 1989). Elvers grow and turn into yellow eels. The growth phase of yellow eels is supposed to last from 3 to 30 years and is longer for females than for males. A remarkable similarity in growth rate of *A. rostrata* from Bermuda (18–28 °C) and the Arctic waters of Greenland has been demonstrated by J. Boëtius (1985). This study was based on correlation of body length and scale zones; scale zones are assumed to indicate age in years after scales develop. In other studies otolith rings have been used; they seem to be fairly reliable until metamorphosis, but after that very difficult to interpret (I. Boëtius, 1985). When yellow eels turn into silver eels, females are generally larger than males. In Danish waters yellow (or silver) eels at a body length over 50 cm are always females, and silver eels less than 42 cm are always males (Bruun *et al.*, 1949). In silver eels the testes (less than 0.1% of the body mass) contain spermatogonia and a few spermatocytes, and the ovaries (about 2% of the body mass) contain oocytes which are previtellogenic or in the early vitellogenic stage. The change from the yellow stage to the silver stage is a typically 'spontaneous', endogenously determined, development, with many characteristics of a metamorphosis, except that the change of form is small. The period constitutes a physiological preparation for the seaward migration.

The silver eels start to migrate towards the Sargasso Sea in autumn, especially on dark nights before new moon. In this phase they do not eat, and the intestine undergoes atrophy, but it is still effective in its role in marine osmoregulation. A few specimens have been captured during their migration in the sea, but never beyond some hundred km from the coast. The eels are supposed to orientate themselves by way of geomagnetic cues, but perhaps also in relation to water currents and olfactory clues (Lecomte-Finiger, 1990). The duration of the migration may be about half a year (from autumn to spring) for Danish eels.

Further development, from the stage of early sexual maturation as seen in silver eels at their downstream migration, we know from eels induced to mature by treatment with hormones (see pp. 93–95).

Fat and protein

Much interest has been paid to a minimum, critical amount of fat as a trigger for metamorphosis (for instance in lampreys) or sexual maturation (e.g. women and female rats). The idea is that the animal in the following phase needs a certain amount of energy in the form of fat to withstand a period of starvation (natural or imposed by lack of food). The lamprey is not able to feed during metamorphosis; human beings and rats can only meet the energy demands of pregnancy and lactation with a large supply of energy, and energy reserves will assure success of the young in periods with low food supply. The silver eel, which does not feed, needs a lot of energy for its migration to the Sargasso Sea (also protein may be used for that purpose) and a lot of energy plus fat and protein to build up gonads, especially ovaries. The ovaries represent about 2% of the body mass when migration starts, and judged from the maximum size in hormonally induced sexual maturation, they will reach 40–60% of the final body mass. The idea of a 'critical fat mass' is discussed in a recent paper; Larsson *et al.*, (1990) examined the percentage of lipid in "muscular tissue" (15–20 g tail, including skin) in yellow and silver eels (all females). In the size range from 70 to 340 g, the fat % increased from about 8% to about 30%. In eels above 340 g (both yellow and silver eels) the fat % did not increase further. This material was taken to indicate that a fat % of 28 (N.B. in the tail) may be a prerequisite for the change from the yellow to the silver stage; however a number of large yellow eels had fat % above 28, and a number of silver eels had fat % below 28 (but not below 20). So a certain fat % is not the 'triggering factor', or only one among others, or just correlated with the triggering internal factor(s) or the triggering situation (external factor(s)).

Boëtius and Boëtius (1985) (Fig. 3.5) examined both males and females. In the figure they express fat in relation to the total energy of the body. When their

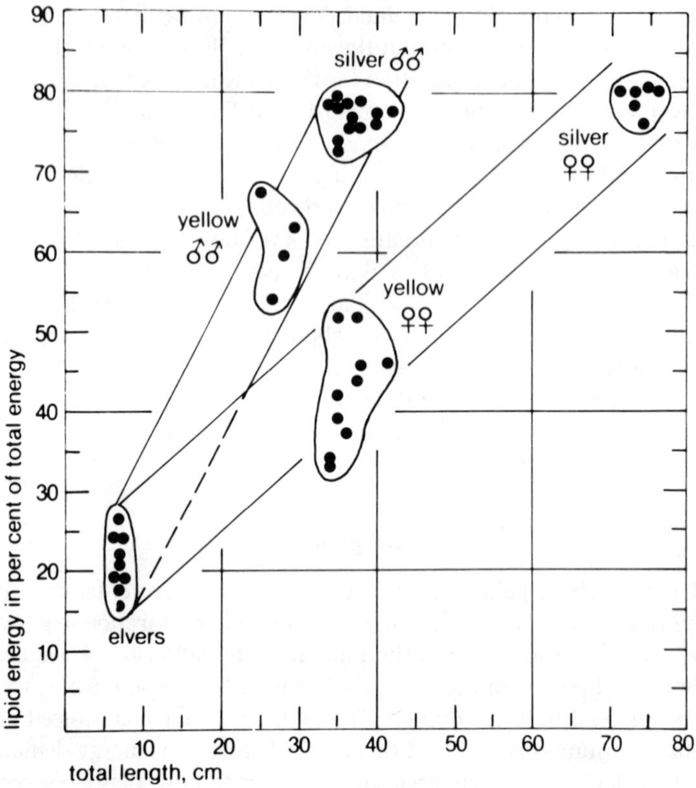

Fig. 3.5 Relationship between fat and body length in *Anguilla anguilla*. The energy represented by lipid was calculated as percentage of the total energy content of the body and plotted against body length. Reproduced with permission from Boëtius and Boëtius (1985).

tabulated data and those of Larsson *et al.*, are combined, it may be concluded that silver eels, regardless of sex, rarely have a fat % lower than 20. Furthermore the data tabulated by Boëtius and Boëtius show that yellow females have higher levels of protein and lower levels of lipid than yellow males. In considerations about the future need of the animals, interest has often focused on energy, and fat represents a good supply of that. However, when gonadal growth (or foetal growth, placental growth, and milk production) is considered, not only fat but also protein is needed, and the fat and protein may need to be of a special quality in order to provide the necessary raw material for growth of tissues or organs or for production of milk. No physiological mechanism has yet been found to explain the correlations between fat (or protein) and lamprey metamorphosis, mammalian puberty or eel silvering.

A very careful analysis of the energy budget of migratory eels, *A. anguilla*, has been made on the basis of analyses of energy content (not %, but amount is needed for the calculations) of whole eels or parts of eels before and after hormone-induced sexual maturation, supplemented with measurements of oxygen uptake during swimming activity (Boëtius and Boëtius, 1980, 1985). These calculations showed that even in females, the silver eel contains energy reserves and organic material sufficient for the long migration and the extensive gonadal growth. The experiments further showed that under laboratory conditions, starving silver eels mobilize protein slightly faster than fat, so eels after a long period of starvation have a slightly higher fat % than at the beginning of the starvation period. Further experiments clarified in more detail the synthesis of various types of lipids in gonads and other organs during hormone-induced maturation of male and female eels (Boëtius *et al.*, 1991). The liver produces vitellogenin (a phospholipoprotein) which is incorporated in the ovary, but the mature egg also contains a large oil globule, like a typical pelagic teleost egg.

Body compartments during sexual maturation

The most detailed study of metabolism of eels during hormonally induced sexual maturation has been performed by Chan and co-workers (Chan and Lau, 1990). As demonstrated in Fig. 3.6, the increase in mass of the gonads

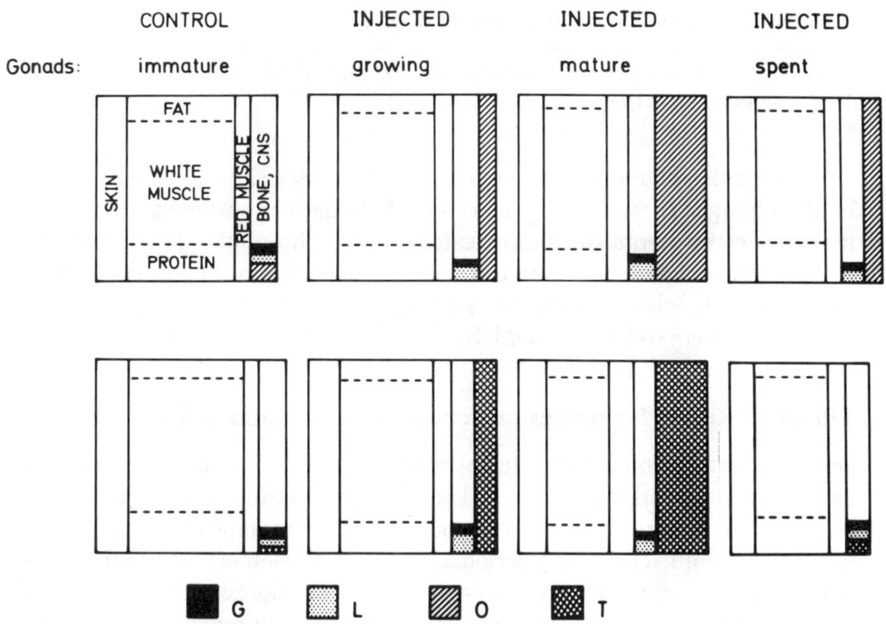

Fig. 3.6 Changes in proportions of body constituents during gonadotropin-induced sexual maturation in *Anguilla japonica*. The figure is based on masses of the whole body and its main constituents, combined with analysis of fat and protein in white muscle samples. Upper row, females; lower row, males. CNS, brain; G, gut; L, liver; O, ovaries; T, testes. (Courtesy D.K.O. Chan.)

Table 3.7 Essential features of gonadotropin-induced sexual maturation*

Yellow eels not responsive; most silver eels responsive (*Anguilla anguilla*)
Males: One dose of human chorionic gonadotropin (HCG) enough (*Anguilla anguilla*)
Females: frequent doses of gonadotropin and pituitary extract necessary (*Anguilla anguilla* and *A. japonica*)
Hypophysectomized males respond rather well to HGG (*Anguilla anguilla* and *A. japonica*)
Hypophysectomized females respond partially to gonadotropin (*Anguilla anguilla*)

* Based on Boëtius and Boëtius (1967, 1980), Fontaine *et al.* (1976), Khan *et al.* (1987), Lau *et al.* (1990), Burzawa-Gerard and Dumas-Vidal (1991).

Table 3.8 Essential features of steroid hormone treatment*

Testosterone implants stimulate testis growth and spermiation (*Anguilla anguilla*)
Testosterone and androstenedione implants stimulate ovarian growth (*Anguilla japonica*)
Oestradiol implants stimulate vitellogenin synthesis and secretion but not vitellogenesis (*Anguilla japonica*)
Oestradiol injections stimulate stainability of several cell types in the pituitary (*Anguilla anguilla*) in ♀ and ♂
Oestradiol and testosterone implants or injections stimulate gonadotropin synthesis in the pituitary (*Anguilla anguilla*) in ♀ and ♂

* Based on Olivereau and Olivereau (1979a,b) Zhang (1988: see Chan and Lau, 1990), Dufour *et al.* (1983; 1989), Boëtius and Larsen (1991), Lin *et al.* (1991 a, b), Quérat *et al.* (1991).

is accompanied by reduction in mass of white muscles, whereas red muscles hardly change. These authors also studied changes in enzyme activities and in blood components, including calcium and phosphate. These ions are mobilized from bones and used in the liver for vitellogenin production in the female; in males phosphate becomes incorporated in the testes. The corpuscles of Stannius seem to be involved when bone salts are mobilized.

Effects of steroid hormones on sexual differentiation and body growth

Sexual differentiation occurs late in the life story of the eel, and the discussion of factors of importance for sex determination is similar to that already presented for lampreys (p. 75). The gonads, at least in some individuals, seem to be undetermined for a long period, and the proportion of males and females may depend on environmental factors. The most impressive demonstration comes from aquaculture, where glass eels are captured and kept under crowded conditions and fed artificial food. Here a large majority of the eels turn out to be males. Since females are larger than males, economic interest has led to attempts at sex manipulation. The most successful result has been obtained by treating eels from 1 to 3 g with various oestradiol preparations added to the food or to the water (Lundshøj, unpublished data). This led not only to higher growth rates but also to nearly 100% female populations. However, this does not allow the interpretation that the larger size of female silver eels is caused by oestradiol. Growth rate is supposed to be the same for both sexes. However, females continue to feed and grow for a longer period than males, before they turn into silver eels and stop feeding and growth; therefore female silver eels are generally considerably larger than male silver eels. Androgens and growth hormone also increased growth rate (Degani, 1986; Degani and Gallagher, 1985).

Hormonally induced sexual maturation

The first successful induction of nearly full sexual maturation was obtained by treatment of males with a preparation from urine of a pregnant woman (Fontaine, 1936). That led to increased testis mass (gonadosomatic index, GSI = mass of gonad as % of body mass), reflecting division of spermatogonia, and further differentiation of spermatogonia to spermatocytes. Since sperm was emitted, spermatids and spermatozoa must also have been formed. Females responded only slightly to a similar treatment. In 1964 M. Fontaine and his collaborators managed to induce marked vitellogenesis in three females, and ovulation and spawning in one female, by treatment with carp pituitary extract; to induce ovulation/spawning, desoxycorticosterone was injected (Fontaine *et al.*, 1964). Later full sexual maturation was induced in *A. japonica* by treatment with 'Synahorin', a combination of human chorionic gonadotropin (HCG) and teleost pituitary extract (Yamamoto and Yamauchi, 1974). In males a single injection may suffice, but often two or more injections are given. Females need treatment for the whole period of development; in the beginning one injection per week is sufficient, but in the last part of the treatment up to two injections per day are sometimes given. Furthermore, the temperature regime is important; a rise in temperature may facilitate final maturation and ovulation in females. In males the maximum testis mass corresponds to a GSI of about 20, but spermiation may occur from GSI 5. In females the corresponding figures are 40–60 and about 40. Tables 3.7 and 3.8 list essential features of hormone experiments.

The experiments cited above show that the lack of spontaneous sexual maturation in eels is not caused by lack of gonadal responsiveness toward gonadotropin.

Hitherto the only other way to obtain complete sexual maturation was by treatment of eels with implants of testosterone. The emission of sperm was an unexpected side effect of a successful attempt to induce growth in eye size, like that observed during gonadotropin-induced sexual maturation (Boëtius and Larsen, 1991; p. 95). Even more surprising is the increase in GSI found in female *A. japonica* implanted with testosterone and androstenedione (Lin *et al.*, 1991a,b). It is noteworthy that implants of oestradiol did not affect ovarian development (Chan and Lau, 1990).

Androgen treatment may cause gonadal growth and development by a direct effect on the gonad and/or by an indirect effect via the pituitary gland (gonadotropin or other hormones). In males testosterone probably stimulates spermatogenesis directly, because this is a phenomenon frequently observed in vertebrates. In females testosterone and androstenedione may act by stimulating endogenous gonadotropin secretion and perhaps also by stimulating vitellogenin production in the liver.

Oestradiol is mainly known for its ability to induce female secondary sex characters (as demonstrated in the lamprey) and vitellogenin synthesis in the liver as a prerequisite for formation of yolk in the ovary (vitellogenesis). In eels, oestradiol was shown to stimulate vitellogenin synthesis in the liver and its secretion to the blood. Gonadotropin is needed for its uptake in the ovary (Burzawa-Gerard and Dumas-Vidal, 1991). Oestradiol also affects pituitary cytology and hormone synthesis. Oestradiol injections were shown to influence markedly the stainability (which reflects accumulated hormones) in several cell types of the pituitary gland of both female and male eels (Olivereau and Olivereau, 1979a,b). The effect on the gonadotropin-cells was later shown to reflect increased accumulation of gonadotropin (Dufour *et al.*, 1989), and the effect is probably exerted on mRNA, at the pretranslational level (Quérat *et al.*, 1991) as seen in female eels treated for 2 weeks with implants of testosterone and oestradiol. Injections of progesterone and cortisol had no or very slight effect, and testosterone was effective only in males, not in females (Dufour *et al.*, 1983). The accumulation of gonadotropin reflected increased synthesis; there were no signs of release, as based on measurement of GSI or of gonadotropin levels in the blood. An interplay between the gonad and the pituitary gland has been demonstrated; in female silver eels treated with carp pituitary extract, both gonadotropin and GnRH content in the pituitary glands increased, but not after ovariectomy (Dufour *et al.*, 1989).

The finding of increased GnRH content in the pituitary gland leads to the next step in analysis of the factors influencing gonadotropin secretion. In lampreys the pars distalis of the pituitary gland is separated from the brain, and GnRH from the brain (if active at all) probably only acts by diffusion. In eels and other teleosts, GnRH-secreting neurones penetrate the pars distalis and release GnRH close to the gonadotropin-producing cells. In tetrapods the GnRH is transferred from the median eminence via portal vessels to the pars distalis.

Attempts to induce final maturation with GnRH are often successful in teleosts. However, in eels in which the gonads are immature, only slight effects have been obtained, and only by a combined treatment with a GnRH-agonist and various dopamine antagonists for many weeks (Dufour *et al.*, 1988, 1991); in oestradiol-pretreated females a GSI of 3.0 (maximum 4.6) was obtained against the control value of 1.8. In females not pretreated with oestradiol, GSI increased slightly, but a marked increase in pituitary content of gonadotropin was found, probably reflecting a markedly increased synthesis and a low rate of secretion. Moreover this pharmacological treatment caused a high level of escape activity in the eels, perhaps reflecting a stimulation of brain centres for migratory behaviour (see also pp. 96–97).

The reason that eels do not spontaneously become sexually mature in captivity may be related to a lack of stimulation by GnRH combined with a dopaminergic inhibition (Dufour *et al.*, 1988; 1991), which has been demonstrated convincingly in goldfish (Peter *et al.*, 1990). If the inhibitory effect is dominant, it might be expected that transplantation of the pituitary gland would stimulate gonadotropin production and secretion; there was no cytological evidence of this (Olivereau, 1970).

From an ecophysiological point of view, it is interesting that the amount of oestradiol present in the blood is higher in seawater than in fresh water, and removal from the blood of injected oestradiol is slower in seawater than in fresh water (Quérat *et al.*, 1987; Petersen and Korsgaard, 1989).

Although the gonads are immature in silver eels and the pituitary gland apparently inactive, hypophysectomy will cause degeneration in the testes (Khan *et al.*, 1987) and degeneration of the ovarian follicle cells (Fontaine *et al.*, 1976).

Eels possess no morphological secondary sex characters that unambiguously distinguish males from females. However, the eyes, which grow somewhat during the transition from yellow to silver stage, grow even larger during hormone-induced sexual maturation (Boëtius and Boëtius, 1967; Pankhurst, 1982). Such growth can also be induced by treatment with testosterone implants (Boëtius and Larsen, 1991) or by oestradiol injections (Dufour, unpublished). Sexually mature females emit odorous substances that may attract male eels (Edel, 1975; Sorensen and Winn, 1984). In this context it is worth noticing that the olfactory system of eels is highly developed, and eels are extremely sensitive towards certain chemicals. This may be of importance, not only in the interplay between the sexes, but also during the migration of silver eels, as cues telling about the route of migration (the same as the one they travelled as larvae, glass eels, elvers, and in some cases as yellow eels; Westin, 1990).

Emission of ovulated eggs or spermiated spermatozoa via the oviducts or sperm ducts may be induced by stripping or may occur during spontaneous movements of the eels or spawning behaviour. The spawning behaviour has been described by Boëtius and Boëtius (1980) and by Sorensen and Winn (1984). Fertilization, embryonic development, hatching, and early larval development have also been described (Yamamoto and Yamauchi, 1974; Yamauchi *et al.*, 1976), but it has not yet been possible to make the larvae survive for more than 3 weeks. It is not clear whether the failure of survival is related to bad quality of gametes or to lack of suitable food or lack of a suitable environment (or a combination thereof).

External factors

Parallel with attempts to induce sexual maturation in silver eels by hormone treatment, attempts were made to change the environment in a way that would permit sexual maturation to occur spontaneously or would induce sexual maturation. Naturally, conditions resembling those met during oceanic migration were tested – seawater, various amounts and quality of light, including darkness, various temperatures, forced swimming in circulating water – but with no effect (Boëtius and Boëtius, 1967). The only treatment with stimulatory effect was gradual submersion of caged female eels to a depth of 450 m for 3 months in the Mediterranean Sea (Fontaine *et al.*, 1985). GSI increased slightly (from 1.6 to 2.2), but pituitary gonadotropin content increased 27 times. In most cases only one condition or factor has been changed at a time; perhaps a special pattern of the tested factors is necessary; perhaps unknown factors are essential (Fontaine and Dufour, 1991). The possibility also exists that eels are especially sensitive to confinement. In this context the experiment mentioned above by J. Boëtius (1967) is relevant. He kept 145 eels in complete darkness from September till June, and twice daily he counted their escape reactions. As shown in Fig. 3.7, there was maximal activity in the two first lunations, then a lunation with no activity, and then an early spring maximum. In all periods, activity was highest when the moon

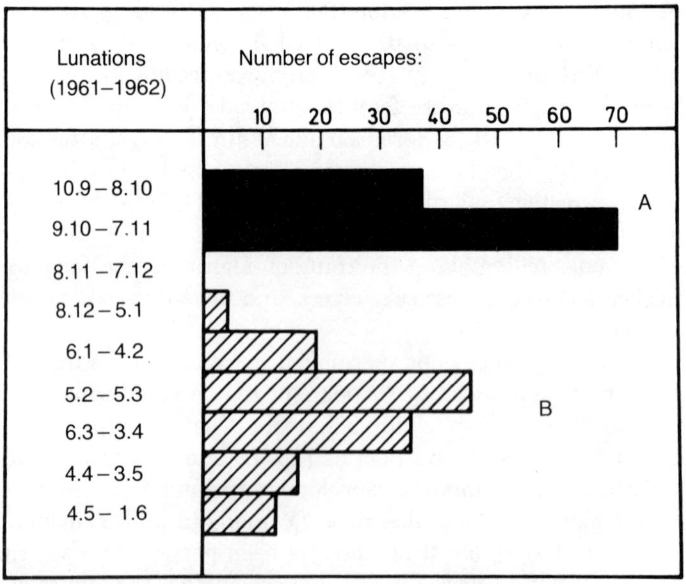

Fig. 3.7 Escape activity of silver eels, *Anguilla anguilla*. Eels (N = 145) were kept in darkness in seawater, at 12–15 °C. They had to leave the water in order to escape. In all 238 escapes were recorded. Escapes occurred in two periods (A,B). Data from Boëtius (1967). Reproduced with permission.

was in its third quarter. All escapes took place between 5pm and 9am. These escape activities are most likely related to the large autumn migration of silver eels, and to the less prominent spring migration known from Danish waters. Since the migratory activity is signalling the beginning of the reproductive period, and since reproduction is inhibited when migration is prevented, the influence of the moon (not via moonlight!) becomes of interest.

Natural death after spawning

For many years it was assumed that eels die after spawning. However, males quite often survived hormonally induced maturation, so this dogma was questioned (Boëtius and Boëtius, 1967). A second partial sexual maturation of male eels was obtained in *A. australis* (one eel) and *A. dieffenbachii* (one eel) (Todd, 1981). In *A. anguilla* even a third maturation was induced in several male eels. Furthermore two feeding and growth periods were interposed between the maturation periods, and the accompanying changes in the

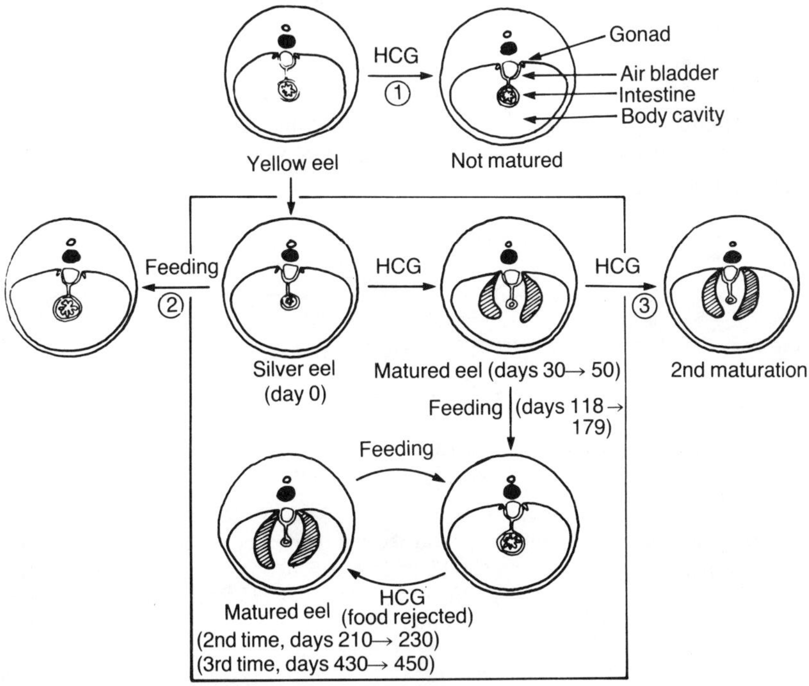

Fig. 3.8 Diagram demonstrating responses of male *Anguilla anguilla* to HCG and feeding. (1) Male yellow eels do not mature when treated with HCG (Boëtius and Boëtius, 1967). (2) Male silver eels will feed after 6 to 7 months in captivity (Boëtius and Boëtius, 1967). (3) Male silver eels can be matured twice with HCG, without intermediate feeding (Boëtius and Boëtius, 1982). In frame: Male silver eels were matured, then fed and matured again twice (Dollerup and Graver, 1985, reproduced with permission).

intestine and in mass of organs and in chemical constitution were followed (Dollerup and Graver, 1985). A survey of the changes between phase of growth and phase of reproduction is shown in Fig. 3.8. Individual growth was followed from day 118 to day 148 and from day 148 to day 179; all combinations in sequences of rapid growth, slow growth, no growth and loss of body mass were encountered.

Females normally do not survive hormonally induced sexual maturation, but one female was fed after spawning, and by hormone treatment a second spawning was induced (Le Belle and Fontaine, 1987).

Conclusion

To summarize, the essential factors demonstrated to be necessary for eel reproduction (Fig. 3.9) are the pituitary with its secretion of gonadotropin and the gonads with their secretion of gonadal steroids. Sexual maturation does not occur spontaneously under the conditions hitherto examined. Probably a situation like that normally experienced by the eel on its way to the supposed spawning area in the Sargasso Sea is essential for GnRH-mediated stimulation of the pituitary gland, via sense organs or via some influence on other cells of the body. An inhibitory influence related to laboratory confinement or confinement in nature (land-locked eels) may also be involved and may act via the dopaminergic inhibition of gonadotropin secretion, as inferred from pharmacological experiments.

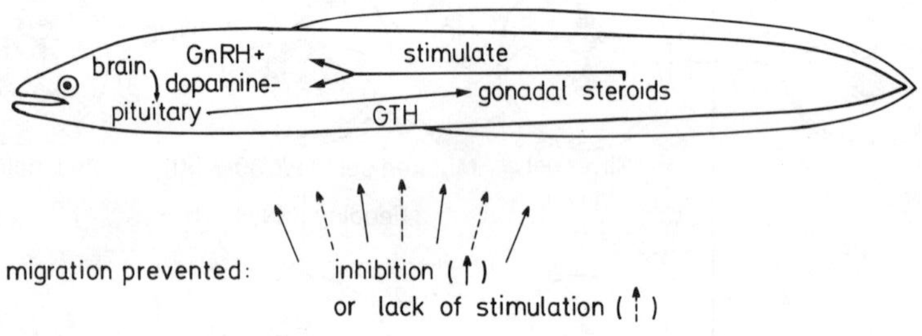

Fig. 3.9 Scheme for neuroendocrine control of eel reproduction. The elements depicted for the lamprey in Fig. 3.4 may be involved in reproduction in all vertebrates (hagfish excepted?), but here they are combined in a way typical of the eel. In the lamprey, endogenous control of *initiation* of sexual maturation seems very important, and the initiation may be directly caused by the genome, as indicated in the figure; but another possibility (Section 3.4) is that the genome early in ontogeny controls the differentiation of the brain in such a way that the brain–soma interactions lead to the later 'programmed changes'. These are thus built into the organism by the genome. The function of the genome in the late stages of ontogeny may rather be controlled by hormones than vice versa. In the eel the environment is decisive for initiation of sexual maturation; unknown stimuli related to time of night and the third quarter of the lunar cycle seem to induce migratory activity; if migration is prevented, sexual maturation does not occur.

3.4 GENERAL DISCUSSION

In the previous sections the physiology and ecology of lampreys and eels were described, especially with regard to growth and reproduction. Focus was on the life story of individual fish. When ecophysiology is considered at the level of populations, one can speak of life cycles, because the life story of individuals is followed by a similar life story of the offspring. In successive generations little change occurs in the structure of the genome, typical of the species, and the generations live in the same environment. In migratory species, individuals are able to adapt physiologically to different types of environment, and it is typical that this adaptation to some extent takes place before migration starts (anticipatorily). This probably reflects a genetically programmed change in morphology, physiology and behaviour; so even if the genome does not change in structure from one generation to the next, it certainly changes in function (repression–expression) during all types of ontogenetic development, and perhaps (but not necessarily) most dramatically when dramatic changes in habitat, morphology, physiology and behaviour take place. In lampreys and eels the transitions between phases of food intake and growth and phases of gonadal development can easily be observed or induced, and they are therefore well suited for studies of how the genetically based subprogrammes for growth and for reproduction shift. Shifts between a phase dominated by food intake and growth and a phase dominated by lack of food intake and by reproduction may be similar to the shift that takes place spontaneously between a phase of food intake and other activities and a phase of sleep in, for instance, human beings. The elements involved and the rapidity with which the shifts occur differ, but the principle may be the same. If this idea is correct, focus should be on the genetic programming of brain activity involved in control of pituitary function and in function of the autonomic nervous system in lampreys and eels. The growth phase is dominated by secretion of, and sensitivity to, growth-promoting hormones, by a drive for food intake, and by a fully functioning digestive tract. Energy and material of the food are used for growth and for maintenance of the already established body. The phase of reproduction is dominated by secretion of, and sensitivity to, gonadotropin, and by a lack of drive for food intake, and a reduced function of the digestive tract. Energy and material for gonadal growth are mobilized from the general body tissues. In human beings the analogous daily shift between food intake and sleep seems to be based on spontaneous shifts in brain activity (centres of sleep and centres of arousal). The changed brain activity also results in different patterns of pituitary function in the two phases. The environment may stimulate or prevent the shift, but it is dominated by endogenous processes.

A deep understanding of the life story of individuals of a species necessitates knowledge of the evolutionary story of that species. Lampreys have attracted interest because they have certain structural features in common with fossil *Cephalospidomorphi* from the Silurian and Devonian periods. It is, however, not certain that lampreys retain more primitive traits from the unknown vertebrate ancestor(s) than any other vertebrate existing today. 'Trait' means any recognizable character in structure or function at the level of molecules, cells, organs or the whole organism; 'function' means the way structures change, move, and interact. A trait may thus range from the amino acid sequence of a protein to the type of spawning behaviour – *A. anguilla* has attracted special interest because of its (supposed) long migrations between the Sargasso Sea and Europe and back again. It is consistent with palaeontological and geological data to assume that eels gradually adapted (genetically) to the geological changes that separated America from Europe (Lecomte-Finiger, 1990).

To really understand the biology of lampreys, eels and other fish – and any other creature – one needs all disciplines of science, united in what one may call integrative biology, in which one attempts to describe how the genetic programme for the whole life story evolved and today interacts with the physiological control systems and with the environment.

ACKNOWLEDGEMENTS

The help of Inge and Jan Boëtius is gratefully acknowledged. They contributed substantially to the chapter with their ample experience with all known parts of the life story of eels, both in the field and in the laboratory, and with their familiarity with eel literature.

REFERENCES

Applegate, V.C. (1950) Natural history of the sea lamprey (*Petromyzon marinus*) in Michigan. *Spec. scient. Rep. U.S. Fish Wildl. Serv., Fish.*, no. 55, 237 pp.

Boëtius, I. (1985) Otholiths of eels of known age. *EIFAC Working Party on Eels. Perpignan, France*, FAO, Rome, 13pp.

Boëtius, I. and Boëtius, J. (1967) Studies in the European Eel, *Anguilla anguilla* (L.). Experimental induction of the male sexual cycle, its relation to temperature and other factors. *Meddr Danm. Fisk.- og Havunders.*, **4**, 339–405.

Boëtius, I. and Boëtius, J. (1980) Experimental maturation of female silver eels, *Anguilla anguilla*. Estimates of fecundity and energy reserves for migration and spawning. *Dana*, **1**, 1–28.

Boëtius, I. and Boëtius, J. (1982) Experimental maturation of European silver eels. *Proc. int. Symp. reprod. Physiol. Fish, Wageningen*, **1982**, 174–6.

Boëtius, I. and Boëtius, J. (1985) Lipid and protein content in *Anguilla anguilla* during growth and starvation. *Dana*, **4**, 1–17.

Boëtius, I. and Boëtius, J. (1989) Ascending elvers, *Anguilla anguilla*, from five European localities. Analyses of pigmentation stages, condition, chemical composition and energy reserves. *Dana*, **7**, 1–12.

Boëtius, I. and Larsen, L.O. (1991) Effects of testosterone on eye size and spermiation in silver eels, *Anguilla anguilla*. *Gen. comp. Endocrinol.*, **82**, 238.

Boëtius, I., Boëtius, J. and Hansen, H.J.M. (1991) Studies on lipid synthesis by

incorporation of ^{14}C-acetate during experimental maturation of silver eels, *Anguilla anguilla*. *Dana*, **9**, 1–14.
Boëtius, J. (1967) Experimental indication of lunar activity in European silver eels, *Anguilla anguilla* (L.). *Meddr Danm. Fisk.- og Havunders.*, **6**, 1–6.
Boëtius, J. (1985) Greenland eels, *Anguilla rostrata* LeSueur. *Dana*, **4**, 41–8.
Boëtius, J. and Harding, E.F. (1985) A re-examination of Johannes Schmidt's Atlantic eel investigations. *Dana*, **4**, 129–62.
Bruun, A.F., Hemmingsen, A.M. and Møller-Christensen, E. (1949) Attempts to induce experimentally maturation of the European eel, *Anguilla anguilla* L. *Acta endocr., Copenh.*, **2**, 212–6.
Burzawa-Gerard, E. and Dumas-Vidal, A. (1991) Effects of 17β-estradiol and carp gonadotropin on vitellogenesis in normal and hypophysectomized European silver female eel (*Anguilla anguilla* L.) employing a homologous radioimmunoassay for vitellogenin. *Gen. comp. Endocrinol.*, **84**, 264–76.
Chan, D.K.O. and Lau, E.Y.L. (1990) Mobilization of body reserves during induced gonadal development in the Japanese eel, *Anguilla japonica* (Temminck & Schlegel): the role of the pituitary gland and corpuscles of Stannius, in *Fish Reproduction, Toxicology and Management* (ed. Robert C. Ryans), U.S. Governmental Protection, Environmental Res. Lab., Athens, GA, pp. 108–27.
Degani, G. (1986) Effect of dietary 17-β-estradiol and 17-α-methyltestosterone on growth and body composition of slow-growing elvers (*Anguilla anguilla* L.). *Comp. Biochem. Physiol.*, **85A**, 243–7.
Degani, G. and Gallagher, M.L. (1985) Effects of dietary 17α-methyltestosterone and bovine growth hormone on growth and food conversion of slow- and normally-growing American elvers (*Anguilla rostrata*). *Can. J. Fish. aquat. Sci.*, **42**, 185–9.
Dollerup, J. and Graver, C.M. (1985) Repeated induction of testicular maturation and spermiation, alternating with periods of feeding and growth in silver eels, *Anguilla anguilla* (L.). *Dana*, **4**, 19–39.
Dufour, S., Belle, N.D.-L. and Fontaine, Y.-A. (1983) Effects of steroid hormones on pituitary immunoreactive gonadotropin in European freshwater eel, *Anguilla anguilla* L. *Gen. comp. Endocrinol.*, **52**, 190–7.
Dufour, S., Lopez, E., Le Menn, F., Le Belle, N., Baloche, S. and Fontaine, Y.-A. (1988) Stimulation of gonadotropin release and of ovarian development, by the administration of a gonadoliberin agonist and of dopamine antagonists, in female silver eel pretreated with estradiol. *Gen. comp. Endocrinol.*, **70**, 20–30.
Dufour, S., Belle, N.D.-L., Baloche, S. and Fontaine, Y.-A. (1989) Positive feedback control by the gonads on gonadotropin (GTH) and gonadoliberin (GnRH) levels in experimentally matured female silver eels, *Anguilla anguilla*. *Fish Physiol. Biochem.*, **7**, 157–62.
Dufour, S., Bassompierre, M., Montero, M., Belle, N.D-L., Baloche, S. and Fontaine, Y.-A. (1991) Stimulation of pituitary gonadotropic function in female silver eel treated by a gonadoliberin agonist and dopamine antagonists, in *Reproductive Physiology of Fish* (eds. A.P. Scott, J.P. Sumpter, D.E. Kime and M.F. Rolfe), Fishsymp. 91, Sheffield, pp. 54–60.
Edel, R.K. (1975) The induction of maturation of female American eels through hormone injections. *Helgoländer wiss. Meeresunters.*, **27**, 131–8.
Evennett, P.J. and Dodd, J.M. (1963) Endocrinology of reproduction in the river lamprey. *Nature, Lond.*, **197**, 715–16.
Farmer, G.J. (1980) Biology and physiology of feeding in adult lampreys. *Can. J. Fish. aquat. Sci.*, **37**, 1751–61.
Fontaine, M. (1936) Sur la maturation complète des organes génitaux de l'anguille mâle et l'émission spontanée de ses produits sexuels. *C.r. Acad. Sci. Paris*, **202**, 1312–15.

Fontaine, M., Bertrand, E., Lopez, E. and Callamand, O. (1964) Sur la maturation des organes génitaux de l'anguille femelle (*Anguilla anguilla* L.) et l'émission spontanée des oeufs en aquarium. *C.r. hebd. Séanc. Acad. Sci. Paris*, **259**, 2907–10.

Fontaine, Y.-A. and Dufour, S. (1991) The eels: from life cycle to reproductive endocrinology. *Bull. Inst. Zool. Acad. Sin.*, Monograph 16, 237–48.

Fontaine, Y.-A., Dufour, S., Alinat, J. and Fontaine, M. (1985) L'immersion prolongée en profondeur stimule la fonction hypophysaire gonadotrope de l'anguille européenne (*Anguilla anguilla* L.) femelle. *C.r. hebd. Séanc. Acad. Sci. Paris*, **300**, 83–7.

Fontaine, Y.-A., Lopez, E., Delerue-Le Belle, N., Fontaine-Bertrand, E., Lallier, F. and Salmon, C. (1976) Stimulation gonadotrope de l'ovarie chez l'anguille (*Anguilla anguilla* L.) hypophysectomisée: morphologie, activités adényl cyclase et phosphodiestérase de l'adénosine monophosphate cyclique. *J. Physiol., Paris*, **72**, 871–92.

Gorbman, A. (1980) Evolution of the brain–pituitary relationship: evidence from the Agnatha. *Can. J. Fish. aquat. Sci.*, **37**, 1680–86.

Hagelin, L.-O. (1959) Further aquarium observations on the spawning habits of the river lamprey (*Petromyzon fluviatilis*). *Oikos*, **10**, 50–64.

Hardisty, M.W. and Baker, B.I. (1982) Endocrinology of lampreys, in *The Biology of Lampreys*, Vol. 4B (eds M.W. Hardisty and I.C. Potter), Academic Press, London, pp. 1–115.

Harris, G.W. and Jacobsohn D. (1952) Functional grafts of the anterior pituitary gland. *Proc. R. Soc.*, **139B**, 263–76.

Khan, I.A., Lopez, E. and Leloup-Hâtey, J. (1987) Induction of spermatogenesis and spermiation by a single injection of human chorionic gonadotropin in intact and hypophysectomized immature European eel (*Anguilla anguilla* L.). *Gen. comp. Endocrinol.*, **68**, 91–103.

Kille, R.A. (1960) Fertilization of the lamprey egg. *Expl Cell Res.*, **20**, 12–27.

Larsen, L.O. (1962) Weight and length in the river lamprey. *Nature, Lond.*, **194**, 1093.

Larsen, L.O. (1970) The lamprey egg at ovulation (*Lampetra fluviatilis* L. Gray). *Biol. Reprod.*, **2**, (1), 37–47.

Larsen, L.O. (1974) Effects of testosterone and oestradiol on gonadectomized and intact male and female river lampreys (*Lampetra fluviatilis* (L.) Gray). *Gen. comp. Endocrinol.*, **24**, 305–13.

Larsen, L.O. (1976) Blood glucose levels in intact and hypophysectomized river lampreys (*Lampetra fluviatilis* L.) treated with insulin, "stress", or glucose, before and during the period of sexual maturation. *Gen. comp. Endocrinol.*, **29**, 1–13.

Larsen, L.O. (1980) Physiology of adult lampreys with special regard to natural starvation, reproduction, and death after spawning. *Can. J. Fish. aquat. Sci.*, **37**, 1762–79.

Larsen, L.O. (1985) The role of hormones in reproduction and death in lampreys and other species which reproduce once and die, in *Current Trends in Comparative Endocrinology* (eds B. Lofts and W.N. Holmes), Hong Kong University Press, Hong Kong, pp. 613–16.

Larsen, L.O. (1987) The role of hormones in initiation of sexual maturation in male river lampreys (*Lampetra fluviatilis* L.): gonadotropin and testosterone. *Gen. comp. Endocrinol.*, **68**, 197–201.

Larsen, L.O. (1990) The role of external and internal factors during the phase of reproduction in lampreys, with special regard to the interplay between gonadal and extragonadal (interrenal) steroids. *Fish Physiol. Biochem.*, **8**, 375–88.

Larsson, P., Hamrin, S. and Okla, L. (1990) Fat content as a factor inducing migratory behavior in the eel (*Anguilla anguilla* L.) to the Sargasso Sea. *Naturwissenschaften*, **77**, 488–90.

Lau, E.Y.L., Lin, H.R. and Chan, D.K.O. (1990) Circulating gonadotropin levels during induced gonadal development in the Japanese eel *Anguilla japonica*. *Soc. Study Endocr. Metab. Reprod. Hong Kong*, **5**, 3.

Le Belle, N. and Fontaine, M. (1987) Toutes les anguilles males et femelles peuvent-elles frayer une seconde fois dans la nature?. *Ichthyophysiol. Acta*, **11**, 7–11.

Lecomte-Finiger, R. (1990) Métamorphose de l'anguille jaune en anguille argentée *Anguilla anguilla* L. et sa migration catadrome. *Annls Biol.*, **29**, 183–94.

Lecomte-Finiger, R. and Yahyaoui, A. (1989) La microstructure de l'otolithe au service de la connaissance du développement larvaire de l'anguille européenne *Anguilla anguilla*. *C.r. hebd. Séanc. Acad. Sci. Paris*, **308**, 1–7.

Lin, H.R., Zhang, M.-L., Zhang, S.-M., Van Der Kraak, G. and Peter, R.E. (1991a) Stimulation of pituitary gonadotropin and ovarian development by chronic administration of testosterone in female Japanese silver eel, *Anguilla japonica*. *Aquaculture*, **96**, 87–95.

Lin, H., Zhang, M., Zhang, S., Shi, H., Lu, M., Van Der Kraak, G. and Peter, R.E. (1991b) Induction of gonadal development and maturation by chronic administration of testosterone and androstenedione in female Japanese silver eel, *Anguilla japonica*, in *Reproductive Physiology of Fish* (eds. A.P. Scott, J.P. Sumpter, D.E. Kime and M.F. Rolfe) 4th int. Symp. reprod. Physiol. Fish, abstract no. 89, p. 281.

Maitland, P.S. (1980) Assessment of lamprey and fish stocks in the Great Lakes in relation to control of the sea lamprey (*Petromyzon marinus*): report from the SLIS assessment measurements task force. *Can. J. Fish. aquat. Sci.*, **37**, 2197–2201.

Malmqvist, B. (1983) Growth dynamics and distribution of a population of the brook lamprey *Lampetra planeri* in a South Swedish stream. *Holarctic Ecol.*, **6**, 404–12.

Malmqvist, B. (1986) Reproductive ecology of lampreys, in *Indo-Pacific Fish Biology: Proc. Second Int. Conf. Indo-Pacific Fishes* (eds T. Uyeno, A.T. Taniuchi and K. Matsuura), Ichthyol. Soc. Japan, Tokyo, pp. 20–30.

Michael, J.H., jun. (1984) Additional notes on the repeat spawning by Pacific lamprey. *Calif. Fish Game*, **70**, 186–8.

Moore, J.W. and Mallatt, J.M. (1980) Feeding of larval lamprey. *Can. J. Fish. Aquat. Sci.*, **37**, 1658–64.

Murdoch, S.P., Docker, M.F., Beamish, F.W.H. (1992) Effect of density and individual variation on growth in sea lamprey (*Petromyzon marinus*) larvae in the laboratory. *Can. J. Zool.*, **70**, 184–8.

Olivereau, M. (1970) Cytologie de l'hypophyse autotransplantée chez l'anguille. Comparaison avec celle de *Poecillia*, in *Neuroendocrinologie* (eds J. Benoit and C. Kordon), Centre Nat. Res. Scient., Paris, pp. 251–60.

Olivereau, M. and Olivereau, J. (1979a) Estradiol-positive feedback on gonadotropic (GTH) cells in freshwater male silver eels. *Gen. comp. Endocrinol.*, **39**, 247–61.

Olivereau, M. and Olivereau, J. (1979b) Effect of estradiol-17β on the cytology of the liver, gonads and pituitary, and on plasma electrolytes in the female freshwater eel. *Cell Tissue Res.*, **199**, 431–54.

Pankhurst, N.W. (1982) Relation of visual changes to the onset of sexual maturation in the European eel *Anguilla anguilla* (L.). *J. Fish Biol.*, **21**, 127–40.

Peter, R.E., Yu, K.-L., Marchant, T.A. and Rosenblum, P.M. (1990) Direct neural regulation of the teleost adenohypophysis. *J. exp. Zool.*, **Supp. 4**, 84–9.

Petersen, I. and Korsgaard, B. (1989) Experimental induction of vitellogenin synthesis in eel (*Anguilla anguilla*) adapted to seawater or freshwater. *Comp. Biochem. Physiol.*, **93B**, 57–60.

Plisetskaya, E. (1985) Some aspects of hormonal regulation of metabolism in agnathans, in *Evolutionary Biology of Primitive Fishes* (eds R.E. Foreman, A. Gorbman, J.M. Dodd and R. Olsson), Plenum, New York, pp. 339–61.

Plisetskaya, E.M. (1990) Recent studies of fish pancreatic hormones: selected topics. *Zool. Sci.*, **7**, 335–53.

Purvis, H.A. (1980) Effects of temperature on metamorphosis and the age and length at metamorphosis in sea lamprey (*Petromyzon marinus*) in the Great Lakes. *Can. J. Fish aquat. Sci.*, **37**, 1827–34.

Quérat, B., Hardy, A. and Fontaine, Y.A. (1991) Regulation of the type-II gonadotrophin α and β subunit mRNAs by oestradiol and testosterone in the European eel. *Jour. molec. Endocrinol.*, **7**, 81–6.

Quérat, B., Nahoul, K., Hardy, A., Fontaine, Y.A. and Leloup-Hâtey, J. (1987) Plasma concentrations of ovarian steroids in the freshwater European silver eel (*Anguilla anguilla* L.): effects of hypophysectomy and transfer to sea water. *J. Endocr.*, **114**, 289–94.

Sorensen, P.W. and Winn, H.E. (1984) The induction of maturation and ovulation in American eels, *Anguilla rostrata* (LeSueur) and the relevance of chemical and visual cues to male spawning behaviour. *J. Fish Biol.*, **25**, 261–8.

Sower, S.A. (1990) Neuroendocrine control of reproduction in lampreys. *Fish Physiol. Biochem.*, **8**, 365–74.

Todd, P.R. (1981) Hormone-induced maturation in male New Zealand freshwater eel (*Anguilla* spp.). *N.Z. J. mar. freshwat. Res.*, **15**, 237–46.

Tsukamoto, K. (1992) Discovery of the spawning area for Japanese eel. *Nature, Lond.*, **356**, 789–91.

Westin, L. (1990) Orientation mechanisms in migrating European silver eel (*Anguilla anguilla*): temperature and olfaction. *Mar. Biol.*, **106**, 175–9.

Wright, G.M. (1986) Immunocytochemical demonstration of growth hormone, prolactin and somatostatin-like immunoreactivities in the brain of larval, young adult and upstream migrant adult sea lamprey, *Petromyzon marinus*. *Cell Tissue Res.*, **246**, 23–31.

Yamamoto, K. and Yamauchi, K. (1974) Sexual maturation of Japanese eel and production of eel larvae in the aquarium. *Nature, Lond.*, **251**, 220–1

Yamauchi, K., Nakamura, M., Takahashi, H. and Takano, K. (1976) Cultivation of larvae of Japanese eel. *Nature, Lond.*, **263**, 412.

Surveys, not cited specifically in the text

Lampreys

Hardisty, M.W. (1979) *Biology of the Cyclostomes*, Chapman and Hall, London, 428 pp.

Hardisty, M.W. and Potter, I.C. (eds) (1971–1982) *The Biology of Lampreys*, Vols. 1,2,3,4A,4B, Academic Press, London.

Holcík, J. (ed.) (1986) *The Freshwater Fishes of Europe. Petromyzontiformes*, Vol. 1, Part I, Aula-Verlag, Wiesbaden, 314 pp.

Leatherland, J.F. and Larsen, L.O. (eds) (1990) Proceedings of the Symposium on Lamprey Endocrinology. *Fish Physiol. Biochem.*, **8**, 355–417.

Smith, B.R. (ed.) (1980) Proceedings of the 1979 Sea Lamprey International Symposium. *Can. J. Fish. aquat. Sci.*, **37**, 1585–2214.

Eels

Bertin, L. (1956) *Eels. A Biological Study*, Cleaver-Hume Press, London, 192 pp.

Moriarty, C. (1978) *Eels, a Natural and Unnatural History*, David and Charles, London, 192 pp.

Sinha, V.R.P. and Jones, J.W. (eds) (1975) *The European Freshwater Eel*, Liverpool Univ. Press, Liverpool, 146 pp.

Tesch, F.-W. (1977) *The Eel. Biology and Management of Anguillid Eels*, Chapman and Hall, London, 434 pp.

Chapter four

Salmonid smolting: a pre-adaptation to the oceanic environment

Gilles Boeuf

4.1 INTRODUCTION

Salmonids are a small family of fishes characterized by high morphological and behavioural homogeneity. These characteristics make it possible, for example, to produce many interspecific hybrids, some of them fertile. However, there is one specific aspect where a clear difference appears, allowing separation of the family into two groups, one remaining in fresh water, the other migrating to the sea. Although at present, no salmonid species is totally marine, and some species remain in fresh water for their entire life, others achieve an active migration from fresh water (FW) to the estuary, and finally to the sea. Later they return to fresh water for spawning.

All salmonids hatch in FW, even pink salmon, *Oncorhynchus gorbuscha*, and chum salmon, *O. keta*, species which leave the river at very early life history stages to go to the sea. At a specific stage of development the young salmonid, called parr, undergoes external and internal transformations which preadapt it for survival and growth in the marine environment. It becomes a smolt, a short and critical FW phase bringing it to the status of a seawater (SW) fish. At the end of this **parr–smolt transformation** or **smoltification** (smolting), the salmon is stimulated by internal and external conditions to begin the active downstream migration to the estuary, remains some time in brackish water to adapt completely, and then goes to the open ocean to find its feeding areas, where growth can be very rapid. This phenomenon is totally different from the observed behaviour of some salmonids migrating transitorily from the river to coastal zones under certain unfavourable circumstances (flood, pollution,

Fish Ecophysiology. Edited by J. Cliff Rankin and Frank B. Jensen. Published in 1993 by Chapman & Hall, London. ISBN 0 412 45920 5.

lack of water or abnormal overheating etc). Under these conditions, the fish do not migrate far and their movements are very limited in both distance and time. They will go back rapidly to FW.

Smolting is a dynamic phenomenon, corresponding to information deeply inscribed in the genome, expressing itself at a specific time in the life of the fish and under internal control (nervous and endocrinological), and external synchronization with ecological factors (mainly photoperiod and temperature: Hoar, 1976, 1988).

In general, salmonids may live in FW from some few weeks to as many as 6 years before smolting, which is usually completed in the spring, and which prepares them for marine life. This phenomenon can be considered as a combination of cytological, morphological, biochemical, physiological, endocrinological and behavioural changes, occurring in FW with the active downstream migration concluding the process. The smolt's physiological status governs the migration behaviour. Some cohorts in a particular strain and year class do not achieve smolt status; they remain in FW, usually for an additional year, until the smolting process is completed.

Among all species, the Atlantic salmon, *Salmo salar*, a circum-Atlantic species, is able to leave rivers from, for example, Southern Europe (some of them very long, > 1000 km) and to reach growing areas in the Davis Strait and return, achieving trips of more than 7000 nautical miles. Pacific salmon or migrating trout like chum or steelhead, *O. mykiss*, can travel 3000 miles.

Various aspects of salmonid smolting have been studied intensively since the 1960s reviews: Fontaine, 1975; Hoar, 1976, 1988; Folmar and Dickhoff, 1980; Wedemeyer *et al.*, 1980). Since 1981, the year of the First Workshop on Smolting Phenomenon at the Scripps Institute, California (Bern and Mahnken, 1982), the research has been better organized and appropriate for applied fields. The study of parr–smolt transformation, of great interest in basic science, has become very important in fisheries management and in the rapidly growing mariculture industry. To the question: "what's a smolt?", the best answer for me is "a wild fish in active migration, just before reaching the estuary of the river". This question of smolt quality is decisive in both basic biology and aquaculture. The study of the basic biochemistry and physiology of the smolt is of great interest because this fish, still living in FW, is in fact transformed into a SW salmon before entering the sea. These changes are therefore not triggered by the external salinity but by a 'specific genetic program', which has resulted during the evolution of the group. Many processes involved in SW adaptation, or pre-adaptation (migration, metabolic capacity for rapid growth, predatory behaviour), are linked to the capacity to grow very rapidly in SW. The stunting phenomenon is commonly observed in mariculture: fish do not grow, or cease growing, in SW, probably owing to incomplete smolt status. This emphasizes the biological significance of true smolting. In both intensive marine culture, consisting of growing salmon in net pens or ponds, and extensive culture, consisting of releasing the fish in rivers or coastal areas and afterwards exploiting their later stages by fishing, good smolt quality is the key to success. There is an acute need for reliable criteria of smolt status.

4.2 CHANGES DURING PARR–SMOLT TRANSFORMATION

General aspects

Parr–smolt transformation is a progressive sequence of events and does not refer only to the last changes occurring at the end of the stay in FW. It is a gradual process starting a long time before completion and readiness to enter the marine environment. In Atlantic salmon, probably the species which exhibits the greatest changes, one of the most interesting findings was the discovery of the existence of two modes in the distribution of length or weight frequency classes in the population (Thorpe, 1977; Thorpe *et al.*, 1982; Boeuf *et al.*, 1985). In Brittany, France, from July to September of the first year, the young salmon of the upper mode already express their capacity to smoltify the following year, growing faster than lower-mode fish of the same age, which will stay in the river an additional year before completing the smolting process (Fig. 4.1, top). This growth function is constantly present during smoltification and bigger fish will migrate downstream earlier. The larger individuals reach the size threshold for smolting, and consequently smoltify and leave the river, before their smaller siblings. Both fast growth and acquisition of euryhalinity occur at the same time. The growth in length is very high near the end of smolting (Fig. 4.2, bottom), and the condition factor ($K = W/L^3 \times 10^5$, W in g and fork length L in mm) decreases at that time, from a range of 1.25–1.40 in the Atlantic salmon pre-smolt to 0.90–1.00 in the smolt. Then it increases again after entry into SW if growth is satisfactory (Fessler and Wagner, 1969; Hoar, 1976).

The smolt looks quite different from the parr. Smolts are characteristically silvery owing to purine (guanine and hypoxanthine) deposition on the scales. However, as pointed out by Gorbman *et al.* (1982), the silvering of the skin has proven easily reversible and not necessarily correlated with size or developing physiological or biochemical features. We could say that a smolt is always silvery, but a silvery salmon is not necessarily a smolt. Probably the relationship between the external appearance and the true physiological smolt status is better understood in Atlantic salmon than in Pacific species, where it is often difficult to separate true smolts from pre-smolts in the population. This helps to explain the high incidence of stunts in coho salmon, *Oncorhynchus kisutch*, culture. Among other external parameters, useful indices of smolting status in coho are emergence and growth of teeth, growth and changes in shape of integumentary folds adjacent to the cloacal opening and the auxiliary appendage of the pelvic fin, as well as growth of scales with respect to radius and number of circuli (Gorbman *et al.*, 1982). Winans and Nishioka (1987) pointed out the relationships between body shape and some physiological criteria of smolting. In Pacific Japanese species, Fujioka (1988)

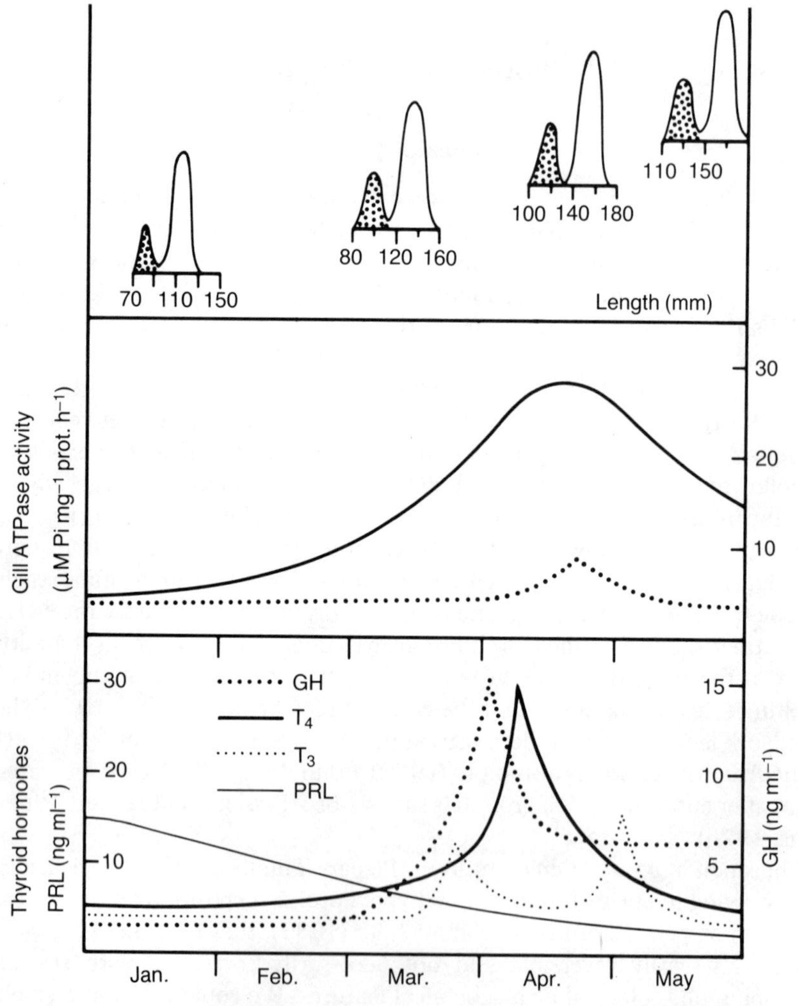

Fig. 4.1 Typical changes in growth and physiology during smolting in Atlantic salmon, *Salmo salar*. (Top) Bimodal length class frequency (%) distribution in the population, with lower (parr, stippled) and upper (pre-smolts or smolts) modes. (Centre) Gill microsomal (Na^+–K^+)-ATPase activity in the two modes (dotted line, parr; continuous line, pre-smolts and smolts). Units of activity are μmol Pi per mg protein per hour. (Bottom) Typical hormonal changes in upper-mode fish in FW. GH, growth hormone; PRL, prolactin; T_3, triiodothyronine; T_4, thyroxine. Data from Boeuf *et al.* (1985, 1989a), Prunet *et al.* (1989). Note different scale for GH.

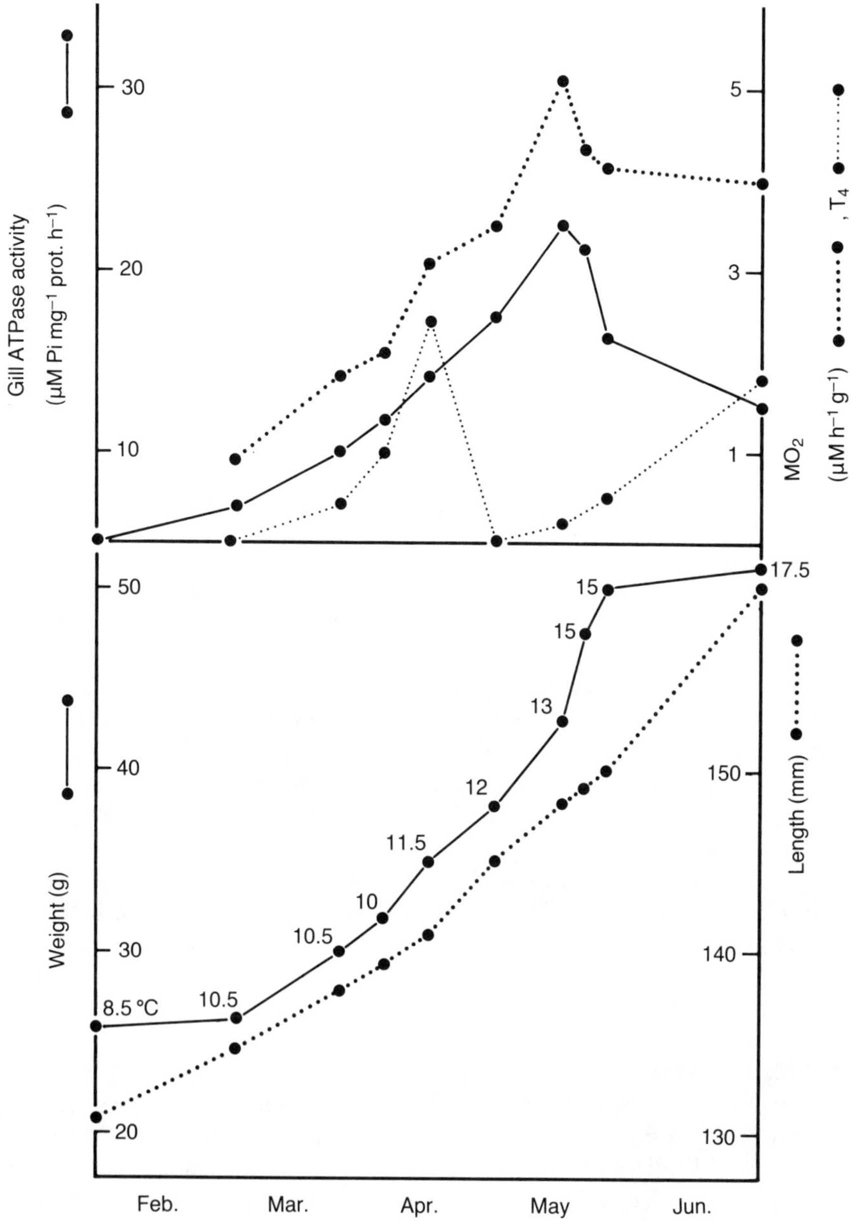

Fig. 4.2 Metabolic and physiological changes in Atlantic salmon pre-smolts and smolts during smoltification in FW. (Top) MO_2 (routine oxygen consumption), gill ATPase activity and blood thyroxine (T_4) level in upper-mode fish. (Bottom) Length and weight changes: they lead to reduced condition factor (*K*) from April (smolt continues to grow in length). Water temperature at time of measurement (numbers on weight curve) remained almost constant from the beginning of March to mid-April (original data).

called attention to changes of body shape, such as body depth and larger eye diameter. Atlantic salmon smolts are usually very slender, with dark margins on their fins and always come from the upper mode of the population (Thorpe, 1977; Boeuf *et al.*, 1985). However, changes in general body shape and coloration are not sufficient by themselves to determine true smolt status.

Apart from these general growth and morphological aspects, the parr–smolt transformation phenomenon involves many metabolic and biochemical changes. Baraduc and Fontaine (1956) demonstrated for the first time the increase in oxygen consumption (MO_2) in smolts compared with parr. Higgins (1985), Wiggs *et al.* (1989) and Maxime *et al.* (1989) reached similar conclusions (Fig. 4.2, top). These last authors showed the existence of a marked endogenous circadian metabolic rhythm, disappearing at the end of smolting if the salmon were kept in FW, and perhaps indicating great physiological disturbances causing the smolt to migrate: about this point, important differences appear between the situation in France and results obtained in Canada, where Atlantic salmon smolts are able to stay in FW and to grow normally in this medium (D. Saunders, 1991, pers. comm.). In France, if we want to keep our salmon in FW after the normal migration time, we lose a high proportion and the survivors grow very poorly, compared with smolts in SW (p. 112). From Fig. 4.2 (top), we also can imagine an involvement of thyroid activity in the beginning of the profound increase in oxygen consumption in smolt, not only dependent on temperature at this time (Fig. 4.2, bottom). Changes in haemoglobin forms appear at the end of the smolting process (review: Hoar, 1988), probably under the influence of thyroid hormones (Sullivan *et al.*, 1985). Red blood cell ATP concentrations increase (Zaugg, 1982).

Condition factor decreases in smolts in association with the intense growth and with the utilization of metabolic reserves (Fig. 4.2, bottom). In 1969, Fessler and Wagner demonstrated the importance of the decrease in total lipid content in steelhead trout (*O. mykiss*), restricted to migrating fish. Concentrations of liver and muscle cholesterol decrease, and triacylglycerol is used extensively during smolting (Sheridan *et al.*, 1983). The smolt is characterized by a high proportion of long-chain polyunsaturated fatty acids which is typical of marine fishes (Sheridan *et al.*, 1985a; Ogata and Murai, 1989). These changes are not dependent on diet quality or water temperature. However, Ogata and Murai (1989) were able to produce better smolts using more linoleate and linolenate in the food. Lipolysis is intense and fatty acid synthesis in the liver (-88%) and mesenteric fat (-81%) decrease markedly (Sheridan *et al.*, 1985b). Increases in tissue lipolytic rates and decreases in rates of lipid synthesis are among the factors responsible for lipid depletion. Lipase-mediated lipid depletion has also been observed following SW exposure. Smolts, with major changes in hormone production and use, appear closer to SW fish than FW salmon (Sheridan, 1989).

A decrease in smolt liver glycogen was first reported by Fontaine and Hatey

(1950) and more recently by Sweeting and McKeown (1989). Rates of glycogenesis and glycogenolysis were estimated by Sheridan *et al.* (1985a) in selected tissues of coho salmon and their results showed a drop of 54% in glycogen synthesis in the liver and a 66% increase in glycogen phosphorylase activity, indicating intense use of available reserves.

Body water content increases during smoltification (Komourdjian *et al.*, 1976a; Farmer *et al.*, 1978; Birt and Green, 1986). In Atlantic salmon, a specific change, not observed in Pacific species, occurs corresponding to a shift of chloride from muscle to plasma (Fontaine, 1951). The hyperconcentration of sodium is still equivocal (Fontaine, 1975; Bradley and Rourke, 1988). These same authors reported that plasma proteins changed during smolting, the liver being involved in the synthesis of new molecules. An increase in the number of low-molecular-weight proteins (< 29 kD) is associated with the profound alterations occurring during smolting, and changes in the quantities and relative proportions of several proteins (57–62 kD) were observed (Johanning and Bradley, 1989). At the same time, plasma NH_4^+ levels decline while ammonia excretion increases (Bradley and Rourke, 1988; Wiggs *et al.*, 1989). General protein content of the body declines in smolts (Fessler and Wagner, 1969). In muscle, an increase in non-proteic nitrogenous compounds, mainly creatine, occurs (review: Fontaine, 1975). Muscle taurine and glycine decrease while levels of threonine and glutamine increase in the brain. There is high production of glutamic acid, rapidly transformed into glutamine, mainly during migration (Fontaine, 1975).

Osmotic and ionic regulation

Small salmon parr are unable to adapt to SW, and after a direct transfer from FW to full salinity mortality is almost 100%; surviving fish do not grow whereas the 'physiological smolt' can adapt perfectly to SW, even after direct transfer, only sometimes suffering a small and transitory osmotic disequilibrium and afterwards growing very fast in this new environment (Parry, 1960; Clarke and Nagahama, 1977; Boeuf *et al.*, 1978; Saunders and Henderson, 1978; Nishioka *et al.*, 1982; Prunet and Boeuf, 1985; Boeuf, 1987; Boeuf *et al.*, 1989a; Nance *et al.*, 1990). 'Physiologically intermediate salmon', like pre-smolt and silvery parr, will develop intermediate responses (Table 4.1).

In the wild, migratory behaviour is controlled by physiological status. Moreover, smolts remain a few days in the estuary, in brackish water, to adapt progressively to increased salinity (Zaugg *et al.*, 1985). In reared salmon, the situation is quite different, because the producers do not know the quality of their fish. Transfer, often directly to seawater, of pre-smolts and post-smolts poses many problems. The fish may survive, but not grow (the stunting phenomenon). For this reason, survival by itself does not provide a good indication of smolt status and acceptable growth and production in mariculture.

Table 4.1 Possible responses of young salmon to direct transfer into full-salinity sea water (> 34‰ parts per thousand) in relation to their physiological status*

Stage			
Parr, P (typical FW fish (hatching to 4–16 months before S)) lower mode	Pre-smolt, PS (4–16 months before S) upper mode	Smolt, S	Post-smolt, PoS (post-migratory status; partial reversion to PS)
Very high mortality in 2–7 days	Low to high mortality (5–40% in 2–10 days)	Very good survival (> 95%)	Low mortality (5–10%)
No feeding activity in surviving fish; dark and thin	Poor feeding activity at the beginning	Immediate feeding activity after seawater contact (2h to 2 days)	Difficult feeding behaviour at the beginning
No growth	Very low growth; stunting	Very fast growth, better than in FW fish	Growth low to normal
–	Fish very susceptible to diseases	–	Possibility of diseases

* These observations are valid at a seawater temperature of 8–14 °C. The higher the salinity and temperature, the more difficult is the adaptation. It is very difficult to specify the limits between the different stages (P, PS, S and PoS); S corresponds to fish at the end of the FW life, already migrating, just before reaching the sea.

The main effector organs which are directly involved in osmoregulation in fishes are the skin, intestine, kidney–urinary bladder system and the gills. Obviously these organs undergo many changes during smolting.

Skin

In all animals the skin represents the fundamental protection against external factors. In aquatic animals it plays the role, among others, of a shield preventing water and electrolyte fluxes from the body to the environment, or vice versa. The skin of salmonids in FW is thick, well coloured and has many mucous cells. It becomes silvery during smolting owing to deposition of purine substances (guanine and hypoxanthine crystals). The scales of smolts are easily dislodged, necessitating careful handling to prevent scale loss and resulting osmotic disequilibrium. Descaled fish do not survive in SW. Johnston and Eales (1970) emphasized the influence of temperature and photoperiod on guanine and hypoxanthine deposition, thyroid hormones being also strongly involved in silvering (Fontaine, 1975; Miwa and Inui, 1983, 1985).

Intestine

In salmonids, the intestine is the most important part of the digestive tract for osmoregulation. Freshwater fish do not drink water and the gut is virtually watertight. Fish in SW drink the ambient water continually and salt water is absorbed across the gut to reach the blood. This process allows the fish to compensate for water losses which mainly occur through the gills. The Atlantic salmon, following SW (30–32‰) exposure, starts to drink at a rate of 2.4 ml kg^{-1} h^{-1}, 4–6 h after transfer, finally reaching 8 ml kg^{-1} h^{-1} in feeding fish (Usher *et al.*, 1988). Intestinal water absorption increases greatly in rainbow trout after transfer to SW (Lahlou *et al.*, 1975), without affecting protein digestion and nutrient absorption. Nonnotte *et al.*, (1987) demonstrated the transient changes in the intestinal absorption of water, sodium and chloride, occurring after entering SW. Transport of Na^+ and Cl^- follows two phases, first an increase in permeability of the epithelium for 1 day, then a transient impermeabilization and increase in protein content of the mucosa and a progressive increase of both unidirectional Na^+ and Cl^- fluxes in association with increased ATPase activity. Polyunsaturated fatty acids play a major role in the transportation of water and electrolytes in the gut through their action on enzyme systems. The most important ultrastructural modifications of the intestinal mucosa occur during the first 2 days in SW, in correlation with important physiological changes following the abrupt increase of environmental salinity (Nonnotte *et al.*, 1986). The situation in salmon smolts can be deduced from these patterns observed in rainbow trout. Transfer to SW could modify the apparent nitrogen digestibility in transferred Atlantic salmon (Usher *et al.*, 1990). During smoltification, water transport capacity pre-establishes in coho salmon in FW; this occurs after the thyroxine surge and coincides with the maximum (Na^+–K^+)-ATPase activity of gill microsomes (Loretz *et al.*, 1982). Intestinal water transport increases greatly and FW salmon come to resemble SW fish (Collie and Bern, 1982).

Kidney and urinary bladder

As with the intestine, few studies have been devoted to the changes in the kidney and urinary bladder during smolting. The FW teleost produces a large volume of dilute urine, while in SW urine is sparse and relatively concentrated: in rainbow trout, for example, it amounts to from 0.344–1.0 ml h^{-1} 100 g^{-1} weight in FW to 0.031 ml h^{-1} 100 g^{-1} in SW (Holmes, 1961; Lahlou *et al.*, 1975). In FW, the kidney and urinary bladder are of considerable importance, their action consisting of eliminating excess water without losing ions; the glomerular filtration rate is very high and tubular reabsorption is low. In SW, the role of these organs is quite different. Ford (1958) showed a reduction in the number of functional glomeruli in SW salmon. During smoltification of steelhead trout, while still in FW, renal water excretion rate declines, accompanied by an increase in urine concentration (Holmes and Stainer, 1966). If the fish is kept in FW, these changes reverse. There is a major change in kidney anatomy and physiology before migration in smoltifying salmonids, these modifications occurring only under the action of external salinity after SW transfer in non-migrating species. A decrease in electrolyte absorption in the urinary bladder of terminal coho smolts occurs in FW (Loretz *et al.*, 1982), predisposing the fish to SW entry. Eddy and Talbot (1985) found an increase in urine production in silvery Atlantic salmon which they interpreted as a consequence of pre-adaptation to increase the water permeability of the gut of the fish still in FW.

Gills

Many studies have been devoted to gill physiology during smolting. Recent reviews present the role of the gill in osmoregulation, this organ becoming of the highest importance in SW: Payan *et al.*, 1984; McCormick *et al.*, 1989b; Avella and Bornancin, 1990; Fig. 4.3. Apart from its role in respiration, nitrogenous excretion, thermal exchange and mucus production, the gill plays a major role in ionic and osmotic regulation through the action of chloride cells (mitochondria-rich cells or ionocytes). These specific cells increase considerably in number and activity in SW compared with FW fish (Conte and Lin, 1967) and they are the sites of active electrolyte transport (Payan *et al.*, 1984).

The number, size and activity of chloride cells in the gills increase in smolts in FW (review: Hoar, 1988). Their proliferation begins well in advance of

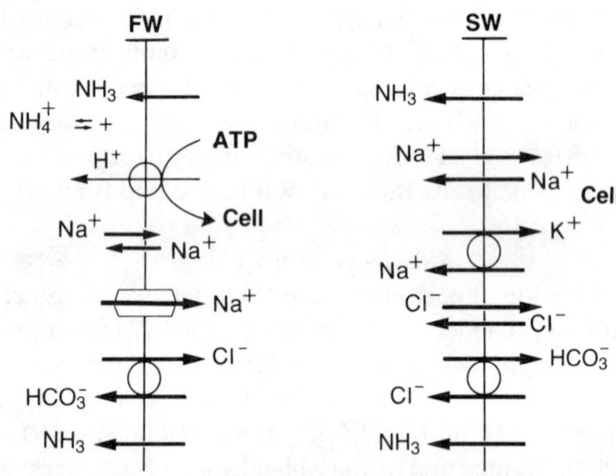

Fig. 4.3 Branchial fluxes and exchange rates of salmonids in fresh water (FW) and seawater (SW). In FW, there is no Na^+/H^+ exchange pump (as in erythrocytes, for example) but a coupling between active H^+ excretion and Na^+ entrance. Cl^-/HCO_3^- exchange is more relevant to acid–base balance than to osmoregulation. The existence of a chloride channel on the apical part of chloride cells has also been imagined. Adapted from Payan *et al.* (1984), Avella and Bornancin (1990) and Randall and Lin (1992). The smolt has to change from the FW to the SW situation.

migration and contact with SW, in correlation with enzymatic changes (Langdon and Thorpe, 1985). Apart from these general modifications, precise ultrastructural and functional changes also occur during smolting (Pisam *et al.*, 1988; Lubin *et al.*, 1989). In a recent paper, Pisam *et al.*, (1990) examined the ultrastructure of chloride cells in both FW and SW fish. The cells of smolts in FW are more similar to those of SW fish than to those of FW salmon (Pisam *et al.*, 1988) with regard to the appearance of accessory cells and the multiplication of shallow apical junctions. Smolts are thus fully and spontaneously pre-adapted to SW life, and contact with SW acts only as a final stimulus in order to achieve the terminal transformation into SW fish. The number of chloride cells on the opercular membrane increases after the thyroxine peak, concomitant with the peak of gill ATPase activity (Collie and Bern, 1982).

Zaugg and McLain (1972) pointed out the increase of a specific enzymatic system activity in gill microsomes on smoltifying salmonids, the (Na^+–K^+)-ATPase. This system is universally known in nervous and excretory cells. Fish in SW have high enzymatic activity in gills, but low activity in kidney and intestine, the opposite to FW fish (Zaugg and McLain, 1972; de Renzis and Bornancin, 1984). Avella and Bornancin (1990) recently reviewed the role of ATPase in gill Na^+ extrusion in SW, ATPase activity being localized in the basolateral plasma membranes and tubular system (Hootman and Philpott, 1979; Pisam *et al.*, 1988).

The level of gill ATPase activity, directly involved in ionic extrusion in SW, can be used as a valuable indicator of SW adaptability in migrating salmonids (reviews: Folmar and Dickhoff, 1980; Wedemeyer *et al.*, 1980; Boeuf, 1987; Hoar, 1988). As already mentioned, only smolts can rapidly adapt to full-salinity SW and grow in the marine environment, and gill ATPase level can represent a predictive criterion of the osmotic quality of the fish in normal rearing conditions (Fig. 4.1 (centre), 4.4; Adams *et al.*, 1973; Boeuf *et al.*, 1978, 1985, 1989a; Saunders and Henderson, 1978; Boeuf and Harache, 1982, 1984; Virtanen and Soivio, 1985; Boeuf, 1987; McCormick *et al.*, 1987; Björnsson *et al.*, 1989a; Zaugg, 1989; Zaugg and Mahnken, 1991). In Atlantic salmon there is a good correlation between high levels of gill (Na^+–K^+)-ATPase activity and the presence of the fish in the upper size mode of the population at the time of completion of smolting (Boeuf *et al.*, 1985): microsomal gill ATPase activity of FW fish increases progressively from December to April. However, the enzyme level is not always sufficient as a single criterion to evaluate the true physiological quality of the smolt (Ewing *et al.*, 1985; Boeuf, 1987; Duston and Saunders, 1990; Saunders and Harmon, 1990; Duston *et al.*, 1991), especially following environmental manipulation during rearing. Anyway, gill ATPase activity measured *in vitro* cannot be used as the sole physiological criterion of osmoregulatory potential; full smolting represents much more.

Results of a recent study enable us to specify the ionic and osmotic capacity of smolt and post-smolt after direct exposure to high-salinity SW, using a new isolated perfused head technique: smolts are able to excrete Na^+ immediately after SW contact, whereas post-smolts are more similar to rainbow trout, needing more time to develop this ability (Nance *et al.*, 1990). There is a good correlation between gill ATPase activity level and ionic regulatory capacity. In another study, Primmett *et al.* (1988) demonstrated a seasonal net loss of Na^+ in FW smolts.

Other enzymatic changes occur in the gill during smolting, e.g. in succinic dehydrogenase, cytochrome C oxidase, citrate synthetase and carbonic anhydrase activities (Langdon and Thorpe, 1985; Hoar, 1988; McCormick *et al.*, 1989a,b). On the other hand, although it increases after SW transfer, alkaline phosphatase activity does not change in FW smolts (Gasser and Kirschner, 1987). In general, the metabolic capacity of gill and kidney does not increase consistently during smolting (McCormick *et al.*, 1989b). However, the lipid composition of gills is affected by parr–smolt transformation, shown mainly by decreasing levels of triglycerides (Takeuchi *et al.*, 1990).

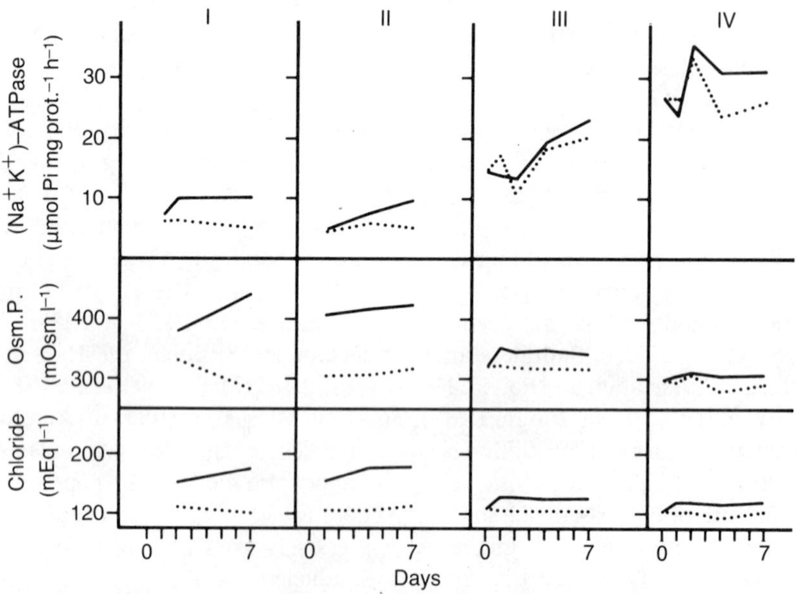

Fig. 4.4 Osmoregulatory capacity of young Atlantic salmon after transfer to full-salinity SW (35‰, continuous lines; dotted lines denote fish in FW). (I) 25 Oct. (9 months old); 116 mm length; low gill ATPase level, low T_4; 45% mortality after 7 days in SW; no growth in surviving fish. (II) 7 Jan. (12 months old); 122 mm; low ATPase and T_4; 25% mortality; no growth. (III) 25 Mar. (14 months old); 160 mm; very high T_4 level, medium ATPase; 6% mortality; medium growth. (IV) 22 Apr. (15 months old); 162 mm; highest level of ATPase, low T_4; 2% mortality; fast growth. Adapted from Boeuf (1987) and Boeuf *et al.* (1989a).

Endocrine changes

An extensive review of endocrine changes during smolting will not be presented here, but some specific points will be underlined. The reader is referred to recent reviews on this topic (Barron, 1986; Boeuf, 1987; Hoar, 1988). All the changes associated with smolting, already presented here, are influenced by environmental factors which act to synchronize the population through nervous and endocrine systems which are directly involved. The fish can be considered to be pre-programmed and will enter into parr–smolt transformation in association with a physiological maturity which enables it to react to external factors. Many physiological functions are involved or affected by smoltification, i.e. respiration, circulation, excretion, osmoregulation and growth. Consequently, many hormones are involved in the phenomenon. An outline of the known endocrine involvement is given in Fig. 4.5 and is summarized below.

Adenohypophysis

This endocrine complex is intimately involved in physiological functions and smoltification; hypophysectomized fish are not able to smoltify (Nishioka *et al.*,

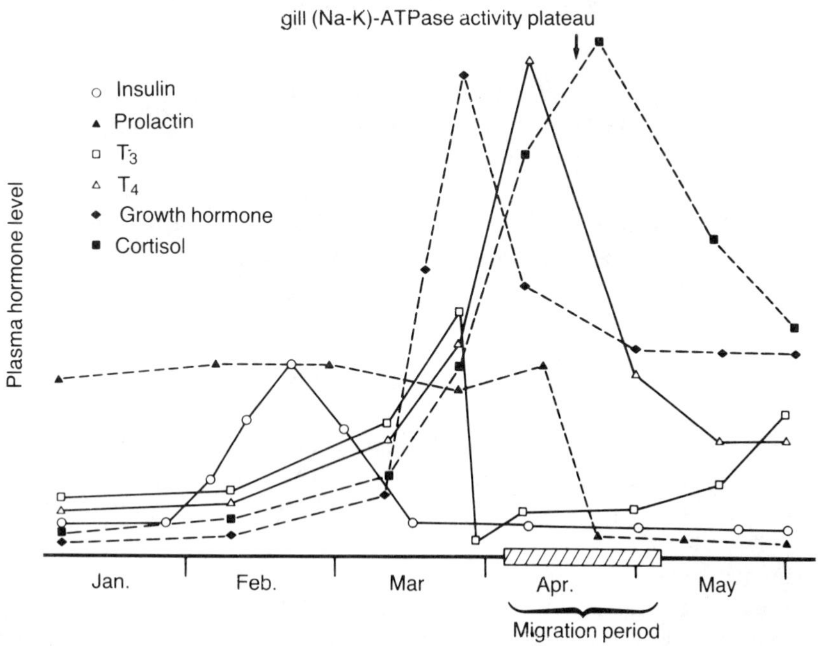

Fig. 4.5 Schematic presentation of hormone changes during smolting of salmonids. Adapted from Boeuf (1987) and Dickhoff *et al.* (1990). Data from Plisetskaya *et al.* (1988: coho insulin), Prunet and Boeuf (1989: Atlantic salmon prolactin), Boeuf *et al.* (1989a: Atlantic salmon T_3, T_4 and growth hormone) and Specker (1982: coho cortisol).

1987). Pituitary cytology is clearly stimulated during parr–smolt transformation (Olivereau, 1954; Komourdjian *et al.*, 1976a,b; Nishioka *et al.*, 1982). Unfortunately, there is no study on the role of the neurohypophysis during smoltification.

Thyroid stimulating hormone (TSH)

TSH stimulates the thyroid gland to produce and release thyroxine (T_4). A degranulation of delta cells suggests an increase of TSH production during smolting (Fontaine, 1975), and the pre-smolt becomes increasingly receptive to mammalian TSH treatment until the T_4 peak (Specker and Schreck, 1984) or later (Swanson and Dickhoff, 1987). Stunted coho salmon in SW have less active or fewer delta cells (Nishioka *et al.*, 1982).

Adrenocorticotropic hormone (ACTH)

ACTH clearly stimulates interrenal tissue to produce cortisol in salmonids (Specker, 1982). So far there is no purified salmon ACTH, and therefore it is difficult to specify how ACTH may be acting, beyond the knowledge that the steroid-producing cells in the interrenal are stimulated during smoltification (Fontaine, 1975; Nishioka *et al.*, 1982). Langdon *et al.*, (1984) reported a stimulation of gill (Na^+–K^+)-ATPase activity provoked by mammalian ACTH treatment.

Prolactin (PRL)

PRL is well known in fish for its specific role in osmoregulation in FW (Loretz and Bern, 1982; Hirano *et al.*, 1987). Some very sensitive hypophysectomized fish species can survive in SW but not in FW. Although levels are variable in salmonids (Nishioka *et al.*, 1987), circulating PRL in the blood clearly decreases during parr–smolt transformation (Prunet and Boeuf, 1985, 1989; Prunet *et al.*, 1989; Young *et al.*, 1989a). Unlike Atlantic salmon, coho salmon develop a PRL peak before the final drop (Young *et al.*, 1989a). However, in both species the highest gill ATPase values occur when the PRL level is low. Values in parr are much lower than in smolts (Prunet and Boeuf, 1989). Cortisol appears to be an important regulator of PRL cells (Kelley *et al.*, 1990).

Growth hormone (GH)

Komourdjian *et al.* (1976a,b) and Clarke *et al.* (1977) demonstrated the action of mammalian or fish GH, not only on growth, but also on osmoregulation and SW adaptability in salmonids. A period of rapid growth is usually correlated with an increase of euryhalinity during the final stages of natural smoltification (Duston and Saunders, 1990). Recent purification of salmon GH has made it possible to demonstrate a notable increase in levels of GH in the blood at the end of smolting (Sweeting *et al.*, 1985; Boeuf *et al.*, 1989a; Prunet *et al.*, 1989; Young *et al.*, 1989a). GH levels surge or reach a high plateau, depending on the year (Boeuf *et al.*, 1990b), but smolts always have a high GH level, which remains elevated after the peak. GH increases steadily but transitorily after Atlantic salmon smolts enter SW (Boeuf *et al.*, 1989a). GH often peaks together with T_3 (triiodothyronine) and has been shown to be involved in the control of T_4–T_3 transformation in salmonids (Nishioka *et al.*, 1982; McLatchy and Eales, 1990). Treatment with mammalian GH makes it possible to transfer juvenile brown trout, *Salmo trutta* (Madsen, 1990a), rainbow trout (Collie *et al.*, 1989) or pre-smolt Atlantic salmon (Komourdjian *et al.*, 1976b; Boeuf *et al.*, 1990a) or coho salmon (Richman and Zaugg, 1987) into SW with good results in terms of SW adaptability, survival and subsequent growth in SW. GH treatment clearly increases the gill (Na^+–K^+)-ATPase activity (Richman and Zaugg, 1987; Boeuf *et al.*, 1990a; Madsen, 1990a). Recombinant salmonid GH was recently used to improve SW adaptability in Atlantic salmon with good results (Boeuf *et al.*, in prep.).

Many explanations have been proposed for this apparent direct role of GH in osmoregulation and smoltification, including its similarity with PRL, the growth effect, or the T_4–T_3 or cortisol-activation effect. A more direct effect is now considered to be responsible. Sheridan (1989) noted that the sensitivity of smolts to hormonal treatment was low, and in common with many hormones, parr or pre-smolt respond much more clearly to an exogenous supply (Boeuf *et al.*, in prep.).

In the stunting phenomenon the fish were clearly demonstrated as hypothyroid (Nishioka *et al.*, 1982; Young *et al.*, 1989b) and apparently paradoxically, with very high GH levels (Bolton *et al.*, 1987; Björnsson *et al.*, 1988; Young *et al.*, 1989b; Gray *et al.*, 1990). It is noteworthy that liver sGH (salmon growth hormone) binding is very low in stunts (Gray *et al.*, 1990), probably indicating a down-regulated receptor system.

Thyroid

The thyroid has probably been the most studied endocrine system during salmonid smolting since the first reports of Hoar (1939), Leloup and Fontaine (1960) and Fontaine (1975). This gland is involved in the phenomenon and produces more thyroxine at the end of smoltification in wild Atlantic salmon (Leloup and Fontaine, 1960). The thyroid in fishes acts as in mammals, with high T_4 production and peripherical transformation into T_3. The same receptor binds the two molecules but with a much higher affinity for T_3 (Eales, 1985). Because it is possible to measure precisely the blood circulating levels of both T_3 and T_4 in salmonids, many authors have demonstrated a dramatic surge of T_4 at the end of smolting in FW, a few weeks before the highest level of gill ATPase activity is reached in both Pacific and Atlantic salmon (reviews: Boeuf, 1987; Hoar, 1988; Figs 4.1 (bottom), 4.2 (top), 4.5). In Atlantic salmon sometimes T_3 also peaks (Boeuf and Prunet, 1985; Virtanen and Soivio, 1985; Boeuf *et al.*, 1989a; Prunet *et al.*, 1989) before and after the T_4 surge. We know that these increases in circulating levels of both T_3 and T_4 are not due to changes in properties of plasma binding proteins (Boeuf *et al.*, 1989b) and have physiological significance.

In Pacific salmon, interesting correlations were found between thyroxine (T4) level, growth, SW adaptability and lunar phases (Grau *et al.*, 1982; Farbridge and Leatherland, 1987). Several T4 peaks appear, the surge associated with smoltification timed precisely to coincide with the new moon phase of the lunar cycle (Grau *et al.*, 1982). Blood thyroxine levels can also be modified by changes in FW quality or water flow (Fontaine and Leloup, 1959; Nishioka *et al.*, 1985; Youngson *et al.*, 1986; Boeuf and Gaignon, 1989; Fujioka *et al.*, 1990; Hoffnagle and Fivizzani, 1990).

Attempts were made to correlate thyroid activity and triggering of the gill ATPase stimulation for some years. The enzyme activity level is much more useful for predicting the SW adaptability than is the absolute level of thyroxine (Boeuf, 1987; McCormick *et al.*, 1987; Boeuf *et al.*, 1989a). However, thyroid stimulation is indispensable before the end of smolting to allow the achievement of full parr–smolt transformation and to develop capacities of adaptation and growth in SW (Folmar and Dickhoff, 1981; Boeuf, 1987). The thyroxine surge occurs much earlier in Atlantic salmon in long rivers than in those originating in short streams (Boeuf and Le Bail, 1990). This fact supports the concept of the role of TH in triggering migratory behaviour and memorizing the environment (i.e. imprinting, p. 122).

It is still difficult to specify the exact role of thyroid hormones in osmoregulation and SW adaptation in salmonids. However, such action is likely and feasible (Fontaine, 1975; Dickhoff *et al.*, 1982; Refstie, 1982), even though many studies did not demonstrate improved SW adaptability following treatment with thyroid hormones (Miwa and Inui, 1983, 1985; Ikuta *et al.*, 1985; Saunders *et al.*, 1985; Omeljaniuk and Eales, 1986; Iwata *et al.* 1987; Sullivan *et al.*, 1987; Madsen, 1990b; Boeuf *et al.*, unpublished data).

Present knowledge suggests the involvement of TH in the following functions during smolting:

1. control of growth (Fontaine, 1975; Nishioka *et al.*, 1982, 1985; Refstie, 1982; Saunders *et al.*, 1985; Boeuf and Gaignon, 1989);
2. stimulation of body silvering (Fontaine, 1975; Miwa and Inui, 1983, 1985; Sullivan *et al.*, 1987);
3. stimulation of metabolism (Leloup *et al.*, 1971; Narayansingh and Eales, 1975; Bhattacharya *et al.*, 1985; Madsen and Korsgaard, 1989);
4. appearance of new haemoglobin forms (reviews: Sullivan *et al.*, 1985; Hoar, 1988);
5. stimulation of migratory behaviour (Baggerman, 1963; Godin *et al.*, 1974; Fontaine, 1975; Youngson *et al.*, 1985; Iwata *et al.*, 1989; Boeuf and Le Bail, 1990);
6. stimulation of imprinting and environment-memorizing (Leloup and Fontaine, 1960; Fontaine, 1975; Hoar, 1976; Hasler *et al.*, 1978; Scholz *et al.*, 1985; Morin *et al.*, 1989a, b).

Interrenal tissue

Profound cytological changes occur in the interrenal gland during smolting (Fontaine, 1975; Nishioka *et al.*, 1982). In salmonids this endocrine gland produces cortisol, rather than cortisone, which is probably a degradation peripheral compound (Specker, 1982). Cortisol has been known for a long time as 'SW hormone' able to promote SW adaptability (Specker, 1982; Barron, 1986). This hormone increases greatly during smoltification, reaching a maximum coincident with the highest plateau of the gill (Na^+–K^+)-ATPase activity in both the Pacific and Atlantic salmon (Specker, 1982; Barton *et al.*, 1985; Virtanen and Soivio, 1985). Cortisol treatment stimulates gill (Na^+–K^+)-ATPase activity, chloride cell number (Richman and Zaugg, 1987; Madsen 1990a,b,c) and increases plasma T_3 (Redding *et al.*, 1991). Blood sampling for cortisol is particularly critical, since this hormone is very sensitive to stress. Simpson (1985) even suggested that the steady rise in plasma cortisol might be regarded as a non-specific response to the rising levels of stress caused by the increasing maladaption of smolting fish to FW. Cortisol is very active in processes of gluconeogenesis and plays a role in synergy with ACTH, TSH, T_3 and GH in both the SW adaptation and the energy mobilization needed by the fish at the end of smolting (Fontaine, 1975).

Other endocrine systems

Chromaffin tissue secretes catecholamines: adrenaline and noradrenaline levels increase sharply at the end of smolting (Fontaine, 1975).

Smolts have significantly more cytological activity than parr in the urophysis (caudal neurosecretory system) from which urotensin II is secreted (Nishioka *et al.*, 1982).

Two types of cells constitute the corpuscles of Stannius, and an increase in cell size was observed during smolting (Aida *et al.*, 1980).

Sex hormones, mainly androgens, generally impair SW adaptability and smolting in salmonids (Miwa and Inui, 1986; Lundqvist *et al.*, 1989, 1990; Madsen and Korsgaard, 1989). Sower *et al.* (1984) reported an increase in plasma oestradiol in smolting coho, but Miwa and Inui (1986) were not able to repeat this observation.

A sharp transitory peak in plasma calcitonin levels occurs early in smoltification in coho salmon (Björnsson *et al.*, 1989b).

Levels of plasma insulin increase in pre-smolts at the earliest stages of smolting, probably indicating a role in modification of metabolic activity at that time (Plisetskaya *et al.*, 1988). Highest specific binding to the liver plasma membranes is coincident with the low level of insulin prevailing in smolt (Gutiérrez and Plisetskaya, 1990). At that time, glucagon-like peptide (GLP) elevates plasma insulin levels slightly in coho salmon (Plisetskaya *et al.*, 1989).

Behavioural modifications

Parr have marked territorial behaviour in FW and gradually lose it at smolting, all the fish emigrating together. Spontaneous physical activity increases acutely, probably under the influence of catecholamines (Fontaine, 1975), appetite is stimulated and oxygen consumption rises. Smolts are clearly attracted by increased salinity and by light (Baggerman, 1963; Iwata *et al.*, 1989). Migration is characterized by a reduction of fast swimming activity (Glova and McInerney, 1977) and a diurnal periodicity (Thorpe and Morgan, 1978). Downstream migration is mainly passive (Thorpe *et al.*, 1981), often, but not systematically, occurring during the night. Using forced-swimming tests against constant high flow rates (50 cm s^{-1}), Virtanen and Forsman (1987) showed that Atlantic salmon migrating downstream had markedly lower physiological capacity than parr. A transitory period in the estuary seems to be important in achieving the final adaptation to full salinity (Zaugg *et al.*, 1985; Birt *et al.*, 1990).

In the last weeks before the migration, the smolt develops a very high sensitivity to environmental factors, olfactory sensitivity increasing concomitant with an olfactory learning ability to facilitate olfactory imprinting (Morin *et al.*, 1989b). Thyroid hormones may facilitate olfactory learning by increasing the excitability of the nervous system (Scholtz *et al.*, 1985; Morin *et al.*, 1989a). The process of imprinting occurs only once in the life of the Atlantic salmon; returning adults come back to the river of their first migration (Foster and Schom, 1989).

4.3 ROLE OF ENVIRONMENTAL FACTORS

There is a considerable literature dedicated to the role of environmental factors and the last Workshop on Salmonid Smolting, in Trondheim (Hansen *et al.*, 1989), dealt mainly with this subject (*Aquaculture*, special issue, 1989). Many attempts are being made to produce 0-age smolts in aquaculture and this is done in two different ways, by hormonal treatment or by manipulation of environmental factors. The first approach may be easier but poses the delicate problem of hormone use in fish for human consumption. This is the reason for the great interest in recent studies using environmental manipulation. There is, of course, great interest in the role of ecological variables such as temperature, photoperiod and salinity on the smolting process. In this section I do not intend to review all the papers published about this topic, but refer the reader to reviews (Wedemeyer *et al.*, 1980; Hoar, 1988); I shall make some comments on the role of temperature and photoperiod.

Smoltification is controlled by an endogenous rhythm, itself controlled by the nervous and endocrine systems, synchronized by external factors such as temperature and photoperiod (Hoar, 1976, 1988; Folmar and Dickhoff, 1980;

Wedemeyer *et al.*, 1980; Duston and Saunders, 1990). If downstream-migrating smolts fail to reach the sea, or are prevented from migrating by retention in FW, they lose at least some of their smolt characteristics (Folmar *et al.*, 1982; Nance *et al.*, 1990; Duston *et al.*, 1991).

Generally, increasing temperature at the end of winter hastens the completion of parr–smolt transformation, which will reach its maximum some weeks after the vernal equinox. This timing can change with latitude and strain. Later, rising temperature in late spring will inhibit smolt migration and any smolt remaining in FW (post smolts) will revert to FW status. The true smolt stage, which allows the full adaption to SW followed by good growth, lasts for a very short time and has led to the statement that there was a window(s) of potential adaptation in the year (Boeuf and Harache, 1982; Hoar, 1988). Whereas some Pacific species are able to stay in FW after the normal migration time and to grow (as in introduced coho in the Great Lakes and in the case of FW rearing in Chile, France, etc.), Atlantic species cannot always tolerate such treatment (p. 110).

The ideal thermal range for migrating salmonids is 12–17 °C, depending on species and strain. Ideal thermal conditions for growth are in the upper part of this range (Saunders and Henderson, 1978; Clarke *et al.*, 1981; Johnston and Saunders, 1981). However, it is very difficult to generalize for all species of salmon, and to be called a smolt, the fish must be able not only to live in SW but also to thrive and grow there. For a long time (Adams *et al.*, 1973; Zaugg and McLain, 1976), we have known the inhibiting power of high temperatures on salinity tolerance: > 13 °C in stenothermal fish like steelhead trout, *O. mykiss*, > 16 °C in coho salmon, with gill (Na^+–K^+)-ATPase activity being clearly altered (see also Duston *et al.*, 1991). However, after environmental factor manipulations, we were able to transfer with good results (long-term survival and good growth) Atlantic salmon in Brittany in August at 18 °C (unpublished data). Very low temperatures (< 6°C) also limit SW adaptation (Sigholt and Finstad, 1990). In cold regions (North Canada, Finland), migration occurs very late at the end of June or July, whereas in southern countries (France, Spain) it starts in March–April.

Photoperiod appears to be essential in the control of smoltification (Saunders and Henderson, 1970; Hoar, 1976, 1988) and many studies are presently devoted to effects of various photoperiod regimes. Photoperiod stimulates the light–pituitary axis (Komourdjian *et al.*, 1976a) and triggers many hormonal effects (Björnsson *et al.*, 1989a). Extreme photoperiod disturbances (rearing in total darkness or continuous light) may prevent or delay smoltification and SW adaptation. Extended daylength (advanced summer solstice, continuous light during some months) in rearing allows stimulation of growth and shortening of the FW cycle of reared salmon, but can also lead to production of large fish which are able to survive in SW but not to grow (Saunders *et al.*, 1985). Smolting can therefore occur at different periods of the year, confirming the hypothesis of a basic endogenous rhythm synchronized by external factors. In stable conditions (12 h light, 11 °C), Atlantic salmon expresses a smolting periodicity of 10 months (Eriksson and Lundqvist, 1982). Extended photoperiod can increase the number of fish in the upper mode in *Salmo salar* (Saunders *et al.*, 1989; Stewart *et al.*, 1990), but can also alter thyroid status (Stewart *et al.*, 1990). In the same way Clarke and Shelbourn (1986) and Clarke *et al.* (1989) demonstrated the feasibility of producing 0-age coho smolts by using a delayed photoperiod in FW. Continuous light applied early in ontogeny and maintained throughout the rearing period inhibits osmoregulatory capacities, whereas increasing daylength during winter–spring stimulates smoltification (Saunders *et al.*, 1985; McCormick *et al.*, 1987; Komourdjian *et al.*, 1989). Several experiments carried out on light-induced changes in salmonids have led to uncoupling between gill (Na^+–K^+)-ATPase activity measured *in vitro* and SW adaptability (Saunders and Harmon, 1990; Duston and Saunders, 1990; Duston *et al.*, 1991). This may be explained by smoltification induction which is not total, or by disturbances in cycle (?). Salinity tolerance would be able to develop without a high level of ATPase activity. In the same way, it is the first time (apart from treatment with anabolic hormones) when there is no correlation between growth and acquisition of euryhalinity, these two functions always being linked during natural smoltification.

4.4 CONCLUSION

Parr–smolt transformation truly corresponds to a pre-adaptation to the marine environment, fish in FW having already developed most of the mechanisms they will need to adapt and grow in hyperosmotic surroundings. The different cytological, anatomical, morphological, biochemical, endocrinological and physiological changes studied in FW demonstrate that at the end of the FW stage, the smolt is closer to a marine fish than to a FW salmon. Is the smolt a marine fish surviving in FW (but still hyperosmoregulatory)? One of the different theories advanced to explain the reason for the migration is maladaptation to FW at the end of the smolting phenomenon (Simpson, 1985): the Atlantic salmon is not always able to survive and to grow in good condition if it is kept in FW after the normal migration time. The fish appears so well prepared for life in SW that the prevention of migration can trigger great physiological disturbances in FW: in respiration (Maxime *et al.*, 1989), and in osmoregulation (Nance *et al.*, 1990), such as decreases in osmotic pressure, Na^+ and Cl^- concentrations in blood (Boeuf, unpublished data), high disease susceptiblity and often death. This is not consistent and some populations can perfectly tolerate remaining in FW. In Canada (D. Saunders, 1991, pers. comm.), anadromous populations can survive and grow well if they are kept in FW after the normal migration time. There are many stocks of land-locked Atlantic salmon. However, in France, we tried several times to keep smolts in

FW after smolting and we observed great disturbances and mortality (up to 50%) in stocks from both Brittany and the Loire–Allier river. Coho salmon do not seem to be greatly affected by staying in FW.

In the wild we have seen that migratory behaviour followed the physiological changes but in aquaculture the quality of the smolt appears to be decisive (Mahnken *et al.*, 1982; Nishioka *et al.*, 1982; Zaugg, 1989; Zaugg and Mahnken, 1991). To resolve this very complex problem, many studies have been undertaken in the world, mainly in Canada, the USA, Scandinavia, Scotland, Japan and France. The use of a specific parameter, as useful as it may be, does not allow us to specify the true physiological smolt status. In fact, we must have a deep understanding of both physiological and behavioural aspects of smolting. Dickhoff *et al.* (1990) recently compared smoltification to amphibian metamorphosis and found many similarities. Smolting is a typical phenomenon in migrating salmonids and represents a very critical phase in their life cycle. The entire process is of great interest in both basic and applied fields and many discoveries are to be expected in the future.

ACKNOWLEDGEMENTS

I wish to greatly thank R.L. Saunders, S.S. Madsen and J.C. Rankin for providing important advice and assistance in reviewing the original manuscript; they brought me a lot of indispensable comments, corrected errors and put in correct English the initial text. This work was not feasible without their help.

REFERENCES

Adams, B.L., Zaugg, W.S. and McLain, L.R. (1973) Temperature effect on parr–smolt transformation in steelhead trout as measured by gill sodium-potassium stimulated adenosine triphosphatase. *Comp. Biochem. Physiol.*, **44A**, 1333–9.

Aida, K., Nishioka, R.S. and Bern, H.A. (1980) Changes in the corpuscles of Stannius of coho salmon *Oncorhynchus kisutch* during smoltification and seawater adaptation. *Gen. comp. Endocrinol.*, **41**, 296–304.

Avella, M. and Bornancin, M. (1990) Ion fluxes in the gills of freshwater and seawater salmonid fish, in *Comparative Physiology*, Vol. 6 (eds R.K.H. Kinne, E. Kinne-Saffran and K.W. Beyenbach), Basel, Karger, pp. 1–13.

Baggerman, B. (1963) The effect of TSH and anti-thyroid substances on salinity preference and thyroid activity in juvenile Pacific salmon. *Can. J. Zool.*, **41**, 307–19.

Baraduc, M.M. and Fontaine, M. (1956) Etude comparée du métabolisme respiratoire du jeune saumon sédentaire (parr) et migrateur (smolt). *C. R. Séanc. Soc. Biol. Paris*, **149**, 1327–9.

Barron, M.G. (1986) Endocrine control of smoltification in anadromous salmonids. *J. Endocr.*, **108**, 313–19.

Barton, B.A., Schreck, C.B., Ewing, R.D., Hemmingsen, A.R. and Patino, R. (1985) Changes in plasma cortisol during stress and smoltification in coho salmon, *Oncorhynchus kisutch*. *Gen. comp. Endocrinol.*, **59**, 468–71.

Bern, H.A. and Mahnken, C.V.W. (1982) Salmonid smoltification, *Aquaculture*, **28**, 1–2.

Bhattacharya, S., Plisetskaya, E., Dickhoff, W.W. and Gorbman, A. (1985) The effects of estradiol and triiodothyronine on protein synthesis by hepatocytes of juvenile coho salmon *Oncorhynchus kisutch. Gen. comp. Endocrinol.*, **57**, 103–9.

Birt, T.P. and Green, J.M. (1986) Parr smolt transformation in female and sexually mature male anadromous and nonadromous Atlantic salmon, *Salmo salar. Can. J. Fish. aquat. Sci.*, **43**, 680–86.

Birt, T.P., Green, J. and Davidson, W.S. (1990) Smolting status of downstream migrating Atlantic salmon (*Salmo salar*) parr. *Can. J. Fish. aquat. Sci.*, **47**, 1136–9.

Björnsson, B.T., Ogasawara, T., Hirano, T., Bolton, J.P. and Bern, H.A. (1988) Elevated growth hormone levels in stunted Atlantic salmon, *Salmo salar. Aquaculture*, **73**, 275–81.

Björnsson, B.T., Thorarensen, H., Hirano, T., Ogasawara, T., and Kristinsson, J.B. (1989a) Photoperiod and temperature affect plasma growth hormone levels, growth, condition factor, and hypoosmoregulatory ability of juvenile Atlantic salmon (*Salmo salar*) during parr smolt transformation. *Aquaculture*, **82**, 77–91.

Björnsson, B.T., Young, G., Lin, R.J., Deftos, L. and Bern, H.A. (1989b) Smoltification and seawater adaptation in coho salmon (*Oncorhynchus kisutch*): plasma calcium regulation, osmoregulation, and calcitonin. *Gen. comp. Endocrinol.*, **74**, 346–54.

Boeuf, G. (1987) Contribution à l'étude de l'adaptation à l'eau de mer chez les Poissons Salmonidés. Détermination de critères de smoltification par mesures de l'activité (Na^+–K^+)-ATPasique des microsomes de la branchie et des hormones thyroïdiennes plasmatiques. Thèse de Doctorat d'Etat, Université de Bretagne Occidentale. (369 pp.)

Boeuf, G. and Gaignon, J.L. (1989) Effects of rearing conditions on growth and thyroid hormones during smolting of Atlantic salmon, *Salmo salar* L. *Aquaculture*, **82**, 29–38.

Boeuf, G. and Harache, Y. (1982) Criteria for adaptation of salmonids to high salinity seawater in France. *Aquaculture*, **28**(1–2), 163–76.

Boeuf, G. and Harache, Y. (1984) Adaptation osmotique à l'eau de mer de différentes espèces (*Salmo trutta, Salmo gairdneri, Salvelinus fontinalis*) et hybride (*Salmo trutta femelle* × *Salvelinus fontinalis* mâle) de salmonidés. *Aquaculture*, **40**(4), 343–58.

Boeuf, G. and Le Bail, P.Y. (1990) Growth hormone and thyroid hormones levels during smolting in different populations of Atlantic salmon, in *Progress in Clinical Research* (eds. A. Epple, C.G. Scanes and M.H. Stetson), Wiley-Liss Inc., New York, **342**, 193–7.

Boeuf, G. and Prunet, P. (1985) Measurements of gill (Na^+–K^+)-ATPase activity and plasma thyroid hormones during smoltification in Atlantic salmon, *Salmo salar* L. *Aquaculture*, **45**(1–4), 111–19.

Boeuf, G., Lasserre, P. and Harache, Y. (1978) Osmotic adaptation of *Oncorhynchus kisutch* Walbaum. II. Plasma osmotic and ionic variations and gill Na^+–K^+-adenosine triphosphatase activity of yearling coho salmon transferred to seawater. *Aquaculture*, **15**(1), 35–52.

Boeuf, G., Le Bail, P.Y. and Prunet, P. (1989a) Growth hormone and thyroid hormones during Atlantic salmon, *Salmo salar* L, smolting and after transfer to seawater. *Aquaculture*, **82**(1–4), 257–68.

Boeuf, G., Le Roux, A., Gaignon, J.L. and Harache, Y. (1985) Gill (Na^+–K^+)-ATPase activity and smolting in Atlantic salmon *Salmo salar* L. in France. *Aquaculture*, **45**(1–4), 73–81.

Boeuf, G., Le Roux, A., Sévère, A., Prunet, P. and Le Bail, P.Y. (1990b) The role of growth hormone in the adaptability of Atlantic salmon (*Salmo salar*) to seawater. *Can. tech. Rep. Fish. aquat. Sci.*, **1761**, 125–31.

Boeuf, G., Prunet, P. and Le Bail, P.Y. (1990a) Un traitement à l'hormone de croissance peut-il stimuler la smoltification du saumon atlantique? *C. r. hebd. Séanc. Acad. Sci. Paris,* **310**(III), 75–80.

Boeuf, G., Uin, L.M. and Eales, J.G. (1989b) Plasma levels of free and bound thyroid hormones during parr–smolt transformation in Atlantic salmon, *Salmo salar* L. *Can. J. Zool.,* **67**, 1654–8.

Bolton, J.P., Young, G., Nishioka, R.S., Hirano, T. and Bern, H.A. (1987) Plasma growth hormone levels in normal and stunted yearling coho salmon, *Oncorhynchus kisutch. J. exp. Zool.,* **242**, 379–82.

Bradley, T.M. and Rourke, A.W. (1988) Smoltification in steelhead trout (*Salmo gairdneri*): developmental aspects of plasma constituents. *Fish Physiol. Biochem.,* **5**(1), 39–47.

Clarke, W.C. and Nagahama, Y. (1977) Effect of premature transfer to seawater on growth and morphology of the pituitary, thyroid, pancreas and interrenal in juvenile coho salmon *Oncorhynchus kisutch. Can. J. Zool.,* **55**, 1620–30.

Clarke, W.C., and Shelbourn, J.E. (1986) Delayed photoperiod produces more uniform growth and great seawater adaptability in underyearling coho salmon *Oncorhynchus kisutch. Aquaculture,* **56**, 287–99.

Clarke, W.C., Shelbourn, J.E. and Brett, J.R. (1981) Effect of artificial photoperiod cycles, temperature and salinity on growth and smolting in underyearling coho (*Oncorhynchus kisutch*), chinook (*Oncorhynchus tsawytsha*) and sockeye (*Oncorhynchus nerka*) salmon. *Aquaculture,* **22**, 105–16.

Clarke, W.C., Farmer, S.W. and Hartwell, K.M. (1977) Effect of teleost pituitary growth hormone on growth of *Tilapia mossambica* and on growth and seawater adaptation of sockeye salmon *O. nerka. Gen. comp. Endocrinol.,* **33**, 174–8.

Clarke, W.C., Shelbourn, J.E., Ogasawara, T. and Hirano, T. (1989) Effects of initial daylength on growth, seawater adaptability and plasma growth hormone levels in underyearling coho, chinook and chum salmon. *Aquaculture,* **82**(1–4), 51–62.

Collie, N.L. and Bern, H.A. (1982) Changes in intestinal fluid transport associated with smoltification and seawater adaptation in coho salmon *Oncorhynchus kisutch* Walbaum. *J. Fish Biol.,* **21**, 337–48.

Collie, N.L., Bolton, J.P., Kawauchi, H. and Hirano, T. (1989) Survival of salmonids in seawater and the time-frame of growth hormone. *Fish Physiol. Biochem.,* **7**, 315–21.

Conte, F.P. and Lin, D.H. (1967) Kinetics of cellular morphogenesis in gill epithelium during salt water adaptation of *Oncorhynchus* Walbaum. *Comp. Biochem. Physiol.,* **23**, 945–57.

Dickhoff, W.W., Brown, C.L., Sullivan, C.V. and Bern, H.A. (1990) Fish and amphibian models for developmental endocrinology. *J. exp. Zool.* (Supp.) **4**, 90–97.

Dickhoff, W.W., Folmar, L.C. Mighell, J.L. and Mahnken, C.W. (1982) Plasma thyroid hormones during smoltification of yearling and underyearling coho salmon and yearling chinook salmon and steelhead trout. *Aquaculture,* **28**(1–2), 39–48.

Duston, J. and Saunders, R.L. (1990) Control of the timing of smoltification in Atlantic salmon: endogenous rhythms and environmental factors. *Can. Tech. Rep. Fish. aquat. Sci.,* **1761**, 99–106.

Duston, J., Saunders, R.L. and Knox, D.E. (1991) Effects of increases in freshwater temperatures on loss of smolt characteristics in Atlantic salmon (*Salmo salar*). *Can. J. Fish. aquat. Sci.,* **48**, 164–9.

Eales, J.G. (1985) The peripheral metabolism of thyroid hormones and regulation of thyroidal status in poikilotherms. *Can. J. Zool.,* **60**, 304–9.

Eddy, F.B. and Talbot, C. (1985) Urine production in smolting Atlantic salmon *Salmo salar* L. *Aquaculture,* **45**(1–4), 67–72.

Eriksson, L.O. and Lundqvist, H. (1982) Circannual rhythms and photoperiod regulation of growth and smolting in Baltic salmon *Salmo salar* L. *Aquaculture*, **28**(1–2), 113–20.

Ewing, R.D., Hemmingsen, A.R., Evenson, A.D. and Lindsay, R.L. (1985) Gill (Na^+–K^+)-ATPase activity and plasma thyroxine concentrations do not predict time of release of hatchery coho *Oncorhynchus kisutch* and chinook salmon *Oncorhynchus tshawytscha* for maximum adult returns. *Aquaculture*, **45**(1–4), 359–73.

Farbridge, K.J. and Leatherland, J.F. (1987) Lunar cycles of coho salmon, *Oncorhynchus kisutch*. II. Scale amino acid uptake, nucleic acids, metabolic reserves and plasma thyroid hormones. *J. exp. Biol.*, **129**, 179–89.

Farmer, G.J., Ritter, J.A. and Ashfield, D. (1978) Seawater adaptation and parr smolt transformation of juvenile Atlantic salmon *Salmo salar*. *J. Fish. Res. Bd Can.*, **35**, 93–100.

Fessler, J.L. and Wagner, H.H. (1969) Some morphological and biochemical changes in steelhead trout *Salmo gairdneri* during parr smolt transformation. *J. Fish. Res. Bd Can.*, **26**, 2823–41.

Folmar, L.C. and Dickhoff, W.W. (1980) The parr–smolt transformation (smoltification) and seawater adaptation in salmonids. A review of selected literature. *Aquaculture*, **21**(1), 1–37.

Folmar, L.C. and Dickhoff, W.W. (1981) Evaluation of some physiological parameters as predictive indices of smoltification. *Aquaculture*, **23**, 309–24.

Folmar, L.C., Dickhoff, W.W., Mahken, C.V.W. and Waknitz, F.W. (1982) Stunting and parr-reversion during smoltification of coho salmon (*Oncorhynchus kisutch*). *Aquaculture*, **28**, 91–104.

Fontaine, M. (1951) Sur la diminution de la teneur en chlore du muscle des jeunes saumons lors de la migration d'avalaison. *C.r. hebd. Séanc. Acad. Sci. Paris*, **232D**, 2477–9.

Fontaine, M. (1975) Physiological mechanisms in the migration of marine and amphihaline fish. *Adv. mar. Biol.*, **13**, 241–55.

Fontaine, M. and Hatey, J. (1950) Variations de la teneur du foie en glycogène chez le jeune saumon *Salmo salar* L. au cours de la smoltification. *C. r. Séanc. Soc. Biol.*, **144**, 953–5.

Fontaine, M. and Leloup, J. (1959) Influence de la nage à contre courant sur le métabolisme de l'iode et le fonctionnenment thyroïdien chez la truite arc-en-ciel, *Salmo gairdneri* Richarson. *C.r. hebd. Séanc. Acad. Sci. Paris*, **249**, 343–7.

Ford, P. (1958) Studies on the development of the kidney of the Pacific pink salmon *Oncorhynchus gorbuscha* Walbaum II. Variation in glomerular count of the kidney of the Pacific pink salmon. *Can. J. Zool.*, **36**, 45–7.

Foster, J.R. and Schom, C.B. (1989) Imprinting and homing of Atlantic salmon (*Salmo salar*) kelts. *Can. J. Fish. aquat. Sci.*, **46**, 714–19.

Fujioka, Y. (1988) Morphological changes associated with growth and parr–smolt transformation in Biwa salmon *Oncorhynchus rhodurus* and Amgao salmon *Oncorhynchus rhodurus*. *Nippon Suisan Gakkaishi*, **54**, 77–86.

Fujioka, Y., Fushiki, S., Tagawa, M., Ogasawara, T. and Hirano, T. (1990) Downstream migratory behaviour and plasma thyroxine levels of Biwa salmon *Oncorhynchus rhodurus*. *Nippon Suisan Gakkaishi*, **54**, 1773–9.

Gasser, K.W. and Kirschner, L.B. (1987) The response of alkaline phosphatase to osmoregulatory changes in the trout, *Salmo gairdneri*. *J. comp. Physiol.*, **157B**, 469–75.

Glova, G.J. and McInerney, J.E. (1977) Critical swimming speeds of coho salmon *Oncorhynchus kisutch* fry to smolt stages in relation to salinity and temperature. *J. Fish. Res. Bd Can.*, **34**, 151–4.

Godin, J.G., Dill, P.A. and Drury, D.E. (1974) Effects of thyroid hormones on behaviour of yearling Atlantic salmon *Salmo salar* L. *J. Fish. Res. Bd Can.*, **31**, 1787–90.

Gorbman, A., Dickhoff, W.W., Mighell, J.L., Prentice, E.F. and Waknitz, F.W. (1982) Morphological indices of developmental progress in the parr smolt coho salmon *Oncorhynchus kisutch. Aquaculture*, **28**(1–2), 1–20.

Grau, G.E., Prunet, P., Gross, T., Nishioka, R.S. and Bern, H.A. (1982) Factors determining the occurrence of the surge in thyroid activity in salmon during smoltification. *Aquaculture*, **28**(1–2), 49–58.

Gray, E.S., Young, G. and Bern, H.A. (1990) Radioreceptor assay for growth hormone in coho salmon (*Oncorhynchus kisutch*) and its applications to the study of stunting. *J. exp. Zool.*, **256**, 290–96.

Gutiérrez, J. and Plisetskaya, E.M. (1990) Insulin binding to liver plasma membranes of coho salmon during smoltification. *Gen. comp. Endocrinol.*, **82**, 466–75.

Hansen, L.P., Clarke, W.C., Saunders, R.L. and Thorpe, J.E. (1989) Salmonid Smoltification III. *Aquaculture*, **82**, 1–4.

Hasler, A.D., Scholz, A.T. and Horrall, R. (1978) Olfactory imprinting and homing in salmon. *Am. Scient.*, **66**, 347–55.

Higgins, P.J. (1985) Metabolic differences between Atlantic salmon *Salmo salar* parr and smolts. *Aquaculture*, **45**(1–4), 33–53.

Hirano, T., Ogasawara, T., Bolton, J.P., Collie, N.L., Hasegawa, S. and Iwata, M. (1987) Osmoregulatory role of prolactin in lower vertebrates, in *Comparative Physiology of Environmental Adaptations* (eds R. Kirsch and B. Lahlou), Karger, Basel, pp. 112–24.

Hoar, W.S. (1939) The thyroid gland of the Atlantic salmon. *J. Morph.*, **65**, 257–95.

Hoar, W.S. (1976) Smolt transformation: evolution, behavior and physiology. *J. Fish. Res. Bd Can.*, **33**, 1233–52.

Hoar, W.S. (1988) The physiology of smolting salmonids, in *Fish Physiology*, Vol. 11B (eds W.S. Hoar and D.J. Randall), Academic Press, New York, pp. 275–343.

Hoffnagle, T.L. and Fivizzani, A.J. (1990) Stimulation of plasma thyroxine levels by novel water chemistry during smoltification in chinook salmon (*Oncorhynchus tshawytscha*). *Can. J. Fish. aquat. Sci.*, **47**, 1513–17.

Holmes, R.M. (1961) Kidney function in migrating salmonids. *Rep. Challenger Soc.*, **3**, 23–8.

Holmes, W.N. and Stainer, I.M. (1966) Studies on the renal excretion of electrolytes by the trout *Salmo gairdneri. J. exp. Biol.*, **44**, 33–46.

Hootman, S.R. and Philpott, C.W. (1979) Ultracytochemical localisation of Na^+, K^+, activated ATPase in chloride cells from the gills of a euryhaline teleost. *Anat. Rec.*, **193**, 99–130.

Ikuta, K., Aida, K., Okumoto, N. and Hanyu, I. (1985) Effects of thyroxine and methyltestosterone on smoltification of masu salmon *Oncorhynchus masou. Aquaculture*, **45**(1–4), 289–303.

Iwata, M., Komatsu, S., Hasegawa, T. and Hirano, T. (1987) Inconsistent effect of thyroid hormone alteration on seawater adaptability of fry of the chum salmon *Oncorhynchus keta. Nippon Suisan Gakkaishi*, **53**, 1969–73.

Iwata, M., Yamanome, T., Tagawa, M., Ida, H. and Hirano, T. (1989) Effects of thyroid hormones on phototaxis of chum and coho salmon juveniles. *Aquaculture*, **82**(1–4), 329–38.

Johanning, K.M. and Bradley, T.M. (1989) Plasma protein changes in Atlantic salmon (*Salmo salar*) during parr smolt transformation. *Comp. Biochem. Physiol.*, **92B**, 555–60.

Johnston, C.E. and Eales, J.G. (1970) Influence of body size on silvering of Atlantic salmon *Salmo salar* L. at parr smolt transformation. *J. Fish. Res. Bd Can.*, **27**, 983–7.

Johnston, L.E. and Saunders, R.L. (1981) Parr smolt transformation of yearling Atlantic salmon *Salmo salar* L. at several rearing temperatures. *Can. J. Fish. Aquat. Sci.*, **38**, 1189–98.

Kelley, K.M., Nishioka, R.S. and Bern, H.A. (1990) In vitro effect of osmotic pressure and cortisol on prolactin cell physiology in the coho salmon (*Oncorhynchus kisutch*) during the parr–smolt transformation. *J. exp. Zool.*, **254**, 72–82.

Komourdjian, M.P., Fenwick, J.C. and Saunders, R.L. (1989) Endocrine-mediated photostimulation of growth in Atlantic salmon. *Can. J. Zool.*, **67**, 1505–9.

Komourdjian, M.P., Saunders, R.L. and Fenwick, J.C. (1976a) Evidence for the role of growth hormone as a part of a "light pituitary axis" in growth and smoltification of Atlantic salmon *Salmo salar* L. *Can. J. Zool.*, **54**, 544–51.

Komourdjian, M.P., Saunders, R.L. and Fenwick, J.C. (1976b) The effect of porcine somatotropin on growth and survival in seawater of Atlantic salmon *Salmo salar* parr. *Can. J. Zool.*, **54**, 531–5.

Lahlou, B., Crenesse, D., Bensahla-Talet, A. and Porthe-Nibelle, J. (1975) Adaptation de la truite d'élevage à l'eau de mer. Effets sur les concentrations plasmatiques, les échanges branchiaux et le transport intestinal du sodium. *J. Physiol., Paris*, **70**, 593–603.

Langdon, J.S. and Thorpe, J.E. (1985) The ontogeny of smoltification: developmental patterns of gill Na^+/K^+-ATPase and chloride cells in juvenile Atlantic salmon *Salmo salar* L. *Aquaculture*, **45**(1–4), 83–95.

Langdon, J.S., Thorpe, J.E. and Roberts, R.J. (1984) Effects of cortisol and ACTH on gill Na^+/K^+ ATPase, SDH and chloride cells in juvenile Atlantic salmon (*Salmo salar* L.). *Comp. Biochem. Physiol.*, **77A**, 9–12.

Leloup, J. and Fontaine, M. (1960) Iodine metabolism in lower vertebrates. *Ann. N. Y. Acad. Sci.*, 316–53.

Leloup, J., Long, P. and Fontaine, M. (1971) Influence des hormones thyroïdiennes sur la synthèse des protéines dans le cerveau des téléostéens. *Gen. comp. Endocrinol.*, **18**, 603.

Loretz, C.A. and Bern, H.A. (1982) Prolactin and osmoregulation in vertebrates. *Neuroendocrinology*, **35**, 292–330.

Loretz, C.A., Collie, N.L., Richman, N.H. and Bern, H.A. (1982) Osmoregulatory changes accompanying smoltification in coho salmon. *Aquaculture*, **28**(1–2), 67–74.

Lubin, R.T., Rourke, A.W. and Bradley, T.M. (1989) Ultrastructural alterations in branchial chloride cells of Atlantic salmon, *Salmo salar*, during parr–smolt transformation and early development in sea water. *J. Fish Biol.*, **34**, 259–72.

Lundqvist, H., Berglund, I., Mayer, I. and Borg, B. (1990) Seawater adaptability in Baltic salmon, *Salmo salar*, immature smolt and mature male parr: lack of effect of springtime castration. *Can. J. Zool.*, **68**, 2181–4.

Lundqvist, H., Borg, B. and Berglund I. (1989) Androgens impair seawater adaptability in smolting salmon (*Salmo salar*). *Can. J. Zool.*, **67**, 1733–6.

McCormick, S.D., Saunders, R.L., Henderson, E.B. and Harmon, P.R. (1987) Photoperiod control of parr smolt transformation in Atlantic salmon (*Salmo salar*): changes in salinity tolerance, gill $Na^+–K^+$ ATPase activity and plasma thyroid hormones. *Can. J. Fish. aquat. Sci.*, **44**, 1462–8.

McCormick, S.D, Saunders, R.L. and McIntyre, A.R.D. (1989a) Mitochondrial enzyme and $Na^+–K^+$ ATPase activity, and ion regulation during parr smolt transformation of Atlantic salmon (*Salmo salar*). *Fish Physiol. Biochem.*, **6**, 231–4.

McCormick, S.D., Moyes, C.D. and Ballantyne, J.S. (1989b) Influence of salinity on the energetics of gill and kidney of Atlantic salmon (*Salmo salar*). *Fish Physiol. Biochem.*, **6** 243–54.

McLatchy, D.L. and Eales, J.G. (1990) Growth hormone stimulates hepatic thyroxine 5′-monodeiodinase activity and 3,5,3′-iodothyronine levels in rainbow trout (*Salmo gairdneri*). *Gen. comp. Endocrinol.*, **78**, 164–72.

Madsen, S.S. (1990a) The role of cortisol and growth hormone in seawater adaptation and development of hypoosmoregulatory mechanisms in sea trout parr (*Salmo trutta trutta*). *Gen. comp. Endocrinol.*, **79**, 1–11.

Madsen, S.S. (1990b) Effect of repetitive cortisol and thyroxine injections on chloride cell number and Na^+–K^+ ATPase activity in gills of freshwater acclimated rainbow trout, *Salmo gairdneri. Comp. Biochem. Physiol.*, **95A**, 171–5.

Madsen, S.S. (1990c) Cortisol treatment improves the development of hypoosmoregulatory mechanisms in sea trout parr (*Salmo trutta trutta*). *Fish Physiol. Biochem.*, **8**, 45–52.

Madsen, S.S. and Korsgaard, B. (1989) Time-course effects of repetitive estradiol-17β and thyroxine injections on the natural spring smolting of Atlantic salmon, *Salmo salar* L. *J. Fish Biol.*, **35**, 119–28.

Mahnken, C., Prentice, E., Waknitz, W., Monan, G., Sims, C. and Williams, J. (1982) The application of recent smoltification research to public hatchery coho salmon *Oncorhynchus kisutch. Aquaculture*, **28**(1–2), 251–68.

Maxime, V., Boeuf, G., Pennec, J.P. and Peyraud, C. (1989) Comparative study of the energetic metabolism of Atlantic salmon (*Salmo salar*) parr and smolts. *Aquaculture*, **82**(1–4), 155–62.

Miwa, S. and Inui, Y. (1983) Effects of thyroxine and thiourea on the parr smolt transformation of amago salmon *Oncorhynchus rhodurus. Bull. nat. Res. Inst. Aquacult.*, **4**, 41–52.

Miwa, S. and Inui, Y. (1985) Effects of L-thyroxine and ovine growth hormone on smoltification of amago salmon *Oncorhynchus rhodurus. Gen. comp. Endocrinol.*, **58**, 436–42.

Miwa, S. and Inui, Y. (1986) Inhibitory effect of 17α methyltestosterone and estradiol 17β on smoltification of sterilized amago salmon *Oncorhynchus rhodurus. Aquaculture*, **53**, 21–39.

Morin, P.P., Dodson, J.J. and Doré, F.Y. (1989a) Thyroid activity concomitant with olfactory learning and heart rate changes in Atlantic salmon, *Salmo salar*, during smoltification. *Can. J. Fish. aquat. Sci.*, **46**, 131–6.

Morin, P.P., Dodson, J.J. and Doré, F.Y. (1989b) Cardiac responses to a natural odorant as evidence of a sensitive period for olfactory imprinting in young Atlantic salmon, *Salmo salar. Can. J. Fish. aquat. Sci.*, **46**, 122–30.

Nance, J.M., Bornancin, M., Sola, F., Boeuf, G. and Dutil, J.D. (1990) Study of transbranchial Na^+ exchange in *Salmo salar* smolts and post-smolts directly transferred to sea water. *Comp. Biochem. Physiol.*, **96A**(2), 303–8.

Narayansingh, T. and Eales, J.G. (1975) Effects of thyroid hormones on *in vivo* I-^{14}C L-leucine incorporation into plasma and tissue protein of brook trout, *Salvelinus fontinalis* and rainbow trout, *Salmo gairdneri. Comp. Biochem. Physiol.*, **52B**, 399–405.

Nishioka, R.S., Bern, H.A., Lai, K.V., Nagahama, Y. and Grau, E.G. (1982) Changes in the endocrine organs of coho salmon during normal and abnormal smoltification. An electron microscope study. *Aquaculture*, **28**(1–2), 21–38.

Nishioka, R.S., Richman, N.H., Young, G., Prunet, P. and Bern, H.A. (1987) Hypophysectomy of coho salmon (*Oncorhynchus kisutch*) and survival in freshwater and seawater. *Aquaculture*, **65**, 343–52.

Nishioka, R.S., Young, G., Bern, H.A., Jochimsen, W. and Hiser, C. (1985) Attempts to intensify the thyroxine surge in coho and king salmon by chemical stimulation. *Aquaculture*, **45**(1–4), 215–25.

Nonnotte, G., Aubree-Pelletier, A. and Colin, D.A. (1987) Transient changes of the intestinal absorption of sodium and chloride in the rainbow trout after abrupt transfer into seawater. *Comp. Biochem. Physiol.*, **88A**, 437–41.

Nonnotte, L., Nonnotte, G. and Leray, C. (1986) Morphological changes in the middle intestine of the rainbow trout, *Salmo gairdneri*, induced by a hyperosmotic environment. *Cell Tissue Res.*, **243**, 619–28.

Ogata, H. and Murai, T. (1989) Effects of dietary fatty acid composition on growth and smolting of underyearling masu salmon, *Oncorhynchus masou*. *Aquaculture*, **82**(1–4), 181–90.

Olivereau, M. (1954) Hypophyse et glande thyroïde chez les poissons. Etude histophysiologique de quelques corrélations endocriniennes, en particulier chez *Salmo salar*. *Ann Inst océanogr., Monaco*, **29**, 95–296.

Omeljaniuk, R.J. and Eales, J.G. (1986) The effect of 3,5,3′-triiodo-L-thyronine on gill Na^+–K^+-ATPase activity of rainbow trout *Salmo gairdneri* in freshwater. *Comp. Biochem. Physiol.*, **84A**, 427–9.

Parry, G. (1960) The development of salinity tolerance in the salmon *Salmo salar* and some related species. *J. Exp. Biol.*, **37**, 425–34.

Payan, P., Girard, J.P. and Mayer-Gostan, N. (1984) Branchial ion movements in teleost: the roles of respiratory and chloride cells, in *Fish Physiology*, Vol. 10B (eds W.S. Hoar and D.J. Randall), Academic Press, New York, pp. 39–63.

Pisam, M., Boeuf, G., Prunet P. and Rambourg, A. (1990) Ultrastructural features of mitochondria-rich cells in stenohaline freshwater and seawater fishes. *Am. J. Anat.*, **187**, 21–31.

Pisam, M., Prunet, P., Boeuf, G. and Rambourg, A. (1988) Ultrastructural features of chloride cells in the gill epithelium of the Atlantic salmon, *Salmo salar*, and their modifications during smoltification. *Am. J. Anat.*, **183**, 235–44.

Plisetskaya, E.M., Ottolenghi, C., Sheridan, M.A., Mommsen, T.P. and Gorbman, A. (1989) Metabolic effects of salmon glucagon and glucagon-like peptide in coho and chinook salmon. *Gen. comp. Endocrinol.*, **73**, 205–16.

Plisetskaya, E.M., Swanson, P., Bernard, M.G. and Dickhoff, W.W. (1988) Insulin in coho salmon (*Oncorhynchus kisutch*) during the parr to smolt transformation. *Aquaculture*, **72**, 151–64.

Primmett, D.R.N., Eddy, F.B., Miles, M.S., Talbot, C. and Thorpe, J.E. (1988) Transepithelial ion exchange in smolting Atlantic salmon (*Salmo salar* L.). *Fish Physiol. Biochem.*, **5**(4), 181–6.

Prunet, P. and Boeuf, G. (1985) Plasma prolactin level during transfer of rainbow trout *Salmo gairdneri* and Atlantic salmon *Salmo salar* from fresh water to seawater. *Aquaculture*, **45**(4), 167–76.

Prunet, P. and Boeuf, G. (1989) Plasma prolactin levels during smolting in Atlantic salmon, *Salmo salar*. *Aquaculture*, **82**, 297–305.

Prunet, P., Boeuf, G., Bolton, J.P. and Young, G. (1989) Smoltification and seawater adaptation in Atlantic salmon (*Salmo salar*): plasma prolactin, growth hormone and thyroid hormones. *Gen. comp. Endocrinol.*, **74**, 355–64.

Randall, D.J. and Lin, H. (1993) Effects of varying water pH on gas and ion transfer across fish gills, in *Fish Ecophysiology* (eds J.C. Rankin and F.B. Jensen), Chapman and Hall, London, 265–75.

Redding, J.M., Patino, R. and Schreck, C.B. (1991) Cortisol effects on plasma electrolytes and thyroid hormones during smoltification in coho salmon *Oncorhynchus kisutch*. *Gen. Comp. Endocrinol.*, **81**, 373–82.

Refstie, T. (1982) The effect of feeding thyroid hormones on salwater tolerance and growth rate of Atlantic salmon. *Can. J. Zool.*, **60**, 2706–12.

Renzis, G. de and Bornancin, M. (1984) Ion transport and gill ATPases, in *Fish Physiology*,

Vol. 10B (eds W.S Hoar and D.J. Randall), Academic Press, New York, pp. 65–104.

Richman, N.H. and Zaugg, W.S. (1987) Effects of cortisol and growth hormone on osmoregulation in pre and post smoltified coho salmon *Oncorhynchus kisutch. Gen. comp. Endocrinol.*, **65**, 189–98.

Saunders, R.L. and Harmon, P.R. (1990) Influence of photoperiod on growth of juvenile Atlantic salmon and development of salinity tolerance during winter–spring. *Trans. Am. Fish. Soc.*, **119**, 689–97.

Saunders, R.L. and Henderson, E.B. (1970) Influence of photoperiod on smolt development and growth of Atlantic salmon *Salmo salar. J. Fish. Res. Bd Can.*, **27**, 1295–1311.

Saunders, R.L. and Henderson, E.B. (1978) Changes in gill ATPase activity and smolt status of Atlantic salmon *Salmo salar. J. Fish. Res. Bd Can.*, **35**(12), 55–66.

Saunders, R.L., Henderson, E.B. and Harmon, P.R. (1985) Effects of photoperiod on juvenile growth and smolting of Atlantic salmon and subsequent survival and growth in sea cages. *Aquaculture*, **45**, 55–6.

Saunders, R.L., Specker, J.L. and Komourdjian, M.P. (1989) Effects of photoperiod on growth and smolting in juvenile Atlantic salmon (*Salmo salar*). *Aquaculture*, **82**, 103–17.

Scholz, A.T., White, R.J., Muzi, M. and Smith, T. (1985) Uptake of radiolabelled triiodothyronine in the brain of steelhead trout (*Salmo gairdneri*) during parr–smolt transformation: implications for the mechanisms of thyroid activation of olfactory imprinting. *Aquaculture*, **45**(1–4), 199–214.

Sheridan, M.A. (1989) Alterations in lipid metabolism accompanying smoltification and seawater adaptation of salmonid fish. *Aquaculture*, **82**(1–4), 191–204.

Sheridan, M.A., Allen, W.V. and Kerstetter, T.H. (1983) Seasonal variation in the lipid composition of the steelhead trout *Salmo gairdneri* Richardson, associated with the parr smolt transformation. *J. Fish Biol.*, **23**(2), 125–34.

Sheridan, M.A. Allen, W.V. and Kerstetter, T.H. (1985a) Changes in the fatty acid composition of steelhead trout *Salmo gairdneri* Richardson associated with parr smolt transformation. *Comp. Biochem. Physiol.*, **80B**, 671–6.

Sheridan, M.A., Woo, N.Y.S. and Bern, H.A. (1985b) Changes in the rates of glycogenesis, glycogenolysis, lipogenesis and lipolysis in selected tissues of the coho salmon *Oncorhynchus kisutch* associated with parr smolt transformation. *J. exp. Zool.*, **236**, 35–44.

Sigholt, T. and Finstad, B. (1990) Effect of low temperature on seawater tolerance in Atlantic salmon (*Salmo salar*) smolts. *Aquaculture*, **84**, 167–72.

Simpson, T.H. (1985) Epilogue. *Aquaculture*, **45**(1–4), 395–8.

Sower, S.A., Sullivan, C.V. and Gorbman, A. (1984) Changes in plasma estradiol and effects of triiodothyronine on plasma estradiol during smoltification of coho salmon *Oncorhynchus kisutch. Gen comp. Endocrinol.*, **54**, 486–92.

Specker, J.L. (1982) Interrenal function and smoltification. *Aquaculture*, **28**(1–2), 59–66.

Specker, J.L. and Schreck, C.B. (1984) Thyroidal response to mammalian thyrotropin during smoltification of coho salmon *Oncorhynchus kisutch. Comp. Biochem. Physiol.*, **78A**, 441–4.

Stewart, M.W., Saunders, R.L. and Wiggs, A.J. (1990) Effects of extended daylength on autumn growth dynamics of juvenile Atlantic salmon, Salmo salar. *Can. J. Fish. aquat. Sci.*, **47**, 755–9.

Sullivan. C.V., Darling, D.S. and Dickhoff, W.W. (1987) Effects of triiodothyronine and propylthiouracil on thyroid function and smoltification of coho salmon (*Oncorhynchus kisutch*). *Fish Physiol. Biochem.*, **4**, 121–35.

Sullivan, C.V., Dickhoff, W.W., Mahnken, C.V.W. and Hershberger, W.K. (1985) Changes in the haemoglobin system of the coho salmon *Oncorhynchus kisutch* during smoltification and triiodothyronine and propylthiouracil treatment. *Comp. Biochem. Physiol.*, **81A**, 807–13.

Swanson, P. and Dickhoff, W.W. (1987) Variation in thyroid response to thyroid-stimulating hormone in juvenile coho salmon (*Oncorhynchus kisutch*). *Gen. comp. Endocrinol.*, **68**, 473–85.

Sweeting, R.M. and McKeown, B.A. (1989) Changes in plasma growth hormone and various metabolic factors during smoltification of coho salmon, *Oncorhynchus kisutch*. *Aquaculture*, **82**, 279–95.

Sweeting, R.M., Wagner, G.F. and McKeown, B.A. (1985) Changes in plasma glucose, amino acid nitrogen and growth hormone during smoltification and seawater adaptation in coho salmon, *Oncorhynchus kisutch*. *Aquaculture*, **45**(1–4), 185–97.

Takeuchi, T., Sampekalo, J., Nomura, T. and Watanabe, T. (1990) Lipid contents and classes in gill of masu salmon *Oncorhynchus masou* during the parr–smolt transformation. *Nippon Suisan Gakkaishi*, **56**, 1527–8.

Thorpe, J. (1977) Bimodal distribution of length of juvenile Atlantic salmon *Salmo salar* L. under artificial rearing conditions. *J. Fish Biol.*, **11**, 175–84.

Thorpe, J.E. and Morgan, R.I.G. (1978) Periodicity in Atlantic salmon, *Salmo salar* L. smolts migration. *J. Fish Biol.*, **12**, 541–8.

Thorpe, J.E., Ross, L.G., Struthers, G. and Watts, W. (1981) Tracking Atlantic salmon smolts, *Salmo salar* L., through Loch Voil, Scotland. *J. Fish Biol.*, **19**, 519–37.

Thorpe, J.E., Talbot, C. and Villarreal, C. (1982) Bimodality of growth and smolting in Atlantic salmon *Salmo salar* L. *Aquaculture*, **28**(1–2), 123–32.

Usher, M.L., Talbot, C. and Eddy, F.B. (1988) Drinking in Atlantic salmon smolts transferred to seawater and the relationship between drinking and feeding. *Aquaculture*, **73**, 237–46.

Usher, M.L., Talbot, C. and Eddy, F.B. (1990) Effects of transfer to seawater on digestion and gut function in Atlantic salmon smolts (*Salmo salar* L.). *Aquaculture*, **90**, 85–96.

Virtanen, E. and Forsman, L. (1987) Physiological responses to continuous swimming in wild salmon (*Salmo salar*) parr and smolt. *Fish Physiol. Biochem.*, **4**(3), 157–163.

Virtanen, E. and Soivio, A. (1985) The patterns of T3, T4, cortisol and Na^+–K^+ ATPase during smoltification of hatchery reared *Salmo salar* and comparison with wild smolts. *Aquaculture*, **45**(1–4), 97–109.

Wedemeyer, A.H., Saunders, R.L. and Clarke, W.L. (1980) Environmental factors affecting smoltification and early marine survival of anadromous salmonids. *Mar. Fish. Rev.*, **42**, 1–14.

Wiggs, A.J., Henderson, E.B., Saunders, R.L. and Kutty, M.N. (1989) Activity, respiration and excretion of ammonia by Atlantic salmon (*Salmo salar*) smolt and postsmolt. *Can. J. Fish. aquat. Sci.*, **46**, 790–95.

Winans, G.A. and Nishioka, R.S. (1987) A multivariate description of change in body shape of coho salmon (*Oncorhynchus kisutch*) during smoltification. *Aquaculture*, **66**, 235–45.

Young, G., Björnsson, B.T., Prunet, P., Lin, R.J. and Bern, H.A. (1989a) Smoltification and seawater adaptation in coho salmon (*Oncorhynchus kisutch*): plasma prolactin, growth hormone, thyroid hormones and cortisol. *Gen. comp. Endocrinol.*, **74**, 335–45.

Young, G., Prunet, P., Ogasawara, T., Hirano, T. and Bern, H.A. (1989b) Growth retardation (stunting) in coho salmon: plasma hormones levels in stunts in seawater and after transfer to freshwater. *Aquaculture*, **82**, 269–78.

Youngson, A.F., McLay, H.A. and Olsen, T.C. (1986) The responsiveness of the thyroid system of Atlantic salmon (*Salmo salar* L.) smolts to increased water velocity. *Aquaculture*, **56**, 243–55.

Youngson, A.F., Scott, D.C.B., Johnstone, R. and Pretswell, D. (1985) The thyroid system's role in the downstream migration of Atlantic salmon (*Salmo salar* L.) smolts. *Aquaculture*, **45**, 392–3.

Zaugg, W.S. (1982) Some changes in smoltification and seawater adaptability of salmonids resulting from environmental and other factors. *Aquaculture*, **28**(1–2), 143–52.

Zaugg, W.S. (1989) Migratory behavior of underyearling *Oncorhynchus tshawytscha* and survival to adulthood as related to prerelease gill (Na^+–K^+)-ATPase development. *Aquaculture*, **82**, 339–53.

Zaugg, W.S. and McLain, L.R. (1972) Changes in gill ATPase activity associated with parr smolt transformation in steelhead trout *Salmo gairdneri*, coho *Oncorhynchus kisutch*, and spring chinook salmon *O. tshawytscha*. *J. Fish. Res. Bd Can.*, **29**, 167–71.

Zaugg, W.S. and McLain, L.R. (1976) Influence of water temperature on gill sodium potassium ATPase activity in juvenile coho *Oncorhynchus kisutch*. *Comp. Biochem. Physiol.*, **54A**, 419–21.

Zaugg, W.S. and Mahnken, C.V.W. (1991) The importance of smolt development to successful marine ranching of Pacific salmon, in *Marine Ranching* (ed. R.S. Svrjcek), *NOAA Tech. Rep. NMFS*, **102**, 89–97.

Zaugg, W.S., Prentice, E.F. and Waknitz, F.N. (1985) Importance of river migration to the development of seawater tolerance in Columbia River anadromous salmonids. *Aquaculture*, **51**, 33–47.

Chapter five

Role of peptide hormones in fish osmoregulation

Yoshio Takei

5.1 INTRODUCTION

The blood of teleost fishes, which has an osmolality of some 300 mOsm kg^{-1}, is separated by only a thin respiratory epithelium from environmental media, which have an osmolality range of a few to 1000 mOsm kg^{-1}. Thus, an ability to osmoregulate is essential for their survival in either freshwater (FW) or seawater (SW) environments. This problem is particularly severe for euryhaline, migratory fishes which experience both FW and SW in their life span. In order to osmoregulate, fishes have to adjust and balance the gain and loss of water and electrolytes, especially of Na^+ and Cl^-. As is the case for terrestrial vertebrates, water and Na^+ are gained by drinking and feeding, and subsequent absorption by the intestine, and they are lost through the kidney as urine. Fishes additionally have the gill and urinary bladder as osmoregulatory organs, the former serving as another route for water and Na^+ gain or loss, depending on the salinity of their environmental media. As far as Na^+ and Cl^- are concerned, the gill of teleost fishes can also act as an important osmoregulatory organ by actively absorbing or excreting the ions against the concentration gradient between blood and environmental media. Marine elasmobranchs accumulate urea and trimethylamine oxide in their plasma to maintain a blood osmolality slightly higher than that of SW. Thus elasmobranch osmoregulation is primarily by ionoregulation, which is the principal function of the rectal gland. Marine cyclostomes like the hagfish *Myxine glutinosa*, have plasma ionic concentrations almost identical to SW, thus, as in marine invertebrates, they scarcely need to osmoregulate and have only a limited ability largely related to volume regulation.

We can depict a general scheme for water and Na^+ economy in teleost fishes. In FW, fishes scarcely drink, absorb little water by the intestine, and

Fish Ecophysiology. Edited by J. Cliff Rankin and Frank B. Jensen. Published in 1993 by Chapman & Hall, London. ISBN 0 412 45920 5.

excrete large volumes of water by the kidney to balance the copious osmotic gain by the gill. The diffusional loss of Na^+ by the gill and the inevitable Na^+ loss with the high rate of urine production are compensated by dietary intake and, if necessary, the active uptake of Na^+ from the media, principally by the gill. In contrast, fishes in SW drink large amounts and absorb most of the imbibed water across the intestine to replace the water lost osmotically across the body surfaces, principally the gill epithelium. The diffusional gain of Na^+ and Cl^- by the gill and ion uptake in the gut are balanced by the active extrusion by the chloride cells of the gill or operculum. The divalent ions that enter the body with imbibed SW are excreted in high concentrations in the small amount of urine produced (Rankin and Davenport, 1981).

Humoral factors are known to play a crucial role in the control of these processes of water and Na^+ economy. Established among these are prolactin and cortisol, which are important for FW and SW adaptation, respectively (Fig. 5.1). Extensive reviews describe the role of prolactin (Nicoll, 1981; Brown and Brown, 1987; Hirano *et al.*, 1987) and cortisol (Balment *et al.*, 1987; Henderson and Kime, 1987) in fish osmoregulation, so these hormones are not dealt with in this chapter. In addition to prolactin and cortisol, a number of peptide hormones are implicated in fish osmoregulation. These include growth hormone (GH), angiotensin II (ANGII), arginine vasotocin (AVT), urotensins, vasoactive intestinal peptide (VIP), and atrial natriuretic peptides (ANP) (Fig. 5.1). ANGII, AVT, and urotensins have been the subject of research for more than a decade, and reviews have been written on their involvement in osmoregulation (Larson and Bern, 1987; Nishimura, 1987; Perks, 1987). However, their roles are far from being established, and new evidence is still

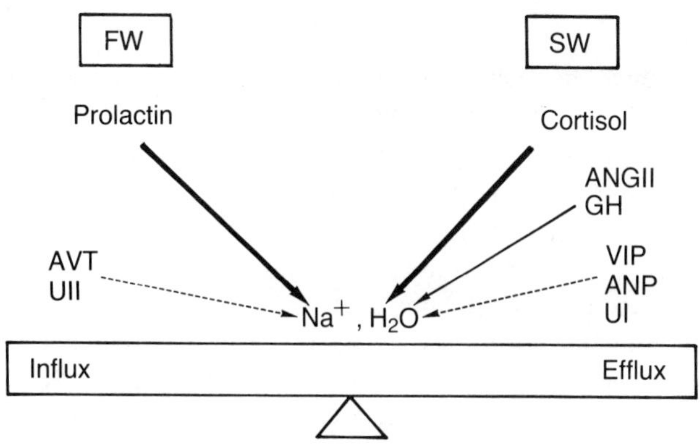

Fig. 5.1 Diagram to show how hormones interact to adjust influx and efflux of Na^+ and water to enable fishes to adapt in either fresh water (FW) or seawater (SW). The relative contribution of each hormone to the adaptation, which can be assessed from the data accumulated thus far, is expressed by the thickness of the line, and broken lines indicate that the role of the hormone is far from being established. ANGII, angiotensin II; ANP, atrial natriuretic peptides; AVT, arginine vasotocin; GH, growth hormone; UI, urotensin I; UII, urotensin II; VIP, vasoactive intestinal peptide.

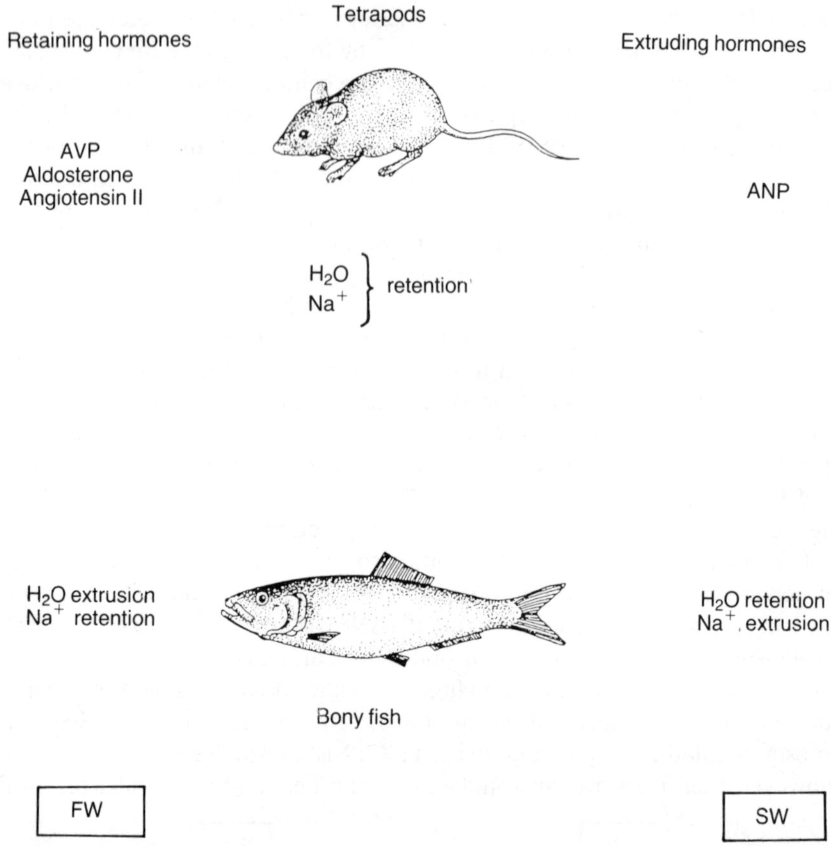

Fig. 5.2 Diagram to show differences in water and Na^+ economy between terrestrial tetrapods and bony fishes in fresh water or seawater.

emerging. In this chapter, therefore, I shall not describe each hormone in detail but add only recent data which are related to osmoregulation. ANP and VIP will be described extensively because these rather new hormones have not been reviewed in detail in relation to their roles in fish osmoregulation, except for a recent review for ANP (Evans, 1990).

It seems that peptide hormones can be divided into two categories according to their mode of action. One is the slow-acting and long-effecting hormones which may be responsible for the development and maintenance of the osmoregulatory system. Protein hormones, such as prolactin and GH, belong to this category. The other category is the fast-acting and short-effecting hormones whose actions may be to allow rapid response to sudden changes in environmental salinity. Oligopeptide hormones, such as ANGII, urotensin II (UII) and AVT, are in this latter category. Peptide hormones of an intermediate size, such as ANP, VIP, and UI, may be placed in the middle, but we know too little about their effects.

The peptide hormones which appear to participate in various aspects of fish osmoregulation do not act independently, but influence one another at the levels of secretion and action to ensure water and electrolyte homeostasis. In terrestrial animals such as mammals, ANGII, vasopressin (or AVT), and aldosterone act in concert to conserve water and Na^+ in the body, whereas ANP appears to be the sole circulating hormone to counteract these retaining hormones (Fig. 5.2). The dominance of retaining hormones may be accounted for by the fact that retention of water and Na^+, i.e. conservation of blood volume, is essential for the survival of terrestrial species (Fig. 5.2). However, the situation is different in fishes: bony fishes in FW have to extrude water and retain Na^+, but those in SW have to retain water and extrude Na^+ (Fig. 5.2). Therefore, the relative importance of these retaining and extruding hormones may be different between tetrapods and bony fishes, or more precisely, among tetrapods, FW fishes and SW fishes. As a first step in our assessment of the role of peptide hormones in fish osmoregulation, it is perhaps appropriate to define them as either predominantly for FW adaptation or SW adaptation, although the available information does not seem to be sufficient for clear definition in many cases.

5.2 GROWTH HORMONE (GH)

GH is an adenohypophysial hormone which has been implicated in promotion of growth in mammals. In fishes, however, since Smith first reported the positive effect of GH on survival of brown trout in SW (Smith, 1956), evidence has been accumulated, though slowly, which indicates that GH is important for SW adaptation, particularly for salmonid fishes (Fig. 5.1; Hirano *et al.*, 1987; Bern *et al.*, 1991). Recently, GH has been isolated and sequenced in several species of teleosts (Fig. 5.3; Kawauchi and Yasuda, 1989). Studies using homologous hormones and homologous radioimmunoassay (RIA) systems show that transfer of salmonids to hypertonic media is associated with increases in plasma GH levels and its metabolic clearance rate. Administration of homologous GH to the immature rainbow trout, *Oncorhynchus mykiss*, or chum salmon, *Oncorhynchus keta*, prevents an increase in plasma Na^+ concentration which otherwise occurs after transfer to SW (Hasegawa *et al.*, 1987; Ogasawara *et al.*, 1989; Sakamoto *et al.*, 1990, 1991).

More recent reports show the presence of specific, saturable GH receptors in the osmoregulatory organs of the rainbow trout (Sakamoto and Hirano, 1991; Yao *et al.*, 1991). An especially high concentration is evident for hepatic receptors, and the concentration increases after adaptation to 80% SW (Sakamoto and Hirano, 1991). It is likely that the effects of GH in salmonids are mediated by the release of somatomedin C/insulin-like growth factor I from the liver as is the case for its growth-promoting effect (Bern *et al.*, 1991). Supportive data have been reported by McCormick *et al.* (1991), who showed

```
                     10                  30                  50                  70                  90
Blue Shark   YPLLPLSDLFAKAVHRAQHLHLVAAETTKDFERKYIPEEQRHS-HKSSPSAFCQSETIPAPTGKEDAQQRSDRELLLYSLLLIQSWLNPIQNLS---
Sturgeon     YPMIPLSSLFTNAVLRAQYLHQLAADIYKDFERTYMPNEQRHS-SKNSPSAFCYSETIPAPTGKDEAQQRSDVELLQFSLALIQSWISPLQSLSRVF
Salmon          IENQRLFNIAVSRVQHLHLLAQKMFNDFDGTLLPDERRQS-NKIFLLDFCNSDSIVSPVDKHETQKSSVLKLLHISFRLIESWEYPSQTL--II
Eel         VEPI-SLYNLFTSAVNRAQHLHTLAAEIYKEFERSIPPEAHRQL-SKTSPLAGCYSDSIPTPTGKDETQEKSDGYLLRISSALIQSWVYPLKTLSDAF
Cod          HPLIDSQRLFSIAVNRIQHLHMLAERIFSELFSSLQIEEQRQL-NKIFLQDFCNSDSIISPIDKHETQRSSVLRLLTVSYRLIESWEFPSQSLP---
Flounder     QPITENQRLFSIAVGRVQYLHLVAKKLFSDFENSLQLEDQRLL-NKIASKEFCHSDNFLSPIDKHETQGSSVQKLLSVSYRLIESWEFFSRFL----
Bullfrog     FPQMSLSNLFTNAVIRAQHLHQMVADTYRDYERTYIPEDQRFS-NKHSYSVYCYSETIPAPTDKDNTHQKSDIELLRFSLLLLQSWMNPIQIVNRVF
Sea Turtle  AFPAMPLSSLFANAVLRAQHLHLLAADTYKEFERTYIPEEQRHS-NKISQSASCYSETIPAPTGKDDAEQKSDMELLRFSLILIQSWLNPVQFLSRVF
Chicken     TFPAMPLSNLFANAVLRAQHLHLLAAETYKEFERTYIPEDQRYT-NKNSQAAFCYSETIPAPTGKDDAQQKSDMELLRFSLVLIQSWLTPVQYLSKVF
Human        FPTIPLSRLFDNAMLRAHRLHQLAFDTYQEFEEAYIPKEQKYSFLQNPQTSLCFSESIPTPSNREETQQKSNLELLRISLLLIQSWLEPVQFLRSVF

                   110                 130                 150                 170                 190
Blue Shark  ----AFRTSD-RVYDKLRDLEEGIFALMKTLEDGGSSQGFAWLK--FSYERFDGNL-SEEALMKNYGLLACFKKDMHKVETYLKVMNCKRFAESNCTV
Sturgeon    TNSLVFLTSD-RVFEKLKDLEEGIVALMRDLGEGGFGSSTLL-K--LTYDKFDVNLRNDDALFKNYGLLSCFKKDMHKVETYLKVMKCRRFVESNCTL
Salmon      SNSLMVRNAN-QISEKLSDLKVGINLLITGSQDGVLSLDDNDSQQLPPYGNYYQNLGGDGNVRRNYELLACFKKDMHKVETYLTVAKCRKSLEANCTL
Eel         SNSLMFGTSD-GIFDKLEDLNKGINELMKVVGDGGIYI-EDV-R-NLRYENFDVHLRNDAGLMKNYGLLACFKKDMHKVETYLKVTKCRRFVESNCTL
Cod         ---GGSVLRN-QISPKLSELKNGIHILIRTSQ-GAG-DALVEADPMSPYGGYYQALGGDGSLRKSYEMLACFKKDMHKVETYLTVAKCRLSPEDNCTL
Flounder    ---VASFART-QVTSKLSELKMGLLKLIEANQDGAGGFSESSVLQLTPYGN---------------SELFACFKKDMHKVETYLTVAKCRLFPEANCTL
Bullfrog    GNNQVFGNID-RVYDRLRDLDEGLHILIRELDDGNVRNYGVL-T--FTYDKFDVNLRSEEGRAKNYGLLSCFKKDMHKVETYLKLMKCRRFVESNCTF
Sea turtle  TNSLVFGTSD-RVYEKLRDLEEGIQALMRELEDGSLRGFQVL-R--PTYDKFDINLRNEDALLKNYGLLSCFKKDLHKVETYLKVMKCRRFGESNCTI
Chicken     TNNLVFGTSD-RVFEKLKDLEEGIQALMRELEDRSPRGPQLLRPTYDKFDIHLRN---EDALLKNYGLLSCFKKDLHKVETYLKVMKCRRFGESNCTI
Human       ANSLVYGASDSNVYDLLKDLEEGIQTLMGRLEDGSPRTGQIFKQTYSKFDTNSHN---DDALLKNYGLLYCFRKDMDKVETFLRIVQCR-SVEGSCGF
```

Fig. 5.3 Comparison of amino acid sequences of growth hormone from various vertebrate classes. References: blue shark, Yamaguchi *et al.* (1989); sturgeon, Yasuda *et al.* (1991); salmon, Sekine *et al.* (1985); eel, Yamaguchi *et al.* (1987); cod, Rand-Weaver *et al.* (1991); flounder and human, Watahiki *et al.* (1989); bullfrog, Kobayashi *et al.* (1991); sea turtle and chicken, Yasuda *et al.* (1989). Gaps are introduced to optimize matching. Identical sequences across different species are boxed.

that both ovine GH and bovine insulin-like growth factor I facilitate the adaptation of juvenile rainbow trout to hypertonic media.

GH appears to play a less significant role in SW adaptation in the eel, *Anguilla japonica*, than in salmonids. Neither plasma GH levels, measured by homologous RIA, nor its metabolic clearance rate, were altered after transfer from FW to SW (Kishida and Hirano, 1988; Duan and Hirano, 1991). However, this may be a characteristic specific to the eel, because it can survive a direct exposure from FW to SW even after hypophysectomy. Hypophysectomized coho salmon, *Onchorynchus kisutch*, are unable to survive such transfers (Nishioka *et al.*, 1987).

5.3 ANGIOTENSIN II (ANGII)

ANGII is a principal component of the renin–angiotensin system which is formed in blood by enzymatic actions of renin and converting enzyme. Nishimura (1987) recently produced a comprehensive review on the osmoregulatory role of the renin–angiotensin system in non-mammalian vertebrates, and fish studies are concisely summarized therein. Thus, we now only add some data which were not included in the review.

The renin–angiotensin system appears to have evolved first in the primitive bony fish and to be absent in elasmobranchs and cyclostomes. This idea is based on the facts that Bowie-positive granules, which are thought to be renin-containing granules of juxtaglomerular cells, are absent in the kidneys of the latter two groups and that a vasopressor substance is not produced after incubation of plasma with kidney extracts when assayed in the rat (Nishimura, 1987). However, mammalian ANGII exhibits biological actions in elasmobranchs such as increases in arterial pressure and stimulation of corticosteroid release (spiny dogfish, *Squalus acanthias*, Opdyke and Holcombe, 1976; European dogfish, *Scyliorhinus canicula*, Hazon and Henderson, 1985; O'Toole *et al.*, 1990), and the incubation of plasma with kidney extracts of European dogfish produces vasopressor substances when assayed in the homologous animal (N. Hazon, personal communication). Furthermore, a substance immunoreactive to the antibody against mammalian ANGII is identified in the plasma, brain, kidney and rectal gland of nurse shark, *Gynclimostoma cirratum*, and the plasma level is altered after the manipulation which is known to change the activity of the renin–angiotensin system (S.M. Galli-Phillips, personal communication). All those results indicate, although they are yet indirect, that the renin–angiotensin system appears to be present in elasmobranchs.

ANGI has been isolated and sequenced in three species of teleost fishes (Fig. 5.4). The common characteristic of fish angiotensin is the replacement of aspartic acid at position 1 with asparagine in addition to various substitutions at position 9 as observed in other vertebrates. This replacement increases the

Man, Pig, Rat, Horse	D R V Y I H P F H L
Ox, Sheep	D R V Y V H P F H L
Fowl, Quail	D R V Y V H P F S L
Snake	X-D R V Y V H P F Y L *
Turtle	D R V Y V H P F H L
Alligator	D R V Y V H P F A L
Bullfrog	D R V Y V H P F N L
Goosefish	N R V Y V H P F H L
Salmon	N R V Y V H P F N L
Eel	N R V Y V H P F G L

Fig. 5.4 Comparison of amino acid sequences of angiotensin I from various vertebrate classes. References: mammals, fowl, bullfrog, and snake, Sakakibara *et al.* (1985); quail, Takei and Hasegawa (1990); turtle, Hasegawa *et al.* (1984); alligator, Takei *et al.*, unpublished data; eel, Hasegawa *et al.* (1983); salmon, Takemoto *et al.* (1983); goosefish, Hayashi *et al.* (1978). *The N-terminal amino acid could be N. X = unidentified molecule. Identical sequences across different species are boxed.

vasopressor potency in the eel (Takei, 1987), although the replacement profoundly decreases the potency in mammals and birds (Sakakibara *et al.*, 1985; Takei and Hasegawa, 1990).

In mammals, ANGII stimulates thirst and sodium appetite, and increases the intestinal absorption of ingested water and Na^+ (Davies *et al.*, 1970; Nishimura, 1987). Physiological concentrations of ANGII decrease glomerular filtration rate and increase tubular reabsorption of water and Na^+ (Nishimura, 1985). All these effects are consistent in view of the conservation of water and Na^+ in the blood. ANGII stimulates the secretion of vasopressin and aldosterone, which indirectly further promotes water and Na^+ conservation. Compared with the ample data in mammals, the data in fishes are still insufficient.

ANGII stimulates drinking in most fishes so far examined, but it was apparently not dipsogenic in elasmobranchs, which scarcely need to conserve water because of their slightly hypertonic plasma (Kobayashi *et al.*, 1983). However, recent studies, which utilize a more sensitive method to measure the volume drunk, indicate that exogenously administered ANGII, as well as procedures which are known to augment endogenous ANGII, result in increased drinking in the European dogfish (Hazon *et al.*, 1989).

In contrast to mammals and birds, in which ANGII acts on the forebrain, in the eel, the site of dipsogenic action of ANGII may be in the medulla oblongata. This idea is based on our data that the removal of forebrain and most of the midbrain of the eel had little effect on drinking induced by peripheral injection of ANGII, and that transection of the vagus and glossopharyngial nerves, which are responsible for swallowing, abolished that drinking (Takei *et al.*, 1979). We also have preliminary data showing that ANGII injected into the cerebral ventricle near the hypothalamus of the eel is dipsogenic at much lower doses than for peripheral injection (Takei, unpublished data). In mammals and birds, ablation of a portion of the forebrain, the subfornical organ, abolished drinking induced by peripheral injection of ANGII (Nishimura, 1987). Aquatic fishes can drink surrounding water simply by the reflex of swallowing, but terrestrial animals have to feel thirsty, search for water, and take water into their mouth before they swallow water. Since a sensation of thirst, which motivates a series of drinking behaviours, is integrated in the forebrain in mammals, it is likely that dipsogenic receptors for ANGII have evolved in the forebrain of terrestrial animals during the evolutionary process toward land.

There seems to be no report on the effect of ANGII on intestinal absorption of water and Na^+ in fishes. Fishes lack the loop of Henle and are unable to concentrate urinary Na^+ above plasma (Nishimura, 1985). Unlike terrestrial mammals and birds, therefore, fish primarily regulate urine volume by glomerular filtration rate (GFR). Thus, systemic injection of pressor doses of ANGII is usually accompanied by glomerular diuresis. However, subpressor doses of ANGII appear to decrease whole-kidney GFR by reducing the proportion of filtering nephrons (glomerular intermittency) in FW rainbow trout and by decreasing the single-nephron GFR in SW trout (Brown *et al.*, 1980). In fact, in the perfused rainbow trout kidney preparation, ANGII causes antidiuresis if the perfusion pressure is kept constant (Rankin *et al.*, 1984). There seem to be no data demonstrating a direct tubular effect of ANGII in fishes.

Fishes usually do not produce aldosterone, and cortisol, secreted from the interrenal, serves as a mineralocorticoid as well as a glucocorticoid (Henderson and Kime, 1987). However, cortisol is not a Na^+-sparing hormone as is aldosterone in tetrapods, but promotes Na^+ extrusion across the gill and is important for SW adaptation, although both steroids stimulate Na^+–K^+-ATPase. Similar to the renin–angiotensin–aldosterone axis in mammals, ANGII stimulates cortisol secretion in the European flounder, *Platichthys flesus* (Perrott and Balment, 1990). It has been shown that ANGII stimulates ACTH release in the goldfish (Weld and Fryer, 1987), which also leads to increased cortisol secretion. ANGII stimulates the release of 1 α-hydroxycorticosterone, which is a major elasmobranch mineralocorticoid (Henderson and Kime, 1987), in the European dogfish *in vivo* (Hazon and Henderson, 1985) and *in vitro* from the perifused interrenal gland (O'Toole *et al.*, 1990).

Collecting all the described biological actions of ANGII, which are consistent in terms of retaining water and excreting Na^+, together with the data that plasma ANGII increases after exposure of FW eels to SW (Kobayashi *et al.*, 1980), it seems that ANGII is an important hormone for SW adaptation (Fig. 5.1).

5.4 ARGININE VASOTOCIN (AVT)

AVT and isotocin are neurohypophysial hormones in teleost fishes. Although several different types of oxytocin-like hormones are present in lungfish and elasmobranchs, all fishes have AVT as the neurohypophysial hormone of the vasopressin lineage (Acher, 1990). Recently, C-terminally elongated AVT peptides named hydrins have been found in addition to AVT in amphibians (Rouille *et al.*, 1989), but they are not yet identified in fishes. As oxytocin appears to stimulate Na^+ excretion in mammals (Brimble *et al.*, 1991), it is possible that isotocin also has some effect on fish osmoregulation. However, only a few studies have examined the effect of isotocin on water and Na^+ management in teleost fishes; the effects appear weak by comparison with that of AVT (Maetz *et al.*, 1964; Chester-Jones *et al.*, 1969).

The complementary DNAs (cDNAs) encoding AVT and isotocin precursor have been cloned and sequenced in the white sucker, *Catostomus commersoni* (Heierhorst *et al.*, 1989) and chum salmon, *Oncorhynchus keta* (Hyodo *et al.*, 1991). The genomic DNA structure was also determined in the sucker (Morley *et al.*, 1990). Two different cDNAs or genes for each of AVT and isotocin were found in these fishes. Even in the same species, microheterogeneity was found in the nucleotide and amino acid sequence of preproAVT, possibly because of different stocks of chum salmon (Heierhorst *et al.*, 1990; Hyodo *et al.*, 1991).

It seems clear that vasopressin is an antidiuretic hormone which acts physiologically on the renal tubules in mammals, but the role of AVT in fish osmoregulation has not yet been established. As summarized in a recent review (Perks, 1987), AVT was first thought to be a diuretic hormone in fishes which may be associated with its vasopressor effect (Henderson and Wales, 1974; Sawyer *et al.*, 1976). However, if subpressor doses of AVT are administered in the eel (Henderson and Wales, 1974; Babiker and Rankin, 1978), or if systemic blood pressure is held constant in the rainbow trout (Rankin *et al.*, 1984; Dunne and Rankin, personal communication) and in the lungfish (Pang, 1983), AVT produces an antidiuresis by afferent arteriolar constriction. Thus, AVT may be an antidiuretic hormone in fishes as in other vertebrates. Since the preglomerular effect is very weak in fishes, a systemic pressure diuresis may dominate the preglomerular antidiuretic effect when high doses are applied. Absence of the tubular effect of AVT and isotocin is reported in the isolated, perfused distal tubules from the rainbow trout (Nishimura and Imai, 1982). It is of interest to note that pressor doses of AVT produce a diuresis in FW eels which are on the edge of overhydration, but a diuretic effect is weak or absent in SW fishes (Maetz and Lahlou, 1974; Babiker and Rankin, 1978). Although AVT has a long history of research, we know relatively little about its physiological function. Detailed analyses as described for ANGII are awaited to define its actions more clearly in fishes. Recently, studies on the second messenger activity in the gill and kidney of the rainbow trout indicate that AVT inhibits adenylate cyclase activity in the membrane fraction of gill cells (Guibbolini and Lahlou, 1987) but stimulates its activity in the isolated nephron (Sainsbury and Balment, 1992). For other osmoregulatory actions, AVT appears to inhibit drinking in FW eels (Hirano and Hasegawa, 1984) and to have no effect on NaCl absorption by the intestine of SW eels (M. Ando, personal communication).

Very recently, Hyodo and Urano (1991) have shown that the level of AVT mRNA was greater in the FW-acclimated rainbow trout than in SW-acclimated fish by using an *in situ* hybridization technique. This result is consistent with the *in vivo* data that AVT concentrations in plasma and pituitary are greater in the FW-acclimated flounder and rainbow trout than in the SW-acclimated fishes (Perrott *et al.*, 1991). These data, together with the weak antidiuretic effect of AVT, may support a possible role for AVT in FW adaptation (Fig. 5.1). However, since plasma AVT levels increase in the eel after transfer from FW to SW (Henderson *et al.*, 1985), and since administration of vasopressin, which could act as a competitive receptor antagonist of AVT, causes diuresis in SW eels (Babiker and Rankin, 1978), AVT is more likely to be a SW-adapting hormone in the eel. We need to accumulate more biological data to define AVT involvement in fish osmoregulation.

5.5 UROTENSINS (UI and UII)

Most fishes, except for cyclostomes and lungfishes, have a caudal neurosecretory system which synthesizes and secretes at least two peptide hormones, UI and UII (Loretz *et al.*, 1981). UI has been sequenced in several species of fishes (Fig. 5.5). The primary structure of carp UI precursor is also inferred from the cDNA sequence (Ishida *et al.*, 1986). UI has sequence homology to mammalian corticotropin-releasing hormone (CRH) and thus may have some effects on adrenocorticotropic hormone (ACTH) release and, subsequently, cortisol release (Woo *et al.*, 1985). Interestingly, homology exists in cDNAs between UI and CRH even in regions which do not encode the mature hormone (Ichikawa *et al.*, 1986), but each peptide is apparently translated from different mRNA in the sucker (Okawara *et al.*, 1988). UII has been sequenced in four species, and some multiple forms are identified in one species (Fig. 5.5). Recently, cDNA which encodes preproUII was sequenced in the carp (Ohsako *et al.*, 1986). Although the amino acid sequence of UII has some homology with somatostatin, there is no similarity between the two preprohormones (Ichikawa *et al.*, 1986).

As described in recent reviews (Ichikawa *et al.*, 1986; Larson and Bern, 1987), the physiological function of urotensins in fish osmoregulation is not yet determined. Initial implication of urotensins in osmoregulation is based on morphological studies which show that the activity of the urophysis is altered after transfer of fish to different salinities, and that the urophysis has a direct blood drainage link to important osmoregulatory organs such as the kidney and urinary bladder. However, urophysectomy, as well as injection of urophysial extract, do not appear to have consistent effects on osmoregulation. Based on these data, Bern (1969) proposed that urotensins may not be involved in major homeostatic adjustment but only in the fine tuning of the osmoregulatory systems in response to slight fluctuations in environmental

Urotensin I (UI)

Flounder	SEDPP	MSIDL	TFHML	RNMIH	MAKME	GEREQ	AQINR	NLLDE	V-NH_4
Sole	SEEPP	MSIDL	TFHML	RNMIH	RAKME	GEREQ	ALINR	NLLDE	V-NH_4
Carp	NDDPP	ISIDL	TFHLL	RNMIE	MARNE	NQREQ	AGLNR	KYLDE	V-NH_4
Sucker	NDDPP	ISIDL	TFHLL	RNMIE	MARIE	NEREQ	AGLNR	KYLDE	V-NH_4
Pig CRH	SEEPP	ISIDL	TFHLL	REVLE	MARAE	QLAQQ	AHSNR	KLMEN	F-NH_4

Urotensin II (UII)

Flounder	AGTTE	CFWKY	CV
Goby	AGTAD	CFWKY	CV
Sucker A	GSGAD	CFWKY	CV
B	GSNTE	CFWKY	CV
Carp α	GGGAD	CFWKY	CV
β	GGNTE	CFWKY	CV*
γ	GGGAD	CFWKY	CI
Somatostatin	AGCKN	FFWKT	FTSC

Fig. 5.5 Comparison of amino acid sequences of urotensins and related peptides from various vertebrate classes. References for UI: flounder, Conlon *et al.* (1990a); sole, McMaster *et al.* (1988); carp, Ichikawa *et al.* (1982); sucker, Lederis *et al.* (1982). References for UII: flounder, Conlon *et al.* (1990b); goby, Pearson *et al.* (1980); sucker, McMaster and Lederis (1983); carp, Ichikawa *et al.* (1984). *Some have S at the second position. Identical sequences across different species are boxed. :.....: = disulphide bridges.

salinity. Since it has recently become possible to use homologous hormone to examine physiological actions, and to measure plasma urotensin levels by homologous RIA, we may soon be able to define more clearly the osmoregulatory role of urotensins in fishes.

Recently, parallel increases in urophysial UI content and plasma cortisol level have been reported in the flounder during adaptation to SW, and synthetic sucker UI or urophysial extract actually increases plasma cortisol level *in vivo* or stimulates cortisol secretion from perifused interrenal gland *in vitro* (Arnold-Reed and Balment, 1989, and personal communication). Furthermore, UI content in the urophysis has been shown to be elevated in SW-adapted flounder (Arnold-Reed *et al.*, 1991a). *In vitro* effects of UI on epithelial transport are generally consistent with a decrease in loss of water and Na^+ from the body (Loretz *et al.*, 1981). Collecting all those results, it seems that UI may contribute to SW adaptation (Fig. 5.1).

UII stimulates active Na^+ absorption through absorptive surfaces such as the urinary bladder and intestine, and inhibits active Cl^- extrusion from secretory surfaces such as the skin and opercular epithelium in fishes, which facilitate FW adaptation (Larson and Bern, 1987). As the urophysial UII content is elevated in SW-adapted flounder (Arnold-Reed *et al.*, 1991a), it needs to be determined whether this is due to the decreased secretion or increased synthesis, to confirm its role in FW adaptation. As expected from the structural similarity to somatostatin, which is known as an endogenous factor to inhibit prolactin and GH release, UII appears to inhibit prolactin secretion, which is inconsistent with the possible involvement of UII in FW adaptation (Fig. 5.1; Loretz *et al.*, 1981). It is not yet determined whether UII inhibits GH release in fishes.

5.6 VASOACTIVE INTESTINAL PEPTIDE (VIP)

VIP was first isolated from porcine duodenal tissue, and histochemical studies using VIP antisera demonstrated VIP-like immunoreactivity in the endocrine cells of the gut (Rosselin *et al.*, 1982). However, the immunoreactive materials in the endocrine cells were later shown to be not VIP itself, but one of the glucagon-like peptides such as secretin which have high structural similarity to VIP (Fig. 5.6). VIP in the intestine is localized in the peripheral nerve terminals, and VIP is in fact a neuropeptide, with VIP-containing neurones widely distributed in a variety of tissues. Plasma VIP levels, measured by specific RIA in humans, range between 0.6 and 30 fmol ml^{-1} (Rosselin *et al.*, 1982). VIP in the circulation is thought to be derived from gut nerve terminals. Recently, VIP was isolated from the intestine of the fowl, cod, *Gadus morhua*, and European dogfish (Fig. 5.6). The sequence is well conserved in spite of the large phylogenetic distance between mammals and elasmobranchs.

In mammals, VIP exhibits a spectrum of biological actions, some of which

<u>Vasoactive intestinal peptide (VIP)</u>

Pig, Ox, Human, Goat, Dog, Rat	HSDAV	FTDNY	TRLRK	QMAVK	KYLNS	ILN-NH_4
Guinea pig	HSDAL	FTDTY	TRLRK	QMAMK	KYLNS	VLN-NH_4
Fowl	HSDAV	FTDNY	SRFRK	QMAVK	KYLNS	VLT-NH_4
Dogfish	HSDAV	FTDNY	SRIRK	QMAVK	KYINS	LLA-NH_4
Cod	HSDAV	FTDNY	SRFRK	QMAAK	KYLNS	VLA-?
Porcine glucagon	HSQGT	FTSDY	SKYLD	SRRAQ	DFVQW	LMNT
secretin	HSDGT	FTSEL	SRLRD	SARLQ	RLLQG	LV-NH_4

Fig. 5.6 Comparison of amino acid sequences of vasoactive intestinal peptide and related peptides from various vertebrate classes. References: mammals, fowl and cod, Thwaites *et al.* (1989); dogfish, Dimaline *et al.* (1987). Identical sequences across different species are boxed.

are related to osmoregulation. These include inhibition of water and Na^+ absorption and stimulation of Cl^- secretion by the intestine, stimulation of prolactin release from the pituitary, stimulation of renin release by the juxtaglomerular cells, and stimulation of corticosteroid release from the adrenal (Rosselin *et al.*, 1982).

In the spiny dogfish, measured plasma VIP concentrations are in the range of 47–210 fmol ml^{-1}, which is much higher than in mammals, and mammalian VIP increases both water and Cl^- secretion from the isolated, perfused rectal gland (Stoff *et al.*, 1979). The VIP action is mediated by cyclic AMP, and abolished by somatostatin. More recently, it was shown that rat ANP also stimulates rectal gland secretion of Cl^- in the spiny dogfish, and that the ANP effect is mediated by the release of VIP from nerve terminals in the rectal gland (Silva *et al.*, 1987). On the other hand, mammalian VIP is without effect on the rectal gland of European dogfish, and Shuttleworth and Thorndyke (1984) could isolate a peptide from the intestine of European dogfish which stimulated rectal gland secretion via cyclic AMP and which they named 'rectin'. As we now know the primary structure of native VIP of European dogfish, it should be determined whether rectin and VIP are identical or not.

In teleost fish, a similar stimulatory effect of VIP on Cl^- secretion is obtained in the opercular epithelium of SW-adapted tilapia, *Sarotherodon mossambicus* (Foskett *et al.*, 1982). This effect of VIP may also be mediated by cyclic AMP. VIP stimulates electrogenic Na^+ reabsorption by the *Gillichythys* urinary bladder (Loretz *et al.*, 1981). However, VIP inhibits water and Na^+ transport by the anterior intestine of both FW- and SW-adapted tilapia (Mainoya and Bern, 1984). In fact, secretion of water and Na^+ occurs in the SW fish. More recently, it was shown that the stimulatory effect of VIP on Cl^- secretion by the intestine of winter flounder, *Pseudopleuronectes americanus*, is mediated by cyclic AMP (O'Grady and Wolters, 1990). Judging from the similarities to the effects of UI (Loretz *et al.*, 1981), VIP may contribute to SW adaptation (Fig. 5.1), though some biological actions are apparently irrelevant.

5.7 NATRIURETIC PEPTIDES

Natriuretic peptides constitute a family of structurally related peptides (Fig. 5.7) which was only discovered in the last decade. It is likely that they have evolved from a common ancestor molecule by gene duplication analogous to that proposed for a family of gut–brain peptides (Adelson, 1971). The first peptide found in this family was ANP, followed by brain natriuretic peptide (BNP) and C-type natriuretic peptide (CNP) (Fig. 5.7). However, it was later shown that BNP is not a 'brain' peptide but a cardiac peptide circulating in blood, while CNP appears to be an intrinsic brain peptide (Balment and Takei, 1992). Thus we now refer to ANP as A-type natriuretic peptide and BNP as B-type natriuretic peptide. Very recently, a possible fourth type of natriuretic

A-type natriuretic peptide

```
Eel              SKSSSPCFGGKLDRIGSYSGLGCNS-RK
Bullfrog            SSDCFGSRIDRIGAQSGMGC-G-RRF
Man            SLRRSS-CFGGRMDRIGAQSGLGCNSFR-Y
Rat            SLRRSS-CFGGRIDRIGAQSGLGCNSFR-Y
```

B-type natriuretic peptide

```
Fowl           MMRDSG-CFGRRIDRIGSLSGMGCNGSRKN
Pig        SPKTM-RDSG-CFGRRLDRIGSLSGLGCNVLRRY
Dog        SPK-MMHKSG-CFGRRLDRIGSLSGLGCNVLRKY
Cattle      PK-MMRDSG-CFGRRLDRIGSLSGLGCNVLRRY
Man        SPK-MVQGSG-CFGRKMDRISSSSGLGCKVLRRH
Rat        S-K-MAHSSS-CFGQKIDRIGAVSRLGCDGLRLF
```

C-type natriuretic peptide

```
Dogfish        GPSRG-CFGVKLDRIGAMSGLGC
Eel            GWNRG-CFGLKLDRIGSLSGLGC
Killifish      GWNRG-CFGLKLDRIGSMSGLGC
Bullfrog       GYSRG-CFGVKLDRIGAFSGLGC
Fowl           GLSRS-CFGVKLDRIGSMSGLGC
Man            GLSKG-CFGLKLDRIGSMSGLGC
```

Ventricular natriuretic peptide

```
Eel            KSFNS-CFGTRMDRIGSWSGLGCNSL-KNGTKKKIFGN
```

Fig. 5.7 Comparison of amino acid sequences of natriuretic peptides from various vertebrate classes. For references, see text and Suzuki *et al.* (1991). Gaps are introduced to optimize matching. Identical sequences across different species are boxed.

peptide was isolated from the eel cardiac ventricle and named ventricular natriuretic peptide (VNP) (Takei *et al.*, 1991). Judging from the amino acid sequence and the nucleotide sequence of VNP, it is more homologous to mammalian ANP than to mammalian BNP (Fig. 5.7). In addition to VNP, ANP was isolated from the eel atrium (Takei *et al.*, 1989) and CNP was isolated from the killifish, *Fundulus heteroclitus*, and eel brain (Price *et al.*, 1990; Takei *et al.*, 1990a). More recently, we isolated CNP from the heart of the European dogfish (Suzuki *et al.*, 1991). This is the first report for the presence of a CNP in tissues other than the brain. Comparing the structural homology of each group of peptides amongst the vertebrates, it is clear that the structure is most highly conserved for CNP, and the variation is greatest for BNP (Fig. 5.7). It is rather surprising that a homology of greater than 80% was noted between mammalian and dogfish CNP in spite of the great phylogenetic distance.

The physiological stimulus for ANP release in mammals is atrial distension, which is consistent with the historical observation that the atrium acts as a stretch receptor for regulation of blood volume (Brenner *et al.*, 1990). However, the mechanisms of secretion do not seem to be so simple in fishes. In teleost fish, plasma levels of ANP, measured by RIA for mammalian ANP, decrease after sculpin, *Myoxocephalus octodecimspinosus*, and winter flounder are transferred to dilute media (Evans *et al.*, 1989) and increase after smolts of Atlantic salmon, *Salmo salar*, are transferred from FW to SW (Smith *et al.*, 1991). Similar results are obtained in *Gila atraria* (Westenfelder *et al.*, 1988) and rainbow trout (Smith *et al.*, 1991); plasma ANP levels are higher in fishes in hyperosmotic media than in those in hyposmotic media. Since the salmon smolts in the above experiments may have suffered from hypovolaemia and hypernatramia after transfer to SW, plasma ANP level should have decreased if atrial stretch is also the principal stimulus for ANP secretion in fishes. It is possible that ANP secretion in fishes may be regulated by some factors other than blood volume. However, measurements of fish ANP using RIA for mammalian ANP may give misleading results, since the RIA for human ANP, which we utilized in our initial study (Takei *et al.*, 1990b), does not measure eel ANP but eel VNP. Indeed, we have preliminary results showing that plasma ANP, CNP, and VNP levels in FW-adapted eels are consistently higher than in SW-adapted fish, when these hormone levels are measured by homologous RIA specific for each peptide (Table 5.1).

In mammals, ANP appears to be a hormone whose function is to remove excess water and Na^+ from the body, thereby decreasing blood volume (Brenner *et al.*, 1990). Briefly, ANP inhibits oral intake of water and Na^+, and increases the renal excretion of water and Na^+. Both glomerular and tubular (inner medullary collecting duct) actions are noted in the kidney, and cyclic GMP may act as a second messenger for the ANP effects (Cogan, 1990; Zeidel, 1990). ANP also inhibits the renin–angiotensin–aldosterone system to indirectly facilitate Na^+ extrusion (Brenner *et al.*, 1990). The effects of ANP on intestinal water and Na^+ absorption are variable, but recent *in vivo* data show that ANP inhibits the absorption across the jejunum but not across the ileum in the dog (Matsushita *et al.*, 1991). The ANP-specific receptors are localized in those tissues where biological actions of ANP are suggested (Bianchi *et al.*, 1985).

The biological actions of ANP have recently been reviewed in fishes (Evans, 1990) and in non-mammals (Balment and Takei, 1992). Unlike mammals, there has been no report thus far on the effect of ANP on drinking in fishes, but ANP apparently decreases water and Na^+ absorption by the teleost intestine. O'Grady (1989) showed that mammalian ANP inhibits the activity of Na-K-2Cl co-transporter in the intestine of winter flounder. We also showed

Table 5.1 Plasma concentration (fmol ml^{-1}) of natriuretic peptides in eels measured by homologous radioimmunoassay specific to each peptide. Values are means ± SEM; *, $P < 0.05$ (Student *t*-test), compared with the value of FW eel, *Anguilla japonica*

	n	ANP	CNP	VNP
FW eel	8	247 ± 66	468 ± 90	754 ± 49
SW eel	7	123 ± 21	28 ± 10*	106 ± 11*

that eel ANP inhibits water and Na^+ absorption by the intestine of SW-adapted eel (Ando *et al.*, (1992). The native eel ANP is 100-fold more potent than mammalian ANP. Cyclic GMP seems to mediate this ANP effect.

ANP appears to stimulate renal excretion of water and Na^+ in teleost fishes. In rainbow trout in FW and in the toadfish, *Opsanus tau*, in SW, mammalian ANP or heart extract causes diuresis and natriuresis (Duff and Olson, 1986; Lee and Malvin, 1987). Since the latter species is aglomerular, this is the first direct evidence for the possible tubular effect of ANP in vertebrates. The diuretic and natriuretic effect of human ANP are also observed in the perfused trout kidney preparation in which perfusion pressure is held constant, and ANP reverses the antidiuretic and antinatriuretic effect of ANGII (Dunne and Rankin, 1992). However, the ANP effect on the kidney is not so simple in elasmobranchs whose plasma osmolality is slightly higher than that of SW. In the spiny dogfish transferred to 70% SW, which may suffer from hypervolaemia, ANP increases Na^+ excretion from the kidney (Solomon *et al.*, 1989). In the more recent study using the unanaesthetized dogfish, however, mammalian ANP was shown to be antidiuretic and antinatriuretic by decreasing GFR (Benyajati and Yokota, 1990). Such an antidiuretic effect of ANP does not appear to be consistent with the increased secretion of immunoreactive ANP after volume expansion reported in the spiny dogfish (Epstein *et al.*, 1987). Since CNP is found in large amounts in the heart and plasma of two species of dogfish: *Scyliorhinus canicula* and *Triakis scyllia* (Suzuki *et al.*, 1991; Suzuki *et al.*, 1992), it should be determined whether or not CNP mimics the action of ANP as observed in mammals. There has been no published evidence for the renal action of ANP in cyclostomes, although ANP binding sites are localized in the kidney of Atlantic hagfish, *Myxine glutinosa* (Kloas *et al.*, 1988).

In addition to the kidney, fishes have specialized organs or cells for NaCl excretion, gill or opercular chloride cells in teleosts and the rectal gland in elasmobranchs. Scheide and Zadunaisky (1988) showed that rat ANP added to the serosal side of the opercular epithelium of SW killifish increases Cl^- secretion by chloride cells. In three species of SW flatfish, human ANP injected intravenously causes transient increases in Na^+ extrusion into environmental media for 20 min (Arnold-Reed *et al.*, 1991b). There is a preliminary report showing that human ANP added to gill cells isolated from SW-adapted rainbow trout stimulates cGMP production (Balment and Lahlou, 1987). Homologous heart extract or rat ANP also stimulates Cl^- excretion from the rectal gland of the spiny dogfish both *in vivo* and *in vitro* (Solomon *et al.*, 1985). As mentioned earlier, the ANP effect could be mediated by the local release of VIP (Silva *et al.*, 1987), but the direct action is also probable because the stimulatory effect is also evident in the cultured rectal gland cells (Karnaky *et al.*, 1991). Clearly, the stimulation by ANP of NaCl extrusion in these fishes would facilitate fish adaptation to hypertonic media.

In contrast to the inhibitory action of ANP on aldosterone and cortisol secretion in mammals (Naruse *et al.*, 1987), ANP stimulates cortisol secretion *in vivo* in the flounder and from the perifused interrenal tissue of the rainbow trout *in vitro* (Arnold-Reed and Balment, 1991). Interestingly, this stimulatory effect is observed in the interrenal tissue from SW-adapted trout but not from FW trout. ANP also stimulates the secretion of 1-α hydroxycorticosterone from the interrenal tissue of the European dogfish (Hazon *et al.*, 1987).

The majority of biological actions of mammalian ANP and stimuli for secretion as indicated by RIA for mammalian ANP are consistent with ANP acting as a hormone for SW adaptation, though the reported diuretic action and inhibitory action of intestinal water absorption appear inappropriate (Fig. 5.1). However, it is not known whether responses to mammalian ANP reliably represent the action of fish ANP in fishes. At least three peptides with high structural homology–ANP, VNP, and CNP–are present in eel blood, and it is not yet determined whether each has the same effect. It is possible that ANP and VNP might have opposite effects, as do GH and prolactin in fish osmoregulation. Furthermore, we do not know which of the three (or more) peptides were detected in the previously employed RIAs for mammalian ANP. Taking the great variety of osmoregulatory systems within the piscine groups into account, it seems necessary to examine the role of ANP in more than two species of fishes using homologous hormones and homologous RIA systems. We are, therefore, currently far from understanding the roles for ANP, VNP and CNP in fishes.

ACKNOWLEDGEMENTS

The author is indebted to Professor Tetsuya Hirano of the Ocean Research Institute, University of Tokyo, and to Dr Richard J. Balment of the Department of Physiological Sciences, University of Manchester, for their valuable comments and critical reading of the manuscript.

REFERENCES

Acher, R. (1990) Structure evolution and processing adaptation of neurohypophysial hormone–neurophysin precursors, in *Progress in Comparative Endocrinology* (eds A. Epple, C.G. Scanes and M.H. Stetson), Wiley-Liss, New York, pp. 1–9.

Adelson, J.W. (1971) Enterosecretory proteins. *Nature, Lond.*, **229**, 321–5.

Ando, M., Kondo, K. and Takei, Y. (1992) Effects of eel atrial natriuretic peptide on NaCl and water transport across the intestine of the seawater eel. *J. comp. Physiol.*, **162B**, 436–439.

Arnold-Reed, D.E. and Balment, R.J. (1989) Steroidogenic role of the caudal neurosecretory system in the flounder, *Platichthys flesus*. *Gen. comp. Endocrinol.*, **76**, 267–73.

Arnold-Reed, D.E. and Balment, R.J. (1991) Atrial natriuretic factor stimulates in-vivo and in-vitro secretion of cortisol in teleosts. *J. Endocr.*, **128**, R17–20.

Arnold-Reed, D.E., Balment, R.J., McCrohan, C.R. and Hackney, C.M. (1991a) The caudal neurosecretory system of *Platichthys flesus*: general morphology and responses to altered salinity. *Comp. Biochem. Physiol.*, **99A**, 137–43.

Arnold-Reed, D.E., Hazon, N. and Balment, R.J. (1991b) Biological actions of atrial natriuretic factor in flatfish. *Fish Physiol. Biochem.*, **9**, 271–7.

Babiker, M.M. and Rankin, J.C. (1978) Neurohypophysial hormonal control of kidney function in the European eel (*Anguilla anguilla* L.) adapted to sea-water or fresh water. *J. Endocr.*, **76**, 347–58.

Balment, R.J. and Lahlou, B. (1987) Atrial natriuretic peptide (ANP) modulation of second messenger activity in isolated trout gill cells. *J. Endocr.*, **115** (Supp.), 164.

Balment, R.J. and Takei, Y. (1992) Natriuretic factors in non-mammalian vertebrates, in *New Insights in Vertebrate Kidney Function* (eds J.A. Brown, R.J. Balment and J.C. Rankin), Cambridge University Press, Cambridge, pp. 351–85. (in press).

Balment, R.J., Hazon, N. and Perrott, M.N. (1987) Control of corticosteroid secretion and its relation to osmoregulation in lower vertebrates, in *Comparative Physiology of Environmental Adaptations*, Vol. 1 (eds R. Kirsch and B. Lahlou), Karger, Basel, pp. 92–102.

Benyajati, S. and Yokota, S.D. (1990) Renal effects of atrial natriuretic peptide in a marine elasmobranch. *Am. J. Physiol.*, **258**, R1201–6.

Bern, H.A. (1969) Urophysis and caudal neurosecretory system, in *Fish Physiology*, Vol. II (eds W.S. Hoar and D.J. Randall), Academic Press, New York, pp. 399–418.

Bern, H.A., McCormick, S.D., Kelley, K.M., Gray, E.S., Nishioka, R.S., Madsen, S.S. and Tsai, P.I. (1991) Insulin-like growth factors "under water": role in growth and function of fish and other poikilothermic vertebrates, in *Modern Concepts of Insulin-like Growth Factors* (ed. E.M. Spencer), Elsevier, New York, pp. 85–96.

Bianchi, C., Gutkowska, J., Thibault, G., Garcia, R., Genest, J. and Cantin, M. (1985) Radioautographic localization of ^{125}I-atrial natriuretic factor (ANF) in rat tissues. *Histochemistry*, **82**, 441–52.

Brenner, B.M., Ballermann, B.J., Gunning, M.E. and Zeidel, M.L. (1990) Diverse biological actions of atrial natriuretic peptide. *Physiol. Rev.*, **70**, 665–99.

Brimble, M.J., Balment, R.J., Smith, C.P., Windle, R.J. and Forsling, M.L. (1991) Influence of oxytocin on sodium excretion in the anaesthetized Brattleboro rat. *J. Endocr.*, **129**, 49–54.

Brown, P.S. and Brown, S.C. (1987) Osmoregulatory actions of prolactin and other adenohypophysial hormones, in *Vertebrate Endocrinology: Fundamentals and Biomedical Implications*, Vol. 2 (eds P.K.T. Pang and M.P. Schreibman), Academic Press, San Diego, pp. 45–84.

Brown, J.A., Oliver, J.A., Henderson, I.W. and Jackson, B.A. (1980) Angiotensin and single nephron glomerular function in the trout *Salmo gairdneri*. *Am. J. Physiol.*, **239**, R509–14.

Chester-Jones, I., Chan, D.K.O. and Rankin, J.C. (1969) Renal function of European eel (*Anguilla anguilla* L.): effects of the caudal neurosecretory system, corpuscles of Stanius, neurohypophysial peptides and vasoactive substances. *J. Endocr.*, **43**, 21–31.

Cogan, M.G. (1990) Renal effects of atrial natriuretic factor. *A. Rev. Physiol.*, **52**, 699–708.

Conlon, J.M., Arnold-Reed, D.E. and Balment, R.J. (1990a) Urotensin I and its N-terminal franking peptide from the flounder, *Platichthys flesus*. *Peptides*, **11**, 891–5.

Conlon, J.M., Arnold-Reed, D.E. and Balment, R.J. (1990b) Post-translational processing of prepro-urotensin II. *FEBS Lett.*, **266**, 37–40.

Davies, N.T., Munday, K.A. and Parsons, B.J. (1970) The effect of angiotensin on rat intestinal fluid transfer. *J. Endocr.*, **48**, 39–46.

Dimaline, R., Young, J., Thwaites, D.T., Lee, C.M., Shuttleworth, T.J. and Thorndyke, M.C. (1987) A novel vasoactive intestinal peptide (VIP) from elasmobranch intestine has full affinity for mammalian pancreatic VIP receptors. *Biochim. biophys. Acta*, **930**, 97–100.

Duan, C. and Hirano, T. (1991) Plasma kinetics of growth hormone in the Japanese eel, *Anguilla japonica*. *Aquaculture*, **95**, 179–88.

Duff, D.W. and Olson, K.R. (1986) Trout vascular and renal responses to atrial natriuretic factor and heart extract. *Am. J. Physiol.*, **251**, R639–42.

Dunne, J.B. and Rankin, J.C. (1992) Effects of atrial natriuretic peptide and angiotensin II on salt and water excretion by the perfused rainbow trout kidney. *J. Physiol., Lond.*, **446**, 92P.

Epstein, F.H., Clark, B., Taylor, M., Silva, P., Dick, L., Silva, M. and Solomon, R. (1987) Atrial natriuretic peptide (ANP) assayed in plasma of *Squalus acanthias*. *Bull. Mt Desert Isl. biol. Lab.*, **27**, 72–3.

Evans. D.E. (1990) An emerging role for a cardiac peptide hormone in fish osmoregulation. *A. Rev. Physiol.*, **52**, 43–60.

Evans, D.H., Chipouras, E. and Payne, J.A. (1989) Immunoreactive atriopeptin in plasma of fishes: its potential role in gill hemodynamics. *Am. J. Physiol.*, **257**, R939–45.

Foskett, J.K., Hubbard, G.M., Machen, T.E. and Bern, H.A. (1982) Effects of epinephrine, glucagon and vasoactive intestinal polypeptide on chloride secretion by teleost opercular membrane. *J. comp. Physiol.*, **146**, 27–34.

Guibbolini, M.E. and Lahlou, B. (1987) Neurohypophysial peptide inhibition of adenylate cyclase activity in fish gills. *FEBS Lett.*, **220**, 98–102.

Hasegawa, S., Hirano, T., Ogasawara, T., Iwata, M., Akiyama, T. and Arai, S. (1987) Osmoregulatory ability of chum salmon, *Oncorhynchus keta*, reared in fresh water for prolonged period. *Fish Physiol. Biochem.*, **4**, 101–10.

Hasegawa, Y., Nakajima, T. and Sokabe, H. (1983) Chemical structure of angiotensin formed with kidney renin in the Japanese eel *Anguilla japonica*. *Biomed. Res.*, **4**, 417–20.

Hasegawa, Y., Cipolle, M., Watanabe, T.X., Nakajima, T., Sokabe, H. and Zehr, J.E. (1984) Chemical structure of angiotensin in the turtle, *Pseudemys scripta*. *Gen. comp. Endocrinol.*, **53**, 159–62.

Hayashi, T., Nakayama, T., Nakajima, T. and Sokabe, H. (1978) Comparative studies of angiotensins. V. Structure of angiotensin formed by the kidney of Japanese goosefish and its identification by Dansyl method. *Chem. pharmacol. Bull.*, **26**, 215–19.

Hazon, N. and Henderson, I.W. (1985) Factors affecting secretory dynamics of 1α-hydroxycorticosterone in the dogfish *Scyliorhinus canicula*. *Gen. comp. Endocrinol.*, **59**, 50–5.

Hazon, N., Decourt, C., O'Toole, L.B., Lahlou, B. and Henderson, I.W. (1987) Vascular and steroidogenic effects of ANF and angiotensin II in elasmobranch fish. *J. Endocr.*, **115**, 161.

Hazon, N., Balment, R.J., Perrott, M. and O'Toole, L.B. (1989) The renin–angiotensin system and vascular and dipsogenic regulation in elasmobranchs. *Gen. comp. Endocrinol.*, **74**, 230–6.

Heierhorst, J., Morley, S.D., Figueroa, J., Krentler, C., Lederis, K. and Richter, D. (1989) Vasotocin and isotocin precursors from the white sucker, *Catostomus commersoni*: cloning and sequence analysis of the cDNAs. *Proc. natn. Acad. Sci. U.S.A.*, **86**, 5242–6.

Heierhost, J., Mahlmann, S., Morley, S.D., Coe, I.R., Sherwood, N.M. and Richter, D. (1990) Molecular cloning of two distinct vasotocin precursor cDNA from chum salmon (*Oncorhynchus keta*) suggest an ancient gene duplication. *FEBS Lett.*, **260**, 301–4.

Henderson, I.W. and Kime, D.E. (1987) The adrenal cortical steroids, in *Vertebrate Endocrinology: Fundamentals and Biomedical Implications*, Vol. 2 (eds P.K.T. Pang and M.P. Schreibman), Academic Press, San Diego, pp. 121–42.

Henderson, I.W. and Wales, N.A.M. (1974) Renal diuresis and antidiuresis after injections of arginine vasotocin in the freshwater eel (*Anguilla anguilla* L.). *J. Endocr.*, **61**, 487–500.

Henderson, I.W., Hazon, N. and Hughes, K. (1985) Hormones, ionic regulation and kidney function in fishes. *Symp. Soc. exp. Biol.*, **39**, 245–65.

Hirano, T. and Hasegawa, S. (1984) Effects of angiotensins and other vasoactive substances on drinking in the eel, *Anguilla japonica. Zool. Sci.*, **1**, 106–13.

Hirano, T., Ogasawara, T., Bolton, J.P., Collie, N.L., Hasegawa, S. and Iwata, M. (1987) Osmoregulatory role of prolactin in lower vertebrates, in *Comparative Physiology of Environmental Adaptation*, Vol. 1 (eds R. Kirsch and B. Lahlou), Karger, Basel, pp. 112–24.

Hyodo, S. and Urano A. (1991) Changes in expression of provasotocin and proisotocin genes during adaptation to hyper- and hypo-osmotic environments in rainbow trout. *J. comp. Physiol.*, **161B**, 549–56.

Hyodo, S., Kato, Y., Ono, M. and Urano, A. (1991) Cloning and sequence analyses of cDNAs encoding vasotocin and isotocin precursors of chum salmon, *Oncorhynchus keta*: evolutionary relationships of neurohypophysial hormone precursors. *J. comp. Physiol.*, **160B**, 601–8.

Ichikawa, T., Lederis, K. and Kobayashi, H. (1984) Primary structures of multiple forms of urotensin II in the urophysis of the carp, *Cyprinus carpio. Gen. comp. Endocrinol.*, **55**, 133–41.

Ichikawa, T., McMaster, D. and Lederis, K. (1982) Isolation and amino acid sequence of urotensin I, a vasoactive and ACTH-releasing neuropeptide from the carp (*Cyprinus carpio*) urophysis. *Peptides*, **3**, 859–67.

Ichikawa, T., Pearson, D., Yamada, C. and Kobayashi, H. (1986) The caudal neurosecretory system of fishes. *Zool. Sci.*, **3**, 585–98.

Ishida, I., Ichikawa, T. and Deguchi, T. (1986) Cloning and sequence analysis of cDNA encoding urotensin I precursor. *Proc. natn. Acad. Sci. U.S.A.*, **83**, 308–12.

Karnaky, K.J., jr., Valentich, J.D., Currie, M.G., Oehlenschlager, W.F. and Kennedy, M.P. (1991) Atriopeptin stimulates chloride secretion in cultured shark rectal gland cells. *Am. J. Physiol.*, **260**, C1125–30.

Kawauchi, H. and Yasuda, A. (1989) Evolutionary aspects of growth hormones from nonmammalian species, in *Advances in Growth Hormone and Growth Factor Research* (eds E.E. Müller, D. Cocchi and V. Locatelli), Pythagora Press, Rome/Springer-Verlag, Berlin, pp. 51–68.

Kishida, M. and Hirano, T. (1988) Development of radioimmunoassay for eel growth hormone. *Nippon Suisan Gakkaishi*, **54**, 1321–7.

Kobayashi, H., Uemura, H. and Takei, Y. (1980) Physiological role of the renin–angiotensin system during dehydration, in *Avian Endocrinology* (eds A. Epple and M.H. Stetson), Academic Press, New York, pp. 319–30.

Kobayashi, H., Uemura, H., Takei, Y., Itatsu, N., Ozawa, M. and Ichinohe, K. (1983) Drinking induced by angiotensin II in fishes. *Gen. comp. Endocrinol.*, **49**, 295–306.

Kobayashi, T., Yasuda, A., Yamaguchi, K., Kawauchi, H. and Kikuyama, S. (1991) The complete amino acid sequence of growth hormone of the bullfrog (*Rana catesbeina*). *Biochim. Biophys. Acta*, **1078**, 383–7.

Kloas, W., Flügge, G., Fuchs, E. and Stolte, H. (1988) Binding sites for atrial natriuretic peptide in the kidney and aorta of the hagfish (*Myxine glutinosa*). *Comp. Biochem. Physiol.*, **91A**, 685–8.

Larson, B.A. and Bern, H.A. (1987) The urophysis and osmoregulation, in *Vertebrate Endocrinology: Fundamentals and Biomedical Implications*, Vol. 2 (eds P.K.T. Pang and M.P. Schreibman), Academic Press, San Diego, pp. 143–56.

Lederis, K., Letter, A., McMaster, D., Moore, G. and Schlesinger, D. (1982) Complete amino acid sequence of urotensin I, a hypotensive and corticotropin-releasing neuropeptide from *Catostomus. Science*, **218**, 162–4.

Lee, J. and Malvin, R.L. (1987) Natriuretic response to homologous heart extract in aglomerular toadfish. *Am. J. Physiol.*, **252**, R1055–8.

Loretz, C.A., Bern, H.A., Foskett, J.K. and Mainoya, J.R. (1981) The caudal neurosecretory system and osmoregulation in fish, in *Neurosecretion: Molecules, Cells, Systems* (eds D.S. Farner and K. Lederis), Plenum, New York, pp. 319–28.

McCormick, S.D., Sakamoto, T., Hasegawa, S. and Hirano, T. (1991) Osmoregulatory actions of insulin-like growth factor-I in rainbow trout (*Oncorhynchus mykiss*). *J. Endocr.*, **130**, 87–92.

McMaster, D. and Lederis, K. (1983) Isolation and amino acid sequence of two urotensin II peptides from *Catostomus commersoni* urophysis. *Peptides*, **4**, 367–73.

McMaster, D., Rivier, J. and Lederis, K. (1988) Isolation, amino acid sequence and synthesis of urotensin I from *Hippoglossoides elassodon*, in *Peptide Chemistry 1987* (eds T. Shiba and S. Sakakibara), Protein Research Foundation, Osaka, pp. 145–8.

Maetz, J. and Lahlou, B. (1974) Actions of neurophysial hormones in fishes, in *Handbook of Physiology*, Sec. 7, Vol. IV, Part 1 (eds E. Knobil and W.H. Sawyer), Am. Physiol. Soc., Washington, DC, pp. 521–44.

Maetz, J., Bourguet, J., Lahlou, B. and Hourdry, J. (1964) Peptides neurophypophysaires et osmoregulation chez *Carassius auratus*. *Gen. Comp. Endocrinol.*, **4**, 508–22.

Mainoya, J.R. and Bern, H.A. (1984) Influence of vasoactive intestinal peptide and urotensin II on the absorption of water and NaCl by the anterior intestine of the tilapia, *Sarotherodon mossambicus*. *Zool. Sci.*, **1**, 100–5.

Matsushita, K., Nishida, Y., Hosomi, H. and Tanaka, S. (1991) Effects of atrial natriuretic peptide on water and NaCl absorption across the intestine. *Am. J. Physiol.*, **260**, R6–12.

Morley, S.D., Schonrock, C., Heierhorst, J., Figueroa, J., Lederis, K. and Richter, D. (1990) Vasotocin genes of the teleost fish *Catostomus commersoni*: gene structure, exon–intron boundary, and hormone precursor organization. *Biochemistry*, **29**, 2506–11.

Naruse, M., Obana, K., Naruse, K., Yamaguchi, H., Demura, H., Inagami, T. and Shizume, K. (1987) Atrial natriuretic polypeptide inhibits cortisol secretion as well as aldosterone secretion *in vitro* from human adrenal tissue. *J. clin. Endocr. Metab.*, **64**, 10–16.

Nicoll, C.S. (1981) Role of prolactin in water and ion balance in vertebrates, in *Prolactin* (ed. R.B. Jaffe), Elsevier, New York, pp. 127–66.

Nishimura, H. (1985) Endocrine control of renal handling of solutes and water in vertebrates. *Renal Physiol.*, **8**, 279–300.

Nishimura, H. (1987) Role of the renin–angiotensin system in osmoregulation, in *Vertebrate Endocrinology: Fundamentals and Biomedical Implications*, Vol. 2 (eds P.K.T. Pang and M.P. Schreibman), Academic Press, San Diego, pp. 157–87.

Nishimura, H. and Imai, T. (1982) Control of renal function in freshwater and marine teleosts. *Fedn Proc. Fedn Am. Socs exp. Biol.*, **41**, 2355–60.

Nishioka, R.S., Richman, N.H. III., Young, G., Prunet, P. and Bern, H.A. (1987) Hypophysectomy of coho salmon (*Oncorhynchus kisutch*) and survival in fresh water and seawater. *Aquaculture*, **45**, 167–76.

Ogasawara, T., Hirano, T., Akiyama, T., Arai, S. and Tagawa, M. (1989) Changes in plasma prolactin and growth hormone concentration during freshwater adaptation of juvenile chum salmon (*Oncorhynchus keta*) reared in seawater for a prolonged period. *Fish Physiol. Biochem.*, **7**, 309–13.

O'Grady, S.M. (1989) Cyclic nucleotide-mediated effects of ANF and VIP on flounder intestinal ion transport. *Am. J. Physiol.*, **256**, C142–6.

O'Grady, S.M. and Wolters, P.J. (1990) Evidence for chloride secretion in the intestine of the winter flounder. *Am. J. Physiol.*, **258**, C243–7.

Ohsako, S., Ishida, I., Ichikawa, T. and Deguchi, T. (1986) Cloning and sequence analysis of cDNAs encoding precursors of urotensin II-α and -γ. *J. Neurosci.*, **6**, 2730–35.

Okawara, Y., Morley, S.D., Burzio, L.O., Zwiers, H., Lederis, K. and Richter, D. (1988) Cloning and sequence analysis of cDNA for corticotropin-releasing factor precursor from the teleost fish *Catostomus commersoni*. *Proc. natn. Acad. Sci. U.S.A.*, **85**, 8439–43.

Opdyke, D.F. and Holcombe, R. (1976) Response to angiotensin I and II and to AI-converting enzyme inhibitor in a shark. *Am. J. Physiol.*, **231**, 1750–3.

O'Toole, L.B., Armour, K.J., Decourt, C., Hazon, N., Lahlou, B. and Henderson, I.W. (1990) Secretory patterns of 1α-hydroxycorticosterone in the isolated perifused interrenal gland of dogfish *Scyliorhinus canicula*. *J. molec. Endocrinol.*, **5**, 55–60.

Pang, P.K.T. (1983) Evolution of control of epithelial transport in vertebrates. *J. exp. Biol.*, **106**, 283–99.

Pearson, D., Shively, J.E., Clark, B.R., Geschwind, I.I., Barkley, M., Nishioka, R.S. and Bern, H.A. (1980) Somatostatin-like peptide in the caudal neurosecretory system of fishes. *Proc. natn. Acad. Sci. U.S.A.*, **77**, 5021–4.

Perks, A.M. (1987) The neurohypophysis: the place of comparative studies in biology and medicine, in *Vertebrate Endocrinology: Fundamentals and Biomedical Implications*, Vol. 2 (eds P.K.T. Pang and M.P. Schreibman), Academic Press, San Diego, pp. 9–44.

Perrott, M.N. and Balment, R.J. (1990) The renin–angiotensin system and the regulation of plasma cortisol in the flounder, *Platichthys flesus*. *Gen. comp. Endocrinol.*, **78**, 414–20.

Perrott, M.N., Carrick, S. and Balment, R.J. (1991) Pituitary and plasma arginine vasotocin levels in teleost fish. *Gen. comp. Endocrinol.*, **83**, 68–74.

Price, D.A., Doble, K.E., Lee, T.D., Galli, S.M., Dunn, B.M., Parten, B. and Evans, D.H. (1990) The sequencing, synthesis, and biological actions of an ANP-like peptide isolated from the brain of the killifish *Fundulus heteroclitus*. *Biol. Bull. mar. biol. Lab., Woods Hole*, **178**, 279–85.

Rand-Weaver, M., Noso, T. and Kawauchi, H. (1991) The complete amino acid sequences of two variants of growth hormone from Atlantic cod (*Gadus morhua*). *Gen. comp. Endocrinol.*, **81**, 39–50.

Rankin, J.C. and Davenport, J.A. (1981) *Animal Osmoregulation*, Blackie, Glasgow, 202 pp.

Rankin, J.C., Wahlqvist, I. and Wallace, B. (1984) Antidiuretic actions of angiotensin II, catecholamines, and neurohypophysial hormones in the *in situ* perfused rainbow trout kidney. *Gen. comp. Endocrinol.*, **53**, 442.

Rosselin, G., Maletti, M., Besson, J. and Rostene, W. (1982) A new neuromodulator: the vasoactive intestinal peptide or VIP. *Molec. cell. Endocrinol.*, **27**, 243–62.

Rouille, Y., Michel, G., Chauvet, M.T., Chauvet, J. and Acher, R. (1989) Hydrins, hydrosmotic neurohypophysial peptides: osmoregulatory adaptation in amphibians through vasotocin precursor processing. *Proc. natn. Acad. Sci. U.S.A.*, **86**, 5272–5.

Sainsbury, R.J. and Balment, R.J. (1992) Arginine vasotocin (AVT) modulation of cyclic AMP production in isolated nephroi and dispersed gills cells of SW and FW rainbow trout. *J. Physiol., Lond.*, **131**, (suppl.), 123P.

Sakakibara, S., Sokabe, H., Nakajima, T., Khosla, M.C., Kumagaye, S. and Watanabe, T.X. (1985) Evolution of angiotensins and some problems in their synthetic preparations, in *Current Trends in Comparative Endocrinology* (eds B. Lofts and W.H. Holmes), Hong Kong Univ. Press, Hong Kong, pp. 1157–9.

Sakamoto, T. and Hirano, T. (1991) Growth hormone receptors in the liver and osmoregulatory organs of rainbow trout: characterization and dynamics during adaptation to seawater. *J. Endocr.*, **130**, 425–33.

Sakamoto, T., Ogasawara, T. and Hirano, T. (1990) Growth hormone kinetics during adaptation to a hypertonic environment in rainbow trout. *J. comp. Physiol.*, **160B**, 1–6.

Sakamoto, T., Iwata, M. and Hirano, T. (1991) Kinetic studies of growth hormone and prolactin during adaptation of coho salmon, *Oncorhynchus kisutch*, to different salinities. *Gen. comp. Endocrinol.*, **82**, 184–91.

Sawyer, W.H., Blair-West, J.R., Simpson, P.A. and Sawyer, M.K. (1976) Renal responses of Australian lungfish to vasotocin, angiotensin II, and NaCl infusion. *Am. J. Physiol.*, **231**, 593–602.

Scheide, J.I. and Zadunaisky, J.A. (1988) Effect of atriopeptin II on isolated opercular epithelium of *Fundulus heteroclitus*. *Am. J. Physiol.*, **254**, R27–32.

Sekine, S., Mizukami, T., Nishi, T., Kuwana, Y., Saito, A., Sato, M., Ito, S. and Kawauchi, H. (1985) Cloning and expression of cDNA for salmon growth hormone in *Escherichia coli*. *Proc. natn. Acad. Sci. U.S.A.*, **82**, 4306–10.

Shuttleworth, T.J. and Thorndyke, M.C. (1984) An endogenous peptide stimulates secretory activity in the elasmobranch rectal gland. *Science*, **225**, 319–21.

Silva, P., Stoff, J.S., Solomon, R.J., Lear, S., Kniaz, D., Greger, R. and Epstein, F.H. (1987) Atrial natriuretic peptide stimulates salt secretion by shark rectal gland by releasing VIP. *Am. J. Physiol.*, **252**, F99–103.

Smith, D.C.W. (1956) The role of endocrine organs in the salinity tolerance of trout. *Mem. Soc. Endocrinol.*, **5**, 83–101.

Smith, N.F., Eddy, F.B., Struthers, A.D. and Talbot, C. (1991) Renin, atrial natriuretic peptide and blood plasma ions in parr and smolts of Atlantic salmon *Salmo salar* L. and rainbow trout *Oncorhynchus mykiss* (Walbaum) in fresh water and after short-term exposure to sea water. *J. exp. Biol.*, **157**, 63–74.

Solomon, R., Solomon, G., Silva, P. and Epstein, F.H. (1989) Renal function changes in *Squalus acanthias* during acute seawater dilution: the importance of hemodynamic changes. *Bull. Mt Desert Isl. biol. Lab.*, **28**, 20–1.

Solomon, R., Taylor, M., Dorsey, D., Silva, P. and Epstein, F.H. (1985) Atriopeptin stimulation of rectal gland function in *Squalus acanthias*. *Am. J. Physiol.*, **249**, R348–54.

Stoff, J.S., Rosa, R., Hallac, R., Silva, P. and Epstein, F.H. (1979) Hormonal regulation of active chloride transport in the dogfish rectal gland. *Am. J. Physiol.*, **237**, F138–44.

Suzuki, R., Takahashi, A. and Takei, Y. (1992) Different molecular forms of C-type natriuretic peptide isolated from the brain and heart of an elasmobranch, *Triakis scyllia*. *J. Endocr.*, 134, 149–53.

Suzuki, R., Takahashi, A., Hazon, N. and Takei, Y. (1991) Isolation of high-molecular-weight C-type natriuretic peptide from the heart of a cartilaginous fish (European dogfish, *Scyliorhinus canicula*). *FEBS Lett.*, **282**, 321–5.

Takei, Y. (1987) In vivo characterization of angiotensin II receptors and converting enzymes in the quail and eel. *Zool. Sci.*, **4**, 1088.

Takei, Y. and Hasegawa, Y. (1990) Vasopressor and depressor effects of native angiotensins and inhibition of these effects in the Japanese quail. *Gen. comp. Endocrinol.*, **79**, 12–22.

Takei, Y., Kobayashi, H. and Hirano, T. (1979) Angiotensin and water intake in the Japanese eel, *Anguilla japonica*. *Gen. comp. Endocrinol.*, **38**, 466–75.

Takei, Y., Takahashi, A., Watanabe, T.X., Nakajima, K. and Sakakibara, S. (1989) Amino acid sequence and relative biological activity of eel atrial natriuretic peptide.

Biochem. Biophys. Res. Commun., **164**, 537–43.

Takei, Y., Takahashi, A., Watanabe, T.X., Nakajima, K., Sakakibara, S., Takao, T. and Shimonishi, Y. (1990a) Amino acid sequence and relative biological activity of a natriuretic peptide isolated from eel brain. *Biochem. Biophys. Res. Commun.*, **170**, 883–91.

Takei, Y., Tamaki, H. and Ando, K. (1990b) Identification and partial characterization of immunoreactive and bioactive atrial natriuretic peptide from eel heart. *J. comp. Physiol.*, **160B**, 119–26.

Takei, Y., Takahashi, A., Watanabe, T.X., Nakajima, K. and Sakakibara, S. (1991) A novel natriuretic peptide isolated from eel cardiac ventricles. *FEBS Lett.*, **282**, 317–20.

Takemoto, Y., Nakajima, T., Hasegawa, Y., Watanabe, T.X., Sokabe, H., Kumagaye, S. and Sakakibara, S. (1983) Chemical structures of angiotensins formed by incubating plasma with the kidney and corpuscles of Stanius in the chum salmon *Oncorhynchus keta*. *Gen. comp. Endocrinol.*, **51**, 219–27.

Thwaites, D.T., Young, J., Thorndyke, M.C. and Dimaline, R. (1989) The isolation and chemical characterization of a novel vasoactive intestinal peptide-related peptide from a teleost fish, the cod, *Gadus morhua*. *Biochim. Biophys. Acta*, **999**, 217–20.

Watahiki, M., Yamamoto, M., Yamakawa, M., Tanaka, M. and Nakashima, K. (1989) Conserved and unique amino acid residues in the domains of the growth hormones. *J. biol. Chem.*, **264**, 312–16.

Weld, M.M. and Fryer, J.N. (1987) Stimulation by angiotensins I and II of ACTH release from goldfish pituitary columns. *Gen. comp. Endocrinol.*, **68**, 19–27.

Westenfelder, C., Birch, F.M., Baranowski, R.L., Rosenfeld, M.J., Shiozawa, D.K. and Kablitz, C. (1988) Atrial natriuretic factor and salt adaptation in the teleost fish *Gila atraria*. *Am. J. Physiol.*, **255**, F1281–6.

Woo, N.Y.S., Hontela, A., Fryer, J.N., Kobayashi, Y. and Lederis, K. (1985) Activation of hypothalamo-hypophysial–interrenal system by urophysectomy in goldfish. *Am. J. Physiol.*, **238**, R197–201.

Yamaguchi, K., Yasuda, A., Kishida, M., Hirano, T., Sano, H. and Kawauchi, H. (1987) Primary structure of eel (*Anguilla japonica*) growth hormone. *Gen. comp. Endocrinol.*, **66**, 447–53.

Yamaguchi, K., Yasuda, A., Lewis, U.J., Yokoo, Y. and Kawauchi, H. (1989) The complete amino acid sequence of growth hormone of an elasmobranch, the blue shark (*Prionace glauca*). *Gen. comp. Endocrinol.*, **73**, 252–9.

Yao, K., Niu, P.-D., Le Gac, F. and Le Bail, P.-Y. (1991) Presence of specific growth hormone binding sites in rainbow trout (*Oncorhynchus mykiss*) tissues: characterization of the hepatic receptor. *Gen. comp. Endocrinol.*, **81**, 72–82.

Yasuda, A., Yamaguchi, K., Papkoff, H., Yokoo, Y. and Kawauchi, H. (1989) The complete amino acid sequence of growth hormone from the sea turtle (*Chelonia mydas*). *Gen. comp. Endocrinol.*, **73**, 242–51.

Yasuda, A., Yamaguchi, K., Noso, T., Papkoff, H., Polenov, A.L., Nicoll, C.S. and Kawauchi, H. (1991) The complete amino acid sequence of growth hormone from sturgeon (*Acipencer guldensitadti*). *Biochim. biophys. Acta*, **1120**, 299–304.

Zeidel, M.L. (1990) Renal actions of atrial natriuretic peptide: regulation of collecting duct sodium and water transport. *A. Rev. Physiol.*, **52**, 747–59.

Chapter six

Environmental perturbations of oxygen transport in teleost fishes: causes, consequences and compensations

Frank B. Jensen, Mikko Nikinmaa and Roy E. Weber

6.1 INTRODUCTION

Aquatic environments exhibit large spatial and temporal variations in physico-chemical factors that influence the physiology of fishes. Environmental changes in O_2 and CO_2 tensions, temperature, salinity, acidity, and input of toxic substances can be of both man-made and natural origin. Oxygen-depleted waters, for instance, occur naturally in habitats with heavy degradation of organic matter. Anthropogenic activity has expanded the number of environments with 'extreme conditions', and has created new environmental problems. Periodic or chronic oxygen depletion resulting from eutrophication is now a well-known phenomenon in many freshwater and coastal areas. Combustion of fossil fuels (emitting nitrogen and sulphur oxides) has acidified many freshwater ecosystems. Furthermore industrial activity has increasingly contaminated waters with toxic substances.

In ecophysiological studies it is important to gain insight into both the acute mechanisms of environmental perturbations and the compensatory mechanisms that are activated to counteract the internal disturbance. One key physiological process that is extremely sensitive to environmental conditions is the transport of gases (O_2 and CO_2) between environment and cells.

Fish Ecophysiology. Edited by J. Cliff Rankin and Frank B. Jensen. Published in 1993 by Chapman & Hall, London. ISBN 0 412 45920 5.

The transport of oxygen from environment to tissue cells consists of four steps: (1) ventilation of gills (water convection); (2) diffusion of O_2 from water to blood; (3) blood gas transport (convection); and (4) diffusion of O_2 from blood to tissue cells. The steady state flux of oxygen ($\dot{M}O_2$) is the same through each step, and at each individual step the transport can be described as a product of a conductance component (highlighted in bold below) and a tension difference (Dejours, 1975)*:

Ventilation:

$$\dot{M}O_2 = \mathbf{\dot{V}w} \times \boldsymbol{\beta}\mathbf{wo_2} \times (PIO_2 - PEO_2) \quad (6.1)$$

Diffusion across gills:

$$\dot{M}O_2 = (\mathbf{K} \times \mathbf{A/l}) \times \Delta PO_2 \text{ (water – capillary blood)} \quad (6.2)$$

Blood oxygen transport:

$$\dot{M}O_2 = \mathbf{\dot{Q}h} \times \boldsymbol{\beta}\mathbf{bo_2} \times (PaO_2 - PvO_2) \quad (6.3)$$

Diffusion in tissues:

$$\dot{M}O_2 = (\mathbf{K'} \times \mathbf{A'/l'}) \times \Delta PO_2 \text{ (capillary blood – tissue)} \quad (6.4)$$

Environmental changes can affect either tension or conductance components. If one component is reduced. (e.g. ΔP in hypoxia), the other component (i.e. conductance) must increase if an unchanged flux of oxygen is to be safeguarded. The present chapter describes how environmental changes affect gas transport, with primary focus on blood oxygen transport and the physiological mechanisms that are available for acclimation/adaptation to environment. Recent reviews related to gas transport in fishes, but with alternative foci and objectives, include: Weber (1982), Randall and Daxboeck (1984), Weber and Jensen (1988), Perry and Wood (1989), Nikinmaa (1990), Jensen (1991).

6.2 HYPOXIA

When a fish is exposed to hypoxia, the simplest way to avoid the problem is behavioural, namely escape! Shift from water breathing to air breathing at low water PO_2 is a sophisticated version of the 'escape strategy', and is used especially by various tropical species which have developed accessory breathing organs. Often, however, escape is not feasible (e.g. in a hypoxic lake) and physiological mechanisms must be used to compensate for the O_2 shortage.

When water PO_2 decreases, the internal O_2 tensions and the O_2 tension differences at each gas transport step also tend to decrease. This is exemplified in Fig. 6.1 for the 'I - E' (cf. Equation 6.1) and 'a - v' (Equation 6.3) transport in

*$\dot{V}w$ is ventilation volume per unit of time; βO_2 is the O_2 capacitance coefficient (= dCO_2/dPO_2); PO_2 is oxygen tension; K is Kroghs diffusion constant; A is diffusion area, l is diffusion distance; $\dot{Q}$h is cardiac output. Indices w, I, E, b, a and v refer to 'water', 'inspired', 'expired', 'blood', 'arterial' and 'venous', respectively.

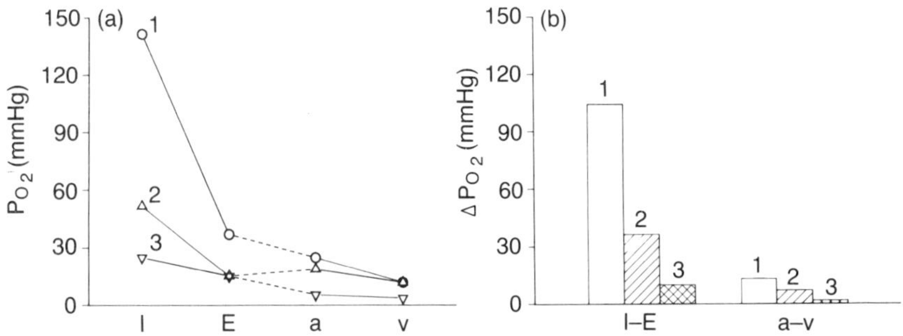

Fig. 6.1 (a) Oxygen tensions in inspired (I) and expired (E) water and arterial (a) and venous (v) blood in carp under normoxia (1) and at two levels of hypoxia (2 and 3). (b) The oxygen tension difference between inspired and expired water, and between arterial and venous blood, at the three levels of water oxygenation depicted in (a). 1, n = 12; 2, n = 6, 3, n = 8. Based on data from Itazawa and Takeda (1978).

carp, *Cyprinus carpio*. The decrease in ΔP_{O_2} can be compensated for by reducing the oxygen demand and/or by raising the conductance components in one or more of the above equations. A decrease in routine O_2 consumption ($\dot{M}_{O_2}$) during hypoxia is observed in many fish species (e.g. Lomholt and Johansen, 1979). In heterothermal environments this can be achieved by behavioural hypothermia, i.e. by selecting a lower ambient temperature (Wood, 1991). At constant temperature, a decreased O_2 demand can be achieved by a decreased activity level. Staying inactive is, however, not possible in the long run, given the needs for feeding, escape reactions or reproduction. Thus, acclimation to hypoxia requires an increased efficiency in the O_2 flux from environment to cells.

Ventilation is immediately increased under hypoxia, elevating the water convection conductance (Equation 6.1). The elevated gill water flow ($\dot{V}_W$) is mediated both through increases in ventilatory frequency and ventilatory stroke volume ($\dot{V}_W = f \times V_S$). Thus the reduced O_2 content of the water is compensated for by bringing more water into contact with the gills per time unit. Most fishes breathe continuously. Some fishes, like carp, however, are intermittent breathers under normoxic conditions. In hypoxia, carp increases breathing frequency by increasing the number of breathing episodes per time unit, leading eventually to a continuous breathing pattern with an elevated tidal volume (Lomholt and Johansen, 1979; Glass *et al.*, 1990). The gill O_2 diffusion conductance (cf. Equation 6.2) may increase in hypoxia via lamellar recruitment (Booth, 1979), which increases the diffusion area (A), and as a result of a reduced diffusion distance (l) (Soivio and Tuurala, 1981). Together, the improved ventilation and gill diffusion conductance will limit the decrease in arterial P_{O_2} in hypoxia. In carp, arterial P_{O_2} is only marginally affected by moderate hypoxia (e.g. Glass *et al.*, 1990), illustrating the efficient compensations at the first two steps in the gas transport process. In severe hypoxia, however, arterial P_{O_2} decreases.

A reduced arterial blood Po_2 tends to decrease the arterial oxygen content. The O_2 flux to the tissues can be secured by increasing either cardiac output ($\dot{Q}h$) or the blood O_2 capacitance coefficient (βbo_2). Both cardiac activity and circulatory patterns change in hypoxia, but cardiac output typically does not increase (Randall and Daxboeck, 1984; Fritsche and Nilsson, 1993). Thus, to safeguard blood O_2 transport in hypoxia, the blood capacitance coefficient βbo_2 (Equation 6.3) must increase. βbo_2 equals dCo_2/dPo_2, which is the slope of the concentration versus tension O_2 equilibrium curve. Several mechanisms and strategies are involved in adjusting and optimizing βbo_2 (reviewed recently by Weber and Jensen, 1988; Jensen, 1991).

βbo_2 can be increased by elevating the blood haemoglobin (Hb) concentration. Fish red cells are nucleated and possess the ability to synthesize Hb during a large part of their life span (Speckner *et al.*, 1989), but it is unknown if hypoxia acts as a signal to increase Hb production in individual cells. Erythropoiesis (formation of new red cells) in fish is activated by hypoxia (Härdig *et al.*, 1978) and may contribute to elevate [Hb] in prolonged (i.e. 1–2 weeks) hypoxia. More rapid elevations of the blood Hb concentration occur by catecholamine-stimulated release of erythrocytes from the spleen (Holmgren and Nilsson, 1975; Yamamoto *et al.*, 1985), or by water shifts (e.g. to lactate-loaded muscles). The Hb concentration increase in hypoxia is quite variable, being substantial in eel, trout, and yellowtail, and almost insignificant in plaice, tench and carp (Weber and Jensen, 1988).

The capacitance coefficient βbo_2 also depends on the O_2 affinity of the blood. This in turn depends on the intrinsic O_2 affinity of the haemoglobin and on the intraerythrocytic chemical environment. The tetrameric haemoglobin molecule shifts between two main structural conformations depending on oxygenation, the tense T ('deoxy') structure, which has a low O_2 affinity, and the relaxed R ('oxy') structure, which has a high O_2 affinity (Perutz, 1983). Erythrocytic hydrogen ions, organic phosphates (ATP and GTP in fish – Weber, 1982) and chloride bind preferentially to the T structure, stabilizing it, and thereby decrease the overall oxygen affinity of the Hb by shifting the allosteric equilibrium between the two structures towards the low O_2 affinity T state (e.g. Weber *et al.*, 1987).

Hypoxia-tolerant fish species have higher blood O_2 affinities than less tolerant species (Krogh and Leitch, 1919). Species differences in blood O_2 affinity can reside both in the functional properties of the haemoglobin and in the chemical milieu inside the red cells. Figure 6.2 provides an example comparing the hypoxia-tolerant tench, *Tinca tinca*, with the rainbow trout, *Oncorhynchus mykiss*, which requires well-oxygenated waters. The blood O_2 equilibrium curve of tench is considerably left-shifted compared with that of trout, allowing tench to obtain high arterial O_2 saturations even at very low oxygen tensions. Oxygen equilibrium curves of the purified haemoglobins at red cell pH values corresponding with those prevailing for the blood curves

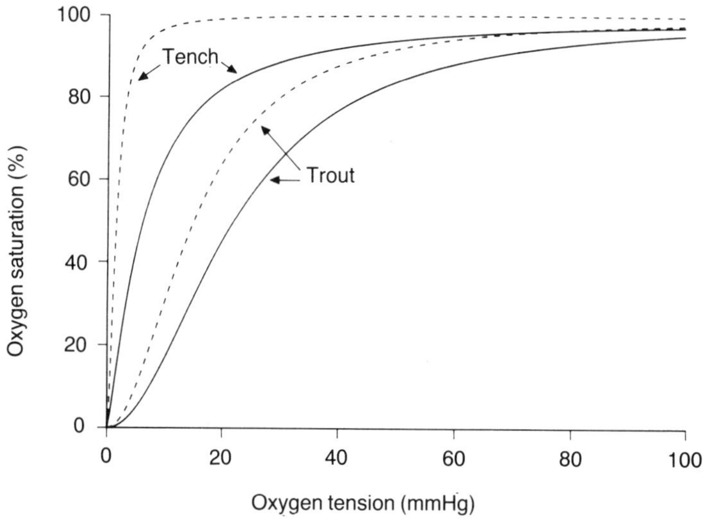

Fig. 6.2 Oxygen equilibrium curves of blood (solid curves) and purified haemoglobin (broken curves) from normoxic rainbow trout and tench at 15 °C. See text for further explanation. Based on data from Jensen and Weber (1982), and Weber *et al.* (1976).

show a similar difference between tench and trout. Thus the difference between these two species resides mainly in species-specific differences in globin structure (amino acid sequences) evolved in the haemoglobins. The difference between the haemoglobin and blood curves for each species is primarily due to the presence of organic phosphates and other anions inside the red cells, which decrease the blood O_2 affinity.

Many fish possess multiple Hbs with mutual differences in intrinsic functional properties (Weber, 1990). This raises the possibility that during hypoxia young red cells (with a capacity to synthesize Hb) could switch to produce components that are better suited for hypoxia. Hypoxia indeed seems to elicit a change in Hb component pattern in some fishes (Weber and Jensen, 1988; Marinsky *et al.*, 1990), whereas in others Hb multiplicity is unaffected by hypoxia (Weber and Lykkeboe, 1978; Jensen and Weber, 1982). The physiological significance of a change in isoHb pattern remains to be documented.

The general intraspecific strategy of adjusting Hb–O_2 affinity in fish to hypoxia is by changing the concentrations of intracellular cofactors, mainly H^+ (pH) and organic phosphates (ATP and GTP).

The hypoxia-induced hyperventilation rapidly causes a respiratory alkalosis (rise in pH through a decrease in P_{CO_2}) in the blood (Holeton and Randall, 1967; Dejours, 1973; Thomas and Hughes, 1982 – cf. 'carp' in Fig. 6.3). This increases O_2 affinity through the pH sensitivity (Bohr effect) of the Hb, given that a concomitant metabolic acidosis (from lactic acid production) does not counter the alkalosis (as happens in severe hypoxia). Deoxygenation of the haemoglobin also leads to large red cell pH increases in teleosts (Albers *et al.*,

1983; Jensen, 1986). This is related to the large Haldane effect (oxygenation-linked H^+ binding) and low buffer values of many teleost haemoglobins (Jensen, 1986, 1989). This characteristic may secure CO_2 transport in the face of a relatively slow red cell HCO_3^-/Cl^- exchanger (Jensen, 1989, 1991). Concerning O_2 transport, the major effect of a large oxygenation-linked red cell pH change (that can be 10 times larger than in mammals) is to decrease the cooperativity of O_2 binding in whole blood (Jensen, 1991).

Another rapid mechanism for increasing red cell pH is adrenergic stimulation of Na^+/H^+ exchange across the red cell membrane. A decade ago, catecholamines were shown to cause swelling and a selective pH increase in rainbow trout red cells, which was suggested to increase HbO_2 affinity in acute hypoxia (Nikinmaa, 1982, 1983). This was confirmed by experiments *in vivo* (Tetens and Christensen, 1987). The mechanism underlying the response has been subject to several recent reviews (Motais and Garcia-Romeu, 1988; Nikinmaa and Tufts, 1989; Nikinmaa, 1990; Fievet and Motais, 1991). In hypoxia, catecholamines are liberated to the blood stream in response to the reduced blood O_2 content (Fievet *et al.*, 1987; Tetens and Christensen, 1987; Perry *et al.*, 1989). The catecholamines bind to the β-adrenergic receptors on the cell membrane, with noradrenaline having a higher affinity than adrenaline (Tetens *et al.*, 1988). The consequent accumulation of cAMP inside the red cells is associated with an activation of the Na^+/H^+ exchange. The efflux of H^+ from the red cells increases intracellular pH (leading to secondary HCO_3^- efflux in exchange for Cl^-) and decreases extracellular pH, even in the presence of functional anion exchange, because the rate of passive equilibration of protons across the red cell membrane is limited by the uncatalysed extracellular dehydration of carbonic acid to CO_2 (Motais *et al.*, 1989; Nikinmaa *et al.*, 1990). Because the dehydration reaction is initially at equilibrium, the proton extrusion will exceed the bicarbonate efflux for a considerable period of time before the rate of dehydration of carbonic acid to CO_2 increases enough to titrate protons at the same rate as they are extruded. At this time the net H^+ efflux will stop (although H^+ and HCO_3^- will continue to be extruded). Apart from increasing red cell pH, the mechanism causes red cell swelling due to the net Na^+ and Cl^- influx to the cells. The activity of the Na^+/H^+ exchanger increases with decreases in pH and P_{O_2} (Borgese *et al.*, 1987; Motais *et al.*, 1987), making the response uniquely suited in controlling red cell pH under exhaustive exercise and acute hypoxia. The P_{O_2} dependence is particularly critical in cyprinids. In carp the response is only elicited at low P_{O_2} values (or at very low pH) and it is absent under normoxic conditions (Nikinmaa *et al.*, 1987; Salama and Nikinmaa, 1988). The same applies to the tench (Jensen, 1987 and unpublished).

The benefit of the β-adrenergic red cell swelling is the associated rapid increase in Hb–O_2 affinity. This affinity increase is primarily due to the red cell pH increase (Nikinmaa, 1983), but minor contributions may also come from a small metabolic decrease in red cell ATP concentration, and swelling per se (which dilutes Hb and ATP and GTP, decreasing complexing). The β-adrenergic response is notably important in acute hypoxia. In chronic hypoxia, prolonged elevation of blood catecholamine levels may lead to a desensitization of the red cell adrenergic response (Thomas *et al.*, 1991), highlighting the need for additional mechanisms to modulate O_2 affinity.

A general mechanism for adjusting O_2 affinity during hypoxia at an intermediate rate (effective within a day or so), is a reduction in the cellular ATP or GTP content (Wood and Johansen, 1972; Weber and Lykkeboe, 1978; Soivio *et al.*, 1980; Jensen and Weber, 1985; Tetens and Lykkeboe, 1985). This increases the blood O_2 affinity, both via reduced allosteric interaction of the phosphates with the haemoglobin, and via an increase in red cell pH. The latter effect is associated with the changed distribution of H^+ across the red cell membrane, which is caused by the reduced amount of non-permeable negative charge carried by the organic phosphates (Wood and Johansen, 1973). The effect of GTP on O_2 affinity is larger than that of ATP. Moreover, in species that have both phosphate cofactors, GTP concentrations decrease more drastically and its effect on O_2 affinity is less inhibited by red cell magnesium (Weber and Lykkeboe, 1978). It appears that fish that encounter hypoxia in their natural habitats (e.g. carp, tench, lungfish, eel) have relatively high red cell concentrations of GTP and that they selectively use this phosphate for O_2 affinity modulation (Weber and Jensen, 1988). Fish living in well-oxygenated waters (e.g. salmonids) appear to use mainly ATP.

The O_2 affinity increase elevates the arterial O_2 saturation in hypoxia and secures a high βb_{O_2} at low P_{O_2} values. Together with improved diffusion conditions in the gills, it moreover contributes to reduce the ventilatory requirement and to increase O_2 extraction from the water (e.g. Lomholt and Johansen, 1979; Malte and Weber, 1987), which gives an obvious energetic advantage during long-term hypoxia.

The Antarctic fish, *Pagothenia borchgrevinki*, which is unlikely ever to encounter environmental hypoxia, responds to hypoxia with an increase in blood [Hb] and a decrease in red cell [ATP] as do fishes that frequently experience hypoxia, suggesting that these compensatory responses are features indigenous to the physiology of fishes (Wells *et al.*, 1989).

The tolerance of different fish species towards hypoxia is very variable. This basically reflects an evolutionary (genetically coded) adaptation of the functional properties of body proteins (including haemoglobin) to life under different conditions. Physiological adjustments fine-tune these basic properties. This is illustrated by the comparison between the hypoxic-tolerant tench and the less tolerant trout (Fig. 6.2). Although blood O_2 affinity in trout increases during hypoxia, by adjustments in the chemical milieu inside the red cells, it cannot become as high as in tench.

The last step in the O_2 transport scheme is diffusion and consumption of

O_2 in the tissues. The tissue O_2 requirement is an essential part of the transport, since it sets the demands on the other transport steps. The increased blood O_2 affinity in fish acclimated to hypoxia tends to decrease off-loading P_{O_2} values, although this decrease may be small due to the increased βb_{O_2}. During long-term acclimation of tench to hypoxia, the capillary supply to muscle fibres is decreased (Johnston and Bernard, 1982). Thus the muscle tissue diffusion conductance is decreased rather than increased. This change may contribute to set $\dot{M}_{O_2}$ at the lower level, to which the other transport steps adjust. The muscle fibres still sustain their energy production aerobically at rest (lactate levels are not elevated in hypoxia-acclimated tench), but the switch to anaerobic energy production probably occurs at a lower activity level in hypoxia than in normoxia.

Severe hypoxia reduces survival in less tolerant species. This can be due to a limited capacity to adjust and safeguard the essential gas transport (notably during periods of increased activity, such as during feeding or escape). Additionally, hypoxia-stress may reduce the fitness of populations by decreasing tolerance to disease and infection.

6.3 COMBINED HYPOXIA–HYPERCAPNIA

Whereas most research concerning environmental O_2 lack has dealt with pure hypoxia, O_2-depleted habitats often tend to be both hypoxic and hypercapnic, since the processes that consume the water O_2 (e.g. microbial degradation of organic matter) also produce CO_2. This applies particularly to habitats where the water surface is covered by floating aquatic plants (or ice) that hinder gas exchange between water and air. In hypoxic–hypercapnic waters the elevated water P_{CO_2} will induce a blood respiratory acidosis (Fig. 6.3; Jensen and Weber, 1982, 1985; Thomas, 1983), which opposes the respiratory alkalosis caused by hyperventilation in acute normocapnic-hypoxia (Fig. 6.3). Thus, in combined hypoxia–hypercapnia, hyperventilation will not cause an increase in blood O_2 affinity. On the contrary, the respiratory acidosis tends to decrease O_2 affinity and (in extreme cases) the O_2 carrying capacity (through the Root effect). The acute extracellular acidosis is slowly compensated by accumulation of bicarbonate, which is taken up from the water across the gills in exchange for chloride, as in fish exposed to pure hypercapnia (Jensen and Weber, 1982, 1985; Thomas, 1983; Heisler, 1986). The pH restoration is relatively fast in trout (about 1 day – Thomas, 1983), but requires several days (or even weeks) in tench (Fig. 6.3), where the higher control pH demands accumulation of larger amounts of bicarbonate (Jensen and Weber, 1985). Blood pH is, however, eventually completely recovered when the fish is in hard water (i.e. water with high HCO_3^- and Ca^{2+} concentrations) via a phenomenal elevation of plasma [HCO_3^-] (Fig. 6.3; Jensen and Weber, 1982).

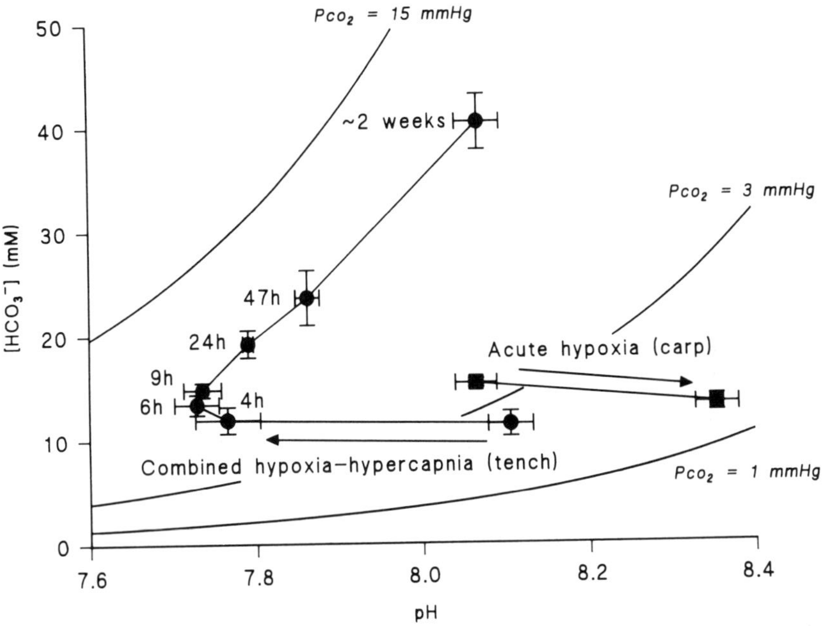

Fig. 6.3 Plasma [HCO_3^-] v. pH diagram illustrating the different acid–base changes in arterial blood during acute hypoxia and combined hypoxia–hypercapnia in the two closely related cyprinids carp, *Cyprinus carpio*, and tench, *Tinca tinca*. In carp, acid–base status was measured in dorsal aortic blood during normoxia and 20 min following a change in water P_{O_2} from 135 mmHg to 25 mmHg (F.B. Jensen, N.A. Andersen and N. Heisler, unpublished). The acid–base status of tench at different times following transfer from normoxia (water P_{O_2} = 140 mmHg and P_{CO_2} = 0.7 mmHg) to combined hypoxia–hypercapnia (water P_{O_2} = 28 mmHg and P_{CO_2} = 7.5 mmHg) are based on the data of Jensen and Weber (1982, 1985).

Whereas the extracellular acidosis is long lasting in hypoxic–hypercapnic tench, the intraerythrocytic pH decreases only transiently, then increases even above the control values (Jensen and Weber, 1985). This is due to swelling of the red cells, a decreased HbO_2 saturation, and a reduced red cell GTP content. Thus, as is the case in pure hypoxia, tench acclimated to combined hypoxia and hypercapnia have a blood O_2 equilibrium curve shifted to the left from the normoxic curve (Jensen and Weber, 1982).

6.4 TEMPERATURE

As in other ectotherms, almost all processes in fish are affected by environmental temperature. The metabolic rate typically increases by a factor of 2–3 for every 10 °C rise in temperature. In parallel with this exponential rise in $\dot{M}_{O_2}$ the solubility of oxygen in water (βw_{O_2}) decreases (e.g. by 35% between

5 and 25 °C). Thus, as is evident from Equation 6.1, the ventilatory flow rate must increase with temperature to bring sufficient amounts of O_2 into contact with the gill.

To safeguard blood O_2 transport, the heart rate and cardiac output increase with temperature (Stevens *et al.*, 1972; Heath and Hughes, 1973). The blood O_2 carrying capacity in different species can increase (De Wilde and Houston, 1967; Cameron, 1970), remain unchanged (Houston and Smeda, 1979) or even decrease (Nikinmaa *et al.*, 1980) with an increase in temperature, suggesting that changes in [Hb] play only a minor role in adjusting oxygen transport to temperature changes. Similarly, the blood volume appears to be relatively insensitive to changes in temperature (Nikinmaa *et al.*, 1981).

The reaction between haemoglobin and oxygen is exothermic, whereby an increase in temperature decreases oxygen affinity (Wyman, 1948). This affinity decrease can be advantageous or disadvantageous, depending on oxygen tensions at the sites of loading and unloading. As long as the arterial oxygen saturation is secured, a decreased affinity increases O_2 unloading in tissue capillaries in parallel with the increased call for oxygen. If, however, O_2 loading in the gills is significantly reduced by the decrease in O_2 affinity, the O_2 transport to the tissues can be hampered. The temperature sensitivity of Hb–O_2 equilibria is dependent on heterotropic effectors like H^+ and organic phosphates. During oxygenation both protons (via the Bohr effect) and ATP or GTP are released from their binding sites on the Hb molecule. This release of heterotropic ligands is endothermic, and tends to mask the heat released in the exothermic oxygenation process, reducing the average heat of oxygenation, $-\Delta H_{av}$ (and thereby the change in O_2 affinity). Thus, under conditions simulating natural red cell conditions (with respect to pH and NTP/Hb ratio) the $-\Delta H_{av}$ value for tench Hb is strongly reduced compared with the intrinsic value of the Hb in the absence of heterotropic ligands (Jensen and Weber, 1987b). A reduced temperature dependence may be critical by avoiding inordinate O_2 affinity decreases in warm waters, which are more liable to hypoxia (due to higher microbial degradation rates). In tuna Hb the release of Bohr protons is nearly zero for the first two O_2 molecules bound to the Hb, but substantial for the last two oxygenation steps. In this way oxygenation is exothermic at low O_2 saturation and endothermic at the high saturation, whereas the average ΔH for the oxygenation reaction is close to zero (Ikeda-Saito *et al.*, 1983). Thus, tuna Hb has an inverse temperature effect at high saturations and virtually none at P_{50}. This peculiarity may be an adaptation to prevent gas bubble formation in blood vessels when cold efferent blood enters the warmer body core (Hochachka and Somero, 1973), just as it contributes to conserve heat in the muscle.

Depending on the temperature range, changes in red cell NTP content during temperature acclimation may either augment or attenuate the temperature-induced decrease in O_2 affinity. Data are scarce and somewhat conflicting, but it appears that as long as the arterial O_2 loading in gills is not compromised, then red cell NTP increases with temperature (Nikinmaa *et al.*, 1980; Dobson and Baldwin, 1982; Tetens *et al.*, 1984; Laursen *et al.*, 1985), resulting in increased O_2 unloading at a given unloading P_{O_2}. At high temperatures, where the temperature-induced decrease in O_2 affinity causes a significant reduction in arterial loading, it appears that this is counteracted by a decrease in red cell [NTP] (Albers *et al.*, 1983; Laursen *et al.*, 1985). A similar effect can be caused by an increased red cell Mg/NTP ratio at elevated temperatures (Houston and Koss, 1984). Magnesium complexes with NTPs, thus reducing binding of NTPs to the Hb and increasing O_2 affinity.

In addition to the effects of temperature per se, seasonal effects may modify gas transport. In autumn, erythropoiesis is switched off in salmon (Härdig and Höglund, 1984). In winter, the methaemoglobin concentrations of temperate fish are higher than in summer (Graham and Fletcher, 1986), and the β-adrenergic responsiveness of trout red cells is reduced (Nikinmaa and Jensen, 1986; Cossins and Kilbey, 1989). These responses limit the scope and flexibility of oxygen transport in winter. In spring, production of new erythrocytes increases, methaemoglobin levels fall and the β-adrenergic responses of the red cells are enhanced, whereby blood O_2 transport can respond to large increases in activity, brought about by the increases in ambient temperature, and to the fluctuations in oxygen levels characteristic of the summer months.

6.5 SALINITY CHANGE

The movement between fresh water and seawater of anadromous fish like salmonids or catadromous fish like the eel demands several osmoregulatory adjustments, which have been subject to extensive research. Gas transport problems associated with changing salinity have, however, received little attention. A transfer from fresh water to seawater typically causes a blood acidosis, whereas the reverse transfer elicits a blood alkalosis (e.g. Maxime *et al.*, 1990; Nonnotte and Truchot, 1990). Together with a changing internal ion composition (e.g. chloride) this predictably changes blood O_2 affinity via the pH and anion sensitivities of the haemoglobin. Indeed, transfer of salmon from seawater to fresh water increased blood O_2 affinity (Maxime *et al.*, 1990).

Fish moving between fresh water and seawater experience variable water O_2 and CO_2 capacitance coefficients. The lower βw_{O_2} in seawater than in fresh water could form a stimulus for an elevated ventilation (Equation 6.1). Salinity changes also appear to affect the diffusion conductance of the gills. Bath and Eddy (1979) observed a large decrease in arterial P_{O_2} when fresh water rainbow trout were acutely exposed to seawater, which may reflect a decreased diffusion conductance. Data on salmon point to an increased branchial diffusion conductance upon the reverse transfer from seawater to fresh water (Maxime *et al.*, 1990).

6.6 EFFECTS OF TOXICANTS

Fish may be exposed to a large variety of natural and man-made toxicants in their environments. Interest has increasingly focused on the ecophysiological and ecotoxicological consequences of toxicant exposures in aquatic animals. At the same time, use of toxic perturbants in physiological studies has proved a valuable means of gaining new insights into basic physiological mechanisms. A few examples are given in the present context.

Nitrite

Nitrite concentrations can be elevated in some natural habitats with large amounts of decaying organic matter as well as in recirculated aquaculture systems due to imbalance in nitrification or denitrification processes (Eddy and Williams, 1987). Fish accumulate nitrite in their blood plasma to concentrations far above that in the environment. Nitrite penetrates the red cell

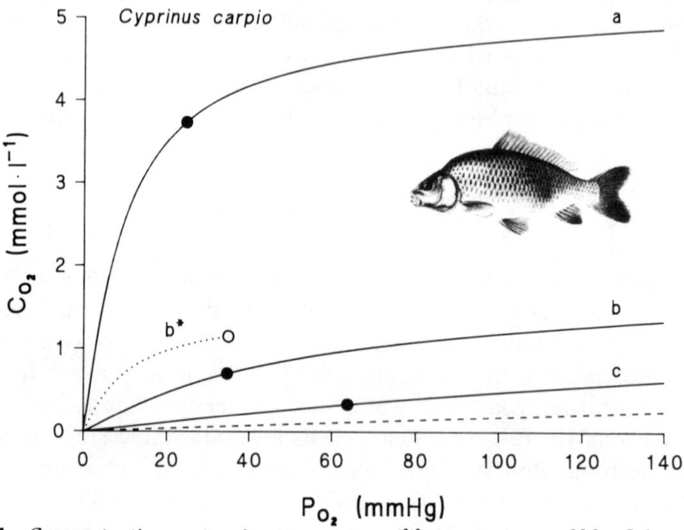

Fig. 6.4 Concentration v. tension oxygen equilibrium curves of blood from carp in normoxic water (curve a) and after 24 h (curve b) and 48 h (curve c) of 1 mM nitrite exposure. Arterial points are shown by filled circles. The slope of the curves below these points reflects the value of the *in vivo* blood O_2 capacitance coefficient under the three conditions. Methaemoglobin levels are 4.9% of the total Hb for curve a, 64.6% for curve b, and 83.3% for curve c. Apart from inducing increased MetHb levels, nitrite exposure is associated with a decreased blood O_2 affinity. Curve b* is a theoretical curve with the same MetHb level as for curve b, but with an unchanged O_2 affinity as in curve a. The vertical distance between the filled circle on curve a and the open circle on curve b* gives the decrease in arterial O_2 content caused by the increased methaemoglobin level after 24 h, and the vertical distance between the open circle on curve b* and the filled circle on curve b gives the additional decrease in arterial O_2 content caused by the O_2 affinity decrease. The dashed line at the bottom is the curve for physically dissolved O_2. Based on Jensen *et al.* (1987).

membrane and oxidizes haemoglobin to methaemoglobin, which cannot transport oxygen. This drastically decreases the arterial O_2 content and the blood capacitance βb_{O_2} (Fig. 6.4) (Jensen *et al.*, 1987). A rise in arterial P_{O_2} (Fig. 6.4) and ΔP_{O_2} between arterial and venous blood may partially compensate blood O_2 transport, but eventually tissue O_2 shortage occurs, and lactic acid is produced (Jensen *et al.*, 1987). Blood O_2 transport is also affected by a major decrease in oxygen affinity of remaining functional haem groups (Jensen *et al.*, 1987; Jensen, 1990). This contrasts with the traditional idea that methaemoglobinaemia is associated with an increased affinity of non-oxidized haem groups. In carp, nitrite causes red cell K^+ loss and shrinkage, which (partly by elevating the red cell nucleoside triphosphate concentration) is a major reason for the O_2 affinity decrease (Jensen, 1990). This affinity decrease is opposite to the affinity increase in swollen cells (cf. Section 6.2), and contributes to lower the arterial O_2 concentration in nitrite-exposed carp (Fig. 6.4). Gas transport may be further compromised by the fact that even a moderate methaemoglobinaemia in rainbow trout strongly inhibits catecholamine-stimulated red cell H^+ efflux (Nikinmaa and Jensen, 1992). Thus, the benefit of β-adrenergic red cell responses cannot be fully exploited if additional stress (e.g. hypoxia) occurs together with the methaemoglobinaemia.

Hydrogen sulphide

Depletion of dissolved oxygen is often associated with microbial generation of hydrogen sulphide in the sediment (Jørgensen, 1980). When a low-pressure weather front moves over such a location, incidents of H_2S release to the water column may occur. More extended periods with elevated water sulphide levels may also develop, such as are frequently registered in the deep basins of the Baltic Sea (Fonselius, 1981). In such situations, adaptations in the blood gas transport to cope with hypoxia are of limited help. When oxygenated Hb is exposed to H_2S, O_2 is liberated and sulphaemoglobin is formed (e.g. Van Assendelft, 1970). But even more critically, H_2S inhibits cytochrome (c) oxidase in the mitochondria of tissue cells, thus blocking the last step in the respiratory chain, where O_2 is the terminal electron acceptor. Incidents with hydrogen sulphide release in significant amounts may accordingly cause extinction of local populations, with a subsequent recolonization when conditions improve.

Acid water/aluminium

Combustion of fossil fuels releases nitrogen oxides and sulphur dioxide, which in the atmosphere are oxidized to nitric acid and sulphuric acid, respectively. This acidifies rain, and subsequently soil and freshwater habitats with poor buffer capacity (i.e. where the calcium and bicarbonate content is low).

Acidification of soft freshwater habitats in parts of Scandinavia, Canada and the USA has been associated with reduced species diversity and extinction of fish populations. The major reason for this seems to be the combination of acidification and elevated aluminium concentrations (aluminium is washed out from soil by acid rain), which obstruct gill functions. Acid water leads to a net ion loss from the fish to the water, but in the absence of aluminium this effect is often only critical when pH values are below those prevailing in natural acid waters. At pH values where fish-kill episodes are observed in nature (i.e. around pH 5), aluminium-containing acid water is more toxic than pure acid water. This is related to the fact that in the presence of aluminium, osmoregulatory as well as respiratory functions of the gills are impeded (Muniz and Leivestad, 1980; Wood, 1989). There appear to be several reasons for the respiratory disturbance. Thus, aluminium compounds precipitate on the gill surface, the mucus production increases and gill lamellae fuse and thicken (e.g. McDonald and Wood, 1993). This reduces the gill diffusion conductance (by decreasing diffusion area and increasing diffusion distance – Equation 6.2), and causes a severe internal hypoxia–hypercapnia. This condition may worsen with time, so even though blood respiratory parameters show adaptive changes similar to those observed during environmental hypoxia or hypoxia–hypercapnia (Jensen and Weber, 1987a), death may be the ultimate result. Acclimation to metal exposure is, however, possible if conditions are not too severe (McDonald and Wood, 1993).

6.7 CONCLUDING REMARKS

The above selection of examples illustrates that the spectrum of environmental perturbations of oxygen transport is large, and that different steps in the gas transport process can be affected. In many circumstances it is possible to compensate these acute disturbances by physiological mechanisms at the organismic, cellular and molecular levels, allowing fish to acclimate to the environmental change. Environmental changes may, however, be so severe that the acute tolerance limits are surpassed or that the capacity of compensatory mechanisms becomes exceeded. Similarly, some environmental changes cause critical, irreversible changes that cannot be rectified.

Most studies on the effect of environmental changes on fish physiology have concentrated on only one environmental parameter, since in order to get insight into the effects of individual environmental factors, only one should be varied experimentally. Environmental changes, however, commonly involve a change in several parameters, and tolerance limits for a given environmental factor often depend on the levels of other environmental factors. Thus, hypoxia may be more frequent in summer where temperature and metabolic rates are higher, making the impact of hypoxia more severe. Similarly, as stated above, hypoxic habitats are often also hypercapnic and more prone to H_2S production and elevated nitrite concentrations than normoxic ones. A detailed understanding of the physiological and ecological impact of environmental changes calls for an intensification of research involving multifactorial changes.

ACKNOWLEDGMENTS

Financial support from the Carlsberg Foundation and the Danish Natural Science Research Council (FBJ) is kindly acknowledged.

REFERENCES

Albers, C., Goetz, K.-H. and Hughes, G.M. (1983) Effect of acclimation temperature on intraerythrocytic acid–base balance and nucleoside triphosphates in the carp, *Cyprinus carpio*. *Respir. Physiol.*, **54**, 145–59.

Bath, R.N. and Eddy, F.B. (1979) Ionic and respiratory regulation in rainbow trout during rapid transfer to seawater. *J. comp. Physiol.*, **134**, 351–7.

Booth, J.H. (1979) The effects of oxygen supply, epinephrine, and acetylcholine on the distribution of blood flow in trout gills. *J. exp. Biol.*, **83**, 31–9.

Borgese, F., Garcia-Romeu, F. and Motais, R. (1987) Ion movements and volume changes induced by catecholamines in erythrocytes of rainbow trout: effect of pH. *J. Physiol., Lond.*, **382**, 145–57.

Cameron, J.N. (1970) The influence of environmental variables on the hematology of pinfish (*Lagodon rhomboides*) and striped mullet (*Mugil cephalus*). *Comp. Biochem. Physiol.*, **32**, 175–92.

Cossins, A.R. and Kilbey, R.V. (1989) The seasonal modulation of Na^+/H^+ exchanger activity in trout erythrocytes. *J. exp. Biol.*, **144**, 463–78.

Dejours, P. (1973) Problems of control of breathing in fishes, in *Comparative Physiology* (eds L. Bolis, K. Schmidt-Nielsen and S.H.P. Maddrell), North-Holland/American Elsevier, Amsterdam, pp. 117–33.

Dejours, P. (1975) *Principles of Comparative Respiratory Physiology*, North-Holland/American Elsevier, Amsterdam, 254 pp.

De Wilde, M.A. and Houston, A.H. (1967) Hematological aspects of the thermoacclimatory process in the rainbow trout, *Salmo gairdneri*. *J. Fish. Res. Bd Can.*, **24**, 2267–81.

Dobson, G.P. and Baldwin, J. (1982) Regulation of blood oxygen affinity in the Australian blackfish *Gadopsis marmoratus*. II. Thermal acclimation. *J. exp. Biol.*, **99**, 245–54.

Eddy, F.B. and Williams, E.M. (1987) Nitrite and freshwater fish. *Chem. Ecol.*, **3**, 1–38.

Fievet, B. and Motais, R. (1991) Na^+/H^+ exchanges and red cell functions in fish, in *Advances in Comparative and Environmental Physiology*, Vol. 8 (ed. R. Gilles), Springer-Verlag, Berlin, pp. 79–104.

Fievet, B., Motais, R. and Thomas, S. (1987) Role of adrenergic-dependent H^+ release from red cells in acidosis induced by hypoxia in trout. *Am. J. Physiol.*, **252**, R269–75.

Fonselius, S. (1981) Oxygen and hydrogen sulphide conditions in the Baltic Sea. *Mar. Poll. Bull.*, **12**, 187–94.

Fritsche, R. and Nilsson, S. (1993) Cardiovascular and ventilatory control during hypoxia, in *Fish Ecophysiology* (eds J.C. Rankin and F.B. Jensen), Chapman and Hall, London, pp. 180–206.

Glass, M.L., Andersen, N.A., Kruhøffer, M., Williams, E.M. and Heisler, N. (1990) Combined effects of environmental PO_2 and temperature on ventilation and blood gases in the carp *Cyprinus carpio* L. *J. exp. Biol.*, **148**, 1–17.

Graham, M.S. and Fletcher, G.L. (1986) High concentrations of methemoglobin in five species of temperate marine teleosts. *J. exp. Zool.*, **239**, 139–42.

Härdig, J. and Höglund, L.B. (1984) Seasonal variations in blood components of reared Baltic salmon, *Salmo salar* L. *J. Fish Biol.*, **24**, 565–79.

Härdig, J., Olsson, L.A. and Höglund, L.B. (1978) Autoradiography on erythrokinesis and multihemoglobins in juvenile *Salmo salar* L. at various respiratory gas regimes. *Acta physiol. scand.*, **103**, 240–51.

Heath, A.G. and Hughes, G.M. (1973) Cardiovascular and respiratory changes during heat stress in rainbow trout (*Salmo gairdneri*). *J. exp. Biol.*, **59**, 323–38.

Heisler, N. (1986) Acid–base regulation in fishes, in *Acid–Base Regulation in Animals* (ed. N. Heisler), Elsevier, Amsterdam, pp. 309–56.

Hochachka, P. and Somero, G.N. (1973) *Strategies in Biochemical Adaptation*, Saunders, Philadelphia, 358 pp.

Holeton, G.F. and Randall, D.J. (1967) The effect of hypoxia upon the partial pressure of gases in the blood and water afferent and efferent to the gills of rainbow trout. *J. exp. Biol.*, **46**, 317–27.

Holmgren, S. and Nilsson, S. (1975) Effects of some adrenergic and cholinergic drugs on isolated spleen strips from the cod, *Gadus morhua*. *Eur. J. Pharmacol.*, **32**, 163–69.

Houston, A.H. and Koss, T.F. (1984) Erythrocytic haemoglobin, magnesium and nucleoside triphosphate levels in rainbow trout exposed to progressive heat stress. *J. therm. Biol.*, **9**, 159–64.

Houston, A.H. and Smeda, J.S. (1979) Thermoacclimatory changes in the ionic microenvironment of haemoglobin in the stenothermal rainbow trout (*Salmo gairdneri*) and eurythermal carp (*Cyprinus carpio*). *J. exp. Biol.*, **80**, 317–40.

Ikeda-Saito, M., Yonetani, T. and Gibson, Q.H. (1983) Oxygen equilibrium studies on hemoglobin from the bluefin tuna (*Thunnus thynnus*). *J. mol. Biol.*, **168**, 673–86.

Itazawa, Y. and Takeda, T. (1978) Gas exchange in the carp gills in normoxic and hypoxic conditions. *Respir. Physiol.*, **35**, 263–9.

Jensen, F.B. (1986) Pronounced influence of Hb-O_2 saturation on red cell pH in tench blood in vivo and in vitro. *J. exp. Zool.*, **238**, 119–24.

Jensen, F.B. (1987) Influences of exercise-stress and adrenaline upon intra- and extracellular acid–base status, electrolyte composition and respiratory properties of blood in tench (*Tinca tinca*) at different seasons. *J. comp. Physiol.*, **157B**, 51–60.

Jensen, F.B. (1989) Hydrogen ion equilibria in fish haemoglobins. *J. exp. Biol.*, **143**, 225–34.

Jensen, F.B. (1990) Nitrite and red cell function in carp: control factors for nitrite entry, membrane potassium ion permeation, oxygen affinity and methaemoglobin formation. *J. exp. Biol.*, **152**, 149–66.

Jensen, F.B. (1991) Multiple strategies in oxygen and carbon dioxide transport by haemoglobin, in *Physiological Strategies for Gas Exchange and Metabolism* (eds A.J. Woakes, M.K. Grieshaber and C.R. Bridges), Cambridge Univ. Press, Cambridge, pp. 55–78.

Jensen, F.B. and Weber, R.E. (1982) Respiratory properties of tench blood and hemoglobin. Adaptation to hypoxic-hypercapnic water. *Molec. Physiol.*, **2**, 235–50.

Jensen, F.B. and Weber, R.E. (1985) Kinetics of the acclimational responses of tench to combined hypoxia and hypercapnia. I and II. *J. comp. Physiol.*, **156B**, 197–211.

Jensen, F.B. and Weber, R.E. (1987a) Internal hypoxia–hypercapnia in tench exposed to aluminium in acid water: effects on blood gas transport, acid-base status and electrolyte composition in arterial blood. *J. exp. Biol.*, **127**, 427–42.

Jensen, F.B. and Weber, R.E. (1987b) Thermodynamic analysis of precisely measured oxygen equilibria of tench (*Tinca tinca*) hemoglobin and their dependence on ATP and protons. *J. comp. Physiol.*, **157B**, 137–43.

Jensen, F.B., Andersen, N.A. and Heisler, N. (1987) Effects of nitrite exposure on blood respiratory properties, acid–base and electrolyte regulation in the carp (*Cyprinus carpio*). *J. comp. Physiol.*, **157B**, 533–41.

Johnston, I.A. and Bernard, L.M. (1982) Routine oxygen consumption and characteristics of the myotomal muscle in tench: effects of long-term acclimation to hypoxia. *Cell Tissue Res.*, **227**, 161–77.

Jørgensen, B.B. (1980) Seasonal oxygen depletion in the bottom waters of a Danish fjord and its effects on the benthic community. *Oikos*, **34**, 68–76.

Krogh, A. and Leitch, I. (1919) The respiratory function of the blood in fishes. *J. Physiol., Lond.*, **52**, 288–300.

Laursen, J.S., Andersen, N.A. and Lykkeboe, G. (1985) Temperature acclimation and oxygen binding properties of blood of the European eel, *Anguilla anguilla. Comp. Biochem. Physiol.*, **81A**, 79–86.

Lomholt, J.P. and Johansen, K. (1979) Hypoxia acclimation in carp – how it affects O_2 uptake, ventilation, and O_2 extraction from water. *Physiol. Zool.*, **52**, 38–49.

McDonald, D.G. and Wood, C.M. (1993) Branchial mechanisms of acclimation to metals in freshwater fish, in *Fish Ecophysiology* (eds J.C. Rankin and F.B. Jensen), Chapman and Hall, London, pp. 297–321.

Malte, H. and Weber, R.E. (1987) The effect of shape and position of the oxygen equilibrium curve on extraction and ventilation requirement in fishes. *Respir. Physiol.*, **70**, 221–8.

Marinsky, C.A., Houston, A.H. and Murad, A. (1990) Effect of hypoxia on hemoglobin isomorph abundances in rainbow trout, *Salmo gairdneri. Can. J. Zool.*, **68**, 884–8.

Maxime, V., Peyraud-Waitzenegger, M., Claireaux, G. and Peyraud, C. (1990) Effects of rapid transfer from sea water to fresh water on respiratory variables, blood acid–base status and O_2 affinity of haemoglobin in Atlantic salmon (*Salmo salar* L.). *J. comp. Physiol.*, **160B**, 31–9.

Motais, R. and Garcia-Romeu, F. (1988) Effects of catecholamines and cyclic nucleotides on Na^+/H^+ exchange, in *Na^+/H^+ Exchange* (ed. S. Grinstein), CRC Press, Boca Raton, FL, pp. 255–70.

Motais, R., Garcia-Romeu, F. and Borgese, F. (1987) The control of Na^+/H^+ exchange by molecular oxygen in trout erythrocytes. A possible role of hemoglobin as a transducer. *J. gen. Physiol.*, **90**, 197–207.

Motais, R., Fievet, B., Garcia-Romeu, F. and Thomas, S. (1989) Na^+–H^+ exchange and pH regulation in red blood cells: role of uncatalyzed H_2CO_3 dehydration. *Am. J. Physiol.*, **256**, C728–35.

Muniz, I.P. and Leivestad, H. (1980) Toxic effects of aluminium on the brown trout, *Salmo trutta* L., in *Ecological Impact of Acid Precipitation* (eds D. Draplos and A. Tollan), Johs. Grefslie Press, Mysen, Norway, pp. 320–21.

Nikinmaa, M. (1982) Effects of adrenaline on red cell volume and concentration gradient of protons across the red cell membrane in the rainbow trout, *Salmo gairdneri. Molec. Physiol.*, **2**, 287–97

Nikinmaa, M. (1983) Adrenergic regulation of haemoglobin oxygen affinity in rainbow trout red cells. *J. comp. Physiol.*, **152**, 67–72.

Nikinmaa, M. (1990) *Vertebrate Red Blood Cells. Adaptation of Function to Respiratory Requirements*, Springer-Verlag, Berlin, 262 pp.

Nikinmaa, M. and Jensen, F.B. (1986) Blood oxygen transport and acid–base status of stressed trout (*Salmo gairdneri*): pre- and postbranchial values in winter fish. *Comp. Biochem. Physiol.*, **84A**, 391–6.

Nikinmaa, M. and Jensen, F.B. (1992) Inhibition of adrenergic proton extrusion in rainbow trout red cells by nitrite-induced methaemoglobinaemia. *J. Comp. Physiol.* **162B**, 429.

Nikinmaa, M. and Tufts, B.L. (1989) Regulation of acid and ion transfer across the membrane of nucleated erythrocytes. *Can. J. Zool.*, **67**, 3039–45.

Nikinmaa, M., Cech, J.J., jun., Ryhünen, E.-L. and Salama, A. (1987) Red cell function of carp (*Cyprinus carpio*) in acute hypoxia. *Exp. Biol.*, **47**, 53–8.

Nikinmaa, M., Tuurala, H. and Soivio, A. (1980) Thermoacclimatory changes in blood oxygen binding properties and gill secondary lamellar structure of *Salmo gairdneri. J. comp. Physiol.*, **140**, 255–60.

Nikinmaa, M., Soivio, A. and Railo, E. (1981) Blood volume of *Salmo gairdneri*: influence of ambient temperature. *Comp. Biochem. Physiol.*, **69A**, 767–9.

Nikinmaa, M., Tiihonen, K. and Paajaste, M. (1990) Adrenergic control of red cell pH in salmonid fish: roles of the sodium/proton exchange, Jacobs–Stewart cycle and membrane potential. *J. exp. Biol.*, **154**, 257–71.

Nonnotte, G. and Truchot, J.-P. (1990) Time course of extracellular acid–base adjustments under hypo- and hyperosmotic conditions in the euryhaline fish *Platichthys flesus. J. Fish Biol.*, **36**, 181–90.

Perry, S.F. and Wood, C.M. (1989) Control and coordination of gas transfer in fishes. *Can. J. Zool.*, **67**, 2961–70.

Perry, S.F., Kinkead, R., Gallaugher, P. and Randall, D.J. (1989) Evidence that hypoxemia promotes catecholamine release during hypercapnic acidosis in rainbow trout (*Salmo gairdneri*). *Respir. Physiol.*, **77**, 351–64.

Perutz, M.F. (1983) Species adaptation in a protein molecule. *Mol. Biol. Evol.*, **1**, 1–28.

Randall, D. and Daxboeck, C. (1984) Oxygen and carbon dioxide transfer across fish gills, in *Fish Physiology*, Vol. XA (eds W.S. Hoar and D.J. Randall), Academic Press, New York, pp. 263–314.

Salama, A. and Nikinmaa, M. (1988) The adrenergic responses of carp (*Cyprinus carpio*) red cells: effects of Po_2 and pH. *J. exp. Biol.*, **136**, 405–16.

Soivio, A. and Tuurala, H. (1981) Structural and circulatory responses to hypoxia in the secondary lamellae of *Salmo gairdneri* gills at two temperatures. *J. comp. Physiol.*, **145**, 37–43.

Soivio, A., Nikinmaa, M. and Westman, K. (1980) The blood oxygen binding of hypoxic *Salmo gairdneri. J. comp. Physiol.*, **136**, 83–7.

Speckner, W., Schindler, J.F. and Albers, C. (1989) Age-dependent changes in volume and haemoglobin content of erythrocytes in the carp (*Cyprinus carpio*). *J. exp. Biol.*, **141**, 133–49.

Stevens, E.D., Bennion, G.R., Randall, D.J. and Shelton, G. (1972) Factors affecting arterial pressures and blood flow from the heart in intact, unrestrained lingcod, *Ophiodon elongatus. Comp. Biochem. Physiol.*, **45A**, 681–95.

Tetens, V. and Christensen, N.J. (1987) Beta-adrenergic control of blood oxygen affinity in acutely hypoxia exposed rainbow trout. *J. comp. Physiol.*, **157B**, 667–75.

Tetens, V. and Lykkeboe, G. (1985) Acute exposure of rainbow trout to mild and deep hypoxia: O_2 affinity and O_2 capacitance of arterial blood. *Respir. Physiol.*, **61**, 221–35.

Tetens, V., Lykkeboe, G. and Christensen, N.J. (1988) Potency of adrenaline and noradrenaline for β-adrenergic proton extrusion from red cells of rainbow trout, *Salmo gairdneri. J. exp. Biol.*, **134**, 267–80.

Tetens, V., Wells, R.M.G. and DeVries, A.L. (1984) Antarctic fish blood: respiratory properties and the effects of thermal acclimation. *J. exp. Biol.*, **109**, 265–79.

Thomas, S. (1983) Changes in blood acid–base balance in trout (*Salmo gairdneri* Richardson) following exposure to combined hypoxia and hypercapnia. *J. comp. Physiol.*, **152**, 53–7.

Thomas, S. and Hughes, G.M. (1982) A study of the effects of hypoxia on acid–base status of rainbow trout blood using an extracorporeal blood circulation. *Respir. Physiol.*, **49**, 371–82.

Thomas, S., Kinkead, R., Walsh, P.J., Wood, C.M. and Perry, S.F. (1991) Desensitization of adrenaline-induced red blood cell H^+ extrusion *in vitro* after chronic exposure of rainbow trout to moderate environmental hypoxia. *J. exp. Biol.*, **156**, 233–48.

Van Assendelft, O.W. (1970) *Spectrophotometry of Haemoglobin Derivatives*, Royal Vangorcum, Assen, Netherlands, 152 pp.

Weber, R.E. (1982) Intraspecific adaptation of hemoglobin function in fish to oxygen availability, in *Exogenous and Endogenous Influences on Metabolic and Neural Controls* (eds A.D.F. Addink and N. Spronk), Pergamon Press, Oxford, pp. 87–101.

Weber, R.E. (1990) Functional significance and structural basis of multiple hemoglobins with special reference to ectothermic vertebrates, in *Animal Nutrition and Transport Processes. 2. Transport, Respiration and Excretion: Comparative and Environmental Aspects* (eds. J.-P. Truchot and B. Lahlou), Karger, Basel, pp. 58–75.

Weber, R.E. and Jensen, F.B. (1988) Functional adaptations in hemoglobins from ectothermic vertebrates. *A. Rev. Physiol.*, **50**, 161–79.

Weber, R.E. and Lykkeboe, G. (1978) Respiratory adaptations in carp blood. Influences of hypoxia, red cell organic phosphates, divalent cations and CO_2 on hemoglobin oxygen affinity. *J. comp. Physiol.*, **128**, 127–37.

Weber, R.E., Jensen, F.B. and Cox, R.P. (1987) Analysis of teleost hemoglobin by Adair and Monod–Wyman–Changeux models. Effects of nucleoside triphosphates and pH on oxygenation of tench hemoglobin. *J. comp. Physiol.* **157B**, 145–52.

Weber, R.E., Wood, S.C. and Lomholt, J.P. (1976) Temperature acclimation and oxygen binding properties of blood and multiple haemoglobins of rainbow trout. *J. exp. Biol.*, **65**, 333–45.

Wells, R.M.G., Grigg, G.C., Beard, L.A. and Summers, G. (1989) Hypoxic responses in a fish from a stable environment: blood oxygen transport in the Antarctic fish *Pagothenia borchgrevinki*. *J. exp. Biol.*, **141**, 97–111.

Wood, C.M. (1989) The physiological problems of fish in acid waters, in *Acid Toxicity and Aquatic Animals* (eds R. Morris, E.W. Taylor, D.J.A. Brown and J.A. Brown), Cambridge Univ. Press, Cambridge, pp. 125–52.

Wood, S.C. (1991) Interactions between hypoxia and hypothermia. *A. Rev. Physiol.*, **53**, 71–85.

Wood, S.C. and Johansen, K. (1972) Adaptation to hypoxia by increased HbO_2 affinity and decreased red cell ATP concentration. *Nature New Biol.*, **237**, 278–9.

Wood, S.C. and Johansen, K. (1973) Organic phosphate metabolism in nucleated red cells: influence of hypoxia on eel HbO_2 affinity. *Neth. J. Sea Res.*, **7**, 328–38.

Wyman, J. (1948) Heme proteins. *Adv. Protein Chem.*, **4**, 407–531.

Yamamoto, K., Itazawa, Y. and Kobayashi, H. (1985) Direct observation of fish spleen by an abdominal window and its application to exercised and hypoxic yellowtail. *Jap. J. Ichthyol.*, **31**, 427–33.

Chapter seven

Cardiovascular and ventilatory control during hypoxia

Regina Fritsche and Stefan Nilsson

7.1 INTRODUCTION

The cardiovascular and respiratory physiology of fish has been an area of major interest in comparative physiology for several decades, but the systems that control these functions and the nature of the nervous reflexes that regulate cardiovascular and respiratory functions have received relatively little interest. Research on cardiovascular control systems has been largely restricted to adrenergic vasomotor control – adrenergic neurones and circulating catecholamines – and of course also the double antagonistic cholinergic and adrenergic innervation of the teleost heart. In addition, recent studies have discussed whether catecholamines are involved in ventilatory control in teleosts (Aota *et al.*, 1990; Playle *et al.*, 1990; Kinkead and Perry, 1990, 1991; Perry *et al.*, 1991).

Fish breathe either seawater or fresh water, but some also breathe air, and the two media show important differences in their physical properties. The amount of oxygen per unit volume is small in water compared with air and, together with the high density and viscosity of the water, this results in a relatively large energy expenditure for water breathing. Unidirectional, continuous flow of water over the gills aids in optimizing the oxygen uptake.

Fishes can be exposed to low oxygen levels in nature on both a short-term and a long-term basis. Ecological success of fishes in hypoxic habitats depends on among other things, their ability to sense O_2 changes, and to use such sensory information to make proper and quick adjustments in cardiovascular and ventilatory activity. Some of these changes may be the direct effect of

Fish Ecophysiology. Edited by J. Cliff Rankin and Frank B. Jensen. Published in 1993 by Chapman & Hall, London. ISBN 0 412 45920 5.

lowered oxygen availability for the vascular and ventilatory muscle involved, others may be due to indirect effects, e.g. hypoxia-induced release of vasoactive compounds from vascular endothelia (Burnstock, 1990; Mione *et al.*, 1990), and finally there are the nervous reflexes, involving peripheral and/or central oxygen receptors.

7.2 CHEMORECEPTORS

Central chemoreceptors

Studies of ventilatory control demonstrate a major role for centrally located oxygen receptors in the control of breathing in most fish species studied (Davis, 1971; Dejours, 1973; Bamford, 1974; Holeton, 1977; Wilkes *et al.*, 1981). The function of such central oxygen receptors in cardiovascular responses to hypoxia is, however, not well understood, although it is clear that certain reflexes, e.g. the increase in arterial blood pressure, are triggered by oxygen receptors which are not located within the gills. The receptors responsible for this reflex may be situated in the central nervous system (CNS).

The respiratory centre

Several investigations on central rhythm generation of breathing provide evidence for a separate site of respiratory rhythmogenesis in the brain stem in fish (Ballintijn and Roberts, 1976; Cohen, 1981; Barrett and Taylor, 1984; Shelton *et al.*, 1986; Milsom, 1991). In mammals, the respiratory centre consists of three groups of neurones located in the medulla oblongata, responsible for the basic control of respiratory rhythmicity (Shelton *et al.*, 1986). Respiratory rhythmic neurones are also found in the medulla in fish, more specifically in the reticular formation and in the trigeminal, facial, glossopharyngeal and vagal motor nuclei, which all drive the breathing muscles (Shelton, 1959, 1961; Ballintijn and Roberts, 1976; Ballintijn, 1982). The central rhythm or pattern generator (CRG or CPG) is responsible for initiating the movements of ventilation (Milsom, 1991).

In contrast to the rhythm-generating neurones in mammals, those in fish appear to be more diffusively arranged (Ballintijn and Roberts, 1976). There is no doubt that the respiratory centre is subject to strong modulation from many sources, both peripheral (mechano- and chemoreceptors) and central (higher brain centres) (Burleson and Smatresk, 1989). At least two different opinions exist concerning the operation of the CRG. One view is that the respiratory centre in the medulla functions independently of external stimuli. Recordings from isolated brainstem spinal cord preparations reveal rhythmic discharge in respiratory motor neurones (Milsom, 1991). These preparations receive no afferent sensory information and hence the respiratory rhythmogenesis functions without external input. Transection experiments where the fore- and midbrain have been removed do not stop breathing in lampreys (Kawasaki, 1979), elasmobranchs (Satchell, 1959) and teleosts (Shelton, 1959), giving further support for this view.

Contrasting results concerning rhythmogenesis have been obtained in experiments where muscle relaxants were given to the fish before recording from respiratory neurones. In the carp, *Cyprinus carpio*, two groups of neurones were found that showed no discharge pattern after paralysis and obviously depended on external input for activity. Another group of neurones also stopped firing during paralysis, but artificial, continuous movements of the jaws induced rhythmical activity in these neurones. This response suggests that the neurones receive information from the central oscillator and from proprioceptive sources. The oscillator input may not be sufficient to bring the cell to its threshold but could modulate its sensitivity to other stimuli (Ballintijn and Roberts, 1976). There seems to be little doubt that the presumed respiratory centre of fish is strongly modulated from peripheral receptors, chemical factors and higher centres, and that all aspects of respiratory control result from the integration of central and peripheral processes.

Factors affecting the respiratory centre

In mammals, the three groups of central respiratory neurones previously described are not directly sensitive to changes in CO_2 or H^+. Instead, there exists, located in the ventrolateral surfaces of the medulla, a very responsive chemosensitive area situated close to the entry of the glossopharyngeal and vagal nerves. This area is very sensitive to changes in either CO_2 or H^+. In fact, CO_2 has a very small direct effect on the neurones in this area, but a potent indirect effect. The reason is that CO_2 in contrast to H^+, readily penetrates the blood–brain barrier. Once CO_2 enters the cerebrospinal and interstitial fluid it reacts with water to form carbonic acid, which in turn dissociates to form H^+ and bicarbonate. The protons have a potent effect on stimulating the neurones in the chemosensitive area, which excites the other portions of the respiratory centre. Oxygen has no direct effect on the respiratory centre in mammals but instead exerts its effect upon the peripheral receptors.

It is well known that fish show ventilatory responses to changes in external and internal oxygen levels. Graham *et al.* (1990) showed that hypercapnia induced a pronounced and persistent increase in ventilation in the Atlantic big skate, *Raja ocellata*. The blood and environmental oxygen levels were maintained constant, thus it appeared that the effect on ventilation was mediated through changes in CO_2 or acid–base status. There is little information available on central v. peripheral effects of changes in gaseous composition in fish. Wood *et al.* (1990) showed that CO_2 penetrates freely through the blood–brain barrier of the skate, whereas H^+ and bicarbonate did not equilibrate across the barrier. The pH of the cerebrospinal fluid was 0.2 units above the pH of arterial blood during exposure to hypercapnia, suggesting an active acid–base regulation of the cerebrospinal fluid. On the other hand, intracranial perfusions of hypercapnic and low pH solutions failed to elicit any ventilatory responses in the bowfin, *Amia calva*, indicating the lack of central chemoreceptors in this species (Hedrick *et al.*, 1991).

It is known from studies on mammals that a variety of neurotransmitters also play a role both in peripheral and in central chemoreception. Microinjections into the central respiratory area have revealed specific effects of different drugs on the respiratory neurones. Cholinergic and monoaminergic neurones have been demonstrated in the respiratory medullary areas of the brain (Dempsey *et al.*, 1986). The effects of acetylcholine on medullary neurones are inconsistent, and both excitatory and inhibitory responses have been described (Dempsey *et al.*, 1986). Noradrenaline and dopamine are the predominant aromatic monoamines in the brain, whereas adrenaline is found in very low concentrations (Hsiao *et al.*, 1989). Axons containing serotonin have been found close to the chemosensitive area. Most studies conclude that the direct central effect of noradrenaline, adrenaline and dopamine is a hypoventilation (Dempsey *et al.*, 1986). Unfortunately, since catecholamines have a variety of actions on the overall chemoreceptor mechanisms, general conclusions are difficult to make.

Aromatic monoamines do not pass the blood–brain barrier in mammals whereas they may do so in fish (Peyraud-Waitzenegger *et al.*, 1979). The responses to systemically administered monoamines in mammals are therefore probably due to peripheral chemoreceptors, whereas the same type of experiment in fish would be difficult to interpret. Nevertheless, several studies on fish have employed pharmacological injections of catecholamines to elucidate any role of these substances in the control of breathing (Peyraud-Waitzenegger, 1979; Metcalfe and Butler, 1988; Aota *et al.*, 1990; Kinkead and Perry, 1990, 1991; Playle *et al.*, 1990; Perry *et al.*, 1992). The conflicting results obtained, some suggesting a hyperventilatory and others a hypoventilatory effect, may result from difficulties in distinguishing between peripheral and central effects (review, Perry *et al.*, 1992).

Peripheral chemoreceptors

Organization of peripheral chemoreceptors

The peripheral chemoreceptors in mammals are located in the carotid and aortic bodies. The afferent, sensory nerve fibres from the carotid bodies run in Hering's nerves to join the glossopharyngeal nerves, and continue into the respiratory centre of the medulla, while the afferent fibres from the aortic bodies run in the vagi.

The gills are considered to be the phylogenetic antecedents of the carotid and aortic regions in mammals. Peripheral chemoreceptors have been localized to the gill region in both elasmobranchs (Butler *et al.*, 1978) and teleosts

(Bamford, 1974; Smith and Jones, 1978; Fritsche and Nilsson, 1989; Burleson and Smatresk, 1990a), and the response characteristics of these receptors indicate that they are homologous to the oxygen receptors of the mammalian carotic and aortic bodies. Other suggested peripheral localization sites in fish include the mouth and pharynx area and the vasculature (De Kock, 1963; Freihofer, 1978; Vasilevskaya and Polyakova, 1979, 1981; Walker *et al.*, 1981). Regardless of the specific site of localization, the pattern of ventilatory responses to aquatic hypoxia suggests that the receptors monitor changes in both the external and the internal environment (Burleson and Samtresk, 1990 a,b).

Mechanistic studies of O_2 chemoreception have only been performed in mammals. Figure 7.1 shows the presumed mammalian 'chemoreceptor unit'. The unit consists of two types of cells, Type I (glomus cells) and Type II (sustentacular cells), and also nerve terminals (both sensory and motor) and fine blood vessels. The Type I cells are characterized by numerous dense-core vesicles (containing neurotransmitters) while the Type II cells have a supportive ('glial') function. Nerve endings are in synaptic contact with the glomus cells, and the fine network of blood vessels that supply the chemoreceptive elements have a rich adrenergic and possibly also cholinergic innervation (Jones and Milsom, 1982).

The exact mechanism of O_2 chemoreception is still unknown, but two main theories exist.

1. The afferent nerve endings are directly sensitive to low oxygen levels and their sensitivity can be modified by neurotransmitters released from the glomus cells.

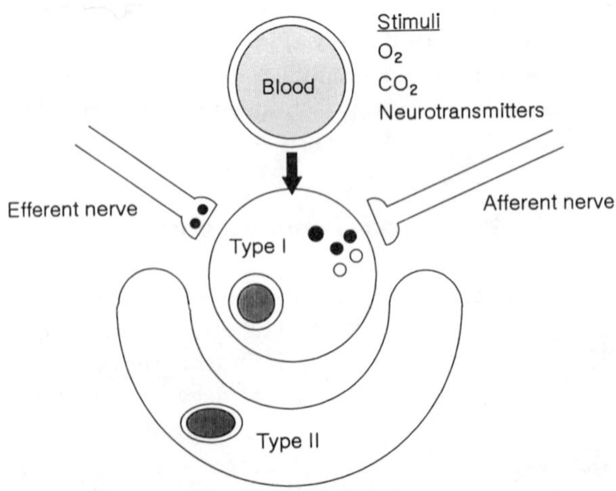

Fig. 7.1 Simplified diagram of the presumed mammalian O_2 chemoreceptive unit showing Type I and Type II cells, afferent and efferent nerve endings and blood vessels. Reproduced with permission from Burleson (1991).

2. The glomus cells are the actual receptor site, releasing neurotransmitters when stimulated by hypoxia/hypoxaemia, which in turn stimulates the afferent nerve endings.

In either case, the afferent signal is transmitted to the respiratory and cardiovascular centres of the medulla, where the information is integrated and hypoxic reflexes are coordinated to correct for the low arterial oxygen content.

Factors affecting the peripheral chemoreceptors

An increase in either CO_2 or H^+ ions in air-breathing vertebrates is known to excite peripheral chemoreceptors, causing an increase in respiration. However, the direct effect of these factors on the respiratory centre is so powerful that the peripheral effects need be of little consequence. Reduced oxygen levels, on the other hand, can affect respiration by acting through the peripheral receptors.

It has been suggested that, in fish, variation in external or internal oxygen levels is the predominant factor affecting ventilation. Reduction in inspired oxygen stimulates ventilation in most teleosts studied. On the other hand, CO_2/H^+-responsive structures involved in the control of breathing in fish have received little attention. Exposure to hypercapnia during hyperoxic conditions stimulates breathing both in elasmobranchs (Heisler, 1989; Graham *et al.*, 1990; Wood *et al.*, 1990) and in rainbow trout, *Oncorhynchus mykiss* (Kinkead, 1990). In contrast, infusion of ammonium bicarbonate (NH_4HCO_3), which produces unchanged arterial pH and oxygen content but increased carbon dioxide tension, did not stimulate ventilation in the bowfin, *Amia calva*. Furthermore, HCl infusion during hyperoxia lowered arterial pH but did not reduce oxygen content and no hyperventilation was observed (McKenzie *et al.*, 1991). The importance of oxygen in respiratory control in fish is not questioned, but present results also suggest that CO_2/H^+ may play a more important role in this regard than previously thought (Thomas *et al.*, 1983; Graham *et al.*, 1990; Wood *et al.*, 1979). Neurotransmitters can affect the chemosensory output, and a number of substances (dopamine, noradrenaline, serotonin and acetylcholine) have been identified in the mammalian carotid body, specifically in the Type I cells and in afferent and efferent nerve endings. A few peptides and amino acids have also been found (Chiocchio *et al.*, 1971).

Neuroepithelial cells in fish gills resemble the Type I cells in mammals and they have been shown to contain serotonin (Dunel-Erb *et al.*, 1982; Laurent, 1984; Bailly *et al.*, 1989). The fact that aquatic hypoxia causes release of serotonin from these cells in trout has been interpreted as a chemoreceptive role of these cells (Laurent, 1984).

Many experiments have been performed to elucidate the role of neurotransmitters in chemoreception in mammals. Acetylcholine stimulates respiration through a nicotine-like action on peripheral chemoreceptors (Dempsey *et al.*, 1986). Noradrenaline excites the chemoreceptors and causes an increase of respiration (Milsom and Sadig, 1983). This could result either from a direct action on the carotid body or from a secondary effect caused by a local vasoconstriction, which will result in hypoxaemia and hence increased activity. Dopamine is selectively depleted from the Type I cells during hypoxia (Hanbauer and Hellström, 1978; Fitzgerald *et al.*, 1983), and injections/infusions of dopamine lead to inhibition of chemoreceptor activity and respiration (Nishino and Lahiri, 1981; Donnely *et al.*, 1981; Folgering *et al.*, 1982; Milsom and Sadig, 1983; Chow *et al.*, 1986). Any general effect of serotonin on carotid body discharge has not been established, but most results suggest a stimulation of chemoreceptor activity of this indolamine (Dempsey *et al.*, 1983). In recent experiments on an isolated, perfused gill arch preparation from rainbow trout, the effects of different putative neurotransmitters on chemoreceptor activity have been investigated. Acetylcholine was found to be the most potent factor in stimulating the discharge. Both dopamine and serotonin caused increased activity followed by a decrease, while noradrenaline and adrenaline had no significant effect on the chemoreceptor discharge (Burleson, 1991). Surprisingly, propranolol inhibited O_2-receptor discharge during both normoxia and hypoxia, which implies a β-adrenergic mechanism involved in O_2 chemoreceptor (Burleson and Milsom, 1990). These findings suggest that neurotransmitters are important in modifying the chemoreceptor output in fish as well as in mammals.

Interaction between central and peripheral receptor systems

Most neurones in the respiratory centre have inherent rhythmicity. However, input from peripheral chemo- and mechanoreceptors will continuously influ-

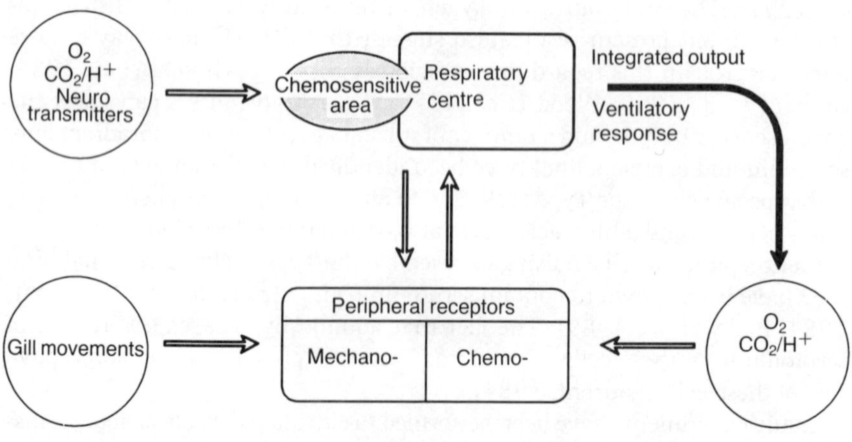

Fig. 7.2 Diagram illustrating the chemosensitive units responsible for ventilatory control in response to changes in P_{O_2}, P_{CO_2}, pH and neurotransmitters. Note the interaction between central and peripheral chemoreceptors.

ence the respiratory centre as the chemical composition of the blood (water) fluctuates. At the same time efferent fibres (sympathetic and parasympathetic) innervating the peripheral chemoreceptors will modulate their activity (O'Regan and Majcherczyk, 1982). Figure 7.2 summarizes the chemosensitive elements responsible for mediating ventilatory adjustments in response to changes in P_{O_2}, P_{CO_2} and H^+ in the external or internal environment.

7.3 MECHANORECEPTORS

Several types of receptors sensitive to mechanical stimuli occur in the branchial region of fish (Nilsson, 1984). Inflation of the pharynx region in the elasmobranch the spiny dogfish, *Squalus acanthias*, leads to inhibition of the respiratory movements, a reflex which shows many similarities with the mammalian 'Hering–Breuer' reflex (Satchell, 1959). Nerve discharges from the afferent branchial fibres in response to mechanical deflection of the branchial chamber, in fish pretreated with muscle relaxant, show a discharge rate proportional to the logarithm of the amplitude of the mechanical stimuli. This suggests a role of these receptors in controlling the respiratory rate in relation to the amplitude of each respiratory cycle (Satchell and Way, 1962).

Mechanoreceptors with afferent nerve fibres in the IXth and Xth cranial nerves have been demonstrated in teleost fish. One set of receptors is activated when the gill filaments are displaced, and the second group is associated with the gill rakers (Sutterlin and Saunders, 1969). In the carp, changes in the electromyograms of the respiratory muscles, induced by loading of the respiratory pumps, suggested a direct mechanosensitive muscle control (Ballintijn and Alink, 1977).

7.4 CATECHOLAMINE RELEASE

The catecholamines adrenaline (epinephrine) and noradrenaline (nor-epinephrine) are synthesized and stored in both adrenergic nerves and chromaffin tissue. Release of the amines normally occurs due to activity in preganglionic autonomic (cholinergic) neurones which reach the adrenergic (postganglionic) neurone, or chromaffin cell. Physical disturbance (handling, exposure to air, 'asphyxia') produces a marked release of catecholamines from chromaffin stores in representatives of all fish groups studied. Although it is clear that hypoxia is a potent stimulus for catecholamine release (Butler *et al.*, 1978; Perry *et al.*, 1991), it is often difficult to assess the role of the ensuing hypoxia, compared with other 'stress factors', in the response (Mazeaud, 1971; Abrahamsson, 1979; Abrahamsson *et al.*, 1979; Wahlqvist and Nilsson, 1980).

The mechanism of catecholamine release in fish has been best studied in teleosts, where the chromaffin tissue resides in the walls of the large veins within the anterior part of the kidney (head kidney). In the Atlantic cod, *Gadus morhua*, electrical stimulation of the autonomic innervation of the tissue causes a catecholamine release into the veins, which is large enough to affect both the heart and branchial vasculature (Nilsson, 1976; Nilsson *et al.*, 1976; Holmgren, 1977; Wahlqvist, 1981).

A gradually (over 30 min) progressing hypoxia produced a graded increase in plasma adrenaline levels in the Atlantic cod, while the noradrenaline levels remained largely unchanged. A more rapid drop in water oxygen levels induced release of both amines, and there is clear evidence that adrenaline was released as a direct effect of hypoxaemia, unrelated to blood acidosis, and independent of the innervation of the chromaffin tissue (Perry *et al.*, 1991).

7.5 METHODS OF STUDYING VENTILATION IN WATER-BREATHING ANIMALS

Studies of fish respiration have employed a variety of techniques, either to get an indication of respiratory water flow (indirect techniques), or directly to measure the volume of water pumped over the gills per unit time ($\dot{V}_W$). It may be invalid to make comparisons of the observed respiratory responses to various stimuli between studies employing different methods for measuring $\dot{V}_W$.

The **indirect techniques** used to estimate branchial water flow require relatively small surgical involvement. One frequently used approach is to measure the opercular or buccal pressure, simply by inserting cannulae in these cavities (Saunders and Sutterlin, 1971; Burleson and Smatresk, 1989, 1990a,b; Hedrick *et al.*, 1991; Fritsche *et al.*, 1992). This gives values for breathing frequency (f_V, but the amplitude of the signal is not necessarily proportional to the ventilatory stroke volume (V_{SV}). Buccal or opercular pressure recordings used together with the van Dam technique, give accurate information of both $\dot{V}_W$ and f_V (Aota *et al.*, 1990).

Another approach is the use of external electrodes for detecting the electrobranchiogram (EBG) (Peyraud-Waitzenegger, 1979; Berschick *et al.*, 1987; Gehrke and Fielder, 1988).

While the described techniques all allow the animal to move freely and the surgical procedure is minimal, these methods give no direct measure of branchial water flow, but merely indications of changes in ventilation. Nevertheless, in studies where minimal disturbance of the fish is crucial, these methods are of considerable importance.

The most frequently used technique for accurate **direct measurements** of $\dot{V}_W$ is the one first described by van Dam (1938). The technique involves the separation of expired from inspired water by a rubber membrane fitted around the jaws of the fish. The membrane is then attached to a divider of a box so that two compartments are obtained and the movements of water from the cephalic to the caudal compartment reflects the ventilatory water flow. The major disadvantage of this technique is that the membrane may restrict the breathing, and of course also movements, of the fish (Cameron and Davis, 1970). Moreover, the total $\dot{V}_W$ without any information of f_V and V_{SV} is obtained, necessitating the simultaneous use of opercular or buccal cannulae.

Electromagnetic flow probes have been used for direct $\dot{V}_W$ measurements (Lomholt and Johansen, 1979; Rantin and Johansen, 1984; Kinkead *et al.*, 1991) (Fig. 7.3). The probes can be placed either in front of the mouth (Rantin and Johansen, 1984; Kinkead *et al.*, 1991; Fritsche, 1992), reflecting inspired $\dot{V}_W$, or behind the operculum, measuring expired $\dot{V}_W$. In neither case will the fish be restrained from moving, and another benefit of this method is the continuous recording obtained (compared with the van Dam technique) and the possibility to monitor f_V from the pulsatile flow signal (Fig. 7.3).

7.6 VENTILATORY RESPONSES TO HYPOXIA

All water breathers show either behavioural or ventilatory responses to aquatic hypoxia. Fish swim away from the oxygen-poor area (avoidance reactions) and/or show ventilatory responses. Depending on their ventilatory responses, fish are considered to belong to either of two groups: 'oxygen conformers' or 'oxygen regulators'. Oxygen regulators are able to maintain or even increase their oxygen uptake during hypoxia until the environmental oxygen tension falls below a critical level, whereas oxygen conformers immediately reduce their oxygen uptake as P_{O_2} falls. These strategies reflect the compromise between oxygen uptake and the energetic cost of breathing. Cyclostomes and teleosts are generally oxygen regulators, and increase ventilation during hypoxia, whereas most elasmobranchs are oxygen conformers.

Elasmobranchs

Most dogfish species studied (*Mustelus* sp., *Scyliorhinus* sp., *Squalus* sp.) show very slight changes in breathing rate and amplitude in response to hypoxia or anoxia (Ogden, 1945; Satchell, 1961; Piiper *et al.*, 1970). *Scyliorhinus canicula* responds with a small increase in total $\dot{V}_W$ and f_V during exposure to moderate (95 mmHg) hypoxia. As hypoxia becomes more severe (54 mmHg), no change in $\dot{V}_W$ and a depression of f_V is observed (Piiper *et al.*, 1970). In the dogfish, *Squalus acanthias*, only a small decrease in f_V followed by an increase occurs. Destruction of the pseudobranch (which has been suggested as a possible O_2-receptor location site) did not abolish the f_V response to anoxia, whereas deafferentation of the branchial nerves eliminated the fall in f_V during the first minutes of anoxia but not the subsequent increase. It is suggested that the secondary 'speeding' is the result of a rise in CO_2, affecting the CNS (Satchell, 1961).

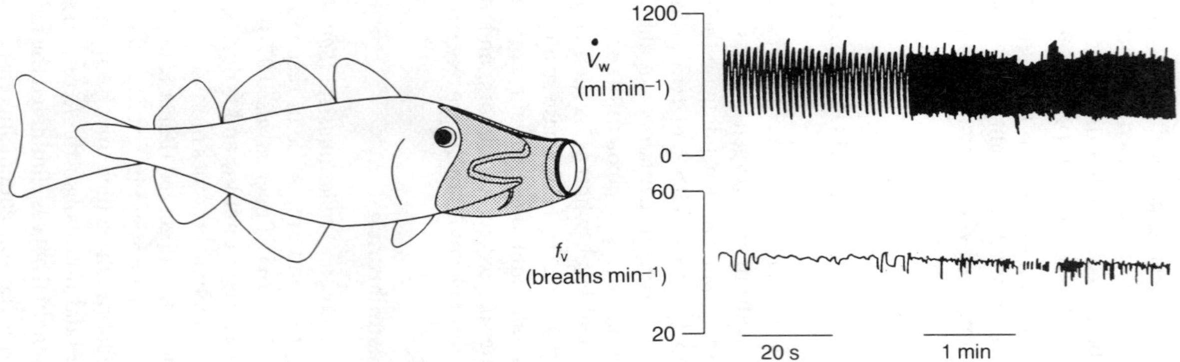

Fig. 7.3 Schematic drawing of a fish equipped with a mask carrying an electromagnetic flow probe, and an example of an original trace of $\dot{V}_W$ and f_V obtained using this technique.

The stationary, bottom-dwelling species, *Torpedo marmorata*, which may well, in contrast to other elasmobranch species, encounter hypoxia naturally, shows a marked increase in frequency and amplitude of breathing during hypoxia. Extreme hypoxia can be endured for many hours although ventilation stops after a while. The onset of the respiratory response to hypoxia is much slower in *Torpedo* than in most teleost species studied, and this has been used to argue against externally located oxygen receptors responsible for ventilatory adjustments in this species (Hughes, 1978).

Teleosts

The reflex responses of teleost fish to hypoxia have been examined in great detail. However, information about the location and innervation of the receptors mediating these reflexes is scarce. Most species studied increase ventilation during hypoxia/hypoxaemia to maintain a sufficient gradient of oxygen between the water and the blood (Randall and Shelton, 1963; Saunders and Sutterlin, 1971; Randall and Jones, 1973; Smith and Jones, 1982; Rantin and Johansen, 1984; Berschick *et al.*, 1987; Gehrke and fielder, 1988; Burleson and Smatresk, 1989, 1990a; Aota *et al.*, 1990; Bushnell and Brill 1991; Kinkead *et al.*, 1991), but a few species reduce ventilation or show no response at all (Burggren and Randall, 1978). It seems that in oxygen regulators, ventilation increases to a certain level of P_{O_2}, and when the cost of ventilation exceeds the gain of the oxygen uptake, $\dot{V}_W$ instead decreases.

The gill region has received by far most attention as a possible O_2-receptor site. Although the receptors involved in the cardiac response to hypoxia have been much studied, information about receptors involved in the reflex control of breathing is scarce. The receptors responsible for the hypoxic bradycardia are most likely external, facing the water, whereas the receptors involved in the ventilatory reflex seem to be both external and internal. Experiments where the aquatic P_{O_2} was changed without any accompanying change in arterial P_{O_2}, stimulated ventilation (Saunders and Sutterlin, 1971) and both external and internal injections of NaCN in the channel catfish, *Ictalurus punctatus*, elicited hyperventilation (Burleson and Smatresk, 1990). Moreover, when the blood oxygen content is experimentally reduced (internal hypoxia/hypoxaemia) by carbon monoxide (Holeton, 1971) or anaemia (Wood *et al.*, 1979; Smith and Jones, 1982), ventilation is also stimulated.

Although these results suggest the existence of receptors monitoring changes in oxygen in both the environment and the internal milieu, it tells nothing about the location and innervation of the receptors. Experiments with branchial nerve sectioning and recording of branchial afferent nerve activity have attempted to localize the receptors, but several nerve section studies have failed to abolish ventilatory responses to hypoxia. The gills and branchial pump muscles are innervated by branches of cranial nerves V, VII, IX and X (Nilsson, 1984). Sectioning of V and VII stops regular breathing movements through destruction of the motor supply to the branchial pump musculature, and the IXth and Xth cranial nerves also carry motor fibres to the gills (Saunders and Sutterlin, 1971; Fritsche and Nilsson, unpublished observations). In the sea raven, *Hemitripterus americanus*, breathing responses to hypoxia were similar before and after sectioning of the IXth and Xth cranial nerves. It was suggested that receptors responsible for the increase in breathing are located in the CNS, or in the mouth/pharynx area and innervated by the Vth and VIIth cranial nerves (Saunders and Sutterlin, 1971).

In the Atlantic cod, *Gadus morhua*, partial denervation had little effect on the ventilatory response to hypoxia, and not even after sectioning the IXth and all of the branchial branches of the Xth, were the ventilatory responses completely abolished (Fig. 7.4). The fact that hyperventilation in the Atlantic cod starts immediately when water oxygen tension declines (Fig. 7.5), i.e. before any reduction in arterial oxygen tension and oxygen content is likely to occur, argues for externally located receptors (Kinkead *et al.*, 1991; Fritsche, unpublished data).

The pseudobranch has attracted interest as a possible site for oxygen receptors involved in the reflex regulation of breathing (Laurent, 1967, 1969). The pseudobranch receives blood that has already passed through the gills, so the receptors would be in a position to monitor arterial gas tensions. However, denervation of the pseudobranch has failed to abolish respiratory responses to hypoxia (Saunders and Sutterlin, 1971; Randall and Jones, 1973). In the channel catfish, *Ictalurus punctatus*, which lacks pseudobranchs, all ventilatory responses to hypoxia were abolished after branchial denervation (Burleson and Smatresk, 1990a). The receptors responsible for changes in breathing appear to be diffusively arranged on all four pairs of gill arches and perhaps also in the pseudobranch.

The best evidence for external and internal branchial O_2-receptors comes from direct recordings from chemoreceptor afferents. It has been shown in an

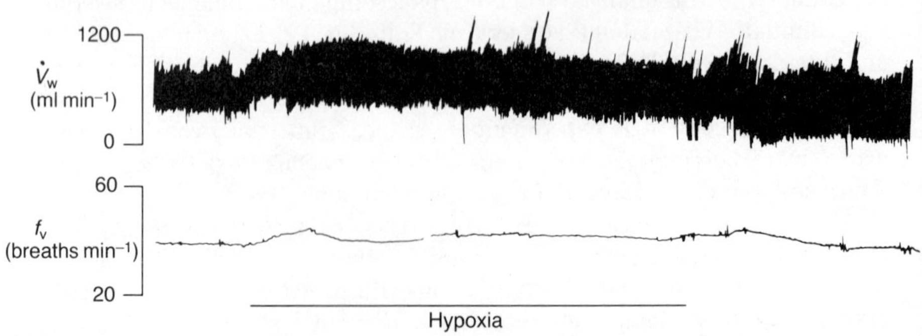

Fig. 7.4 Recordings of $\dot{V}_w$ and f_v in an Atlantic cod, *Gadus morhua*, with the cranial nerves to all pairs of gill arches bilaterally sectioned. The tracings show the response to aquatic hypoxia (80 mmHg). Note the remaining increase in both V_w and f_v.

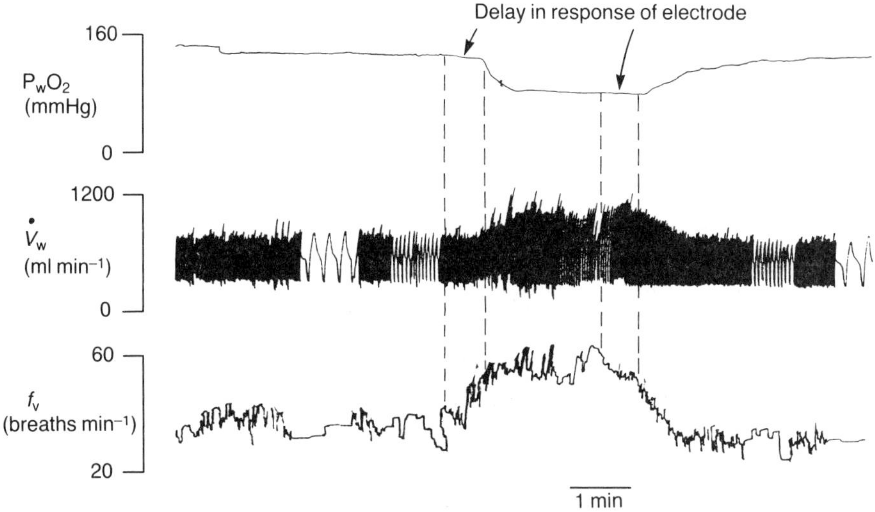

Fig. 7.5 Recordings showing the ventilatory response to mild hypoxia (110 mmHg) in an Atlantic cod. Note the immediate increase in V_W and f_V as water PO_2 is reduced.

isolated perfused gill arch preparation, that submerging the preparation in a hypoxic solution, or perfusing it with hypoxic Ringer's solution, stimulates the discharge from afferent fibres, thus giving evidence for oxygen-sensitive chemoreceptors in this area (Milsom and Brill, 1986; Burleson and Milsom, 1990a, b).

Dipnoans

Aquatic hypoxia may serve as a major selective force in the evolution of air breathing, and several fish groups show (independent) development of air breathing systems. In conditions of aquatic hypoxia, there is a clear advantage to those animals that have the possibility to breathe atmospheric oxygen. The respiratory adjustments used by lungfish during hypoxia include: (1) an increase in the flow of water across the gills or (2) an increase in the rate of surfacing to breathe air. Exposure to hypoxic water evokes different breathing responses in lungfish depending on the relative role of aquatic and air breathing in overall gas exchange.

The Australian lungfish, *Neoceratodus forsteri*, which is more or less exclusively aquatic, reacts to aquatic hypoxia by rapidly increasing gill ventilation (Johansen *et al.*, 1967). However, both the oxygen extraction and overall uptake decrease during the same period. After a few minutes, the number of air breaths increases from zero to more than three breaths in the first half hour of hypoxia. The stimulation of air breathing is generally much slower in onset than that of gill ventilation.

Among the obligate air breathers, the African (*Protopterus* spp.) and South American lungfish (*Lepidosiren paradoxa*), the breathing responses to aquatic hypoxia are different. Aquatic hypoxia causes no ventilatory response in these fish. This implies the absence of external chemoreceptors sensing changes in oxygen in these species (Johansen and Lenfant, 1968). However, introduction of hypoxic gas into the lung of the African lungfish causes a prompt and sharp increase in the rate of air breathing (Johansen and Lenfant, 1968; Lahiri *et al.*, 1970). Injection of hypoxic blood or NaCN (internal hypoxia) into the afferent branchial arteries of the African lungfish elicited hyperventilation, and bilateral sectioning of the nerves to the first, second and third gill arches markedly reduced the response to hypoxaemia and NaCN. These results demonstrate that *Protopterus* possesses internal chemoreceptors in the lungs and the gill vasculature which are responsible for the respiratory responses to hypoxia (Lahiri *et al.*, 1970).

7.7 CIRCULATORY RESPONSES TO HYPOXIA

Teleosts exposed to environmental hypoxia generally show bradycardia and an increase of the arterial blood pressure. While it seems clear that a direct effect of hypoxia on the vasculature, and probably also on the heart, may occur, it is quite clear that both branchial and extra-branchial oxygen receptors trigger reflexes that affect both arterial blood pressure and cardiac functions (Laurent, 1984; Nilsson, 1984). The autonomic nervous system, both the cranial autonomic ('parasympathetic') and the spinal autonomic ('sympathetic') divisions are involved in this control, together with the adrenergic influence of catecholamines released from the chromaffin tissue.

Cardiovascular control systems

Vasomotor nerves in fish have been thought to derive almost exclusively from the spinal autonomic ('sympathetic') division of the autonomic nervous system, and these fibres are very often adrenergic (i.e. the neurotransmitter released is adrenaline and/or noradrenaline). In addition, new evidence suggests that vasomotor control in many organs, notably the gut, depends also on neurones that release non-adrenergic, non-cholinergic (NANC) transmitters. A number of substances, particularly neuropeptides ('regulatory peptides'), but also other amines such as serotonin (5-hydroxytryptamine, 5-HT), may act as neurotransmitters in the vertebrate cardiovascular system, either alone or as co-transmitters (Burnstock and Griffith, 1988; Holmgren, 1989). The control of the vertebrate heart also involves several 'new' substances, many of them peptides, but substantial cardiac control takes place by the 'classical' neurotransmitters: acetylcholine and the catecholamines (Holmgren and Nilsson, 1991; Nilsson and Holmgren, 1991).

Organization of vasomotor and cardiac nerves

The autonomic nervous system provides vasomotor and cardioregulator innervation in practically all the vertebrates, although the cyclostomes (particularly the myxinoids) have a poorly developed autonomic nervous system, and probably lack cardiac innervation (extensive review: Nilsson, 1983). The paravertebral autonomic ganglia of elasmobranchs, which are the equivalent of the 'true' sympathetic chains of the teleosts and tetrapods, are connected to the spinal nerves solely by white rami communicantes. This means that vasomotor fibres to the somatic vasculature, if at all present, do not run in the spinal nerves. Vasomotor fibres to the viscera run in the several splanchnic nerves, and innervate visceral blood vessels. On each side, the most anterior sympathetic ganglia form the axillary bodies, which in addition contain catecholamine-containing and -releasing chromaffin cell masses (Abrahamsson, 1979; Nilsson, 1983). The cardiac innervation is solely vagal, cholinergic and inhibitory, with no contribution of fibres from the paravertebral ('sympathetic') ganglia. There is no vasomotor innervation of the branchial vasculature of elasmobranchs (Metcalfe and Butler, 1984).

Sympathetic chains are well developed in teleosts and other actinopterygians, and these chains continue into the head with sympathetic chain ganglia in connection with the cranial nerves. Fibres of spinal autonomic ('sympathetic') origin can enter (1) the cranial nerves (and run to the branchial vasculature), (2) the spinal nerves (and run to the somatic blood vessels), and finally (3) the anterior splanchnic nerve (and to a lesser degree the posterior splanchnic (vesicular) nerve), providing a vasomotor innervation of the viscera (Nilsson, 1973, 1976, 1983).

Vasomotor control of the teleost branchial vasculature involves cranial autonomic fibres, which are probably mainly cholinergic (extensive reviews: Laurent, 1984; Nilsson, 1984). In addition, histochemical and electron microscopical studies have demonstrated adrenergic nerves in the gill vasculature of several teleosts (Donald, 1984, 1987; Dunel-Erb and Bailly, 1986; Nilsson, 1986), and the function of the adrenergic vasomotor innervation in the gills has been further confirmed in physiological studies (Nilsson, 1984). Recent studies demonstrate the presence of serotonergic nerves and other types of 5-HT-containing cells in the gills of teleosts (Bailly *et al.*, 1989). In most teleosts and in the holostean *Lepisosteus platyrhincus*, spinal autonomic fibres innervate the heart mainly along the vagosympathetic trunks. The fibres are excitatory, and histochemical and ultrastructural studies as well as physiological and pharmacological experiments reveal that they are, at least to a large extent, adrenergic (Govyrin and Leont'eva, 1965; Yamauchi and Burnstock, 1968; Gannon and Burnstock, 1969; Holmgren, 1977; Donald and Campbell, 1982; Nilsson, 1983).

NANC neurotransmitters of the cardiovascular system

One major action of 5-HT on the cardiovascular system in teleosts is a marked

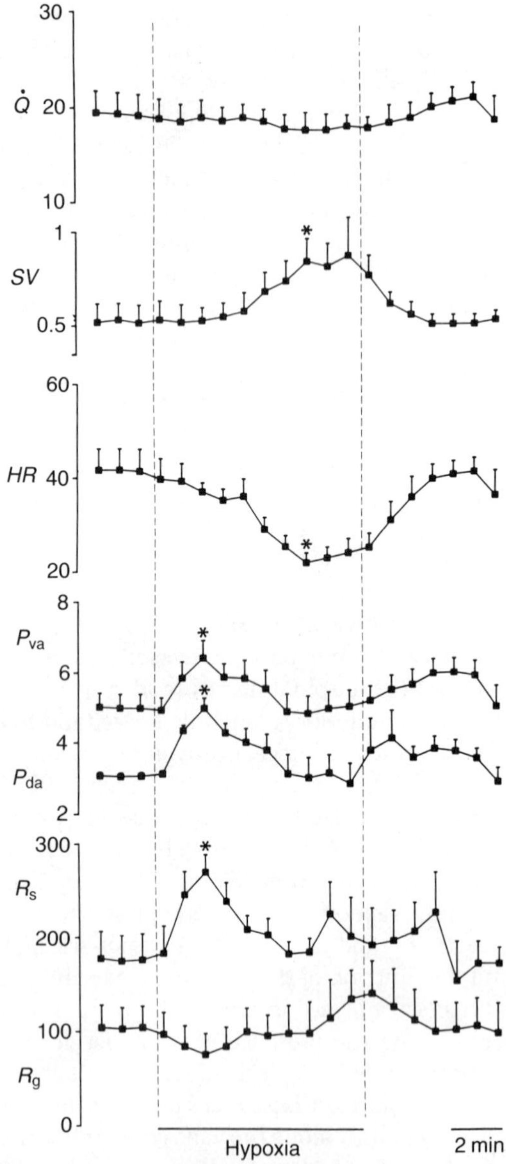

Fig. 7.6 Effects of 8 min hypoxia (P_{WO_2} = 30–40 mmHg) in the Atlantic cod, showing effects on cardiac output ($\dot{Q}$, ml min^{-1} kg^{-1}), cardiac stroke volume (*SV*, ml beat^{-1} kg^{-1}), heart rate (*HR*, beats min^{-1}), ventral and dorsal aortic blood pressure (P_{va} and P_{da}, respectively, kPa) and systemic and branchial vascular resistance (R_s and R_g, respectively, kPa kg min ml^{-1}). Asterisks show statistically significant (Wilcoxon signed-ranks test $P < 0.02$) difference determined in min- or max-points compared with the corresponding pre-hypoxic value. Note offset of the bradycardia by increased stroke volume, and different delays before the minimum in heart rate (about 5–6 min) and the maxima in blood pressures (2 min).

constriction of the branchial vasculature, an effect that can be abolished by the 5-HT antagonist methysergide (Östlund and Fänge, 1962; Reite, 1969; Katchen *et al.*, 1976). 5-HT also caused a decrease of dorsal aortic blood pressure in the five teleost species studied by Reite (1969), and in anaesthetized Atlantic cod and eel in the same experiments, the ventral aortic pressure increased markedly. In the Atlantic cod, injection of 5-HT produced an increase in ventral aortic pressure and heart rate, while mean blood pressure and pulse pressure in the dorsal aorta decreased, again indicating a constriction of the branchial vessels (R. Fritsche, unpublished).

Histochemical studies of rainbow trout gills by Dunel-Erb and co-workers demonstrate the presence of 5-HT in neurones (and nerve fibres), polymorphous granular cells (PGCs) and neuroepithelial cells (NECs) (Dunel-Erb *et al.*, 1982, Bailly *et al.*, 1989). The neurones are of vagal origin, innervating the proximal part of the efferent filament artery (and sphincter), the efferent lamellar arterioles and central venous sinus. It is thus possible that 5-HT, released from nerves or other cells within the gills, causes a constriction of the efferent arteries, thereby increasing lamellar recruitment during conditions such as hypoxia, exercise or stress.

Hypoxic bradycardia

In the elasmobranch *Scyliorhinus canicula*, hypoxia induces a marked bradycardia which is compensated for by an increase of cardiac stroke volume to keep cardiac output more or less stable (Short *et al.*, 1977; Taylor *et al.*, 1977). In the spiny dogfish, *Squalus acanthias*, the effect was only partially abolished by sectioning cranial nerves IX and X, and Satchell (1961) concluded that central oxygen receptors were involved. However, in *Scyliorhinus*, oxygen receptors were also demonstrated in areas innervated by the facial and trigeminal nerves (cranial nerves V and VII). Bilateral sectioning of all four cranial nerves (V, VII, IX and X) abolished the hypoxic bradycardia, and released all vagal cardioinhibitory nervous tonus (Butler *et al.*, 1977), showing lack of centrally located oxygen receptors involved in the hypoxic bradycardia reflex in this species. In teleosts, a reduction in oxygen tension triggers branchial oxygen receptors in the first pair of gill arches, which send fibres in the glossopharyngeal and vagus to a cardiovascular control centre in the medulla (Figs. 7.6, 7.7). The hypoxia may also be detected by receptors located within the central nervous system.

Branchial oxygen receptors activate cardiac vagal fibres (X), which in most teleosts produce hypoxic bradycardia (Randall and Shelton, 1963; Randall and Smith, 1967; Stevens *et al.*, 1972; Bamford, 1974; Holeton, 1977; Daxboeck and Holeton, 1978; Smith and Jones, 1978; Wood and Shelton, 1980; Randall and Daxboeck, 1982, 1984; Fritsche and Nilsson, 1989, 1990; Fritsche, 1990). In some species, e.g. the sea raven, *Hemitripterus americanus* (Saunders and Sutterlin, 1971), the five-bearded rockling, *Ciliata mustela*

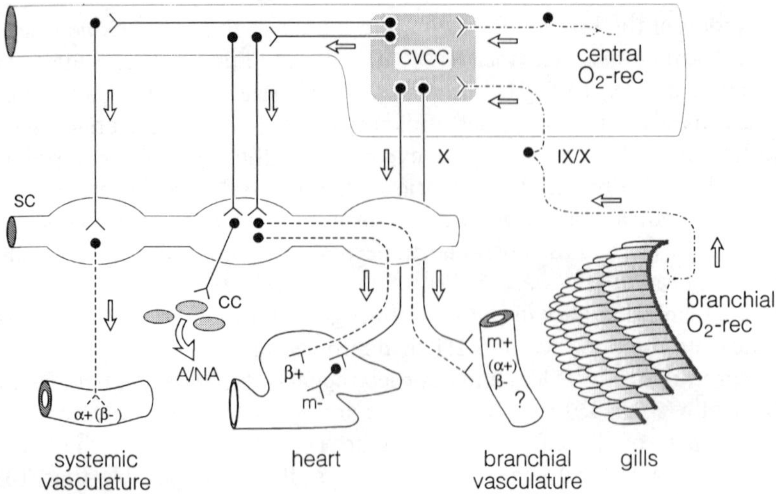

Fig. 7.7 Summary of nervous pathways involved in cardiovascular responses to hypoxia in a teleost fish. α, β, adrenoceptors; A/NA, catecholamines (adrenaline and noradrenaline) released from chromaffin tissue; cc, chromaffin cells of the head kidney; CVCC, cardiovascular control centre; m, muscarinic cholinoceptors; sc, sympathetic chains; IX, glossopharyngeal nerve; X, vagus nerve; + and - refer to excitatory (cardioacceleration or vasoconstriction) and inhibitory (cardioretardation or vasodilatation) effects, respectively. Solid lines represent cholinergic neurones, broken lines represent adrenergic neurones, and dot/dash lines indicate sensory neurones from O_2 receptors.

(Fritsche, 1990), and the rockpool fish, *Gobius cobitis* (Berschick *et al.*, 1987), the bradycardia induced by hypoxia may be weak or even absent. Studies of several teleost species show that both the cholinergic and the adrenergic tonus to the heart increases during hypoxia, but in the intact fish the cholinergic inhibitory component dominates, producing the bradycardia (Fritsche, 1990; Fritsche and Nilsson, 1990).

Whether or not branchial vagal fibres (X), cholinergic and/or serotonergic in nature, to the branchial vasculature are also activated by the hypoxic reflex is unknown.

Hypoxic hypertension

The cyclostome, *Myxine glutinosa*, responds to severe aquatic hypoxia ($P\text{w}O_2$ = 1.5–2.2 kPa) by an increase of both ventral and dorsal aortic blood pressures, but the origin of this reflex remains unknown (Axelsson *et al.*, 1990). In elasmobranchs, hypoxia induces a marked bradycardia (p. 197), and the effect on arterial blood pressure, if any, is a small hypotension (Piiper *et al.*, 1970; Butler and Taylor, 1971). In most teleosts, the central oxygen receptor triggers a second reflex producing an elevated systemic vascular resistance and arterial blood pressure (Holeton and Randall, 1967; Saunders and Sutterlin, 1971; Fritsche and Nilsson, 1989, 1990; Fritsche, 1990). In the Atlantic cod, this effect is relatively rapid, and develops in less than 2 min, compared with the slower (> 6 min) hypoxic bradycardia reflex (Fig. 7.6). The hypertensive reflex is largely abolished by the adrenergic neurone-blocking agent bretylium, showing the involvement of adrenergic vasomotor fibres (Fritsche and Nilsson, 1990).

7.8 GENERAL CONCLUSIONS

Available studies of respiratory and cardiovascular systems in fish, especially the teleosts, show that the responses to hypoxia resemble the responses in the higher vertebrates. At least two distinct locations for chemoreceptors that are sensitive to changes in the oxygen levels can be identified: one in the gills, innervated by cranial nerves IX and X (and, in elasmobranchs, also V and VII), and another probably within the central nervous system. Responses to decreased oxygen levels include increased ventilatory water flow, elevated blood pressure and bradycardia. There is also a dramatic increase in plasma levels of catecholamines, particularly if the rate of decrease of the aquatic oxygen tension is high. These may further affect respiratory mechanisms, such as the oxygen handling by the red blood cells. Although patterns are starting to emerge which identify a number of sensory and motor components that control ventilation and blood circulation, there is still a long way to go before the interaction between all the systems involved is fully understood.

ACKNOWLEDGEMENTS

Our work on cardiovascular reflexes in fish is currently supported by the Swedish Natural Science Research Council.

REFERENCES

Abrahamsson, T. (1979) Phenylethanolamine-N-methyl transferase (PNMT) activity and catecholamine storage and release from chromaffin tissue of the spiny dogfish, *Squalus acanthias*. *Comp. Biochem. Physiol.*, **64C**, 169–72.

Abrahamsson, T., Holmgren, S., Nilsson, S. (1979) On the chromaffin system of the African lungfish, *Protopterus aethiopicus*. *Acta physiol. scand.*, **107**, 135–9.

Aota, S., Holmgren, K.D., Gallaugher, P. and Randall, D.Y. (1990) A possible role for catecholamines in the ventilatory responses associated with internal acidosis or external hypoxia in rainbow trout *Oncorhynchus mykiss*. *J. exp. Biol.*, **151**, 57–70.

Axelsson, M., Farrell, A.P. and Nilsson, S. (1990) Effects of hypoxia and drugs on the cardiovascular dynamics of the Atlantic hagfish, *Myxine glutinosa*. *J. exp. Biol.*, **151**, 297–316.

Bailly, Y., Dunel-Erb, S., Geffard, M. and Laurent, P. (1989) The vascular and epithelial serotonergic innervation of the actinopterygian gill filament with special reference to the trout, *Salmo gairdneri*. *Cell Tissue Res.*, **258**, 349–63.

Ballintijn, C.M. (1982) *Neuronal Control of Respiration in Fishes and Mammals. Exogenous and Endogenous Influences on Metabolic and Neuronal Control*, (eds, A.D.F. Addink and N. Spronk), Pergamon Press, Oxford, pp 127–70.

Ballintijn, C.M. and Alink, G.M. (1977) Identification of respiratory motor neurones in the carp and determination of their firing characteristics and interconnections. *Brain Res., Amst.*, **136**, 261–76.

Ballintijn, C.M. and Roberts, J.L. (1976) Neural control and proprioceptive load matching in reflex respiratory movements of fishes. *Fedn. Proc. Fedn Am. Socs exp. Biol.*, **35**, 1983–91.

Bamford, O.S. (1974) Oxygen reception in the rainbow trout, *Salmo gairdneri*. *Comp. Biochem. Physiol.*, **48A**, 69–76.

Barrett, D.J. and Taylor, E.W. (1984) Changes in heart rate during progressive hypoxia in the dogfish, *Scyliorhinus canicula* L.: evidence for a venous oxygen receptor. *Comp. Biochem. Physiol.*, **78A**, 697–703.

Berschick, P., Bridges, C.R. and Grieshaber, M.K. (1987) The influence of hyperoxia, hypoxia and temperature on the respiratory physiology of the intertidal rockpool fish *Gobius cobitis* Pallas. *J. exp. Biol.*, **130**, 369–87.

Burggren, W.W. and Randall, D.J. (1978) Oxygen uptake and transport during hypoxic exposure in the sturgeon, *Acipenser transmontanus*. *Respir. Physiol.*, **34**, 171–83.

Burleson, M.L. (1991) Oxygen sensitive chemoreceptors and cardiovascular and ventilatory control in rainbow trout, PhD thesis, University of British Columbia. (152 pp.)

Burleson, M.L. and Milsom, W.K. (1990a) Pharmacology of trout gill O_2 sensitive chemoreceptors. *Bull. Can. J. Zool.*, **21**, (2), 32.

Burleson, M.L. and Milsom, W.K. (1990b) Propranolol inhibits O_2-sensitive chemoreceptor activity in trout gills. *Am. J. Physiol.*, **258**, R1089–91.

Burleson, M.L. and Smatresk, N.J. (1989) The effect of decerebration and anesthesia on the reflex responses to hypoxia in catfish. *Can. J. Zool.*, **67**, 630–35.

Burleson, M.L. and Smatresk, N.J. (1990a) Effects of sectioning cranial nerves IX and X on cardiovascular and ventilatory reflex responses to hypoxia and NaCN in channel catfish. *J. exp. Biol.*, **154**, 407–20.

Burleson, M.L. and Smatresk, N.J. (1990b) Evidence for two oxygen sensitive chemoreceptor loci in channel catfish. *Physiol. Zool.*, **63**, 208–21.

Burnstock, G. (1990) Local mechanisms of blood flow control by perivascular nerves and endothelium. *J. Hypertension*, **8**, S95–106.

Burnstock, G. and Griffith, S.G. (1988) *Non-adrenergic Innervation of Blood Vessels*, CRC Press, Boca Raton, FL, 150 pp.

Burnstock, G., Campbell, G., Bennett, M. and Holman, M. (1963) Inhibition of the smooth muscle of the taenia coli. *Nature, Lond.*, **200**, 581–2.

Burnstock, G., Campbell, G., Bennett, M. and Holman, M. (1964) Innervation of the guinea-pig taenia coli: are there intrinsic inhibitory nerves which are distinct from sympathetic nerves? *Int. J. Neuropharmacol.*, **3**, 163–6.

Bushnell, P.G. and Brill, R.W. (1991) Responses of swimming skipjack (*Katsuwonus pelamis*) and yellowfin (*Thunnus albacores*) tunas to acute hypoxia and a model of their cardiorespiratory function. *Physiol. Zool.*, **64**(3), 787–811.

Butler, P.J. and Taylor, E.W. (1971) Response of the dogfish (*Scyliorhinus canicula* L.) to slowly induced and rapidly induced hypoxia. *Comp. Biochem. Physiol.*, **39A**, 307–23.

Butler, P.J., Taylor, E.W., Capra, M.F. and Davison, W. (1978) The effect of hypoxia on the levels of circulating catecholamines in the dogfish *Scyliorhinus canicula*. *J. comp. Physiol.*, **127**, 325–30.
Butler, P.J., Taylor, E.W. and Short, S. (1977) The effects of sectioning cranial nerves V, VII, IX and X on the cardiac response of the dogfish *Scyliorhinus canicula* to environmental hypoxia. *J. exp. Biol.*, **69**, 233–45.
Cameron, J.N. and Davis, J.C. (1970) Gas exchange in rainbow trout (*Salmo gairdneri*) with varying blood oxygen capacity. *J. Fish. Res. Bd Can.*, **27**, 1069–85.
Chiocchio, S.R., King, M.P., Carballo, L. and Angelakos, E.T. (1971) Monoamines in the carotid body cells of the cat. *J. Histochem. Cytochem.*, **19**, 621–6.
Chow, C.M., Winder, C. and Read, D.J.C. (1986) Influences of endogenous dopamine on carotid body discharge and ventilation. *J. appl. Physiol.*, **60**, 370–75.
Cohen, M.I. (1981) Central determinants of respiratory rhythm. *A. Rev. Physiol.*, **43**, 91–104.
Davis, J.C. (1971) Circulatory and ventilatory responses of rainbow trout (*Salmo gairdneri*) to aritifical manipulation of gill surface area. *J. Fish. Res. Bd Can.*, **28**, 1609–14.
Daxboeck, C. and Holeton, G.F. (1978) Oxygen receptors in the rainbow trout, *Salmo gairdneri*. *Can. J. Zool.*, **56**, 1254–9.
Dejours, P. (1973) Problems of control of breathing in fishes, in *Comparative Physiology* (eds L. Bolis, K. Schmidt-Nielsen and S.H.P. Maddrell), North-Holland-Am. Elsevier, Amsterdam, pp. 117–33.
DeKock, L.L. (1963) A histochemical study of the head region of two salmonids with special reference to pressor- and chemo-receptors. *Acta anat.*, **55**, 39–50.
Dempsey, J.A., Olson, E.B. and Skatrud, J.B. (1986) Hormones and neurochemicals in the regulation of breathing, in *Handbook of Physiology*, Sect. 3, *The Respiratory System*, Vol. II: *Control of Breathing*, Part 1 (eds A.P. Fishman, N.S. Cherniak, J.G. Widdicombe *et al.*), Am. Physiological Soc., Bethesda, MD, pp. 181–221.
Donald, J. (1984) Adrenergic innervation of the gills of brown and rainbow trout, *Salmo trutta* and *S. gairdneri*. *J. Morph.*, **182**, 307–16.
Donald, J. and Campbell, G. (1982) A comparative study of the adrenergic innervation of the teleost heart. *J. comp. Physiol.*, **147**, 85–91.
Donald, J.A. (1987) Comparative study of the adrenergic innervation of the teleost gill. *J. Morph.*, **193**, 63–73.
Donnely, D.F., Smith, E.J. and Dutton, R.E. (1981) Neuronal response of carotid chemoreceptors following dopamine blockade. *J. Appl. Physiol.*, **50**, 172–7.
Dunel-Erb, S. and Bailly, Y. (1986) The sphincter of the efferent filament artery in teleost gills: II. Sympathetic innervation. *J. Morph.*, **187**, 239–46.
Dunel-Erb, S., Bailly, Y. and Laurent, P. (1982) Neuroepithelial cells in fish gill primary lamellae. *J. appl. Physiol.*, **53**, 1342–53.
Fitzgerald, R.S., Garger, P., Hauer, M.C., Raff, H. and Fechter, L. (1983) Effect of hypoxia and hypercapnia on catecholamine content in cat carotid body. *J. appl. Physiol.*, **54**, 1408–13.
Folgering, H., Ponte, J. and Saidig, T. (1982) Adrenergic mechanisms and chemoreception in the carotid body of the cat and rabbit. *J. Physiol., Lond.*, **325**, 1–21.
Forster, M.E., Axelsson, M., Farrell, A.P. and Nilsson, S. (1991) Cardiac function and circulation in hagfish. *Can. J. Zool.*, **69**, 1985–92.
Freihofer, W.C. (1978) Cranial nerves of a percoid fish, *Polycentrus schomburgkii* (Family Nandidae), a contribution to the morphology and classification of the order perciformes. *Calif. Acad. Sci.*, **128**, 1–78.
Fritsche, R. (1990) Effects of hypoxia on blood pressure and heart rate in three marine teleosts. *Fish Physiol. Biochem.*, **8**, 85–92.

Fritsche, R. and Nilsson, S. (1989) Cardiovascular responses to hypoxia in the Atlantic cod, *Gadus morhua*. *Exp. Biol.*, **48**, 153–60.

Fritsche, R. and Nilsson, S. (1990) Autonomic nervous control of blood pressure and heart rate during hypoxia in the cod, *Gadus morhua*. *J. comp. Physiol.*, **160B**, 287–92.

Fritsche, R., Thomas, S. and Perry, S.F. (1992) Effects of serotonin on circulation and respiration in the rainbow trout, *Oncorhynchus mykiss*. *J. exp. Biol.*, in press.

Gannon, B.J. and Burnstock, G. (1969) Excitatory adrenergic innervation of the fish heart. *Comp. Biochem. Physiol.*, **29**, 765–73.

Gehrke, P.C. and Fielder, D.R. (1988) Effects of temperature and dissolved oxygen on heart rate, ventilation rate and oxygen consumption of spangled perch, *Leiopotherapon unicolor* (Günther 1859), (Percoidei, Teraponidae). *J. comp. Physiol.*, **157**, 771–82.

Govyrin, V.A. and Leont'eva, G.R. (1965) Distribution of catecholamines in myocardium of vertebrates. *J. evol. Biochem. Physiol.*, **1**, 38–44.

Graham, J.B., Dewar, H., Lai, N.C., Lowell, W.R. and Arce, S.M. (1990) Aspects of shark swimming performance determined using a large water tunnel. *J. exp. Biol.*, **151**, 175–92.

Greeff, K., Kasperat, H. and Oswald, W. (1962) Paradoxe Wirkungen der elektrischen Vagusreizung am isolierten Magen ond Herzvorhofpräparat des Meerschweinchens sowie deren Beieinflussung durch Ganglienblocker, Sympathicolytica, Reserpin und Cocain. *Naunyn-Schmiedebergs Arch. Exp. Pathol. Pharmakol.*, **243**, 528–45.

Hanbauer, I. and Hellström, S. (1978) The regulation of dopamine and noradrenaline in rat carotid body and its modification by denervation and by hypoxia. *J. Physiol., Lond.*, **282**, 21–34.

Hedrick, M.S., Burleson, M.L., Jones, D.R. and Wilson, W.K. (1991) An examination of central chemosensitivity in an air-breathing fish (*Amia calva*). *J. exp. Biol.*, **155**, 165–74.

Heisler, N. (1989) Acid–base regulation, in *Physiology of Elasmobranch Fishes* (eds. T.J. Shuttleworth), Springer-Verlag, Berlin, pp. 215–52.

Holeton, G.F. (1971) Respiratory and circulatory responses of rainbow trout larvae to carbon monoxide and to hypoxia. *J. exp. Biol.*, **55**, 683–94.

Holeton, G.F. (1977) Constancy of arterial blood pH during CO-induced hypoxia in the rainbow trout. *Can. J. Zool.*. **55**, 1010–13.

Holeton, G.F. and Randall, D.J. (1967) Changes in blood pressure in the rainbow trout during hypoxia. *J. exp. Biol.*, **46**, 297–305.

Holmgren, S. (1977) Regulation of the heart of a teleost, *Gadus morhua*, by autonomic nerves and circulating catecholamines. *Acta physiol. scand.*, **99**, 62–74.

Holmgren, S. (ed.) (1989) *Comparative Physiology of Regulatory Peptides*. Chapman and Hall, London.

Holmgren, S. and Nilsson, S. (1991) Novel neurotransmitters in the autonomic nervous system of nonmammalian vertebrates, in *Novel Peripheral Neurotransmitters*, Chapter 11, *International Encyclopedia of Pharmacology and Therapeutics*, Section 135 (ed. C. Bell), Pergamon Press, New York, pp. 293–328.

Hsiao, C., Lahiri, S. and Mokashi, A. (1989) Peripheral and central dopamine receptors in respiratory control. *Respir. Physiol.*, **76**, 327–36.

Hughes, G.M. (1978) On the respiration of *Torpedo marmorata*. *J. exp. Biol.*, **73**, 85–105.

Johansen, K. (1970) Air breathing in fishes, in *Fish Physiology*, Vol. 4 (eds W.S. Hoar and D.J. Randall), Academic Press, New York, pp. 361–411.

Johansen, K. and Lenfant, C. (1968) Respiration in the African lungfish *Protopterus aethiopicus*. II. Control of breathing. *J. exp. Biol.*, **49**, 453–68.

Johansen, K., Lenfant, C. and Grigg, C.G. (1967) Respiratory control in the lungfish, *Neoceratodus forsteri*. *Comp. Biochem. Physiol.*, **20**, 835–54.

Johansen, K., Lenfant, C. and Hanson, D. (1973) Gas exchange in the lamprey, *Entosphenus tridentatus*. *Comp. Biochem. Physiol.*, **44A**, 107–19.

Jones, D.R. and Milsom, W.K. (1982) Peripheral receptors affecting breathing and cardiovascular function in non-mammalian vertebrates. *J. exp. Biol.*. **100**, 59–91.

Katchen, M.S., Olson, K.R. and Wayne, C. (1976) Effects of histamine and serotonin on isolated perfused gill of rainbow trout (*Salmo gairdneri*). *Fedn. Proc. Fedn Am. Socs exp. Biol.*, **35**, 528(1774).

Kawasaki, R. (1979) Breathing rhythm-generation in the adult lamprey *Entosphenus japonicus*. *Jap. J. Physiol.*, **29**, 327–38.

Kinkead, R. (1990) The role of circulating catecholamines in the regulation of breathing in teleosts, MSC thesis, University of Ottawa. (146 pp.)

Kinkead, R. and Perry, S.F. (1990) An investigation of the role of circulating catecholamines in the control of ventilation during acute moderate hypoxia in rainbow trout (*Oncorhynchus mykiss*). *J. comp. Physiol.*, **160**, 441–8.

Kinkead, R. and Perry, S.F. (1991) The effects of catecholamines on ventilation in rainbow trout during hypoxia or hypercapnia. *Respir. Physiol.*, **84**, 77–92.

Kinkead, R., Fritsche, R., Perry, S.F. and Nilsson, S.F. (1991) The role of circulating catecholamines in the ventilatory and hypertensive responses to hypoxia in the Atlantic cod (*Gadus morhua*). *Physiol. Zool.*

Lahiri, S., Szidon, J.P. and Fishman, A. (1970) Potential respiratory and circulatory adjustments to hypoxia in the African lungfish. *A. Fed. Proc.*, **29**, 1141–8.

Laurent, P. (1967) Neurophysiologie – La pseudobranchie des téléostéens: Preuves électrophysologiques de ses fonctions chémoréceptrice et baroréceptrice. *C.r. Hebd. Séanc. Acad. Sci., Paris*, **264**, 1879–82.

Laurent, P. (1969) Action du pH et de la pO_2 sur le potential de membrane des cellules de l'épithelium récepteur dans la pseudobranchie d'un poisson téléostéen. *Rev. Can. Biol.*, **28**, 149–55.

Laurent, P. (1984) Gill internal morphology, in *Fish Physiology*, Vol. X (eds W.S. Hoar and D.J. Randall), Academic Press, Orlando, FL, pp. 73–183.

Lomholt, J.P. and Johansen, K. (1979) Hypoxia acclimation in carp – how it affects O_2 uptake, ventilation and O_2 extraction from water. *Physiol. Zool.*, **52**, 38–49.

McKenzie, D.J., Aota, S. and Randall, D.J. (1991) Ventilatory and cardiovascular responses to blood pH, plasma PCO_2, Blood O_2 content, and catecholamines in an air-breathing fish, the bowfin (*Amia calva*). *Physiol. Zool.*, **64**, 432–50.

Marvin, D.E. and Heath, A.G. (1968) Cardiac and respiratory responses to gradual hypoxia in three ecologically distinct species of fresh-water fish. *Comp. Biochem. Physiol.*, **27**, 349–55.

Mazeaud, M.M. (1971) *Recherches sur la biosynthése, la sécrétion et le catabolisme de l'adrénaline et de la noradrénaline chez quelques espèces de cyclostomes et de poissons*, PhD thesis, Université de Paris, Paris. (130 pp.)

Metcalfe, J.D. and Butler, P.J. (1984) On the nervous regulation of gill blood flow in the dogfish *Scyliorhinus canicula*. *J. exp. Biol.*, **113**, 253–67.

Metcalfe, J.D. and Butler, P.J. (1988) The effects of alpha- and beta-adrenergic receptor blockade on gas exchange in the dogfish (*Scyliorhinus canicula* L.) during normoxia and hypoxia. *J. comp. Physiol.*, **158**, 39–44.

Metcalfe, J.D. and Butler, P.J. (1989) The use of alpha-methyl-p-tyrosine to control circulating catecholamines in the dogfish *Scyliorhinus canicula*: the effects on gas exchange in normoxia and hypoxia. *J. exp. Biol.*, **141**, 21–32.

Milsom, W. and Brill, R. (1986) Oxygen sensitive afferent information arising from the first gill arch of yellowfin tuna. *Respir. Physiol.*, **66**, 193–203.

Milsom, W.K. (1989) Mechanisms of ventilation in lower vertebrates: adaptations to respiratory and non-respiratory constraints. *Can. J. Zool.*, **67**, 2943–55.

Milsom, W.K. (1991) Intermittent breathing in vertebrates. *A. Rev. Physiol.*, **53**, 87–105.

Milsom, W.K. and Sadig, T. (1983) Interaction between norepinephrine and hypoxia on carotid body chemoreception in rabbits. *J. appl. Physiol. (Respirat. Environ. Exercise Physiol. 6)*, **55**, 1893–8.

Mione, M.C., Ralevic, V. and Burnstock, G. (1990) Peptides and vasomotor mechanisms. *Pharmacol. Ther.*, **46**, 429–68.

Nilsson, S. (1973) Fluorescent histochemistry of the catecholamines in the urinary bladder of a teleost, *Gadus morhua. Comp. gen. Pharmacol.*, **4**, 17–21.

Nilsson, S. (1976) Fluorescent histochemistry and cholinesterase staining of sympathetic ganglia in a teleost, *Gadus morhua. Acta zool., Stockh.*, **57**, 69–77.

Nilsson, S. (1983) *Autonomic Nerve Function in the Vertebrates*, Springer-Verlag, Berlin, 000 pp.

Nilsson, S. (1984) Innervation and pharmacology of the gills, in *Fish Physiology*, Vol. XA (eds W.S. Hoar and D.J. Randall), Academic Press, Orlando, FL, pp. 185–227.

Nilsson, S. (1986) Control of gill blood flow, in *Fish Physiology: Recent Advances* (eds S. Nilsson and S. Holmgren), Croom Helm, Dover, NH, pp. 86–101.

Nilsson, S. and Holmgren, S. (1991) Cardiovascular control by purines, 5-hydroxytryptamine and neuropeptides, in *Fish Physiology*, Vol XX (eds W.S. Hoar, D.J. Randall and A.P. Farrell), Academic Press, New York.

Nilsson, S., Abrahamsson, T. and Grove, D.J. (1976) Sympathetic nervous control of adrenaline release from the head kidney of the cod, *Gadus morhua. Comp. Biochem. Physiol.*, **55C**, 123–7.

Nishino, T. and Lahiri, S. (1981) Effects of dopamine on chemoreflexes in breathing. *J. appl. Physiol.*, **50**, 892–7.

Ogden, E. (1945) Respiratory flow in *Mustelus. Am. J. Physiol.*, **145**, 134–9.

O'Regan, R.G. and Majcherczyk, S. (1982) Role of peripheral chemoreceptors and central chemosensitivity in the regulation of respiration and circulation. *J. exp. Biol.*, **100**, 23–40.

Östlund, E. and Fänge, R. (1962) Vasodilation by adrenaline and noradrenaline and the effects of some other substances on perfused fish gills. *Comp. Biochem. Physiol.*, **5**, 307–9.

Perry, S.F., Fritsche, R., Kinkead, R. and Nilsson, S. (1991) Control of catecholamine release *in vivo* and *in situ* in the Atlantic cod (*Gadus morhua*) during hypoxia. *J. exp. Biol.*, **155**, 549–66.

Perry, S.F., Kinkead, R. and Fritsche, R. (1992) Are circulating catecholamines involved in the control of breathing by fishes? *Rev. Fish Biol. Fish.*, **2**, 65–83.

Peyraud-Waitzenegger, M. (1979) Simultaneous modifications of ventilation and arterial PO_2 by catecholamines in the eel, *Anguilla anguilla* L.: Participation of α and β effects. *J. comp. Physiol.*, **129**, 343–54.

Peyraud-Waitzenegger, M., Savina, A., Laparra, J. and Morfin, R. (1979) Blood brain barrier for epinephrine in the eel *Anguilla anguilla* L. *Comp. Biochem. Physiol.*, **63**, 35–8.

Piiper, J., Baumgarten, D. and Meyer, M. (1970) Effects of hypoxia upon respiration and circulation in the dogfish *Scyliorhinus stellaris. Comp. Biochem. Physiol.*, **36**, 513–20.

Playle, R.C., Munger, R.S. and Wood, C.M. (1990) Effects of catecholamines on gas exchange and ventilation in rainbow trout (*Salmo gairdneri*). *J. exp. Biol.*, **152**, 353–67.

Randall, D.J. and Daxboeck, C. (1982) Cardiovascular changes in the rainbow trout (*Salmo gairdneri* Richardson) during exercise. *Can. J. Zool.*, **60**, 1135–40.

Randall, D.J. and Daxboeck, C. (1984) Oxygen and carbon dioxide transfer across fish gills, in *Fish Physiology*, Vol. XA (eds W.S. Hoar and D.J. Randall), Academic Press, Orlando, FL, pp. 263–314.

Randall, D.J. and Jones, D.R. (1973) The effect of deafferentation of the pseudobranch on the respiratory response to hypoxia and hyperoxia in the trout (*Salmo gairdneri*). *Respir. Physiol.*, **17**, 291–301.

Randall, D.J. and Shelton, G. (1963) The effects of changes in environmental gas concentrations on the breathing and heart rate of a teleost fish. *Comp. Biochem. Physiol.*, **9**, 229–39.

Randall, D.J. and Smith, L.S. (1967) The effect of environmental factors on circulation and respiration in teleost fish. *Hydrobiologia*, **29**, 113–24.

Rantin, F.T. and Johansen, K. (1984) Responses of the teleost *Hoplias malabaricus* to hypoxia. *Env. Biol. Fishes*, **11**, 221–8.

Reite, O.B. (1969) The evolution of vascular smooth muscle responses to histamine and 5-hydroxytryptamine. I. Occurrence of stimulatory actions in fish. *Acta physiol. scand.*, **75**, 221–39.

Satchell, G.H. (1959) Respiratory reflexes in the dogfish. *J. exp. Biol.*, **36**, 62–71.

Satchell, G.H. (1961) The response of the dogfish to anoxia. *J. exp. Biol.*, **38**, 531–43.

Satchell, G.H. and Way, H.K. (1962) Pharyngeal proprioceptors in the dogfish *Squalus acanthias*. *J. exp. Biol.*, **39**, 243–50.

Saunders, R.L. and Sutterlin, A.M. (1971) Cardiac and respiratory responses to hypoxia in the sea raven, *Hemitripterus americanus*, and an investigation of possible control mechanisms. *J. Fish. Res. Bd Can.*, **28**, 491–503.

Shelton, G. (1959) The respiratory centre in the tench (*Tinca tinca* L.). *J. exp. Biol.*, **36**, 191–202.

Shelton, G. (1961) The respiratory centre in the tench (*Tinca tinca*). I. Respiratory neuronal activity in the medulla oblongata. *J. exp. Biol.*, **38**, 79–92.

Shelton, G., Jones, D.R. and Milsom, W.K. (1986) Control of breathing in ecothermic vertebrates, in *Handbook of Physiology*, Section 3, *The Respiratory System*, Vol. II, *Control of Breathing*, Part 2 (eds S.R. Geiger, A.P. Fishman, N.S. Cherniak *et al.*), Waverly Press, Baltimore, MD, pp. 857–909.

Short, S., Butler, P.J. and Taylor, E.W. (1977) The relative importance of nervous, humoral and intrinsic mechanisms in the regulation of heart rate and stroke volume in the dogfish, *Scyliorhinus*. *J. exp. Biol.*, **70**, 77–92.

Smith, F.M. and Jones, D.R. (1978) Localization of receptors causing hypoxic bradycardia in trout, *Salmo gairdneri*. *Can. J. Zool.*, **56**, 1260–65.

Smith, F.M. and Jones, D.R. (1982) The effect of changes in blood oxygen-carrying capacity on ventilation volume in the rainbow trout (*Salmo gairdneri*). *J. exp. Biol.*, **97**, 325–34.

Stevens, D.E., Bennion, G.R. and Randall, D.J. (1972) Factors affecting arterial pressures and blood flow from the heart in intact, unrestrained lingcod *Ophiodon elongatus*. *Comp. Biochem. Physiol.*, **43A**, 681–95.

Sutterlin, A.M. and Saunders, R.L. (1969) Proprioceptors in the gills of a teleost. *Can. J. Zool.*, **47**, 1209.

Taylor, E.W., Short, S. and Butler, P.J. (1977) The role of the cardiac vagus in the response of the dogfish (*Scyliorhinus canicula* L.) to hypoxia. *J. exp. Biol.*, **70**, 57–75.

Thomas, S., Fievet, B., Barthelemy, L. and Degraud (1983) Comparison of the effect of exogenous and endogenous hypercapnia on ventilation and oxygen uptake in the rainbow trout *Salmo gairdneri*. *J. comp. Physiol.*, **151**, 185–90.

van Dam, L. (1938) *On the Utilization of Oxygen and Regulation of Breathing in Some Aquatic Animals*, Volharding, Groningen, Netherlands, 144 pp.

Vasilevskaya, N.E. and Polyakova, N.N. (1979) Comparison of single unit responses in the vagal and facial lobes of the carp medulla to chemoreceptor stimulation. *Neurophysiology (Engl. Transl.)*, **10**, 613–21.

Vasilevskaya, N.E. and Polyakova, N.N. (1981) Unit responses in the vagal lobe of the carp brain to stimulation of chemoreceptors of the skin, mouth and gills. *Neurophysiology (Engl. Transl.)*, **12**, 397–404.

Wahlqvist, I. (1981) Branchial vascular effects of catecholamines released from the head kidney of the Atlantic cod, *Gadus morhua*. *Mol. Physiol.*, **1**, 235–41.

Wahlqvist, I. and Nilsson, S. (1980) Adrenergic control of the cardio-vascular system of the Atlantic cod, *Gadus morhua*, during 'stress'. *J. comp. Physiol.*, **137**, 145–50.

Walker, E.R., Fidler, S.F. and Hinton, D.E. (1981) Morphology of the buccopharyngeal portion of the gill in the fathead minnow *Pimephales promelas* (Rafinesque). *Anat. Rec.*, **200**, 67–81.

Wilkes, P.R.H., Walker, R.L., McDonald, D.G. and Wood, C.M. (1981) Respiratory, ventilatory, acid–base and ionoregulatory physiology of the white sucker *Catostomus commersoni*: the influence of hypoxia. *J. exp. Biol.*, **91**, 239–54.

Wood, C.M. and Shelton, G. (1980) Cardiovascular dynamics and adrenergic responses of the rainbow trout in vivo. *J. exp. Biol.*, **87**, 247–70.

Wood, C.M., McMahon, B.R. and McDonald, D.G. (1979) Respiratory gas exchange in the resting starry flounder, *Platichthys stellatus*: a comparison with other teleosts. *J. exp. Biol.*, **78**, 167–79.

Wood, C.M., Turner, J.D., Munger, R.S. and Graham, M.S. (1990) Control of ventilation in the hypercapnic skate *Raja ocellata*: II. Cerebrospinal fluid and intracellular pH in the brain and other tissues. *Respir. Physiol.*, **80**, 279–98.

Yamauchi, A. and Burnstock, G. (1968) An electronmicroscope study of the innervation of the trout heart. *J. comp. Neurol.*, **132**, 567–88.

Chapter eight

Acid–base regulation in response to changes of the environment: characteristics and capacity

Norbert Heisler

8.1 INTRODUCTION

Fish are challenged much more than terrestrial animals by changes of their environment. This is due to utilization of water rather than air as the respiratory gas exchange medium. The composition of water varies to a much larger extent than that of air, owing to a number of natural or contamination-related factors. Fish frequently encounter large and rapid changes in water oxygen and CO_2 concentrations, as well as changes in temperature and environmental electrolyte concentrations.

The large gill surface area, in combination with the typical countercurrent water/blood flow orientation, provides a very efficient interface for respiratory gas exchange between environment and body fluids. In spite of the much lower oxygen content of water as compared with air, this system efficiently satisfies the oxygen demand of metabolism. This is achieved, however, only by a large specific ventilation (volume per unit oxygen consumption) of the viscous breathing medium, water, which requires considerably more energy for convective gas exchange than in terrestrial animals. Contact of the gill epithelium with a large amount of water (about 0.2 of the body volume per minute at about 17 °C) of considerable thermal capacity subjects the animals to frequent and often large changes in body temperature.

In contrast to air-breathing terrestrial animals with lungs as specifically

Fish Ecophysiology. Edited by J. Cliff Rankin and Frank B. Jensen. Published in 1993 by Chapman & Hall, London. ISBN 0 412 45920 5.

adapted gas exchange structures, the gills of fish are the combined site of respiratory gas exchange and ionic regulation. Under normal conditions, ionic transfer is made easy by the large rate of water flow past the gill structures, with hardly any inspired–expired differences in ionic composition. Accordingly, ions can readily be transferred across the large surface area without the necessity of 'multiplying' concentration mechanisms for an establishment of large gradients between internal (plasma) and external (tubular urine) fluids as in the mammalian kidney system.

A severe disadvantage of the open epithelial surface of the gill system is, however, that changes of the water ionic composition will directly affect ionic transfer processes for osmoregulation and acid–base regulation, and often will also have considerable impact on respiratory gas exchange. Conversely, the kidney system is mechanically protected against changes of the environment and is accordingly less sensitive to environmental disturbances, but carries the disadvantage of a one-way system.

This chapter will briefly describe the characteristics of acid–base regulation in response to environmentally induced disturbances, and delineate the principal mechanisms available to fish for maintenance of homeostasis. It will attempt to delineate common features in response to different challenges, and to describe some factors known to determine the effective capacity of the involved regulatory systems.

8.2. ENVIRONMENTALLY INDUCED CHANGES OF THE ACID–BASE STATUS

Changes in temperature

Fish are subject to slow-rate seasonal changes in water temperature of up to 20 °C or more (e.g. Kato and Cavallo, 1967), but may also experience rapid and large changes in temperature in a diurnal rhythm, during migration or when passing thermoclines during predatory activities. Marine fish, in particular sharks, frequently move from close to the surface to water depths of more than 600 m (e.g. Bullis, 1967), passing temperature gradients of 20 °C or more (e.g. Harvey, 1974) in a short time. Although less active fish tend to avoid thermoclines, abrupt and large temperature changes may be imposed upon animals during predatory activities and emergency conditions. Due to the large branchial interface area and the large caloric capacity of water, any environmental temperature changes are rapidly transferred to the body fluids.

Plasma pH, P_{CO_2} and $[HCO_3^-]$ during steady state are a function of body temperature (reviews: Heisler, 1984b, 1986b). Originally considered as loosely or not at all regulated (review: Heisler, 1986c), pH in heterothermal animals was recognized to be inversely associated with temperature only by Winterstein (1954) and Robin (1962), a proposition later confirmed by numerous studies (review: Heisler, 1986c). Winterstein (1954) and Rahn (1967) independently suggested that changes in pH be expressed by the $[OH^-]/[H^+]$ ratio (constant relative alkalinity), which features a negative correlation of pH with temperature. The imidazole alphastat hypothesis, a model of pH/temperature regulation based on adjustment of pH by ventilation related to the ionization of biological histidine–imidazole moieties (Reeves, 1972; Reeves and Malan,

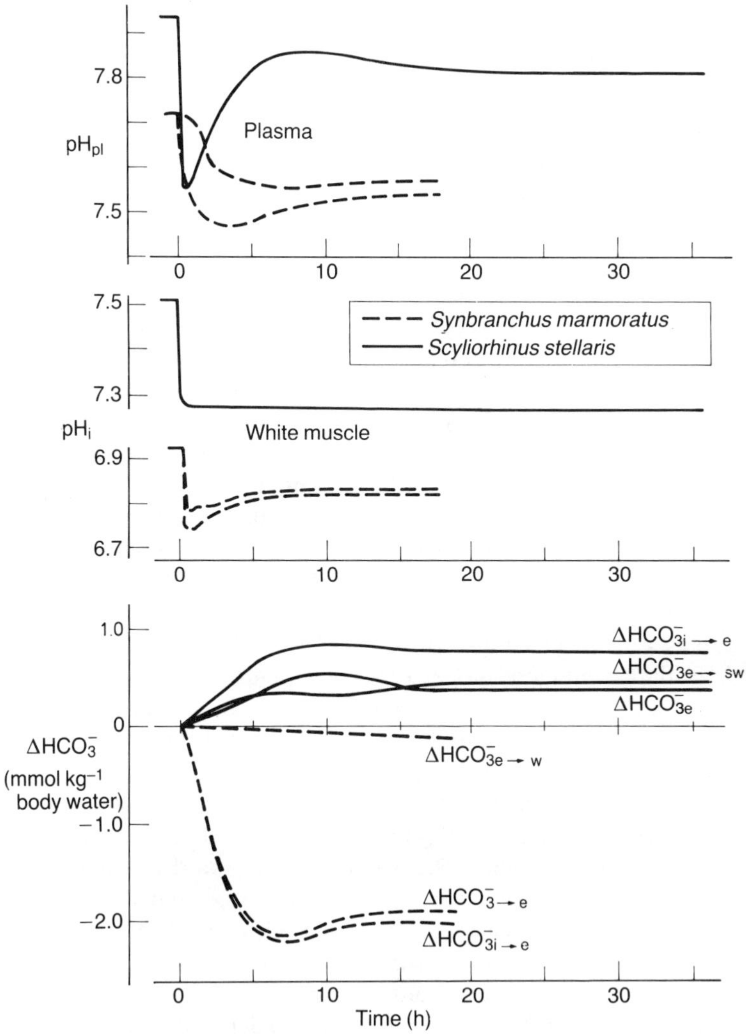

Fig. 8.1 Kinetics of pH adjustment and transfer of bicarbonate equivalent ions between intra- and extracellular spaces (indices 'i' and 'e', respectively), and the environmental water (indices 'w' or 'sw') upon an increase in temperature by 10 °C in two fish species. The time course of adjustment is, except for *Scyliorhinus* white muscle, mainly determined by the adjustment of bicarbonate concentration in the body fluids (see also text).

1976), failed to be confirmed by most of the available experimental data. In fish plasma, pH generally changes much less (average $\Delta pH/\Delta t$: -0.011 U °C^{-1} Heisler, 1986c) than expected from $\Delta pK'/\Delta t$ of imidazole compounds (-0.018 to 0.024 U °C^{-1}, Edsall and Wyman, 1958). Also pH in intracellular compartments and the magnitude of temperature-induced changes in ventilation are generally in contrast with the above hypotheses (Heisler, 1986c).

A central feature of the imidazole alphastat hypothesis was adjustment of pH in body fluids without any ionic transfer (Reeves, 1972; Reeves and Malan, 1976). The limited information on this point falls short of supporting this proposition. Although the results are quite heterogeneous, all three studied fish species, the elasmobranch *Scyliorhinus stellaris*, the freshwater catfish *Ictalurus punctatus* and the tropical freshwater teleost *Synbranchus marmoratus*, exhibit reversible transfer of acid–base relevant ions among intra- and extracellular fluid compartments upon changes in temperature. The first two of the above species utilize ionic transfer also to the environmental water (Heisler, 1978; Cameron and Kormanik, 1982), whereas *Synbranchus* behaves as a whole animal more like a semi-closed buffer system (open for only molecular CO_2), with little and insignificant ionic exchange with the ambient water (Heisler, 1984a).

Adjustment of the acid–base status after step changes in temperature is a time-consuming process (Fig. 8.1). Elevating temperature (10–20 °C) results in an initial plasma pH shift of more than threefold ($\Delta pH/\Delta t \sim$ -0.039 U °C^{-1}) compared with the finally attained steady state change ($\Delta pH/\Delta t \sim$ -0.012 U °C^{-1}) in *Scyliorhinus* (Heisler, 1978, 1984a). Initially, this is primarily due to an overshoot in P_{CO_2}, later adjustment of pH is dominated by slow adjustment of bicarbonate. A new steady state is attained only after more than 12 h. *Synbranchus* exhibits similarly slow acid–base adjustment kinetics in response to an analogous temperature step (20–30 °C) (Heisler, 1984b).

The slow adjustment kinetics of extracellular pH are related to the establishment of new levels of bicarbonate. This is not effected by non-bicarbonate buffering. The product of non-bicarbonate buffer capacity of blood and extravascular extracellular space, and the effective changes in pH, $\Delta pH - \Delta pK$, is negligible in comparison with the changes brought about by bicarbonate-equivalent ion transfer processes in both *Scyliorhinus* and *Synbranchus* (Holeton and Heisler, 1983; Heisler, 1984a,b, 1986a). The pattern of ionic transfer is quite diverse: in *Scyliorhinus*, bicarbonate-equivalent ions are transferred from the intracellular to the extracellular space and a sizable fraction is released to the environmental water, whereas the transmembrane transfer in *Synbranchus* is much larger and in the opposite direction, with little transfer taking place at the epithelial surface (Fig. 8.1; Heisler, 1984a). This pattern is related to differences in temperature-dependent intracellular buffering characteristics and pH changes (Heisler and Neumann, 1980; Heisler, 1984a).

Direct determination of intracellular pH kinetics is technically difficult and has not been possible under physiological conditions *in vivo*. However, first approximations to this point can be acquired by appropriate model calculations on the basis of the rate of transmembrane bicarbonate-equivalent transfer, and tissue-specific buffer characteristics (Heisler and Neumann, 1980; Heisler, 1984a, 1986a). Uncertainties of such estimates are mostly related to assumptions with respect to a homogeneous distribution of transmembrane ionic transfer rates in various tissues (Heisler, 1984a, 1986b). The reliability of such estimates depends on the fractional contribution of the specific tissue to the body weight, and is accordingly high at least for white muscle of *Scyliorhinus* (~ 54%; Heisler, 1978) and *Synbranchus* (~ 75% of body weight; Heisler, 1982a). Adjustment of intracellular pH in *Scyliorhinus* white muscle will be almost complete after about 30 min after a change in temperature (Fig. 8.1), whereas a new steady state will not be attained before 2 h in heart muscle and 8 h in red muscle, assuming the same specific transfer rates for white muscle, red muscle, and heart muscle. In *Synbranchus*, white muscle intracellular pH will be steady only after 7–8 h (Heisler, 1984a).

The observed overall rates of acid–base relevant ion transfer after changes in temperature (Fig. 8.1) are at least one order of magnitude smaller than those achieved during peak perturbations of other acid–base stress conditions (see below, p. 216, p. 221), being comparable to rates observed during mild disturbances, or during approach to control values. Accordingly, temperature-induced pH changes do not appear to affect cell functions to a great extent, a notion supported by relatively wide pH optima (at least in terms of pH changes dealt with in connection with temperature changes) of the energy-producing enzymes (Heisler, 1986c, 1990b).

Environmental hypercapnia

With few exceptions (most notable: air-breathing fishes) the arterial P_{CO_2} in fish is low (1–4 mmHg) as compared with terrestrial animals, particularly homoiotherms (30–45 mmHg). This is related to an about 30 times lower oxygen capacitance of water than air, requiring an equivalently high gill ventilation in order to meet the oxygen demand of metabolism (Heisler, 1989). The capacitance of CO_2 of pure water, however, is similar to that of air (the exact ratio depends on temperature). Moreover, in the range of the normally low environmental P_{CO_2} values (0–1 mmHg), the effective CO_2 capacitance is enhanced in seawater and carbonated fresh water (habitats typical of most fish species) by carbonate- and other non-bicarbonate buffering of CO_2 to values far above those for air (Heisler, 1989). Combined with the oxygen demand-driven high relative rate of ventilation, this may effect inspired-arterial P_{CO_2} differences in fish of well below 1 mmHg (compared with 30–45 mmHg in mammals) (Heisler 1989).

Environmental-arterial P_{CO_2} differences as small as that clearly provide limitations for any compensatory reduction of P_{CO_2} via hyperventilation, which is a well-known mechanism in mammals (Woodbury, 1965). Apart from problems related to energetic cost and to respiratory muscle fatigue during extensive compensatory hyperventilation of the viscous breathing medium, water, the scope for a reduction of arterial P_{CO_2} is generally small. Furthermore, a certain fraction of the inspired-arterial P_{CO_2} difference (which is difficult to quantify) is not related to the rate of gill ventilation, but is a function of other factors like blood or water shunting past the gas exchange surface (Heisler, 1989), or incomplete CO_2 hydration and disequilibrium as a result of insufficient carbonic anhydrase activity on the external side of the gill epithelium (e.g. Henry *et al.*, 1988; Heisler, 1990a).

Hyperventilation will accordingly only reduce arterial P_{CO_2} significantly in fish with relatively large inspired-arterial P_{CO_2} differences. A typical example is carp, *Cyprinus carpio*, which, partly owing to a very high blood oxygen affinity, has an arterial P_{O_2} of only 15–20 mmHg, and a P_{CO_2} of 3–4 mmHg. This species is capable of reducing arterial P_{CO_2} by a factor of 2 upon exposure to hypoxia (75 mmHg), but for this the rate of ventilation has to be elevated by a factor of about 3 (e.g. Glass *et al.*, 1990). These data suggest that at least in carp, a fraction of about 1 mmHg of the inspired-arterial P_{CO_2} difference is related to factors like incomplete CO_2 hydration and/or blood/water shunting. The effect of hyperventilation on arterial P_{CO_2} will be much smaller in animals with a smaller inspired-arterial P_{CO_2} difference.

However, even reduction of P_{CO_2} by a large factor will not result in any appreciable effect, when relatively small changes of environmental P_{CO_2} effect severe acid–base disturbances. A rise of inspired P_{CO_2} by 10 mmHg results in fish in increases of P_{CO_2} by factors of 3–6 (Heisler, 1986b), with deflections of arterial pH of ~ 0.4–0.5 pH units. In mammals, however, the acid–base status will hardly be affected: a slight hyperventilation (+ 30%) will reduce the inspired/alveolar P_{CO_2} difference from 40 to 30 mmHg. Accordingly, hyperventilation is hardly exploited in fish during environmental hypercapnia: the response is generally small and transient without being related to the time course of pH normalization (Dejours, 1975, 1981; Randall *et al.*, 1976).

The range of environmental P_{CO_2} values that water-breathing fishes may encounter is large. Although surface seawater P_{CO_2} usually reflects equilibrium with air (0.15–0.3 mmHg) and is little affected by temperature and biological processes, P_{CO_2} at depths of 200–500 m may rise to values of 5–10 mmHg due to anaerobic metabolism of bacteria and organic debris. Natural freshwater P_{CO_2} may range from extremely low photosynthesis-induced values (less than atmospheric, $< \sim$ 0.26 mmHg) to values as high as 60 mmHg in extreme cases (Heisler *et al.*, 1982), due to inhibition of surface gas exchange by dense surface vegetation, combined with CO_2 production from bicarbonate by anaerobic microbial metabolism and thermostratification. According to the above considerations, the adjustment of gill ventilation cannot be a measure to deal with such disturbances.

Environmental hypercapnia is almost immediately transmitted to the arterial blood by the large gill surface area. The associated elevation in plasma P_{CO_2} results in a general fall in plasma and intracellular pH values. With some exceptions, however, pH recovers towards control values soon after initiation

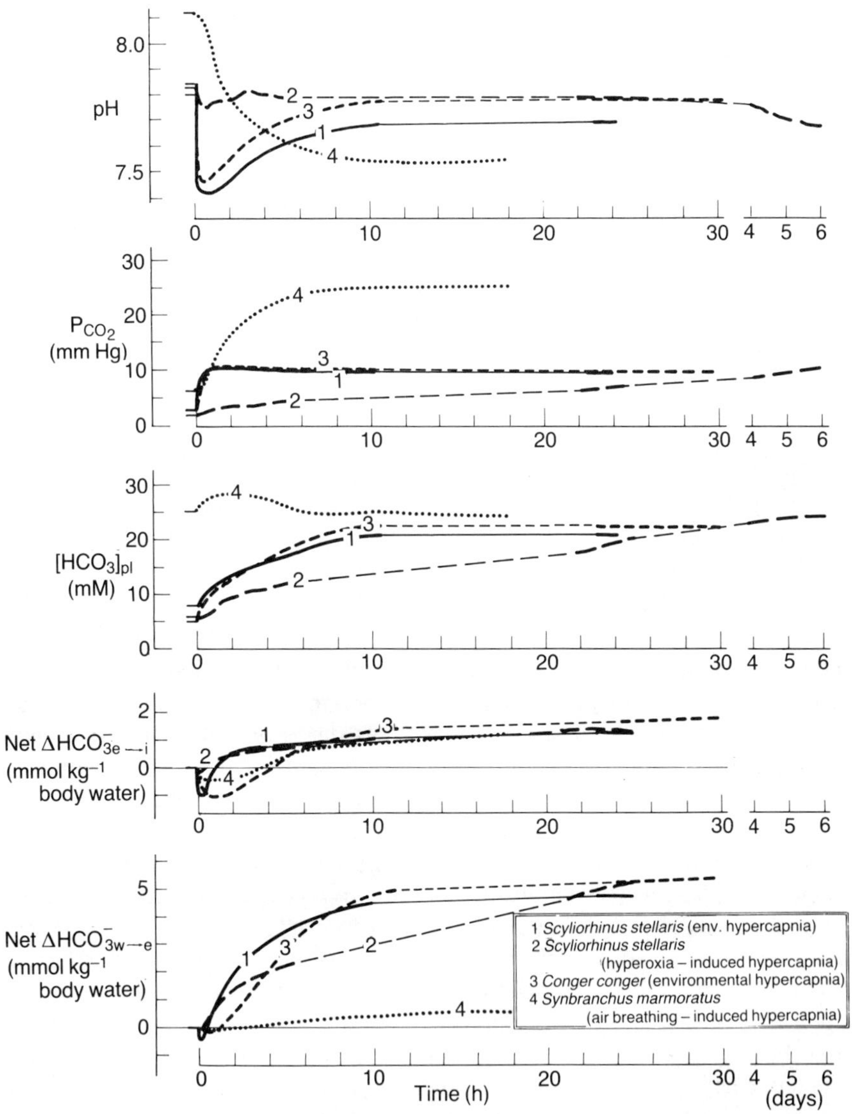

Fig. 8.2 Examples of the acid–base regulatory response of fishes to respiratory acidoses resulting from environmental hypercapnia, hyperoxia-induced hypercapnia, or hypercapnia induced by the transition from water to air breathing. Note differences in the time course of P_{CO_2} elevation, differences in the amount of bicarbonate gained from the environment (see also text).

of hypercapnia by elevation of the bicarbonate concentration, partially or even completely compensating the elevation in P_{CO_2} (Fig. 8.2, '1' and '3'). During the initial phase with large deflections in pH, the rise in extracellular bicarbonate is partially due to non-bicarbonate buffering of CO_2, particularly in well-buffered intracellular muscle compartments. With increasing compensation and return of pH towards control values, the fraction of bicarbonate supplied by this mechanism falls, approaching zero when pH compensation is complete. In fish, mobilization of carbonate from the osteal structures is generally not involved during short-term hypercapnia (directly measured: Cameron, 1985; estimated from changes in plasma [Ca^{2+}] and [Mg^{2+}], N. Heisler and N.A. Andersen, unpublished data).

The main fraction of bicarbonate accumulated during compensation is supplied by acid–base relevant ion transfer mechanisms (Fig. 8.2). The amount gained from the environment during hypercapnia of about 1% CO_2 is in the range of 3 to 6 mmol kg^{-1} body weight (e.g. Heisler *et al.*, 1976; Claiborne and Heisler, 1984, 1986; Heisler, 1986b), much more than the amount present in the animal during normocapnia. The rate of bicarbonate gain is several-fold larger in fish than in Man (Heisler, 1989). However, the uptake rate is apparently linked to ionic concentrations in the environmental water, and may well be reduced considerably under adverse conditions (Heisler, 1984b, 1986b; pp. 223–4). This sensitivity of fishes resides in the direct contact with the environmental water of the branchial epithelium, which is usually responsible for more than 95% of the ion transfer (reviews: Heisler, 1984b, 1986b).

The pattern of acid–base relevant ionic transfer observed upon step changes in environmental P_{CO_2} is characterized by an initial elevation of extracellular bicarbonate due to a transmembrane bicarbonate-equivalent shift from the well-buffered intracellular space (Fig. 8.2; Heisler *et al.*, 1976; Toews *et al.*, 1983), before transepithelial bicarbonate-equivalent uptake mechanisms are sufficiently activated. During this initial phase, there is often a loss of bicarbonate-equivalents to the environment, a maladaptive response expressing still-incomplete adjustment of ion-retaining and resorbing structures to the elevated plasma bicarbonate levels during hypercapnia. The initial release of bicarbonate-equivalents from intracellular to extracellular space (Fig. 8.2) may represent an analogous deficiency of adjustment at the cellular level. When the transepithelial transfer is fully activated, the bicarbonate originally supplied by the intracellular compartment is returned, and bicarbonate gained from the environment is accumulated in both extracellular and intracellular fluid compartments, generally resulting in almost complete pH compensation (Fig. 8.2). Characteristically, intracellular pH is always restored closer to the original controls than extracellular pH (Heisler, 1984b, 1986b, 1988b).

Although the extracellular space is much smaller than the intracellular body fluid compartments (0.2–0.25 v. 0.75–0.8 of the body water, e.g. Heisler,

1978, 1982a), the largest fraction of additional bicarbonate is accumulated in the extracellular space. Nevertheless the small fraction of bicarbonate transferred to the large intracellular space is sufficient for preferential regulation of intracellular pH. This is mainly due to the fact that extracellular control bicarbonate levels are generally larger by a factor of 4–8 (Heisler, 1984b, 1986b). Since the degree of pH compensation is related to the factor by which $[HCO_3^-]$ is elevated (Heisler, 1986c), compensation in the extracellular space is rather costly in terms of accumulated bicarbonate as compared with the small amounts sufficient for compensation on the basis of intracellular bicarbonate levels.

Hyperoxia- and air-breathing-induced hypercapnia

The regulation of ventilation in fish is governed by the oxygen demand of the animals (Dejours, 1975, 1981). When the availability of oxygen is enhanced by environmental hyperoxia or by the switch of facultative air-breathing fishes to air as gas exchange medium, then the ventilatory drive is reduced and P_{CO_2} becomes elevated as compared with control conditions (e.g. Dejours, 1973; Heisler *et al.*, 1981, 1988; Wilkes *et al.*, 1981; Hobe *et al.*, 1984; review: Heisler, 1986b).

In freshwater habitats, P_{O_2} can be extremely variable, with low values resulting from oxygen consumption by animals, plants and microorganisms during dark periods, and extremely high values induced by photosynthesis during daytime (see references in Heisler, 1984b, 1986b). In seawater, the variability of P_{O_2} is much smaller, ranging from 20 to 60 mmHg at water depths of 400–500 m in some particular ocean areas up to values of 150 –190 mmHg in near-surface water layers (Harvey, 1974).

The rise in P_{CO_2} induced by hyperoxia takes place at a much slower rate than that effected by environmental hypercapnia. In some experiments, associated deflections of plasma pH were comparable to those observed during environmental hypercapnia (*Scyliorhinus canicula*, Truchot *et al.*, 1980), whereas others indicated only small initial deviations of plasma pH (~ -0.08, later -0.03 units; *Scyliorhinus stellaris*, Heisler *et al.*, 1981, 1988), but pH started to deviate significantly from control values (-0.1) when P_{CO_2} approached 11 mmHg after 5–6 days of hyperoxia. In these latter experiments, the rise in P_{CO_2} was obviously well matched to the accumulation of plasma bicarbonate, resulting in minimal changes in pH during adaptation to the new environment (Heisler *et al.*, 1988). This shows a switch from primarily oxygen-orientated towards pH-stat regulation (Heisler, 1988a), in which gill ventilation is reduced less than equivalent to the oxygen demand, tracking (with some regulatory oscillations) the accumulation of bicarbonate. During this tracking period, the disparity between oxygen demand and ventilation is evident from the rise of arterial P_{O_2} to more than 200 mmHg during the first day and the subsequent fall to levels below 100 mmHg, close to the normoxic controls, after 5 days of hyperoxia. The plasma [HCO_3^-] steadily rises to values of almost 25 mM, concurrent with an elevation of P_{CO_2} from originally 2 to finally 11 mmHg (Heisler *et al.*, 1988).

The finally attained steady-state acid–base status in *Scyliorhinus stellaris* is very similar with environmental and hyperoxia-induced hypercapnia, with about the same amount of bicarbonate gained from the environment and analogous distribution of additional bicarbonate between intra-and extracellular body compartments (Heisler *et al.*, 1976, 1988). The regulatory mechanisms utilized during environmental and hyperoxia-induced hypercapnia, however, reflect the different time course of P_{CO_2} increase. During the initial stages of environmental hypercapnia, the large deflection of pH leads to a much higher contribution of non-bicarbonate buffering, which is reduced with continuing pH compensation. During hyperoxia, the maximal transepithelial transfer rate of 15 μeq kg^{-1} min^{-1} of bicarbonate equivalent resorption for this species (Heisler *et al.*, 1976; 1977; Holeton and Heisler, 1983; Heisler, 1988b) was attained only during the first hours of hyperoxia, but was reduced to about 5 μmol min^{-1} kg^{-1} when pH recovered to less than -0.03 U after about 3–5 h of hyperoxia, falling further to about 2 μmol min^{-1} kg^{-1} after 21–25 h of hyperoxia. Accordingly it takes at least one order of magnitude longer to accumulate the same quantity of bicarbonate during hyperoxia than during environmental hypercapnia. During this time, ventilation is much higher than required on the basis of the oxygen demand. The capacity of branchial ion transfer mechanisms would allow much faster reduction of ventilation to the final low rate even with pH kept completely compensated. The reason for this type of delayed regulation is unknown, but may be related to the energetics involved (i.e. the additional energetic cost for ventilation would be less than that of a rapid accumulation of bicarbonate), or to the kinetics of readjustment of the respiratory centre. Interestingly, also during hyperoxia-induced hypercapnia, intracellular pH is preferentially regulated as compared with extracellular pH (Heisler *et al.*, 1988).

Air-breathing fish have adapted to frequently or always hypoxic or anoxic environments by switching to air as the gas exchange medium. Many of these fish species are obligate air breathers with reduced gill secondary lamellae or even reduced filaments, utilizing gas exchange organs like modified swim bladders (e.g. *Arapaima gigas*) or the buccal–pharyngeal cavity which largely enhanced surface area and capillarization (e.g. *Electrophorus electricus*). These animals usually have permanently adapted to the higher oxygen content of air as compared with water by appropriate adjustment of ventilation, resulting in higher arterial P_{CO_2} than in water-breathing fish. Little is known about the ionoregulation in those fishes. The reduction of lamellar surface area may not affect the ion-transfer capabilities of the gills, which are thought to be located mainly in mitochondria-rich cells (chloride cells) at the base of the secondary lamellae, or the animals exploit renal mechanisms to a much larger extent than water-breathing fishes.

Very few facultative air-breathing species are capable of covering their oxygen demand completely by exclusively breathing either air or water. In such fishes, the transition from water breathing to air breathing is directly comparable to the exposure of water breathers to hyperoxia. *Synbranchus marmoratus*, one of those species, is well equipped for survival in hypoxic waters. It has a high blood oxygen affinity ($P_{50} \sim 5$ mmHg) and the haematocrit is extremely high for a fish (55%), being in the high range for higher vertebrates (Heisler, 1982a). In tropical floodplains, however, the water is frequently completely anoxic. The animals switch to exclusive air breathing when P_{CO_2} in the water falls below about 16 mmHg, utilizing the same gas exchange structure, the gills, for aerial as well as aqueous gas exchange (Heisler, 1982a).

Upon air breathing, the rate of 'gill' ventilation is reduced considerably and P_{CO_2} rises from about 6 to 25 mmHg within 10 to 72 h, causing plasma pH to fall from 8.15 to 7.5 (Fig. 8.2, '4'). In contrast to water-breathing species, pH remains deflected, with no sign of compensation. After an initial rise in plasma bicarbonate due to blood non-bicarbonate buffering, the bicarbonate concentration levels off at essentially the same level as during water breathing. Also, *Synbranchus* does not gain significant amounts of bicarbonate from the environment during hypercapnia. In contrast to the extracellular space, pH is very well compensated in white and heart muscle intracellular compartments, quite comparable to the degree attained in water breathers. This is achieved by transfer of bicarbonate originally produced by blood non-bicarbonate buffering to the intracellular space (Fig. 8.2; Heisler, 1982a).

The complete lack of extracellular pH compensation may be related to a number of factors. The observed fivefold rise in P_{CO2} would imply an equivalent rise in bicarbonate in order to achieve complete compensation (Heisler, 1986c). On the basis of the high bicarbonate level during water breathing (24 mM), however, an equimolar fall in extracellular [Cl^-] in order to maintain electroneutrality is impossible, and even a fraction of that could hardly be tolerated (Heisler, 1982a). But apart from these considerations, the animals may not be able to take up a sufficient amount of bicarbonate anyway. Ventilation of the buccal cavity with air prevents a continuous irrigation with water, reducing the contact time of the ion-transporting epithelium with the water to short flushes every 5 to 35 min (Heisler, 1982a). Additionally, the environmental conditions for these fish are quite unfavourable with respect to ion exchange with the environment (pp. 223–4). With only a limited amount of additional bicarbonate available, regulation of intracellular homoiostasis has evidently much higher priority. This type of adjustment is very efficient: the small amount of bicarbonate produced by extracellular buffering would not result in any significant compensation of the extracellular space, but is sufficient to serve for an almost complete intracellular pH compensation on the basis of the much lower bicarbonate levels in intracellular fluid compartments.

Tissue hypoxia and anoxia

A disparity between oxygen demand for energy production and oxygen supply at the tissue level may result from a number of mechanisms. Environmental hypoxia as one of the most common factors can at least partially be compensated for by increased gill ventilation. Hypoxia of even moderate extent results in a persistent considerable increase in ventilation (Dejours, 1975, 1981; Glass *et al.*, 1990) with an associated respiratory alkalosis (Glass *et al.*, 1990; Thomas and Motais, 1990). The high ventilation rate in fishes in combination with the high viscosity of water, however, provides energetic limitations. The extraction of oxygen from the water is generally high in fish (60–90%, equivalent to expired P_{O_2} of 60 to 15 mmHg at inspired 150 mmHg, as compared with typical values of 25–35% in mammals; e.g. Wood and Jackson, 1980; Heisler *et al.*, 1988). Any further decrease in expiratory P_{O_2} associated with an increased O_2 extraction will have to be achieved by a reduction of mixed venous P_{O_2} (cf. countercurrent gas exchange system; e.g. Heisler, 1989), with the associated negative effects for the diffusive transfer of oxygen from the capillaries to the site of consumption. Accordingly, any environmental hypoxia exceeding a limited degree will usually result in at least partially anaerobic energy production, expressed by development of a metabolic acidosis (Thomas and Motais, 1990).

Similar problems for the oxygen supply of tissues arise from intoxification with nitrite, a substance recently found to frequently contaminate natural waters. After being taken up, nitrite effects enhanced oxidation of the normal (Fe^{2+})-Hb to methaemoglobin (Fe^{3+}), which cannot reversibly associate oxygen (Jensen *et al.*, 1987; Williams *et al.*, 1989, 1992). The reduced functional oxygen-carrying capacity, a condition equivalent to anaemia induced by other factors, results in principally the same pattern of disturbance of the tissue oxygen supply as does environmental hypoxia. A further condition leading to tissue hypoxia or anoxia is an elevation of energy production associated with extensive muscular activity, such as during predation and emergency conditions. Although ventilation as well as cardiac output can be elevated to a great extent in order to support such activities, the poorly capillarized and perfused white musculature of fish often provides mechanical energy on the basis of anaerobiosis.

The common result of incomplete tissue oxygen supply is breakdown of creatine phosphate and ATP, and the production of intermediary glycolytic metabolic end products, with the most prominent one in vertebrates being lactic acid. After reestablishment of oxic conditions the oxygen debt is quickly repaid with respect to creatine phosphate and ATP, whereas the removal of lactic acid is comparatively delayed. According to the low pK′ value (~ 3.9), lactic acid is completely dissociated at physiological pH values, providing an equimolar stress for the acid–base regulation. A few fish species (e.g. goldfish, *Carassius auratus*, crucian carp, *Carassius carassius*, Shoubridge and Hochachka, 1980; van den Thillart *et al.*, 1983) have developed the capability to produce ethanol as anaerobic end product, which is advantageous by being neutral for the acid–base status. This metabolic strategy is of great ecological importance, supporting the survival of crucian carp in long winters in complete anoxia under the ice cover (Holopainen and Hyvärinen, 1985). However, the benefit of diffusional

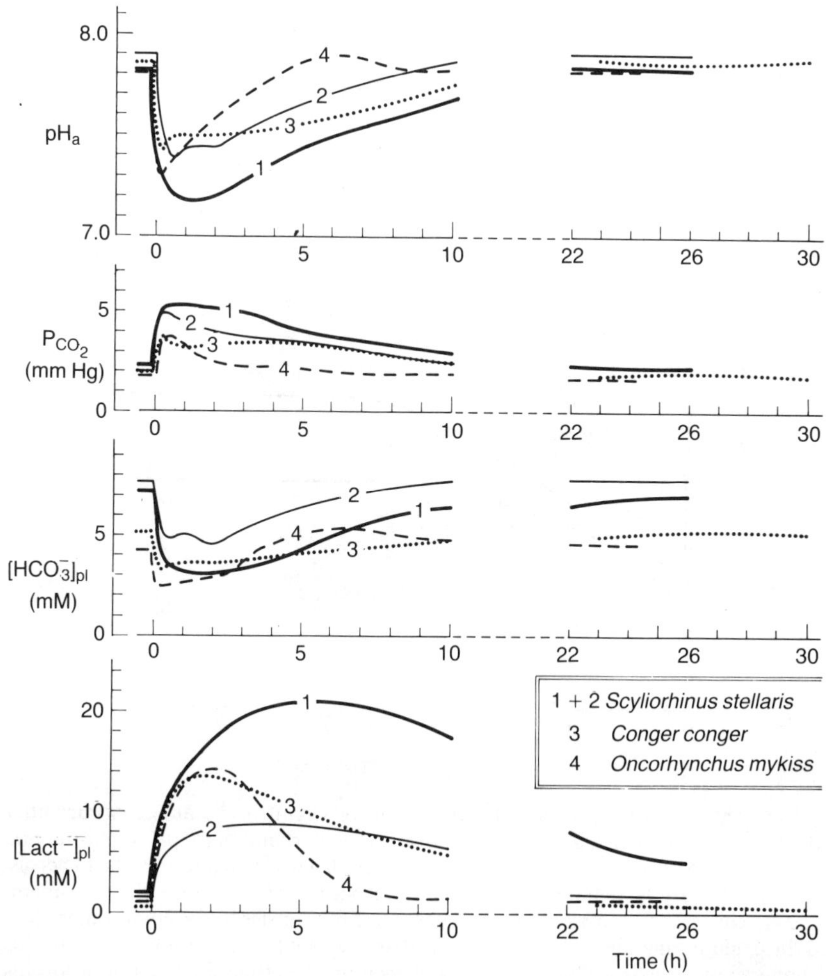

Fig. 8.3 Changes in plasma pH, P_{CO_2}, $[HCO_3^-]$ and lactate concentration during lactacidosis induced by exhausting muscular activity in three selected fish species. Note elevation of P_{CO_2} and the normalization of pH before lactate concentration (see also text).

elimination of a neutral substance through the gill epithelium similar to CO_2 is paid off by the loss of a large amount of inherent energy.

Although lactic acid produced during environmentally induced (water hypoxia, nitrite exposure) hypoxia is more homogeneously distributed through-

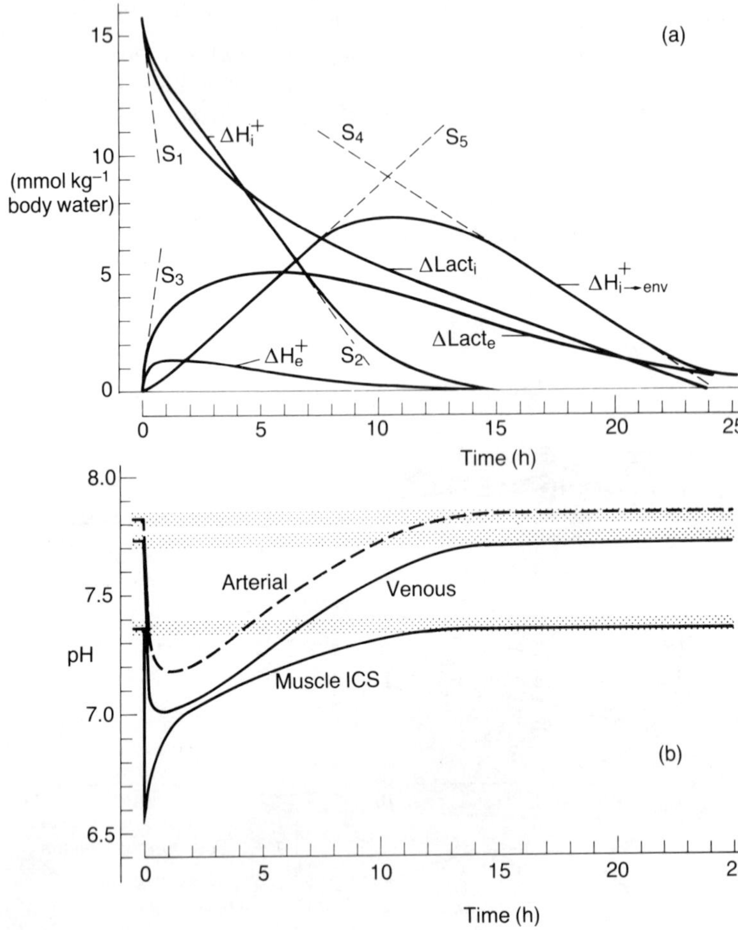

Fig. 8.4 Dynamics of (a) surplus H^+ and lactate ions in intracellular and extracellular body fluid compartments, and the environmental water, and of (b) pH in plasma and muscle cells of *Scyliorhinus stellaris* during recovery from severe muscular lactacidosis. After initially very fast elimination of H^+ ions from muscle cells (S_1), the transmembrane H^+ elimination kinetics (S_2) are determined by transfer to the environment (S_5) and the oxidative metabolization of lactic acid (equivalent to S_4). The amount of surplus H^+ buffered in the extracellular space is always much less than the amount of surplus lactate (equilibrium limitation, see text). After an initial period, intracellular pH is deflected less than extracellular pH in spite of the limitation of transmembrane transfer (see also text). ICS = intracellular space; stippled areas represent range of control values. For for underlying model calculations see Heisler (1984b, 1986a).

out the organisms than during functional hypoxia (muscular activity), the general acid–base regulatory pattern is similar. After production and dissociation of lactic acid during extensive muscular activity, the dissociation products H^+ and lactate$^-$ ions are gradually eliminated from the intracellular space. The time course of elimination is different for the two dissociation products and variable among fish species. Generally, the extracellular acid–base deflections peak much earlier than extracellular lactate (Fig. 8.3). In *Scyliorhinus stellaris*, plasma pH and [HCO_3^-] are maximally reduced 15–30 min after the end of strenuous muscular activity. The time course of the return to control conditions is clearly a matter of the magnitude of the imposed anaerobic activity, but always occurs long before plasma lactate levels are normalized. This is achieved by transient net transfer of H^+ ions to the environmental water until the lactic acid is completely removed by aerobic metabolic processing, whereas lactate$^-$ is never transferred to the environment (Fig. 8.4; Holeton and Heisler, 1983; Holeton *et al.*, 1983).

The intracellular pH of muscle tissues is generally less affected than extracellular pH after an initial efflux period (Fig. 8.4). During this period H^+ ions are eliminated from the cells at a tremendous rate (> 250 μmol min^{-1} kg^{-1}), which is soon reduced due to the establishment of a transient equilibrium between intracellular and extracellular pH. H^+ ions are then further released from the intracellular space only at the rate at which they are transferred to the environmental water ('equilibrium limitation', Holeton and Heisler, 1983). Under these conditions, less than 8% of surplus H^+ ions are buffered extracellularly (Fig. 8.4). Net H^+ transfer to the environment takes place at a rate of about 15 μmol kg^{-1} body water, similar to the maximal rate during hypercapnia. When extracellular pH and [HCO_3^-] approach control conditions, H^+ ions required for aerobic processing of lactate are returned from the environment at the rate of lactic acid metabolization. This advantageous use of the ambient water for transient storage of surplus H^+ ions is found in all studied species (e.g. *Conger conger*, Toews *et al.*, 1983; *Oncorhynchus mykiss*, Holeton *et al.*, 1983).

8.3 CHARACTERISTICS OF REGULATORY RESPONSES

Fish take advantage, at least transiently, of all available mechanisms for acid–base regulation (Heisler, 1989). Modulation of the rate of ventilation, however, usually does not contribute to a great extent. During changes of temperature, readjustment of ventilation is dominated by the requirements for oxygen acquisition: transient oscillations in P_{CO_2} are generally quieted long before the more important ion-transfer processes are complete. During environmental hypercapnia, gill ventilation in water-breathing fishes has been reported to be either little or not at all affected, or only transiently affected to a limited extent (Dejours, 1975, 1981; Randall *et al.*, 1976). Moderate non-respiratory decreases in pH usually result in no or only slight increases in ventilation (Dejours, 1975, 1981), whereas more severe reductions of pH have a more pronounced effect (G.F. Holeton and N. Heisler, unpublished data). This may be related to the normally low oxygen content of water and the resulting primarily oxygen-orientated regulation of ventilation, which relies upon pH as the leading stimulus only during conditions of environmental hyperoxia (Heisler, 1988a; Heisler *et al.*, 1988). Small and transient increases in ventilation during non-respiratory disturbances have little effect on acid–base regulation: P_{CO_2} values typically elevated during lactacidosis indicate that with such conditions ventilation is not the leading parameter in determining the inspired-arterial P_{CO_2} difference (pp. 211–12).

Buffering is a valuable mechanism for transient acid–base regulation in fish, but its role is limited as compared with terrestrial animals. In terrestrial animals, pH is generally normalized only with removal of the original stress factor (e.g. return to environmental normocapnia or aerobic metabolic processing of lactic acid), whereas acid–base regulation in fish is characterized by early recovery of pH. This normalization reduces the role of buffering for the removal of surplus H^+ ions from the body fluids. During changes in temperature, the production of bicarbonate by mainly intracellular non-bicarbonate buffers may actually represent a stress factor: pH shifts are often expected on the basis of semi-closed system buffering in opposite directions from those actually attained by transmembrane and transepithelial ion-transfer processes (Heisler, 1984b).

Reestablishment of control pH in fish is generally performed by transient elimination of H^+ ion equivalents from the body fluids or from certain body fluid compartments before the original stress factor is removed. Ionic transfer at the branchial epithelium is the most important mechanism in this matter. During severe lactacidosis, the kinetics of transmembrane removal of H^+ equivalents from the site of lactic acid production are governed to a great extent by the rate of elimination from the extracellular space by the branchial epithelium by establishment of an equilibrium between intra- and extracellular pH at a reduced level (equilibrium limitation, p. 221; Holeton and Heisler, 1983). The accumulation of additional bicarbonate in the body fluids during hypercapnia is apparently linked to readjustment of bicarbonate thresholds in membranes and epithelia: during the initial stages of environmental hypercapnia, bicarbonate is released from intracellular to extracellular space and to the environment. These 'overflow' phenomena are apparently less pronounced during slowly developing (hyperoxia- and air-breathing-induced) hypercapnia.

Under all studied conditions, regulation of intracellular pH has a higher priority than that of extracellular pH. This is indicated by earlier and closer restoration of pH towards control values. In the extreme case of air-breathing *Synbranchus*, intracellular pH is almost completely, and extracellular pH not at all, compensated. Preferential regulation of intracellular pH is facilitated by initially lower intracellular bicarbonate concentrations, reducing the amount of additional bicarbonate required for a certain degree of compensation. Acid–base regulation in fish must accordingly be considered as a two-line defence mechanism for the internal milieu of the energy-producing structures of the organism.

8.4. CAPACITY OF ACID–BASE RELEVANT ION-TRANSFER MECHANISMS AS A FUNCTION OF ENVIRONMENTAL CONDITIONS

Transfer of acid–base relevant ions between environmental water and extracellular space as a leading mechanism for fish acid–base regulation can be performed at a rate much higher than that of comparable mechanisms in terrestrial animals. The typical overall acid–base relevant transfer rate per unit standard metabolic rate is in the range of 0.13–0.47 (μmol ion transfer per μmol O_2 consumption) in fishes as compared with 0.02–0.03 in typical mammals like the dog and Man (Heisler, 1988b). This is most likely related to the direct contact of the large surface area of the transfer site, the gill epithelium, with a large volume of environmental water (see above). The water flow past the gill epithelium (resting 5000–20 000 ml kg^{-1} h^{-1}, e.g. Randall *et al.*, 1976) being larger than the typical urine flow rate (1–10 ml kg^{-1} h^{-1}, Hunn, 1982) by a factor of 10^3 to 10^4 will certainly facilitate ion-transfer processes to a much greater extent, provided the ionic composition of the environmental water is favourable.

Comparison of literature data first suggested that the ionic composition of the environmental water was an important factor for fish in dealing with environmental hypercapnia (Heisler, 1982b). Rainbow Trout in dilute Vancouver tap water ($[Na^+]$ 0.022–0.087 mM; $[HCO_3^-]$ 0.050 mM) required 72 h to attain approximately the same level of pH compensation (Janssen and Randall, 1975) as that achieved in water of higher ionic content by trout ($[Na^+]$ 0.5 mM; $[HCO_3^-]$ 3 mM) after less than 24 h (Eddy *et al.*, 1977) and *Conger* (seawater, $[Na^+]$ 480 mM; $[HCO_3^-]$ 2.5 mM) after 10 h (Toews *et al.*, 1983). This general pattern was confirmed by studies in *Scyliorhinus stellaris*, where the rate of net bicarbonate gain from the environment was constant in a range of $[HCO_3^-]_{pl}/[HCO_3^-]_{sw}$ = 0.3–4.0, and decreased in an apparently linear fashion at higher $[HCO_3^-]_{pl}/[HCO_3^-]_{sw}$ ratios attaining zero at $[HCO_3^-]_{pl}/[HCO_3^-]_{sw}$ = 13. At even higher ratios bicarbonate-equivalents were lost to the environmental water in spite of continuing hypercapnia. Since all ions except HCO_3^- remained constant during these experiments (except for very small elevations in $[Na^+]$ and $[Cl^-]$ due to seawater titration with $NaHCO_3$ and HCl), the environmental $[HCO_3^-]$ appears to be a limiting factor for trans-epithelial ion transfer, at least in this species (Heisler and Neumann, 1977;

Heisler, 1988a). Comparable data were obtained in rainbow trout *Oncorhynchus mykiss*, exposed to environmental hypercapnia in water of 3, 100 and 300 mM salinity and in carp exposed to hypercapnia in water of 0.1, 1, 10 and 50 mM [HCO_3^-], resulting in significantly faster compensation at higher salinities (Iwama and Heisler, 1991) or higher bicarbonate concentrations (N.A. Andersen and N. Heisler, unpublished data). Much larger acid–base relevant branchial ion-transfer capacities were also reported in seawater as compared with fresh water for genetically identical rainbow trout populations after acid infusion (Tang and Boutilier, 1988) and after exhausting exercise (Tang *et al.*, 1989).

The ion-exchange mechanisms in the branchial epithelium are considered to be best described as active ion-exchange processes of H^+ or NH_4^+ against Na^+, and of HCO_3^- against Cl^- (Evans, 1986; Heisler 1986b). All of these mechanisms have been demonstrated to be present under certain conditions (Evans, 1986). Under physiological conditions, however, not all of them may be involved in the regulatory process. Recent experiments with ionic water concentrations in the mM range have indicated that in carp, Na^+-related mechanisms have hardly any role in the transfer of bicarbonate-equivalent ions for compensation of hypercapnic acidoses (Heisler, 1990a), and unidirectional tracer flux studies have indicated also for trout that acid–base regulation during hypercapnia is to a large fraction performed by Cl^--related mechanisms (Ishimatsu *et al.*, 1991; A. Ishimatsu, G. Iwama and N. Heisler, unpublished data).

Independent of the type of mechanism, sufficient availability of the appropriate counter ions in the environmental water (HCO_3^- and/or Na^+) will be a limiting factor for net transfer of H^+ ions to the environment. The data on bicarbonate-equivalent resorption as a function of water bicarbonate concentration in *Scyliorhinus* (p. 223) suggest that bicarbonate is gained from the environment only with plasma/water bicarbonate ratios of less than 13, but is net lost at higher values. In normal seawater with [HCO_3^-] around 2 mM this pattern will limit the maximal bicarbonate concentration to about 25 mM. Hyperoxia-induced hypercapnia in *Scyliorhinus* is actually completely compensated until plasma bicarbonate is elevated above 20 mM; at higher values the compensation is increasingly incomplete, and even after 6 days of hyperoxia, plasma bicarbonate does not rise above 25 mM (Heisler *et al.*, 1988). A similar limit is attained in this species also during environmental hypercapnia (Heisler *et al.*, 1976).

When literature data about the maximal bicarbonate concentration attained during hypercapnia were compiled, it emerged as a typical pattern that a similar limit for plasma bicarbonate concentration existed for a number of lower vertebrate species in the range of 23 to 33 mM, with fishes covering the lower end, and with reptiles (27–45 mM) and higher vertebrates (40–50 mM) tending to the higher end of this range or exceeding it (Heisler, 1988b). As a consequence of this bicarbonate limit, the degree of compensation (expressed

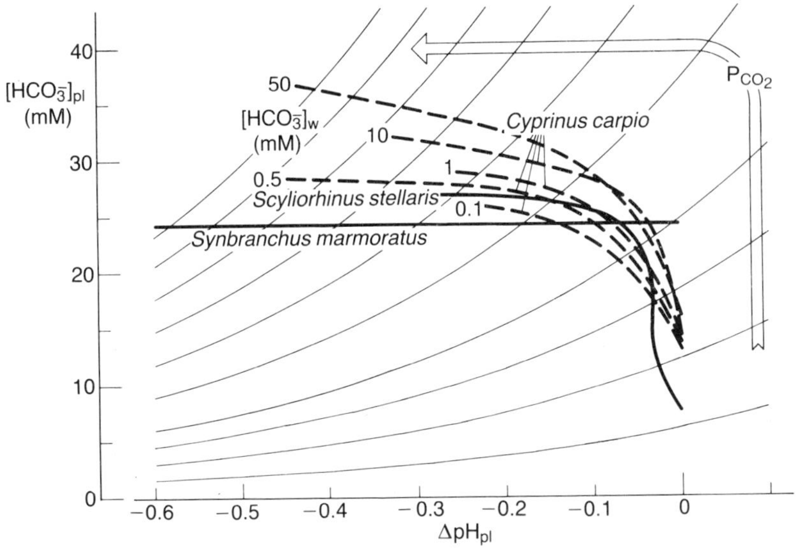

Fig. 8.5 Bicarbonate concentration in three selected fish species during hypercapnia. When certain plasma bicarbonate levels are attained, increasing P_{CO_2} results in a rising degree of decompensation as indicated by the fall in plasma pH (Δ pH). These bicarbonate levels are, at least in *Cyprinus*, a function of the environmental $[HCO_3^-]_w$ (see also text).

as % pH recovery as compared with the pH shift at constant bicarbonate concentration) is generally lower in animals with high control bicarbonate concentration, owing to their inability to achieve an increase of bicarbonate by a factor comparable to the factor of P_{CO_2} increase (Heisler, 1986c). Typical examples for this type of limitation are carp, *Cyprinus carpio*, about 50% (Claiborne and Heisler, 1984, 1986) and the tropical air breather *Synbranchus*, 0% (Heisler, 1982a), with initial bicarbonate concentrations of 13 and 24 mM respectively (Heisler, 1988b). In these animals, the plasma bicarbonate concentration is apparently limited by a certain specific threshold for the retention and net resorption of bicarbonate equivalents; even infused bicarbonate would not be retained, but quantitatively released to the environment, as demonstrated for at least two species (*Cyprinus carpio*, Claiborne and Heisler, 1986; *Siren lacertina*, Heisler *et al.*, 1982). In contrast to this pattern, some experiments demonstrated that much higher bicarbonate concentrations could be attained during compensation of hypercapnia (e.g. Børjeson, 1976, 1977; Jensen and Weber, 1982; Cameron and Iwama, 1987; Dimberg, 1988).

As evident from the bicarbonate equivalent resorption as a function of water bicarbonate in *Scyliorhinus* (p. 223), the bicarbonate threshold must be a function of environmental factors like water bicarbonate concentration. Preliminary data suggest that this holds as well for carp and water bicarbonate (Fig. 8.5), but the phenomenal capacity to compensate hypercapnic acidoses with plasma bicarbonate concentrations of up to 50 mM in for instance the

Uppsala trout (Børjeson, 1976, 1977; Dimberg, 1988), Odense tench (Jensen and Weber, 1982) and *Ictalurus punctatus* (Cameron and Iwama, 1987) is difficult to conceive on the basis of counter-ion concentrations alone. A long-term acclimation effect may be involved, particularly in the Uppsala trout, which were adapted to elevated levels of environmental P_{CO_2} for several months before the experiments, much longer than the 3 weeks of adaptation during which carp did not attain higher levels of plasma bicarbonate than 26 mM (Claiborne and Heisler, 1986). Values of 50 mM, however, were attained in catfish after only 5 days of hypercapnia (Cameron and Iwama, 1987).

One common factor in all those experiments, in which such high plasma bicarbonate levels were achieved, was a rather high environmental $[Ca^{2+}]$ of close to or higher than 2 mM, combined with relatively high levels of bicarbonate in the range of 3–4 mM. The extraordinary capacity to compensate a hypercapnic acidosis in those animals may accordingly be related to a combination of the supportive effect of high environmental bicarbonate with the sealing effect of Ca^{2+} on paracellular pathways and the associated general reduction of unidirectional flux rates, including a reduction in the rate of transepithelial bicarbonate leakage. Even a direct effect of Ca^{2+} on the ion-transporting structures cannot be excluded. The precise interrelationship of these factors is unclear and requires further investigation.

8.5. CONCLUSION

The most important mechanism for fish acid–base regulation is transfer of acid–base relevant ions between intracellular and extracellular body fluid compartments and the environmental water. The involved transfer mechanisms are generally much more effective than equivalent processes in terrestrial animals, facilitated by the large water flow across the exchange site. Because of the exposed site in the branchial epithelium, the performance of these mechanisms is affected to a large extent by the composition of the environmental water, in particular by the availability of appropriate counterions. Additional factors like water $[Ca^{2+}]$ may be involved as well, limiting the maximal transfer rate, and the maximal plasma bicarbonate concentration.

REFERENCES

Børjeson, H. (1976) Some effects of high carbon dioxide tension on juvenile salmon (*Salmo salar* L.). *Acta univ. Ups.*, **383**, 1–35.

Børjeson, H. (1977) Effects of hypercapnia on the buffer capacity and haematological values in *Salmo salar* (L.). *J. Fish Biol.*, **11**, 133–42.

Bullis, H.R., jun. (1967) Depth segregations and distribution of sex-maturity groups in the marbled catshark, *Galeus arae*, in *Sharks, Skates and Rays* (eds W. Gilbert, R.F. Mathewson and D.P. Rall), Johns Hopkins Univ. Press, Baltimore, MD, pp. 141–8.

Cameron, J.N. (1985) The bone compartment in a teleost fish, *Ictalurus punctatus*: size, composition and acid–base response to hypercapnia. *J. Exp. Biol.*, **117**, 307–18.

Cameron, J.N. and Iwama, G.K. (1987) Compensation of progressive hypercapnia in channel catfish and blue crabs. *J. exp. Biol.*, **133**, 183–97.

Cameron, J.N. and Kormanik, G.A. (1982) Intracellular and extracellular acid–base status as a function of temperature in the freshwater channel catfish, *Ictalurus punctatus*. *J. exp. Biol.*, **99**, 127–42.

Claiborne, J.B. and Heisler, N. (1984) Acid–base regulation in the carp (*Cyprinus carpio*) during and after exposure to environmental hypercapnia. *J. exp. Biol.*, **108**, 25–43.

Claiborne, J.B. and Heisler, N. (1986) Acid–base regulation and ion transfers in the carp (*Cyprinus carpio*): pH compensation during graded long- and short-term environmental hypercapnia and the effect of bicarbonate infusion. *J. exp. Biol.*, **126**, 41–61.

Dejours, P. (1973) Problems of control of breathing in fishes, in *Comparative Physiology* (eds. L. Bolis, K. Schmidt-Nielsen and S.H.P. Maddrell), North Holland Publ., Amsterdam, pp. 117–33.

Dejours, P. (1975, 1981) *Principles of Comparative Respiratory Physiology*, North-Holland Publ., Amsterdam, 254 pp. (1975) and 266 pp. (1981).

Dimberg, K. (1988) High blood CO_2 levels in rainbow trout exposed to hypercapnia in bicarbonate-rich hard fresh water – a methodological verification. *J. exp. Biol.*, **134**, 463–6.

Eddy, F.B., Lomholt, J.P., Weber, R.E. and Johansen, K. (1977) Blood respiratory properties of rainbow trout (*Salmo gairdneri*) kept in water of high CO_2 tension. *J. exp. Biol.* **67**, 37–47.

Edsall, J.T. and Wyman, J. (1958) *Biophysical Chemistry*, Academic Press, New York, 700 pp.

Evans, D.H. (1986) The role of branchial and dermal epithelia in acid–base regulation in aquatic animals, in *Acid–Base Regulation in Animals* (ed. N. Heisler), Elsevier Biomedical Press, Amsterdam, pp. 139–72.

Glass, M.L., Andersen, N.A., Kruhøffer, M., Williams, E.M. and Heisler, N. (1990) Combined effects of environmental P_{O_2} and temperature on ventilation and blood gases in the carp *Cyprinus carpio* L. *J. exp. Biol.*, **148**, 1–17.

Harvey, H.W. (1974) *The Chemistry and Fertility of Sea Waters*, Cambridge Univ. Press, London, 240 pp.

Heisler, N. (1978) Bicarbonate exchange between body compartments after changes of temperature in the larger spotted dogfish (*Scyliorhinus stellaris*). *Respir. Physiol.*, **33**, 145–60.

Heisler, N. (1982a) Intracellular and extracellular acid–base regulation in the tropical freshwater teleost fish *Synbranchus marmoratus* in response to the transition from water breathing to air breathing. *J. exp. Biol.*, **99**, 9–28.

Heisler, N. (1982b) Transepithelial ion transfer processes as mechanisms for fish acid–base regulation in hypercapnia and lactacidosis. *Can. J. Zool.*, **60**, 1108–22.

Heisler, N. (1984a) Role of ion transfer processes in acid–base regulation with temperature changes in fish. *Am. J. Physiol*, **246**, R441–51.

Heisler, N. (1984b) Acid–base regulation in fishes, in *Fish Physiology*, Vol. XA, (eds W.S. Hoar and D.J. Randall), Academic Press, New York, pp. 315–401.

Heisler, N. (1986a) Buffering and transmembrane ion transfer processes, in *Acid–base Regulation in Animals* (ed. N. Heisler), Elsevier, Amsterdam, pp. 3–47.

Heisler, N. (1986b) Acid–base regulation in fishes, in *Acid–base Regulation in Animals* (ed. N. Heisler), Elsevier, Amsterdam, pp. 309–56.

Heisler, N. (1986c) Comparative aspects of acid–base regulation, in *Acid–base Regulation in Animals* (ed. N. Heisler), Elsevier Amsterdam, pp. 397–450.

Heisler, N. (1988a) Acid–base regulation in elasmobranch fishes, in *Physiology of Elasmobranch Fishes* (ed. T.J. Shuttleworth), Springer, Heidelberg, pp. 215–52.

Heisler, N. (1988b) Ion transfer processes as mechanisms for acid–base regulation, in *Lung Biology in Health and Disease – Comparative Pulmonary Physiology: Current Concepts* (ed. S.C. Wood), Marcel Decker, New York, pp. 539–83.

Heisler, N. (1989) Acid–base regulation in fishes I. mechanisms, in *Acid toxicity and Aquatic Animals* (eds R. Morris, E.W. Taylor, D.J.A. Brown and J.A. Brown). (Society of Experimental Biology Seminar Series), Cambridge Univ. Press, Cambridge, pp. 85–97.

Heisler, N. (1990a) Acid–base regulation: interrelationships between gaseous and ionic exchange, in *Vertebrate Gas Exchange from Environment to Cell* (ed. R.G. Boutilier) (Advances in Environmental and Comparative Physiology, Vol. 6; (R. Gilles, managing editor), Springer, Heidelberg, pp. 211–51.

Heisler, N. (1990b) Interaction between gas exchange, metabolism and ion transport in animals: an overview. *Can. J. Zool.*, **67**, 2923–35.

Heisler, N. and Neumann, P. (1977) Influence of sea water pH upon bicarbonate uptake induced by hypercapnia in an elasmobranch (*Scyliorhinus stellaris*). *Pflügers Arch. ges. Physiol.*, **368**, Suppl., R19.

Heisler, N. and Neumann, P. (1980) The role of physico-chemical buffering and of bicarbonate transfer processes in intracellular pH regulation in response to changes of temperature in the larger spotted dogfish (*Scyliorhinus stellaris*). *J. exp. Biol.*, **85**, 99–110.

Heisler, N., Weitz, H. and Weitz, A.M. (1976) Hypercapnia and resultant bicarbonate transfer processes in an elasmobranch fish. *Bull. Eur. Physiopathol. Respir.*, **12**, 77–85.

Heisler, N., Holeton, G.F. and Toews, D.P. (1981) Regulation of gill ventilation and acid–base status in hyperoxia-induced hypercapnia in the larger spotted dogfish (*Scyliorhinus stellaris*). *Physiologist*, **24**, 58 (305).

Heisler, N., Forcht, G., Ultsch, G.R. and Anderson, J.F. (1982) Acid–base regulation in response to environmental hypercapnia in two aquatic salamanders, *Siren lacertina* and *Amphiuma means*. *Respir. Physiol.*, **49**, 141–58.

Heisler, N., Toews, D.P. and Holeton, G.F. (1988) Regulation of ventilation and acid–base status in the elasmobranch *Scyliorhinus stellaris* during hyperoxia-induced hypercapnia. *Respir. Physiol.*, **71**, 227–46.

Henry, R.P., Smatresk, N.J. and Cameron, J.N. (1988) The distribution of branchial carbonic anhydrase and the effects of gill and erythrocyte carbonic anhydrase inhibition in the channel catfish, *Ictalurus punctatus*. *J. exp. Biol.*, **134**, 201–18.

Hobe, H., Wood, C.M. and Wheatly, M.G. (1984) The mechanisms of acid–base and ionoregulation in the freshwater rainbow trout during environmental hyperoxia and subsequent normoxia. I. Extra- and intracellular acid–base status. *Respir. Physiol.*, **55**, 139–54.

Holeton, G.F. and Heisler, N. (1983) Contribution of net ion transfer mechanisms to the acid–base regulation after exhausting activity in the larger spotted dogfish (*Scyliorhinus stellaris*). *J. exp. Biol.*, **103**, 31–46.

Holeton, G.F., Neumann, P. and Heisler, N. (1983) Branchial ion exchange and acid–base regulation after strenuous exercise in rainbow trout (*Salmo gairdneri*). *Respir. Physiol.*, **51**, 303–18.

Holopainen, I. J. and Hyvärinen, H. (1985) Ecology and physiology of crucian carp (*Carassius carassius*, L.) in small Finnish ponds with anoxic conditions in winter. *Verh. int. Verein. Limnol.*, **22**, 2566–70.

Hunn, J.B. (1982) Urine flow rate in freshwater salmonids: a review. *Progve. Fish-Cult.*, **44**, 119–25.

Ishimatsu, A., Iwama, G.K., Bentley, T. and Heisler, N. (1991) Role of the secondary circulatory system in acid–base regulation in trout. *Proc. IUBS comp. Physiol. Biochem.*, **3**, 137 (3017).

Iwama, G.K. and Heisler, N. (1991) Effect of environmental water salinity on the acid–base regulation during environmental hypercapnia in the rainbow trout (*Salmo gairdneri*). *J. exp. Biol.*, **158**, 1–18.

Janssen, R.G. and Randall, D.J. (1975) The effect of changes in pH and $P\text{CO}_2$ in blood and water on breathing in rainbow trout, *Salmo gairdneri*. *Respir. Physiol.*, **25**, 235–45.

Jensen, F.B. and Weber, R.E. (1982) Respiratory properties of tench blood and hemoglobin. Adaptation to hypoxic–hypercapnic water. *Mol. Physiol.*, **2**, 235–50.

Jensen, F.B., Andersen, N.A. and Heisler, N. (1987) Effect of nitrite exposure on blood respiratory properties, acid–base and electrolyte regulation in the carp (*Cyprinus carpio*). *J. comp. Physiol.*, **157B**, 533–41.

Kato, S. and Cavallo, A.H. (1967) Shark tagging in the eastern Pacific Ocean, 1962–1965, in *Sharks, Skates and Rays* (eds W. Gilbert, R.F. Mathewson, and D.P. Rall), Johns Hopkins Press, Baltimore, MD, pp. 93–110.

Rahn, H. (1967) Gas transport from the environment to the cell, in *Development of the Lung* (eds A.V.S. de Reuck and R. Porter), Churchill, London, pp. 3–23.

Randall, D.J., Heisler, N. and Drees, F. (1976) Ventilatory response to hypercapnia in the larger spotted dogfish *Scyliorhinus stellaris*. *Am. J. Physiol.*, **230**, 590–4.

Reeves, R.B. (1972) An imidazole alphastat hypothesis for vertebrate acid–base regulation: tissue carbon dioxide content and body temperature in bullfrogs. *Respir. Physiol.*, **14**, 219–36.

Reeves, R.B. and Malan, A. (1976) Model studies of intracellular acid–base temperature responses in ectotherms. *Respir. Physiol.*, **28**, 49–63.

Robin, E.D. (1962) Relationship between temperature and plasma pH and carbon dioxide tension in the turtle. *Nature*, **135**, 249–51.

Shoubridge, E. and Hochachka, P. (1980) Ethanol: novel end-product of vertebrate anaerobic metabolism. *Science*, **209**, 308–9.

Tang, Y. and Boutilier, R.G. (1988) Clearance of lactate and protons following acute lactacidosis: a comparison between seawater- and freshwater-adapted rainbow trout (*Salmo gairdneri*). *Exp. Biol.*, **48**, 41–4.

Tang, Y., McDonald, D.G. and Boutilier, R.G. (1989) Acid–base regulation following exhaustive exercise: a comparison between freshwater- and seawater-adapted rainbow trout (*Salmo gairdneri*). *J. exp. Biol.*, **141**, 407–18.

Thomas, S. and Motais, R. (1990) Acid–base balance and oxygen transport during acute hypoxia in fish, in *Animal Nutrition and Transport Processes, II. Transport, Respiration and Excretion: Comparative and Environmental Aspects* (eds J.P. Truchot and B. Lahlou) (Series Comparative Physiology), Basel, Karger, pp. 76–91.

Thillart, G. Van den, Van Berge-Henegouwen, M. and Kesbeke, F. (1983) Anaerobic metabolism of goldfish, *Carassius auratus* (L.): ethanol and CO_2 excretion rates and anoxia tolerance at 20, 10 and 5°C. *Comp. Biochem. Physiol.*, **76A(2)**, 295–300.

Toews, D.P., Holeton, G.F. and Heisler, N. (1983) Regulation of the acid–base status during environmental hypercapnia in the marine teleost fish *Conger conger*. *J. exp. Biol.*, **107**, 9–20.

Truchot, J.P., Toulmond, A. and Dejours, P. (1980) Blood acid–base balance as a function of water oxygenation: a study at two different ambient CO_2 levels in the dogfish, *Scyliorhinus canicola*. *Respir. Physiol.*, **41**, 13–28.

Wilkes, P.R.H., Walker, R.L., McDonald, D.G. and Wood, C.M. (1981) Respiratory, ventilatory, acid–base and ionoregulatory physiology of the white sucker *Catostomus commersoni*: the influence of hyperoxia. *J. exp. Biol.*, **91**, 239–54.

Williams, E.M., Glass, M.L. and Heisler, N. (1989) The effect of nitrite on carp (*Cyprinus carpio*) blood oxygen affinity and methaemoglobin formation. *Scand. J. Physiol.*, **136**, 16a (C26).

Williams, E.M., Glass, M.L. and Heisler, N. (1992) Blood oxygen transport in carp (*Cyprinus carpio* L.) exposed to hypoxia and methaemoglobinaemia. *Aquaculture and Fisheries Management* (in press).

Winterstein, H. (1954) Der Einfluβ der Körpertemperatur auf das Säure-Basen-Gleichgewicht im Blut. *Arch. exper. Path. u. Pharmakol.*, **223**, 1–18.

Wood, C.M. and Jackson, E.B. (1980) Blood acid–base regulation during environmental hyperoxia in the rainbow trout (*Salmo gairdneri*). *Respir. Physiol.*, **42**, 351–72.

Woodbury, J.W. (1965) Regulation of pH, in *Physiology and Biophysics* (eds T.C. Ruch and H.D. Patton), Saunders, Philadelphia, pp. 899–934.

Chapter nine

Environmental effects on fish gill structure and function

Steve F. Perry and Pierre Laurent

9.1 INTRODUCTION

Aquatic habitats are remarkably diverse with respect to the chemical and physical properties of the water. For example, salinity may vary between full-strength seawater and near distilled water while pH may differ by as much as 6.0 pH units. Furthermore, aquatic environments, especially the freshwater ecosystems, are typically unstable and characterized by marked natural fluctuations of temperature, pH, oxygen, carbon dioxide, and dissolved ions. The ability of fish to inhabit these diverse and oscillating environments arises from a variety of adaptive physiological mechanisms. Owing to the location of the gill between the external and internal environments and its crucial role in gas transfer, acid–base balance, and ionic regulation, adaptive changes in branchial function are especially important. In many instances, the compensatory adjustments of gill function originate from profound morphological changes. Generally, the morphological adjustments to environmental changes are considered to be adaptive, although in certain instances gill function may actually be impaired. Moreover, owing to the multi-functional nature of the gill, morphological amelioration of a particular physiological function might compromise another.

The use of morphological techniques to evaluate the nature of physiological adaptation to environmental change is a powerful approach which essentially links the traditional disciplines of fish gill morphology and physiology. In recent years, several comprehensive reviews have been written on the interrelationships between fish gill morphology and properties of the environment

Fish Ecophysiology. Edited by J. Cliff Rankin and Frank B. Jensen. Published in 1993 by Chapman & Hall, London. ISBN 0 412 45920 5.

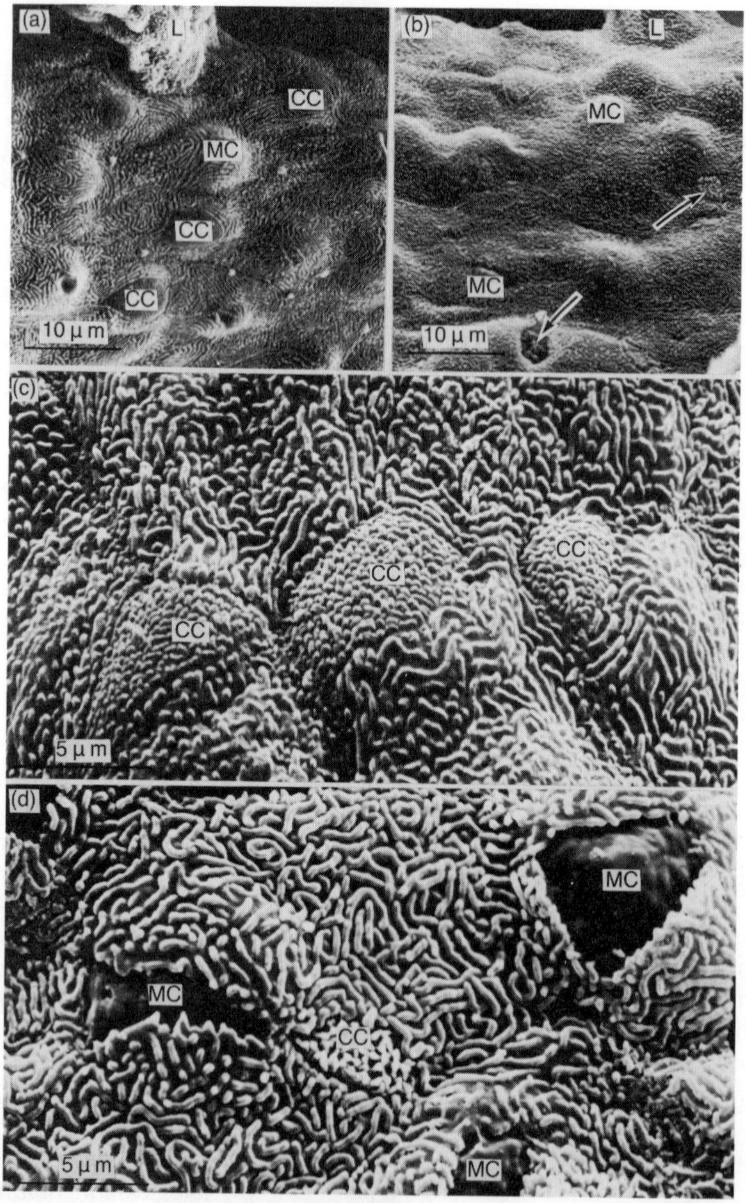

Fig. 9.1 Representative scanning and transmission electron micrographs of rainbow trout, *Oncorhynchus mykiss*, gills from freshwater- and seawater-adapted animals. In fresh water (a, c), the chloride cells (CC) are larger, raised above the surface of the surrounding filamental pavement cells, and in this example, more numerous. In seawater (b, d), the CCs are smaller, often recessed within the filament epithelium, and in this example, less abundant. Panels (e) and (f) illustrate some characteristic ultrastructural differences between freshwater- and seawater-adapted trout. In fresh

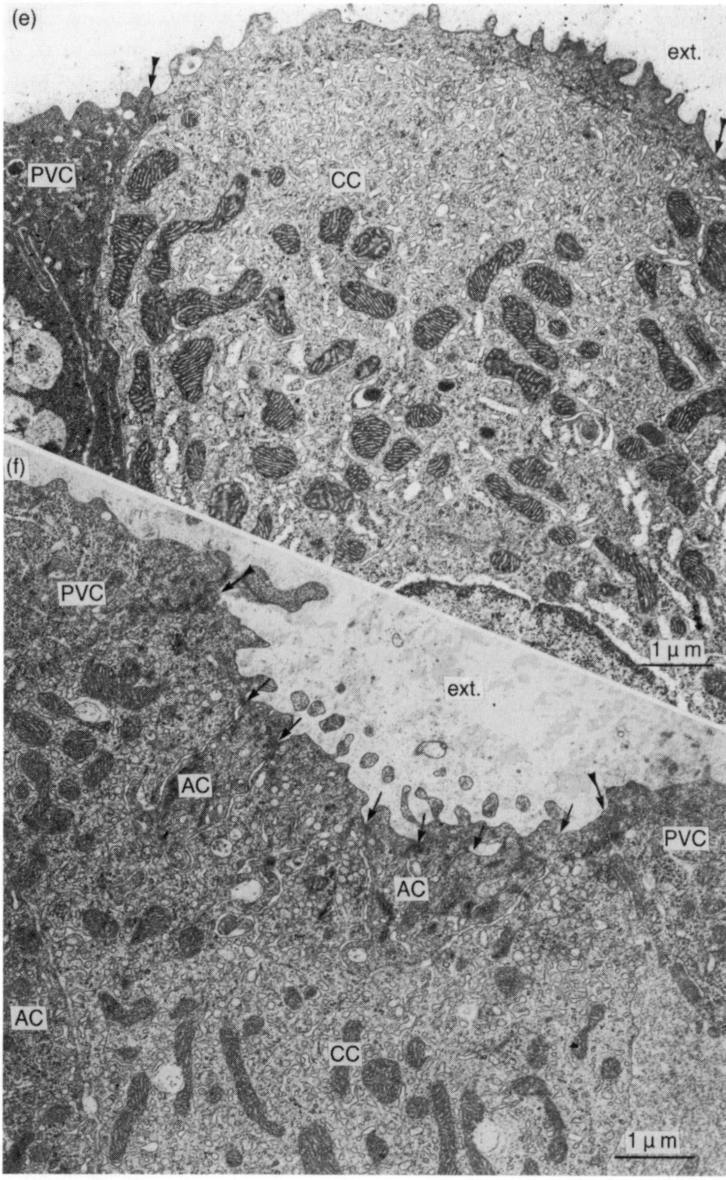

Fig.9.1(continued).
water (e), the CCs are characterized by an expansive, convex, and uninterrupted single apical membrane. In seawater (f), the CCs are characterized by an apical crypt formed from the apical membranes of a central CC and associated interdigitating processes (arrows) of a laterally disposed accessory cell (AC). In each case, the boundaries of the CC apical exposure are indicated by the double arrowheads. MC, mucous cell; PVC, pavement cell; ext. external water.

(Laurent, 1989; Laurent and Hebibi, 1989; Laurent and Perry, 1991). The reader is referred to these, and earlier, reviews (Laurent and Dunel, 1980; Laurent, 1984, 1985) for basic descriptions of fish gill structure and function. In this chapter, our goal is to focus on recent developments pertaining to the effects of selected environmental perturbations as well as to review some of the classical examples of environmental adaptation. Emphasis is placed on those areas where morphological studies have been particularly helpful in establishing or elucidating underlying mechanisms of adaptation. These include adaptations to ionic and acid–base disturbances, and altered environmental oxygen. Particular emphasis is placed on the roles of the three predominant cell types of the lamellar and filament epithelia, the mucous cell, the chloride cell, and the pavement cell.

9.2 ENVIRONMENTAL IONS

Common morphological features in fresh water and seawater

The gill surface epithelia of fish adapted to fresh water (FW) and seawater (SW) share numerous common features. The predominant cell type is the pavement cell (often referred to as the respiratory cell in earlier publications), which generally covers greater than 90% of the filament (often referred to as the primary epithelium in earlier literature) and lamellar (often referred to as the secondary epithelium in earlier literature) epithelia. The filament epithelium typically is multi-layered while the lamellar epithelium is normally composed of a double layer of cells separated by extracellular spaces. The apical membrane of the pavement cell is characterized by projecting microridges (e.g. Olson and Fromm, 1973; Dunel-Erb and Laurent, 1980). The microridges markedly increase the apical surface area of the epithelial pavement cells (Olson and Fromm, 1973) and thus may be important in increasing the diffusive conductance of the gill to respiratory gases. Additionally, the microridges may assist in anchoring mucus to the epithelial surfaces. The ornamentation of the lamellar pavement cells is often less developed or totally absent (Hossler *et al.*, 1979) in comparison with the filament pavement cells. This may be important in decreasing the extent of mucus accumulation on the gas transfer interface. Owing to the probable effect of the mucus layer on impairing oxygen uptake (Ultsch and Gros, 1979), it may be functionally advantageous to limit mucus accumulation on the gas transfer surfaces. At the ultrastructural level, the pavement cell is characterized by a well-developed Golgi apparatus, abundance of rough endoplasmic reticulum, and numerous vesicles of variable shapes and sizes (Laurent and Dunel, 1980). In contrast to the chloride cell (see below), mitochondria are usually only sparsely distributed throughout the cytoplasm. A characteristic ultrastructural property of the pavement cell which has largely been overlooked is the appearance of ordered arrays of rod-shaped particles on the apical membrane of freeze-fractured epithelia (Sardet, 1977). The function of these structures is unknown, although a comparison with other H^+-secreting epithelia (Wade, 1976; Brown *et al.*, 1978, 1987) suggests a possible role in acid–base regulation (p. 252).

Epithelial pavement cells are joined to each other or to adjacent chloride cells by occluding tight junctions (Sardet *et al.*, 1979; Sardet, 1980; Bartels, 1988). These junctions are composed of numerous strands and are impermeable to lanthanum (Sardet *et al.*, 1979). Although measurements of gill epithelial electrical resistance are not possible, it is nevertheless assumed that such tight junctions impart high electrical resistivity and that the paracellular 'channels' between pavement cells, or between pavement cells and chloride cells, are relatively (though not totally (e.g. McDonald *et al.*, 1991a,b)) impermeable to electrolytes.

Mucous cells are located on both the lamellar and filament epithelia, although they are normally most abundant on the efferent side (leading edge) of the filament (e.g. Laurent and Hebibi, 1989). The mucus secreted by these cells is a polyanionic glycoprotein. The function of the mucus coat covering the branchial epithelium is uncertain, although suggested roles include defence against pathogens (Pickering, 1974), prevention of turbulent water flow during swimming (Rosen and Cornford, 1971), and osmotic or ionic regulation (Olivereau and Lemoine, 1971; Handy, 1989).

The chloride cells, although comprising less than 10% of the total gill surface area, have been studied more extensively than the other cell types. The characteristic ultrastructural features of the chloride cell include an abundance of mitochondria, an amplification of the basolateral membrane to form an extensive tubular network within the cell, and a well-developed vesicular system in the apical regions of the cell (review: Laurent and Dunel, 1980). Chloride cells are located on the lamellar and filament epithelia, although they are most abundant on the afferent side (trailing edge) of the filament and in the interlamellar regions. Usually, the chloride cells are restricted to the base of the lamellae, although in certain instances they may also occupy central and distal areas of lamellae (see below).

The seawater environment

The surface and ultrastructural morphology of the gill epithelium differs markedly in SW- and FW-adapted fish. The most obvious and consistently observed difference concerns the structure of the apical membrane of the chloride cell (Fig. 9.1). In SW fish or euryhaline species adapted to SW, the apical membrane is often recessed within the epithelium to form apical crypts or pits which contact the external environment through small pores or through concave apical membranes (Sardet *et al.*, 1979; Dunel-Erb and Laurent, 1980; Laurent and Dunel, 1980; Sardet, 1980; Hossler *et al.*, 1985;

Franklin, 1990). With few exceptions (Thomson and Sargent, 1977; Franklin, 1990), the density of chloride cells is greater in SW-adapted fish (Laurent and Hebibi, 1989; for references see Laurent and Dunel, 1980; Laurent, 1984), owing primarily to proliferation of cells on the afferent side of the filament. Despite the increased density of chloride cells in SW-adapted fish, the area of apical membrane exposed to the external environment is probably reduced owing to the formation of the narrow apical pit. Moreover, in many instances an obvious channel extends from the apical crypt to the opening or pore on the epithelial surface (e.g. Sardet *et al.*, 1979). The invagination of the apical membrane and its relative isolation from the external environment presumably restricts the convection of water in these areas. The functional advantage of such a configuration is questionable, although it is likely to impede passive inward diffusion of electrolytes.

A fascinating and functionally significant difference between SW- and FW-adapted species is a rearrangement of the junctional organization between neighbouring chloride cells. In fresh water, the chloride cells are normally singly distributed and thus are joined on either side to pavement cells by tight junctions. Only rarely are two fully developed chloride cells observed adjacent to each other (Laurent and Dunel, 1980). In SW-adapted fish, the mature chloride cells are flanked on either, or both, sides by accessory cells first described by Dunel and Laurent (1973) (see also Dunel-Erb and Laurent, 1980; Hootman and Philpott, 1980). These accessory or companion cells are unique to SW-adapted fish. The origin of the accessory cell is disputed; one possibility is that this cell is simply a degenerative stage (Hootman and Philpott, 1980) or a developing (young) stage (Sardet *et al.*, 1979; Wendelaar Bonga and van der Meij, 1989) of the mature chloride cell, while another possibility is that it represents a distinctly different type of cell that has differentiated from stem cells (Laurent and Dunel, 1980). Regardless of its origin, the accessory cell sends interdigitations into the apical membrane of the mature chloride cell such that the apical membrane is actually a complex of interconnected cells. These interdigitations, commonly observed in seawater fish, are strikingly absent in the cichlid fish, *Oreochromis mossambicus* (Wendelaar Bonga and van der Meij, 1989).

The junctions between the accessory and chloride cells are shallow, single-stranded, and permeable to lanthanum. On the basis of these morphological observations, the paracellular channels formed by the 'leaky' junctions between accessory and chloride cells are believed to be highly permeable to electrolytes. The physiological significance of this organization is that it allows Na^+ to diffuse from the blood to water though the paracellular channels (reviews: Karnaky, 1980; 1986; Foskett *et al.*, 1983; Zadunaisky, 1984). The outward transepithelial movement of Cl^- occurs transcellularly through the chloride cell (Foskett and Scheffey, 1982). The junctional reorganization of the filament epithelial cells is apparently a prerequisite for successful transfer of euryhaline species from fresh water to seawater (Hwang and Hirano, 1985;

Hwang, 1987). More recently, it was demonstrated (Franklin, 1990) that the ability of sockeye salmon, *Oncorhynchus nerka*, to adapt successfully to seawater was directly related to the prevailing density of filament chloride cells at the time of transfer. A possible explanation for this phenomenon is that the greater number of chloride cells provides for a potentially greater number of leaky junctions to facilitate transepithelial paracellular Na^+ efflux.

Several other morphological features of the chloride cell in SW-adapted fish support the accepted view of its involvement in NaCl efflux. The tubular system originating from the basolateral membranes is generally more extensive, and the tubules themselves are often of larger diameter (e.g. Karnaky *et al.*, 1976). These observations are consistent with the increased activity of Na^+/K^+-ATPase in gills of SW-adapted fish because this enzyme is localized on the basolateral membranes and associated tubules (reviews: DeRenzis and Bornancin, 1984; Evans, 1984). In addition, the size and density of mitochondria may increase (e.g. Karnaky *et al.*, 1976), which is consistent with the observation of increased mitochondrial enzyme activity in SW-adapted fish (Conte and Lin, 1967).

On the basis of radiotracer turnover (Conte and Lin, 1967) and morphological (Wendelaar Bonga and van der Meij, 1989) studies, it is apparent that the rate of renewal of epithelial cells (pavement and chloride cells) is markedly stimulated in SW-adapted fish. In fully acclimated fish, the increased rate of cell turnover is the result of accelerated differentiation and apoptosis (physiologically controlled cell death), whereas in fish acutely transferred from FW to SW, there is an additional component of cellular necrosis (Wendelaar Bonga and van der Meij, 1989).

Numerous studies indicate that the glucocorticoid hormone, cortisol, acting alone or in concert with other hormones, is involved in the morphological and biochemical transformation of the gill following the transfer of euryhaline species to seawater. First, plasma cortisol levels increase after abrupt transfer to SW (Nichols and Weisbart, 1985; Patino *et al.*, 1987; Weisbart *et al.*, 1987) or during smoltification (Barton *et al.*, 1985; Langhorne and Simpson, 1986). Second, metabolic clearance rates of cortisol increase after transfer to seawater (Nichols and Weisbart, 1985) or during the parr–smolt transformation (Patino *et al.*, 1985). Third, chronic pre-treatment of FW-adapted fish with cortisol markedly improves or accelerates the capacity for ionic regulation after transfer into SW (Forrest *et al.*, 1973; Madsen, 1990a) owing to increased activity of branchial Na^+/K^+-ATPase. Fourth, with one notable exception (Eib and Hossner, 1985), repeated injections of FW- or SW-adapted fish with cortisol causes pronounced proliferation of chloride cells on the gill or opercular membrane (Foskett *et al.*, 1981; Perry and Wood, 1985; Perry and Walsh, 1989; Laurent and Perry, 1990). Recently, it was demonstrated (McCormick, 1990) that elevated cortisol levels can directly stimulate differentiation of opercular membrane chloride cells maintained *in vitro*. Other hormones that have been implicated in mediating the morphological transformation of the gill in seawater include growth hormone (Madsen, 1990b) and thyroid hormone (Dange, 1986).

Although elevated levels of hormones such as cortisol, growth hormone, and thyroid hormone are capable of enhancing the levels of branchial Na^+/K^+-ATPase and/or stimulating the proliferation of chloride cells, other unidentified factors are clearly involved in the complex structural reorganization of the surface morphology of the gill upon transfer to seawater. It is important to stress that treatment of FW-adapted fish with cortisol does not cause the surface or ultrastructural morphology of the chloride cell to transform to the unique configuration observed in SW-adapted fish; instead, the typical structure of the FW chloride cell is maintained (Laurent and Perry, 1990).

The effects of transfer to SW on gill mucous cell morphology are highly variable, with no obvious pattern emerging (e.g. Solanki and Benjamin, 1982; Laurent and Hebibi, 1989; Franklin, 1990). The variable results may partially reflect the different techniques used for detecting epithelial mucous cells. Scanning electron microscopy probably yields the best estimate of functional mucous cells because it detects only those cells exposed to the water. On the other hand, transmission electron microscopy and traditional light microscopy probably yield more reliable estimates of total mucous cell numbers, since both the exposed and covered cells are visible.

A role for mucus in osmotic or ionic regulation in seawater fish has not yet been established. Although it is often assumed that the mucus layer impedes osmotic water movements or the passive diffusion of electrolytes, the available evidence in fact shows that mucus is not an effective barrier to Na^+ or Cl^- diffusion (Marshall, 1978) and that it only marginally lessens (< 10%) the overall permeability of the gill epithelium to water (Shephard, 1981). Further, owing to the polyanionic composition of mucus, its presence next to the gill epithelium would seemingly exacerbate the problem of passive Na^+ gain and regulatory paracellular efflux by creating a microenvironment rich in Na^+. Interestingly, fish gill mucus contains the calcium-binding protein, calmodulin (Flik *et al.*, 1984), and thus may play a role in transepithelial Ca^{2+} uptake.

Hagfish (e.g. Bartels, 1988) and elasmobranchs (e.g. Laurent and Dunel, 1980) also possess gill chloride cells, despite an apparent absence of significant branchial ionoregulatory mechanisms in these seawater fish. The hagfish, *Myxine glutinosa*, are particularly interesting because the plasma is essentially isoionic with the external seawater. Recently, Bartels (1988) showed that unlike those in SW teleosts, the chloride cells are not associated with accessory cells. Further, the intercellular junctions between chloride cells and pavement cells have the morphological features of tight occluding junctions and do not differ from the junctions between adjacent pavement cells (Bartels, 1988). These observations are entirely consistent with the notion that the hagfish gill does not participate in regulatory NaCl efflux. On the other hand, it has been suggested that the chloride cells of hagfish may participate in acid–base regulation (Bartels, 1988). A similar argument can be applied to the chloride cells of elasmobranchs, since the gill plays only a limited role in the regulation of internal NaCl levels. A distinctive characteristic of the elasmobranch gill chloride cell is the lack of a tubular system. The tubular system appears to be functionally replaced by extensive infoldings of the basolateral membrane (Laurent and Dunel, 1980).

There is a paucity of data concerning the effects of seawater adaptation on gill morphometrics. Recently, Laurent and Hebibi (1989) demonstrated that branchial diffusing capacity is approximately fourfold greater (five times as great as) in SW-acclimated rainbow trout owing to a decrease in the blood-to-water diffusion distance and an increase in lamellar surface area. Such an increase in diffusing capacity may explain, in part, the increased rates of unidirectional Na^+ and Cl^- fluxes across the gills of seawater teleosts.

The freshwater environment

The surface morphology of the lamellar epithelium of FW-adapted fishes is not noticeably different from that in SW-adapted fishes. In contrast, the morphology of the filament epithelium is markedly different (Fig. 9.1). Except under very unusual circumstances (e.g. Leino *et al.*, 1987), the apical membrane of the chloride cell does not form a recessed apical crypt, but instead is either flush or slightly raised above the adjacent pavement cells. An exception may be the tilapia, *Oreochromis mossambicus*, which displays recessed chloride cells (Wendelaar Bonga and van der Meij, 1989; see also Fig. 9.5 below). Occasionally, the apical membrane may have a convex appearance. The surface structure of the apical membrane can vary tremendously, with some cells possessing apical microvilli extending above the epithelium and others possessing a totally smooth apical membrane (Olson and Fromm, 1973; Franklin and Davison, 1989; Laurent and Perry, 1990; see also Fig. 9.1). These dissimilar morphological appearances may reflect different stages of development of a singular type of chloride cell (Wendelaar Bonga and van der Meij, 1989) or different chloride cell varieties (Pisam *et al.*, 1987). The smooth-surfaced chloride cells may represent degenerative stages of mature chloride cells (Olson and Fromm, 1973) or 'worn-out' chloride cells (Laurent and Perry, 1990). Accessory cells are not present in FW-adapted species, although two or more chloride cells may occasionally exist adjacent to one another (Wendelaar Bonga and van der Meij, 1989). Clusters of adjacent chloride cells are typically observed in FW fish adapted to ion-poor media or chronically treated with cortisol (p. 243). Ultrastructurally, the chloride cells of FW fish or of stenohaline hyperosmotic-adapted fish (e.g. *Oreochromis alcalicus grahami*) transferred to FW (Maina, 1990) have a less developed tubular system and fewer mitochondria.

Mucous cells are often reported to be more abundant in FW-adapted fish, which may reflect the increased secretion of prolactin in FW (e.g. Avella *et al.*,

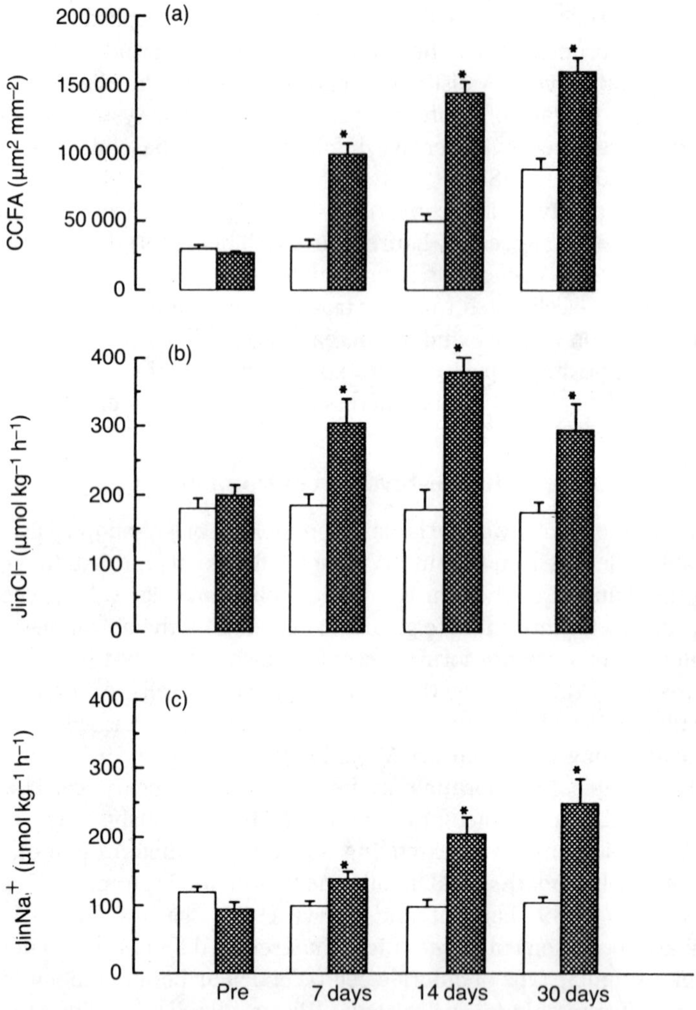

Fig. 9.2 The effects of long-term exposure of rainbow trout, *Oncorhynchus mykiss*, to low [NaCl] water (0.25 μmol l^{-1}; cross-hatched bars) or normal (control) water (0.15 mmol l^{-1} [NaCl]; empty bars) on (a) gill filament chloride cell fractional area (CCFA), (b) whole-body chloride uptake ($JinCl^-$), and (c) whole-body sodium uptake ($JinNa^+$). Values shown are means ± 1 SE; * indicates significant difference from the corresponding control value ($P < 0.05$; Student's *t* test). Data are replotted from Perry and Laurent (1989). See text for further details.

1991; Suzuki and Hirano, 1991) by virtue of the effects of prolactin on stimulating gill mucous cell production (Marshall, 1976). Although mucus is probably not a significant barrier to the diffusion of electrolytes (Marshall, 1978), an increased gill epithelial mucus layer in FW-adapted fish might conceivably be advantageous owing to the ability of mucus to bind Na^+ and

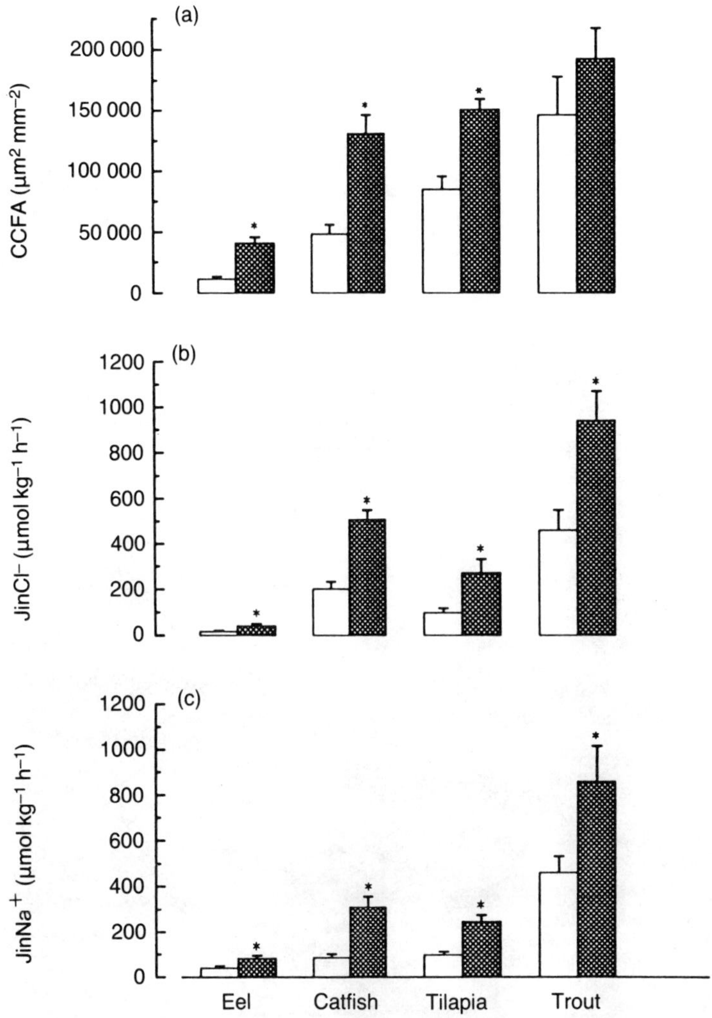

Fig. 9.3 The effects of chronic (10 day) cortisol treatment (daily intramuscular injections of 4 mg kg^{-1} body weight hydrocortisone hemisuccinate; cross-hatched bars) or sham treatment (controls; daily NaCl injections; empty bars) in American eel, *Anguilla rostrata*, brown bullhead catfish, *Ictalurus nebulosus*, tilapia, *Oreochromis mossambicus*, and rainbow trout, *Oncorhynchus mykiss*, on (a) gill filament chloride cell fractional area (CCFA), (b) whole-body chloride uptake ($JinCl^-$), and (c) whole-body sodium uptake ($JinNa^+$). Values shown are means ± 1 SE; * indicates significant difference from corresponding control value ($P < 0.05$; Student's *t* test).

Cl^- (Handy, 1989). Thus, a mucus layer rich in Na^+ and Cl^- next to the branchial epithelium could impede the outward passive diffusion of these ions and also assist in active absorption.

Freshwater ecosystems are considerably more variable in their ionic composition (e.g. Laurent *et al.*, 1985) than the seawater environment, yet

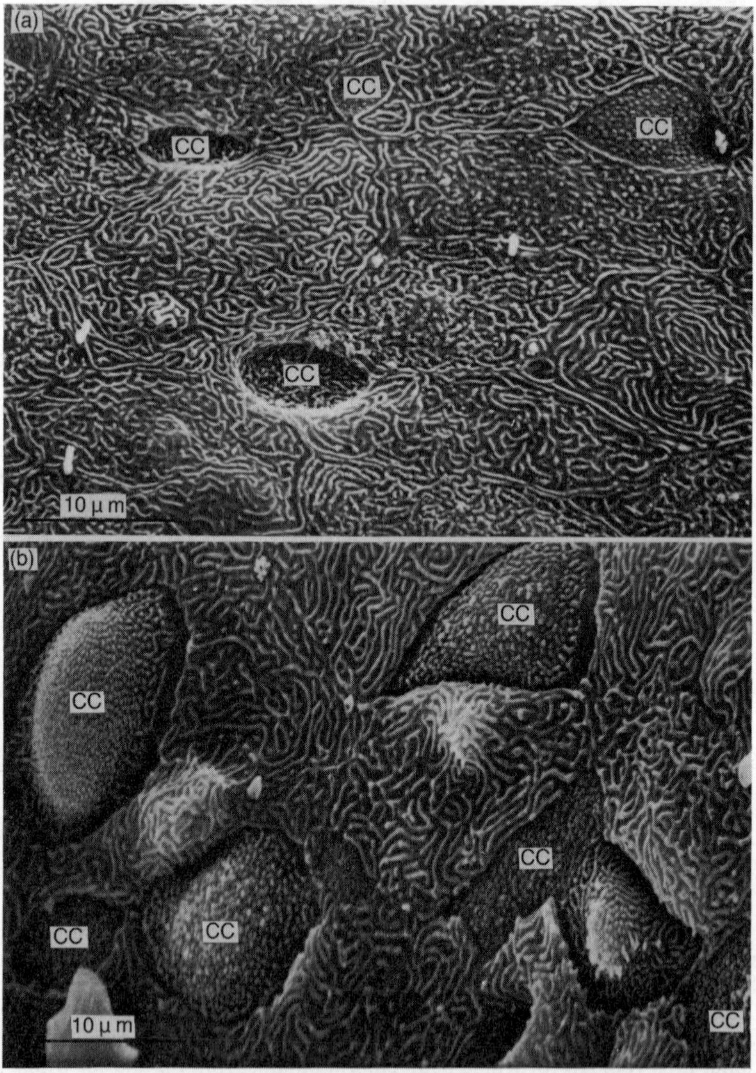

Fig. 9.4 Representative scanning and transmission electron micrographs of rainbow trout gills from animals acclimated to (a, c) normal city of Ottawa tap water (approximately 0.15 mmol l^{-1} [NaCl]) or (b, d) ion-poor water (approximately 0.025 mmol l^{-1} [NaCl]). Acclimation to ion-poor water markedly increases the fractional area of chloride cells (CCs) exposed to the water, owing to the combined effects of increased numbers and enlargement of individual cells (compare a and b). A comparison of c and d reveals the pronounced expansion of the exposed areas of the CC apical membrane (delineated by the arrows). PVC, pavement cell; ext., external water.

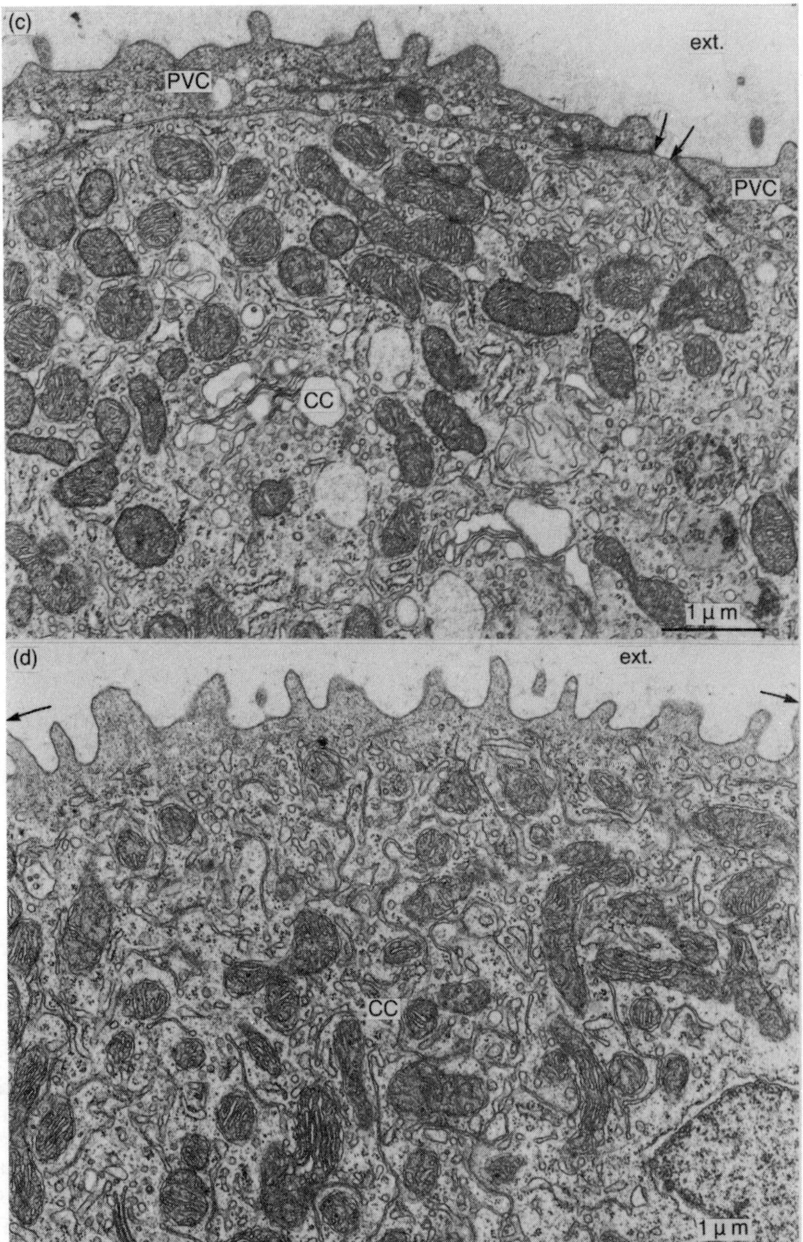
(c)
PVC
ext.
PVC
CC
1 µm
(d)
ext.
CC
1 µm

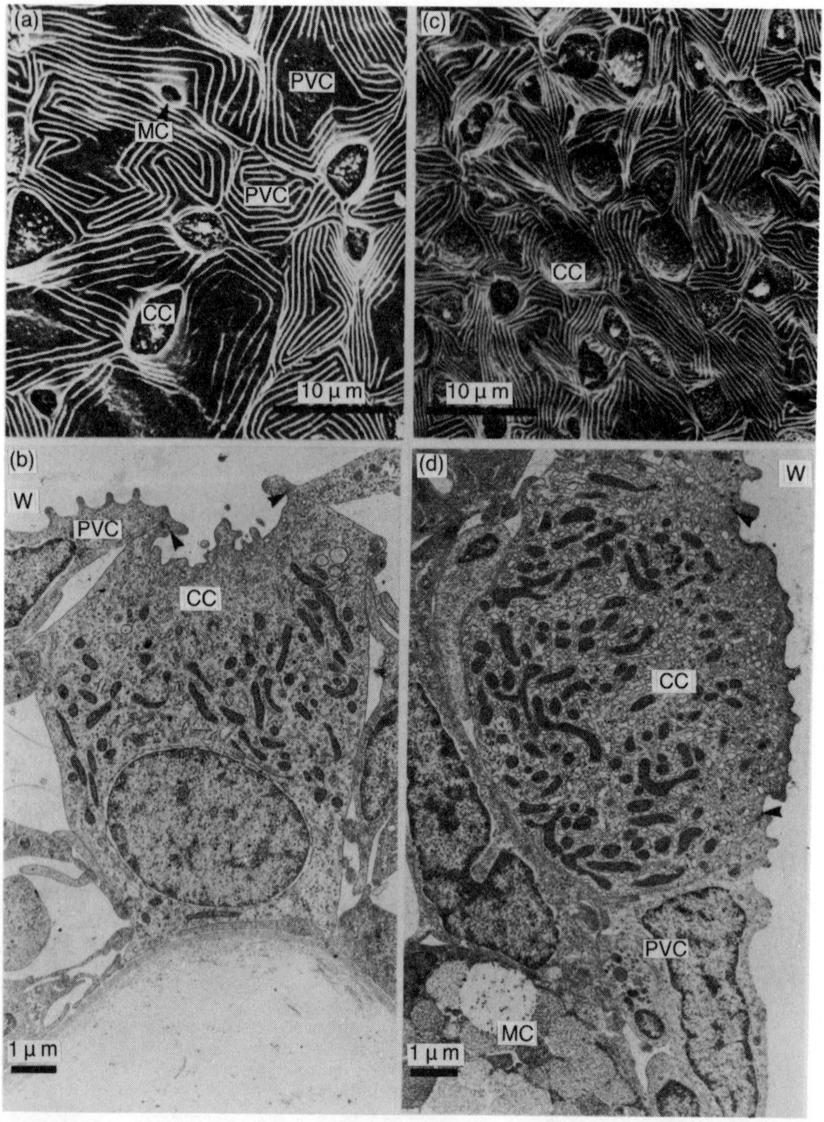

Fig. 9.5 The effects of chronic cortisol treatment (details as in Fig. 9.4) on the epithelial surface and ultrastructural morphology of tilapia, *Oreochromis mossambicus* gill filaments as viewed using scanning or transmission electron microscopy. The sham-treated fish (a and b) displayed the usual array of cell types including the predominant pavement cells (PVC), the less numerous chloride cells (CC) and the sparse mucous cells (MC). Note the characteristic microridges of the PVCs and the microvilli of the recessed CCs; the apical membrane of the CC in sham-treated fish formed an apical crypt (b). The chloride cells were more abundant after cortisol treatment (c), were raised above the adjacent PVCs, and were characterized by expanded apical membranes with an obvious convex appearance (d). W, water; arrowheads in b and d delineate the CC apical membrane.

relatively few studies have addressed the mechanisms by which fish adapt to dilute or ion-poor media. Numerous adaptive responses to exposure to ion-poor water have been elucidated including (i) adjustments of urine flow rate (McDonald and Rogano, 1986), (ii) prolactin-mediated alterations of gill osmotic permeability (Wendelaar Bonga and Van der Meij, 1981), (iii) temporary reductions in gill ionic permeability (McDonald and Rogano, 1986), and (iv) long-lasting increases in the ion-transporting capacity of the gill (Maetz, 1974; Perry and Wood, 1985; McDonald and Rogano, 1986; Avella *et al.*, 1987; Perry and Laurent, 1989). Morphological studies have been especially useful in establishing the underlying mechanism(s) for the increased ion-transporting capacity of the gills in fish maintained in ion-poor water. It is now well established that proliferation of chloride cells on the filament and lamellar epithelia is a physiologically significant response to exposure to ion-poor water (Laurent and Dunel, 1980; Laurent *et al.*, 1985; Perry and Wood, 1985; Avella *et al.*, 1987; Leino *et al.*, 1987; Spry and Wood, 1988; Laurent and Hebibi, 1989; Perry and Laurent, 1989; McDonald *et al.*, 1991a) that may be mediated by increased plasma levels of cortisol (Perry and Laurent, 1989; Laurent and Perry, 1990). Figures 9.2 and 9.3 illustrate the simultaneous increases in the fractional surface area of gill filament chloride cell apical membranes exposed to the water and the whole-body uptake rates of both Cl^- and Na^+ either in rainbow trout transferred to NaCl-deficient water (Fig. 9.2) or in a variety of teleost species chronically injected intramuscularly with cortisol (Fig. 9.3). The overall enlargement in the area of chloride cell apical membrane exposed to the water is caused by the combined effects of increased apical membrane area of individual chloride cells and increased density of these cells (Figs 9.4 and 9.5). Figure 9.5 illustrates the particularly intriguing chloride cell surface morphology and ultrastructure of tilapia before, and after, cortisol treatment. Note in particular the recession of the chloride cell within the filament epithelium and the appearance of an apical crypt in the sham-treated fish. In cortisol-treated tilapia, the apical membrane of the chloride cell is greatly expanded to form a convex structure which is raised above the adjacent pavement cells (Fig. 9.5(d)).

The proliferation of lamellar chloride cells in ion-poor water, although presumably beneficial for increasing the ion-transport capacity of the gill, may compromise respiratory gas transfer. Indeed, Thomas *et al.* (1988) demonstrated pronounced differences in the tolerance of freshwater rainbow trout to hypoxia in fish naturally adapted to widely different ionic environments (0.1 v. 1.0 mmol l^{-1} [NaCl]). The fish adapted to the more dilute medium were less resistant to hypoxia, which was attributed to the preponderance of lamellar chloride cells and consequent reduction of the surface area available for oxygen uptake. On the other hand, Laurent and Hebibi (1989) reported a marked increase in the diffusing capacity of the trout gill after acclimation to ion-poor water owing primarily to a thinning of the blood-to-water diffusion distance. This may be a secondary compensatory response to the initial proliferation of chloride cells.

The proliferation of chloride cells in ion-poor water, together with the excellent correlation between the rates of NaCl uptake and gill chloride cell fractional area, suggest that the chloride cell is an important site of Na^+ and Cl^- uptake in freshwater fish (e.g. Perry and Laurent, 1989; Laurent and Perry, 1990). Indeed, the model proposed by Girard and Payan (1980), in which lamellar pavement cells are presumed to be the sites of Na^+ and Cl^- influx, is now generally discounted. However, the notion that the chloride cell is the sole site of NaCl uptake is probably an oversimplification. Although it is now largely assumed that both Na^+ and Cl^- influx occurs through the same cell type (probably the chloride cell), there is no reason to discount the possibilities that (i) these ions are being transported across both the chloride cells and the pavement cells, or (ii) the influx of Na^+ and Cl^- occurs across different cell types. Although there is compelling evidence that the chloride cell is the single cell type involved in transepithelial Cl^- uptake, the situation for Na^+ uptake may be markedly different. Under normal conditions, the uptake rates of both ions are identically related to gill filament chloride cell surface area. However, during periods of imposed blood acidosis, the relationship holds only for Cl^- uptake (Goss *et al.*, 1992a). The lack of correspondence between Na^+ uptake and gill chloride cell fractional area is most obvious in hypercapnic catfish, *Ictalurus nebulosus* (Goss *et al.*, 1992a), the most extensively studied species. Briefly, during hypercapnia (15 Torr external P_{CO_2}), the chloride cell fractional area is dramatically reduced by about 95% after 48 h and corresponds with a near total reduction of Cl^- uptake. In contrast, Na^+ uptake is maintained or even markedly stimulated, despite the almost total absence of functional epithelial chloride cells. On the basis of these results, it was suggested (Goss *et al.*, 1992a,b) that the pavement cell may play an important role in branchial Na^+ uptake under all conditions, or alternatively may subserve the role of the chloride cell during conditions of chloride cell disappearance. The functional significance of these results for the purposes of acid–base regulation is discussed below (p. 250). We believe that the potential role of the pavement cell in gill Na^+ uptake is certainly an area that merits further investigation.

Traditionally, the gill Na^+ and Cl^- uptake mechanisms are thought to involve electroneutral exchanges with H^+ (and/or NH_4^+) and HCO_3^- (and/or OH^-), respectively (Krogh, 1938). The concept of a Cl^-/HCO_3^- exchanger on the apical membrane of gill chloride cells is supported by numerous empirical studies (e.g. Perry and Randall, 1981; Perry *et al.*, 1981) and is now universally accepted. On the other hand, the notion of an apically located Na^+/H^+ exchanger has recently been challenged (Avella and Bornancin, 1989, 1990), primarily on the basis of thermodynamic criteria and com-

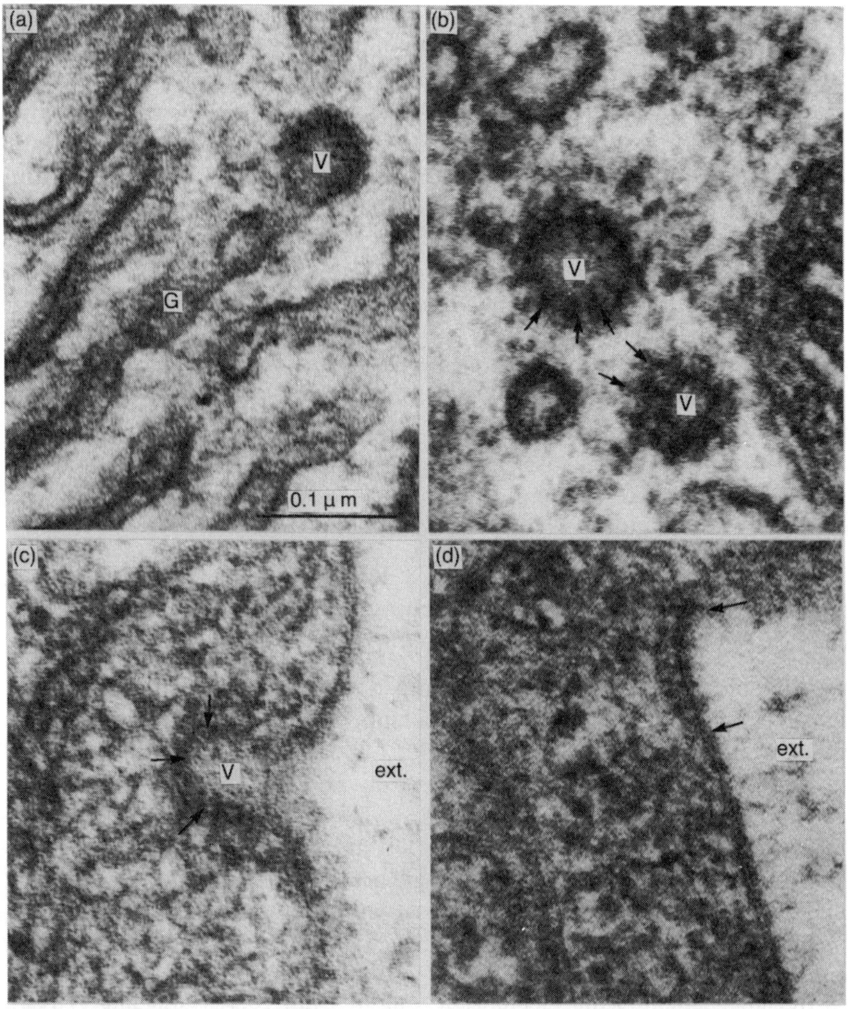

Fig. 9.6 High-magnification transmission electron micrographs of catfish, *Ictalurus nebulosus* gill pavement cells showing (a) the formation of vesicles (V) coated with putative proton pumps from the Golgi, (G), (b), the distribution of these vesicles with their characteristic 'studs' (arrowheads) in the cytoplasm, (c) fusion of the vesicle with the apical membrane, and (d) insertion of the 'studs' (believed to be the structural correlate of the proton pump) into the apical membrane. Ext., external water.

parisons with other Na^+-transporting epithelia. Instead, it has been proposed (Avella and Bornancin, 1989, 1990) that Na^+ uptake is accomplished through conductive Na^+ channels coupled to electrogenic outward H^+ transport (see also Randall and Lin, Chapter 10, this volume). This putative Na^+ uptake mechanism coupled to proton pumping is reminiscent of the pattern of Na^+ uptake in frog skin. There is no unequivocal evidence to support either model of Na^+ uptake across the fish gill. Recently, we (P. Laurent, S.F. Perry and G.G. Goss, unpublished observations) have identified, using high-magnification transmission electron microscopy (Fig. 9.6), characteristic vesicles in pavement cells with essentially identical morphological properties to those of the H^+ pump vesicles described by Brown *et al.* (1987) in toad urinary bladder. Our observations demonstrate that this putative H^+ pump vesicle is restricted to pavement cells (Fig. 9.6). Furthermore, the insertion of these vesicles into the apical membrane (Fig. 9.6) appears to be related to blood acid–base status in a similar fashion to that previously described in acidotic medullary collecting duct cells of rat (Madsen and Tisher, 1983). It is conceivable that the ordered orthogonal arrays of rod-shaped particles viewed by freeze-fracture in teleost (Sardet, 1977) or hagfish (Bartels, 1988) pavement cells also represent the H^+ pump vesicle.

With respect to acid–base regulation (p. 247), the two models of trans-epithelial Na^+ uptake are identical because in each case the inward movement of Na^+ is coupled to the excretion of acidic equivalents.

Calcium

There is considerable indirect evidence to implicate the chloride cell as the site of branchial calcium uptake, including (i) excellent correlations between lamellar chloride cell numbers and calcium uptake in rainbow trout acclimated to low external [Ca^{2+}] (Perry and Wood, 1985), (ii) cortisol-induced proliferation of lamellar chloride cells in trout and accompanying increases in calcium uptake (Perry and Wood, 1985), (iii) stimulation of branchial high-affinity Ca^{2+}-ATPase by cortisol treatment in trout and an associated stimulation of Ca^{2+}-transporting capacity of gill basolateral membrane vesicles (Flik and Perry, 1989), (iv) histological observations of oxalate-induced calcium precipitates specifically localized to the chloride cell in goldfish, *Carassius auratus* (Ishihara and Mugiya, 1987), and (v) specific binding of lanthanum to the apical membrane of chloride cells in trout (Perry and Flik, 1988). Additionally, Payan *et al.* (1981) demonstrated, using a perfused head preparation of trout, that calcium uptake was restricted to the filament epithelium, which implicated the filament chloride cells as the location of branchial calcium uptake. On the other hand, Perry and Wood (1985) reported that in the same preparation, the bulk of gill calcium uptake occurred across the lamellar epithelium and thus proposed that lamellar chloride cells were likely the route of calcium influx. Regardless of the exact site of calcium uptake, the mechanism is thought to involve passive diffusion through apical membrane Ca^{2+} channels followed by active transport (through the high-affinity Ca^{2+}-ATPase: Flik *et al.*, 1983) across the basolateral membranes (Perry and Flik, 1988; review: Fenwick, 1989).

9.3 MORPHOLOGICAL COMPONENT TO ACID–BASE REGULATION

A variety of environmental perturbations are known to induce internal acid–base disorders including hypoxia, hyperoxia, hypercapnia, acidification, salinity changes, and fluctuations of ambient calcium levels. The goal of this chapter is not to review in detail the causes and consequences of the various perturbations, but instead to focus on the common mechanisms of branchial regulation, especially those with a morphological component.

It is generally accepted that the gill is the principal site of acid–base regulation in fishes and that this is achieved by dynamic manipulation of the net transepithelial fluxes of Na^+ and Cl^- (reviews: Heisler, 1984, 1986; Cameron, 1989; Cameron and Iwama, 1989; McDonald *et al.*, 1989; Wood and Goss, 1990; see also Heisler, Chapter 8, this volume). The manipulations of the net fluxes are caused by complex adjustments of both the unidirectional influx and efflux components. Historically, modulation of the influx component has been viewed as the predominant mechanism, although recent studies suggest that adjustments of Na^+ and Cl^- effluxes are also important (Goss and Wood, 1990a,b; Wood and Goss, 1990).

The dominant mechanism for correcting internal acidosis, induced by environmental hypercapnia (e.g. Cameron, 1976) or hyperoxia (review, Wood, 1991), is simultaneous reduction of Cl^- uptake and (to a lesser extent) increased Na^+ uptake. These responses transfer basic equivalents from the water to the fish (equivalent to transfer of acidic equivalents from the fish to the water) and metabolically correct the acid–base disturbance. Traditionally, the reduction of Cl^- uptake is believed to reduce HCO_3^- excretion by virtue of the obligate electroneutral coupling of Cl^- uptake and HCO_3^- efflux, while the increase of Na^+ uptake is thought to enhance H^+ excretion owing to the electroneutral coupling of Na^+ uptake and H^+ efflux. An alternate view is that the relationship between Na^+ uptake and H^+ excretion stems from the dependence of Na^+ uptake on the membrane potential which, in part, is determined by active electrogenic outward H^+ movement.

Although it is well established that modification of Na^+ or Cl^- influx is the primary mode of branchial acid–base regulation, the underlying mechanisms promoting these adjustments are virtually unknown. Circulating catecholamines have been implicated in mediating the adjustments through interaction with β-adrenoceptors (Perry *et al.*, 1984; Vermette and Perry, 1987; Tang *et al.*, 1988, 1989; McDonald *et al.*, 1989), yet in many acid–base disturbances (e.g. hyperoxia, mild or moderate hypercapnia), plasma catecholamine levels do not vary (Perry *et al.*, 1989; Kinkead and Perry, 1991). Furthermore, antagonism of β-adrenoceptors, using propranolol, has produced conflicting results on the importance of a β-adrenergic component to acid–base regulation (Tang *et al.*, 1988; Vermette and Perry, 1988; McDonald *et al.*, 1989). Clearly, other factors are involved. Recently, Goss and Wood (1991; see also review

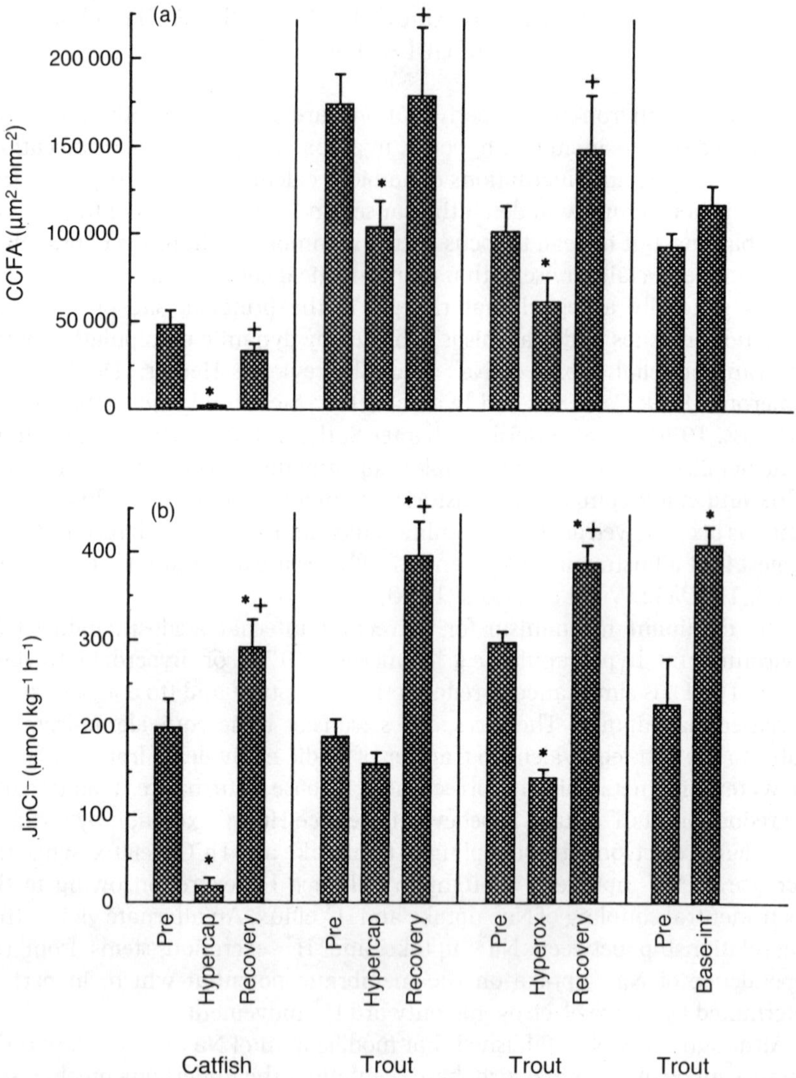

Fig. 9.7 The effects of a variety of experimentally imposed acid–base disturbances including hypercapnic acidosis, hyperoxic acidosis and base (HCO_3^-)-infusion in brown bullhead catfish, *Ictalurus nebulosus*, or rainbow trout, *Oncorhynchus mykiss*, on (a) gill filament chloride cell fractional area (CCFA) and (b) whole-body chloride uptake ($JinCl^-$). * Indicates significant difference from pre (control) value; $^+$ denotes significant difference from the value during the acid–base disturbance (i.e. before recovery). The data for catfish are replotted from Goss *et al.* (1992a).

by Wood and Goss, 1990) proposed that the levels of H^+ and HCO_3^-, themselves, in the blood (and thus also presumably in the gill epithelial cells) could affect the rates of Cl^-/HCO_3^- exchange and Na^+/H^+ exchange (alternatively Na^+ uptake coupled to electrogenic H^+ efflux), and therefore influence

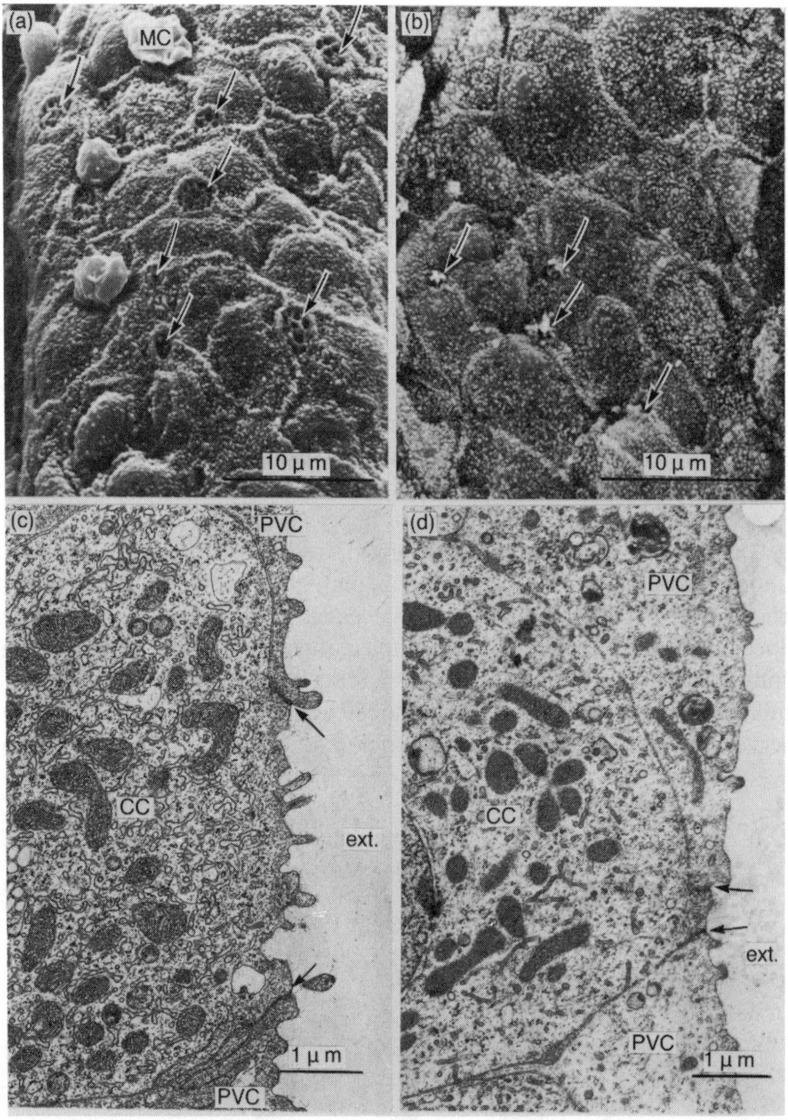

Fig. 9.8 Representative scanning and transmission electron micrographs of brown bullhead catfish, *Ictalurus nebulosus*, gills showing morphological changes associated with prolonged (48 h) hypercapnic acidosis (2% CO_2 in air). In control animals (a, c), there are numerous chloride cells (CCs; arrows) on the filament epithelium with relatively large areas of apical membrane exposed to the external environment (delineated by the arrowheads on the TEMs). After 48 h of hypercapnia (b, d), the area of CCs exposed to the water is markedly reduced owing to a reduction in the number of exposed CCs and a decrease in the size of the apical membrane exposure. Note the scarcity of intracellular organelles in panel d, suggesting a reduced activity. MC, mucous cell; PVC, pavement cell; ext., external water.

acid–base regulation. In their model, increased levels of H^+ directly stimulate Na^+ uptake while increased levels of HCO_3^- directly inhibit Cl^- uptake. These are appropriate responses to correct metabolic acid–base disturbances. On the other hand, the proposed relationship between internal HCO_3^- levels and Cl^- uptake may actually exacerbate the problem of acid–base regulation during respiratory disturbances. For example, elevation of plasma [HCO_3^-] during respiratory acidosis and further elevation during progressive compensation would be expected to cause inappropriate stimulation of Cl^-/HCO_3^- exchange. Similarly, the decrease of plasma [HCO_3^-] during respiratory alkalosis would be expected to inappropriately decrease Cl^-/HCO_3^- exchange. Since Cl^-/HCO_3^- exchange is appropriately modified under such conditions, other factors must be influencing this process.

We suggest that morphological alteration of the gill epithelium is an additional and perhaps crucial mechanism for regulating the rates of Cl^-/HCO_3^- exchange during acid–base disturbances. Figure 9.7 illustrates that a common response of brown bullhead catfish or rainbow trout to the respiratory acidosis induced by hypercapnia or hyperoxia is a reduction of the surface area of chloride cells exposed to the external environment. This is most pronounced in catfish, where an approximate 95% reduction in chloride cell surface area was observed (Fig. 9.7(a), Goss *et al.*, 1992a). In each case, the reduced area of exposed chloride cell apical membranes appears to be caused by physical covering by adjacent pavement cells (Fig. 9.8). The physiological significance of this reduction in chloride cell apical membrane exposure is that it serves to decrease the number of functional Cl^-/HCO_3^- exchange sites and

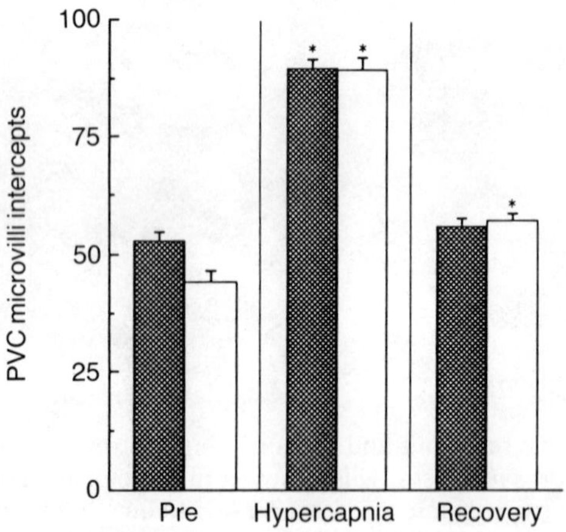

Fig. 9.9 The effects of hypercapnic acidosis (2% CO_2 in air) on the surface microvilli density of lamellar (empty bars) and filamental (cross-hatched bars) pavement cells in brown bullhead catfish, *Ictalurus nebulosus*. Microvilli density was quantified by counting the number of pavement cell microvilli intercepts on a morphometry grid. * Indicates significant difference from the pre-hypercapnia values ($P < 0.05$; Student's t test). Data are re-plotted from Goss *et al.* (1992a).

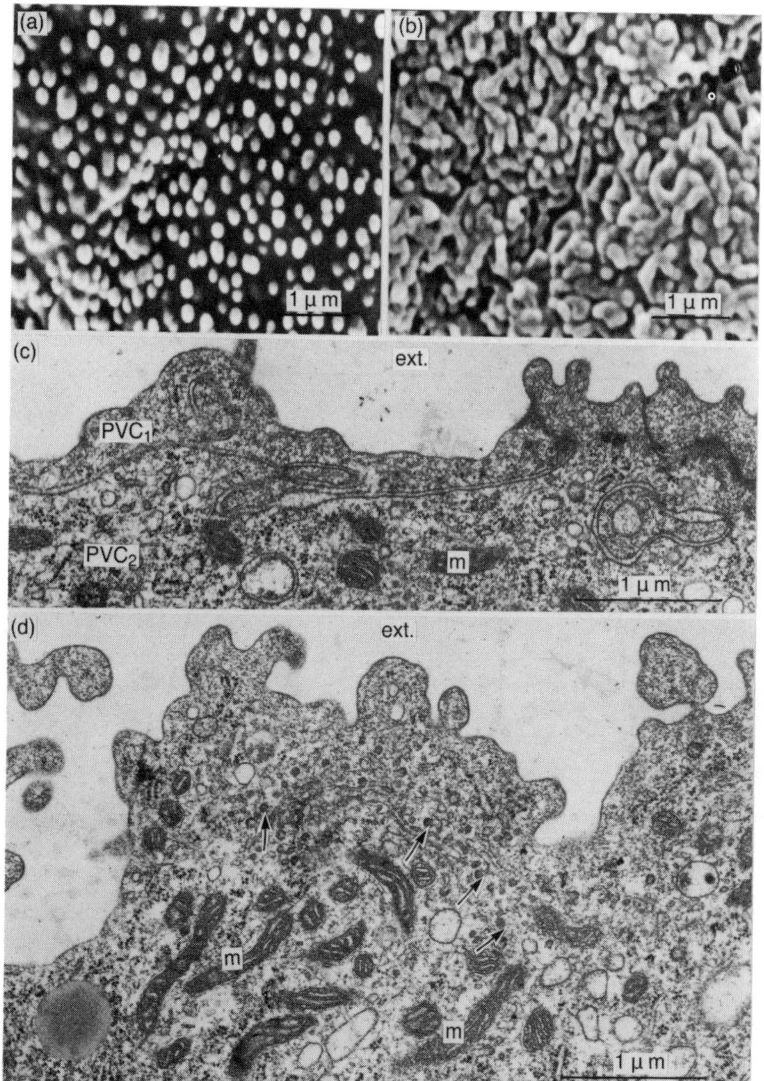

Fig. 9.10 Representative scanning and transmission electron micrographs of brown bullhead catfish, *Ictalurus nebulosus*, gills showing the surface and ultrastructural morphology of pavement cells (PVCs) before (a, c) and after (b, d) prolonged (48 h) exposure to hypercapnia (2% CO_2 in air). Hypercapnia was associated with a pronounced increase in the density of PVC apical membrane microvilli (compare a and b) and an increase in the numbers of mitochondria (m) and coated vesicles (arrows) in many PVCs. Additionally, in panel c note that a cell (PVC_2) below the most superficial PVC (PVC_1) is partly covered and opens to the external water (ext.) through a small surface area. PVC_2 displays a more complex ultrastructure than PVC_1.

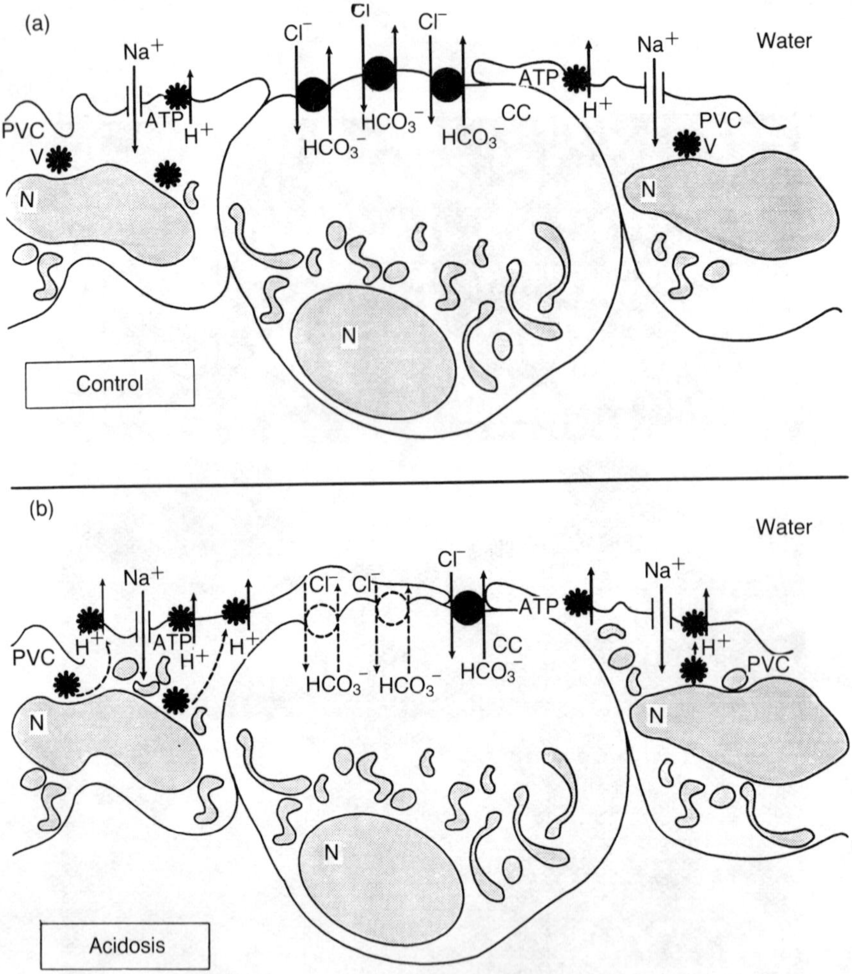

Fig. 9.11 Model illustrating a proposed morphological role for acid–base regulation during conditions of internal acidosis. (a) Under control conditions, the apical membrane of the chloride cell (CC) is largely exposed to the water, with only small areas of CC apical membrane covered by the adjacent pavement cells (PVC). This permits more or less unrestricted functioning of the CC apical membrane Cl^-/HCO_3^- exchangers (black circles) with the rates of Cl^-/HCO_3^- exchange essentially dependent on the levels of external Cl^- and internal HCO_3^-. Proton pumps are located on the apical membrane of the PVCs and additionally proton pump vesicles (V) are located throughout the cytoplasm. Na^+ uptake occurs through PVC apical membrane Na^+ channels, which are coupled to the electrogenic proton pumps through the membrane potential. (b) Under acidotic conditions, the area of the CC apical membrane in contact with the water is reduced, owing to the physical covering by adjacent PVCs. This serves to reduce the number of functional Cl^-/HCO_3^- exchangers, and thus decrease the excretion of basic equivalents. Additionally, the PVCs undergo morphological transformations including hypertrophy and apical polarization of mitochondria, apical microvilli, and increased fusion of proton pump vesicles with the apical membrane. N, nucleus; ATP, adenosine triphosphate.

thereby lower the rate of Cl^-/HCO_3^- exchange (Fig. 9.7(b)). In this way, the rates of Cl^-/HCO_3^- exchange can be persistently decreased during acidosis despite the compensatory increases in plasma HCO_3^- levels. The chloride cells rapidly re-emerge after the environmental perturbation is removed (Fig. 9.7) and assist in clearing the accumulated HCO_3^- because Cl^-/HCO_3^- exchange can now function at very high rates owing to an abundance of Cl^-/HCO_3^- exchange sites and the high levels of HCO_3^- in the blood.

Recent evidence suggests that the stimulation of Na^+ uptake during periods of acidosis also may reflect gill morphological adjustments (Goss *et al.*, 1992a). The physical covering of the chloride cells argues against a role for this cell type in transepithelial Na^+ uptake at these times. On the other hand, Goss *et al.*, (1992a) observed pronounced surface and ultrastructural changes in pavement cell morphology during hypercapnia in catfish, and suggested that the pavement cell might play an important role in Na^+ uptake during acidosis and perhaps during resting conditions as well. The morphological changes included an increase in the density of apical membrane microvilli (Figs 9.9 and 9.10) and in some instances a marked proliferation of mitochondria (Fig. 9.10). These observations, together with the finding of proton pump vesicles in the cytoplasm and areas of proton pump domains on the apical membrane (Fig. 9.6), suggest that the stimulation of Na^+ uptake is associated with increased electrogenic active H^+ extrusion by the pavement cells. This mechanism of acid–base regulation may be more important in fish such as the eel *Anguilla* which possess few chloride cells and have very low rates of Cl^- uptake under resting conditions.

In summary, morphological adjustment of the gill epithelium, especially during acidotic conditions, appears to be an essential mechanism of acid–base regulation to physically adjust the functional number of Cl^-/HCO_3^- exchange sites on the chloride cell apical membrane and the number of presumptive proton pumps on the pavement cell apical membrane. These morphological responses complement, or in some cases diminish, the effects of altered internal HCO_3^- and H^+ levels on branchial Cl^- and Na^+ uptake, respectively. Figure 9.11 is a diagrammatic model illustrating the morphological component to the regulation of acidotic disturbances.

9.4 ENVIRONMENTAL PH AND OXYGEN

Environmental acidification

The exposure of fish to acidified water may impair numerous gill physiological processes including gas transfer, ionic regulation, and acid–base balance (reviews: McDonald, 1983; Wood, 1989). The two most obvious branchial morphological changes associated with exposure of fish to sublethal levels of acidity are hypertrophy of filament and lamellar chloride cells (Leino and

McCormick, 1984; Laurent and Perry, 1990; Wendelaar Bonga *et al.*, 1990) and stimulation of filament mucous cells (Daye and Garside, 1976; Laurent and Perry, 1990).

The increase in chloride cell numbers is thought to be an adaptive response to increase the ion-transport capacity of the gill in the face of the acidification-induced increases in epithelial ionic permeability. In other words, the chloride cell proliferation may increase the ability of fish to absorb NaCl from the water and thus oppose the increased diffusional losses of these ions. It should be stressed, however, that the ion-transport capacity of the gills is most likely related to the number of mature chloride cells in contact with the external environment, rather than to the total number of epithelial chloride cells, which is the sum of immature, degenerating and mature cells (Wendelaar Bonga *et al.*, 1990). Indeed, owing to the increased turnover of chloride cells in acid-exposed tilapia (Wendelaar Bonga *et al.*, 1990), there is a preponderance of immature and degenerative chloride cells while the number of mature chloride cells increases only slightly. This likely explains why the increase in Na^+/K^+-ATPase activity in acid-exposed tilapia is substantially lower than that predicted on the basis of the increase in the total number of epithelial chloride cells (Wendelaar Bonga *et al.*, 1990).

In some species (fathead minnows and yellow perch), the structure of the apical membrane of the chloride cell resembles the apical crypt of SW-adapted fish (Leino *et al.*, 1987). The functional significance of this type of apical membrane arrangement in acid-exposed fish is unknown. An additional unusual morphological feature of the chloride cell apical membrane of acid-exposed fish is the occurrence of chloride cell/accessory cell or chloride cell/chloride cell complexes characterized by 'leaky' junctions similar to those found in SW-adapted fish (Laurent and Perry, 1991). The formation of these 'leaky' junctions may explain, in part, the increased ionic permeability of the gills in fish exposed to low pH. In addition, the depth of the tight junctions between pavement cells and chloride cells is significantly reduced in rainbow trout exposed to pH 4.0 (Freda *et al.*, 1991), which presumably also contributes to the increased diffusional loss of ions.

Traditionally, the increased mucus secretion in acidified water has been viewed as an adaptive response to restrain the diffusional losses of Na^+ and Cl^-. However, in view of the limited barrier that mucus imposes on ionic diffusion (Marshall, 1978), this assumption may not be correct. Indeed, McDonald *et al.* (1991b) recently demonstrated that interspecific differences in the tolerance of fish to external acidification were unrelated to the properties of the mucus layer covering the gill epithelium. Furthermore, exposure of fish to soft water (non-acidified) does not appear to induce hypertrophy of mucous cells (Leino *et al.*, 1987; McDonald *et al.*, 1991b) despite the increased diffusional loss of ions under such conditions. This further substantiates the view that mucus is probably not an effective impediment to ionic diffusion. Alternatively, the increased production of mucus (known to contain the enzyme carbonic anhydrase; Wright *et al.*, 1986; Rahim *et al.*, 1988; Perry and Laurent, 1990) in acid-exposed fish may be important in creating a microenvironment next to the gill epithelium which is abundant in the enzyme carbonic anhydrase (review: Perry and Laurent, 1990). As discussed by Laurent and Perry (1991), it is possible that additional 'external' carbonic anhydrase is required to maintain carbon dioxide excretion in acidified water because the uncatalysed CO_2 hydration reaction is slowed and thus could lower the blood-to-water P_{CO_2} gradients. Finally, we cannot discount the possibility that the increase in mucus secretion serves no adaptive purpose but instead is solely detrimental owing to its probable disruptive effects on branchial gas transfer.

It is now apparent that exposure of fish to sublethal pH levels does not cause substantial structural damage to the gills (Wendelaar Bonga *et al.*, 1990; McDonald *et al.*, 1991b). On the other hand, exposure of fish to lethal levels of acidification causes pronounced structural damage, including exfoliation of the lamellar epithelium, formation of oedematous areas, deformation of lamellae, and damage to the chloride cells.

Hypoxia

Surprisingly, there have been relatively few studies on the effects of environmental hypoxia on gill morphology (Soivio and Hughes, 1978; Soivio and Tuurala, 1981; Tuurala *et al.*, 1984). Obvious consequences of prolonged exposure of trout to hypoxia are (i) a reduction of the blood-to-water diffusion distance, (ii) distension of the gill vascular spaces, and (iii) erection of the lamellae with a substantial decrease in the proportion of lamellae buried within the filament. These changes are important in increasing the diffusive conductance for oxygen and enhancing lamellar perfusion, thus compensating the diminished blood-to-water oxygen diffusion gradient (review: Perry and Wood, 1989).

Hyperoxia

There are only scant data concerning the effects of hyperoxia on gill morphology, despite the facts that hyperoxia can be encountered commonly by fish in the natural environment and that experimental hyperoxia is a widely used tool to induce respiratory acidosis in the laboratory. Aside from conflicting reports on the effects of hyperoxia on chloride cell density (Goss *et al.*, 1992b; Laurent and Perry, 1991; Fig. 9.7), the only other reported response is a proliferation of lamellar pavement cells leading to a multi-layered and thickened lamellar epithelium (Laurent and Perry, 1991). The consequences of this evident thickening of the blood-to-water diffusion distance on gill physiological function have not yet been examined.

ACKNOWLEDGEMENTS

Original work cited in this chapter was supported by NSERC of Canada Operating and Equipment grants to SFP and CNRS grants to PL. We are grateful to G. Sullivan (Ottawa), C. Chevalier (Strasbourg), F. Scheer (Strasbourg) and G. Bombarde (Strasbourg) for excellent technical assistance. We thank Drs C.M. Wood and D.G. McDonald for access to unpublished manuscripts. We especially appreciate the help and constructive criticism of G.G. Goss (Ottawa) in the preparation of this manuscript.

REFERENCES

Avella, M. and Bornancin, M. (1989) A new analysis of ammonia and sodium transport through the gills of the freshwater rainbow trout (*Salmo gairdneri*). *J. exp. Biol.*, **142**, 155–75.

Avella, M. and Bornancin, M. (1990) Ion fluxes in the gills of freshwater and seawater salmonid fish, in *Animal Nutrition and Transport Processes. 2. Transport, Respiration and Excretion: Comparative and Environmental Aspects* (eds J.P. Truchot and B. Lahlou) Karger, Basel, pp. 1–13.

Avella, M., Masoni, A., Bornancin, M. and Mayer-Gostan, N. (1987) Gill morphology and sodium influx in the rainbow trout (*Salmo gairdneri*) acclimated to artificial freshwater environments. *J. exp. Zool.*, **242**, 159–69.

Avella, M., Schreck, C.B. and Prunet, P. (1991) Plasma prolactin and cortisol concentrations of stressed coho salmon, *Oncorhynchus kisutch*, in fresh water or salt water. *Gen. comp. Endocrinol.*, **81**, 21–7.

Bartels, H. (1988) Intercellular junctions in the gill epithelium of the Atlantic hagfish, *Myxine glutinosa. Cell Tissue Res.*, **254**, 573–83.

Barton, B.A., Schreck, C.B., Ewing, R.D., Hemmingsen, A.R. and Patino, R. (1985) Changes in plasma cortisol during stress and smoltification in coho salmon, *Oncorhynchus kisutch. Gen. comp. Endocrinol.*, **59**, 468–71.

Brown, D., Gluck, S. and Hartwig, J. (1987) Structure of the novel membrane-coating material in proton-secreting epithelial cells and identification as an H^+-ATPase. *J. Cell Biol.*, **105**, 1637–48.

Brown, D., Ilic, V. and Orci, L. (1978) Rod-shaped particles in the plasma membrane of the mitochondria-rich cell of the amphibian epidermis. *Anat. Rec.*, **192**, 269–76.

Cameron, J.N. (1976) Branchial ion uptake in Arctic grayling: resting values and the effects of acid–base disturbance. *J. exp. Biol.*, **64**, 711–25.

Cameron, J.N. (1989) Acid–base homeostasis; past and present perspectives. *Physiol. Zool.*, **62**, 845–65.

Cameron, J.N. and Iwama, G.K. (1989) Compromises between ionic regulation and acid–base regulation in aquatic animals. *Can. J. Zool.*, **67**, 3078–84.

Conte, F.P. and Lin, D.H.Y. (1967) Kinetics of cellular morphogenesis in gill epithelium during sea water adaptation of *Oncorhynchus* (Walbaum). *Comp. Biochem. Physiol.*, **23**, 945–57.

Dange, A.D. (1986) Branchial Na^+, K^+-ATPase activity in freshwater or saltwater acclimated tilapia *Oreochromis (Sarotherodon) mossambicus*: effects of cortisol and thyroxine. *Gen. comp. Endocrinol.*, **62**, 341–3.

Daye, P.G. and Garside, E.T. (1976) Histopathologic changes in superficial tissues of brook trout *Salvelinus fontinalis* exposed to acute and chronic levels of pH. *Can. J. Zool.*, **54**, 2140–55.

DeRenzis, G. and Bornancin, M. (1984) Ion transport and gill ATPases, in *Fish Physiology*, Vol. XB (eds W.S. Hoar and D.J. Randall), Academic Press, New York, pp. 65–104.

Dunel, S. and Laurent, P. (1973) Ultrastructure compareé de la pseudobrancie chez les Téléostéens marins et d'eau douce. I. L'épithélium pseudobrancial. *J. Microsc. (Paris)*, **16**, 53–74.

Dunel-Erb, S. and Laurent, P. (1980) Ultrastructure of marine teleost gill epithelia: SEM and TEM study of the chloride cell apical membrane. *J. Morph.*, **165**, 175–86.

Eib, D.W. and Hossner, K.L. (1985) The effects of cortisol and actinomycin D injections on chloride cells and branchial Na^+–K^+-ATPase in rainbow trout (*Salmo gairdneri*). *Gen. comp. Endocrinol.*, **59**, 449–52.

Evans, D.H. (1984) The roles of gill permeability and transport mechanisms in euryhalinity, in *Fish Physiology*, Vol. XB (eds W.S. Hoar and D.J. Randall), Academic Press, New York, pp. 239–83.

Fenwick, J.C. (1989) Calcium exchange across fish gills, in *Vertebrate Endocrinology: Fundamentals and Biomedical Implications*, Vol. 3 (eds P.K.T. Pang and M.P. Schreibman), Academic Press, New York, pp. 319–42.

Flik, G. and Perry, S.F. (1989) Cortisol stimulates whole body calcium uptake and the branchial calcium pump in freshwater rainbow trout. *J. Endocr.*, **120**, 75–82.

Flik, G., Wendelaar Bonga, S.E. and Fenwick, J.C. (1983) Ca^{2+}-dependent phosphatase and ATPase activities in eel gill plasma membranes – I. Identification of Ca^{2+}-activated ATPase activities with non-specific phosphatase activities. *Comp. Biochem. Physiol.*, **76B**, 745–54.

Flik, G., Van Rijs, J.H. and Wendelaar Bonga, S.E. (1984) Evidence for the presence of calmodulin in fish mucus cells. *Eur. J. Biochem.*, **138**, 651–4.

Forrest, J.N. jun., Cohen, A.D., Schon, D.A. and Epstein, F.H. (1973) Na transport and Na-K ATPase in gills during adaptation to seawater: effects of cortisol. *Am. J. Physiol.*, **224**, 709–13.

Foskett, J.K. and Scheffey, C. (1982) The chloride cell: definitive identification as the salt-secreting cell in teleosts. *Science*, **215**, 164–6.

Foskett, J.K., Logsdon, C.D., Turner, T., Machen, T.E. and Bern, H.A. (1981) Differentiation of the chloride extrusion mechanism during seawater adaptation of a teleost fish, the cichlid *Sarotherodon mossambicus*. *J. exp. Biol.*, **93**, 209–24

.Foskett, J.K., Bern, H.A., Machen, T.E. and Conner, M. (1983) Chloride cells and the hormonal control of teleost fish osmoregulation. *J. exp. Biol.*, **106**, 255–81.

Franklin, G.E. (1990) Surface ultrastructure changes in the gills of sockeye salmon (Teleostei: *Oncorhynchus nerka*) during seawater transfer: comparison of successful and unsuccessful seawater adaptation. *J. Morph.*, **206**, 13–23.

Franklin, G.E. and Davison, W. (1989) S.E.M. observations of morphologically different chloride cells in freshwater-adapted sockeye salmon, *Oncorhynchus nerka*. *J. Fish Biol.*, **34**, 803–4.

Freda, J., Sanchez, D. and Bergman, H.L. (1991) Shortening of branchial tight junctions in acid exposed rainbow trout. *Can. J. Fish. aquat. Sci.*, **48**, 2028–33.

Girard, J.P. and Payan, P. (1980) Ion exchanges through respiratory and chloride cells in freshwater- and seawater-adapted teleosteans. *Am. J. Physiol.*, **238**, R260–68.

Goss, G.G. and Wood, C.M. (1990a) Na^+ and Cl^- uptake kinetics, diffusive effluxes and acidic equivalent fluxes across the gills of rainbow trout. I. Responses to environmental hyperoxia. *J. exp. Biol.*, **152**, 521–47.

Goss, G.G. and Wood, C.M. (1990b) Na^+ and Cl^- uptake kinetics, diffusive effluxes and acidic equivalent fluxes across the gills of rainbow trout. II. Responses to bicarbonate infusion. *J. exp. Biol.*, **152**, 549–71.

Goss, G.G. and Wood, C.M. (1991) Two-substrate kinetic analysis: a novel approach linking ionic and acid–base transport in freshwater trout. *J. comp. Physiol.*, **161**, 635–46.

Goss, G.G., Laurent, P. and Perry, S.F. (1992a) Evidence for a morphological component in the regulation of acid–base balance in hypercapnic catfish (*Ictalurus nebulosus*). *Cell Tissue Res.*, **268**, 539–52.

Goss, G.G., Perry, S.F., Wood, C.M. and Laurent, P. (1992b) Relationships between ion and acid–base regulation in freshwater fish. *J. exp. Zool.*, **263**, 143–59.

Handy, R.D. (1989) The ionic composition of rainbow trout body mucus. *Comp. Biochem. Physiol.* **93A**, 571–5.

Heisler, N. (1984) Acid–base regulation in fishes, in *Fish Physiology*, Vol. XA (eds W.S. Hoar and D.J. Randall), Academic Press, New York, pp. 315–401.

Heisler, N. (1986) Acid–base regulation in fishes, in *Acid–base Regulation in Animals* (ed. N. Heisler) Elsevier, Amsterdam, pp. 309–56.

Hootman, S.R. and Philpott, C.W. (1980) Accessory cells in teleost branchial epithelium. *Am. J. Physiol.*, **238**, R199–206.

Hossler, F.E., Musil, G., Karnaky, K.J. jun. and Epstein, F.H. (1985) Surface ultrastructure of the gill arch of the killifish, *Fundulus heteroclitus*, from seawater and freshwater, with special reference to the morphology of the apical crypts of chloride cells. *J. Morph.*, **185**, 377–86.

Hossler, F.E., Ruby, J.R. and McIlwain, T.D. (1979) The gill arch of the mullet, *Mugil cephalus* I. Surface ultrastructure. *J. exp. Zool.*, **208**, 379–98.

Hwang, P.P. (1987) Tolerance and ultrastructural responses of branchial chloride cells to salinity changes in the euryhaline teleost *Oreochromis mossambicus*. *Mar. Biol.*, **94**, 643–9.

Hwang, P.P. and Hirano, R. (1985) Effects of environmental salinity on intracellular organization and junctional structure of chloride cells in early stages of teleost development. *J. exp. Zool.*, **236**, 115–26.

Ishihara, A. and Mugiya, Y. (1987) Ultrastructural evidence of calcium uptake by chloride cells in the gills of the goldfish, *Carassius auratus*. *J. exp. Biol.*, **242**, 121–9.

Karnaky, K.J. jun. (1980) Ion-secreting epithelia: chloride cells in the head region of *Fundulus heteroclitus*. *Am. J. Physiol.*, **238**, R185–98.

Karnaky, K.J. jun. (1986) Structure and function of the chloride cell of *Fundulus heteroclitus* and other teleosts. *Am. Zool.*, **26**, 209–24.

Karnaky, K.J. jun., Ernst, S.A. and Philpott, C.W. (1976) Teleost chloride cell I. Responses of pupfish *Cyprinodon variegatus* gill Na, K-ATPase and chloride cell fine structure to various high salinity environments. *J. Cell Biol.*, **70**, 144–56.

Kinkead, R. and Perry, S.F. (1991) The effects of intra-arterial infusion of catecholamines on ventilation in rainbow trout (*Oncorhynchus mykiss*) exposed to external hypoxia or hypercapnia. *Respir. Physiol.*, **84**, 77–92.

Krogh, A. (1938) The active absorption of ions in some freshwater animals. *Z. vergl. Physiol.*, **25**, 335–50.

Langhorne, P. and Simpson, T.H. (1986) The interrelationship of cortisol, gill (Na+K) ATPase, and homeostasis during the parr–smolt transformation of Atlantic salmon (*Salmo salar* L.). *Gen. comp. Endocrinol.*, **61**, 203–13.

Laurent, P. (1984) Gill internal morphology, in *Fish Physiology*, Vol. XA (eds W.S. Hoar and D.J. Randall), Academic Press, New York, pp. 73–183.

Laurent, P. (1985) A survey of the morpho-functional relationships between epithelia, vasculature and innervation within the fish gill, in *Forschritte der Zoologie*, Band 30 (Dunker/Fleischer, eds.), *Vetebrate Morphology*, Gustav Fischer Verlag, Stuttgart, pp. 340–51.

Laurent, P. (1989) Gill structure and function: fish, in *Comparative Pulmonary Physiology: Current Concepts* (ed. S.C. Wood), Dekker, New York, pp. 69–120.
Laurent, P. and Dunel, S. (1980) Morphology of gill epithelia in fish. *Am. J. Physiol.*, **238**, R147–59.
Laurent, P. and Hebibi, N. (1989) Gill morphometry and fish osmoregulation. *Can. J. Zool.*, **67**, 3055–63.
Laurent, P. and Perry, S.F. (1990) The effects of cortisol on gill chloride cell morphology and ionic uptake in the freshwater trout, *Salmo gairdneri*. *Cell Tissue Res.*, **259**, 429–42.
Laurent, P. and Perry, S.F. (1991) Environmental effects on fish gill morphology. *Physiol. Zool.*, **64**, 4–25.
Laurent, P., Hobe, H. and Dunel-Erb, S. (1985) The role of environmental sodium chloride relative to calcium in gill morphology of freshwater salmonid fish. *Cell Tissue Res.*, **240**, 675–92.
Leino, R.L. and McCormick, J.H. (1984) Morphological and morphometrical changes in chloride cells of the gills of *Pimephales promelas* after chronic exposure to acid water. *Cell Tissue Res.*, **236**, 121–8.
Leino, R.L., McCormick, J.H. and Jensen, K.M. (1987) Changes in gill histology of fathead minnows and yellow perch transferred to soft water or acidified soft water with particular reference to chloride cells. *Cell Tissue Res.*, **250**, 389–99.
McCormick, S.D. (1990) Cortisol directly stimulates differentiation of chloride cells in tilapia opercular membrane. *Am. J. Physiol.*, **259**, R857–63.
McDonald, D.G. (1983) The effects of H^+ upon the gills of freshwater fish. *Can. J. Zool.*, **61**, 691–703.
McDonald, D.G. and Rogano, M.S. (1986) Ion regulation by the rainbow trout, *Salmo gairdneri*, in ion-poor water. *Physiol. Zool.*, **59**, 318–31.
McDonald, D.G., Tang, Y. and Boutilier, R.G. (1989) Acid and ion transfer across the gills of fish: mechanisms and regulation. *Can. J. Zool.*, **67**, 3046–54.
McDonald, D.G., Cavdek, V. and Ellis, R. (1991a) Gill design in freshwater fishes – interrelationships among gas exchange, ion regulation, and acid–base regulation. *Physiol. Zool.*, **64**, 103–23.
McDonald, D.G., Freda, J., Cavdek, V., Gonzalez, R. and Zia, S.H. (1991b) Interspecific differences in gill morphology of freshwater fish in relation to tolerance of low pH environments. *Physiol. Zool.*, **64**, 124–44.
Madsen, K.M. and Tisher, C.C. (1983) Cellular response to acute respiratory acidosis in rat medullary collecting duct. *Am. J. Physiol.*, **245**, F670–79.
Madsen, S.S. (1990a) Cortisol treatment improves the development of hypoosmoregulatory mechanisms in the euryhaline rainbow trout, *Salmo gairdneri*. *Fish Physiol. Biochem.*, **8**, 45–52.
Madsen, S.S. (1990b) Enhanced hypoosmoregulatory response to growth hormone after cortisol treatment in immature rainbow trout, *Salmo gairdneri*. *Fish Physiol. Biochem.*, **8**, 271–9.
Maetz, J. (1974) Aspects of adaptation to hypo-osmotic and hyper-osmotic environments, in *Biochemical and Biophysical Perspectives in Marine Biology* (eds D.C. Malin, and J.R. Sargent), Academic Press, London, pp. 1–167.
Maina, J.N. (1990) A study of the morphology of the gills of an extreme alkalinity and hyperosmotic adapted teleost *Oreochromis alcalicus grahami* (Boulenger) with particular emphasis on the ultrastructure of the chloride cells and their modifications with water dilution. *Anat. Embryol.*, **181**, 83–98.
Marshall, W.S. (1976) Effects of hypophysectomy and ovine prolactin on the epithelial mucus-secreting cells of the Pacific staghorn sculpin, *Leptocottus armatus* (Teleostei: Cottidae). *Can. J. Zool.*, **54**, 1604–9.

Marshall, W.S. (1978) On the involvement of mucous secretion in teleost osmoregulation. *Can. J. Zool.*, **56**, 1088–91.

Nichols, D.J. and Weisbart, M. (1985) Cortisol dynamics during seawater adaption of Atlantic salmon *Salmo salar*. *Am. J. Physiol.*, **248**, R651–9.

Olivereau, M. and Lemoine, A.M. (1971) Action de la prolactine chez l'anguille intacte et hypophysectomisee. VII. Effet sur la teneur en acide sialique (N-acetyl neuraminique) de la peau. *Z. vergl. Physiol.*, **73**, 34–43.

Olson, K.R. and Fromm, P.O. (1973) A scanning electron microscope study of secondary lamellae and chloride cells of rainbow trout (*Salmo gairdneri*). *Z. Zellforsch. Mikrook. Amat.*, **143**, 439–49.

Patino, R., Schreck, C.B. and Redding, J.M. (1985) Clearance of plasma corticosteroids during smoltification of coho salmon, *Oncorhynchus kisutch*. *Comp. Biochem. Physiol.* **82A**, 531–5.

Patino, R., Redding, J.M. and Schreck, C.B. (1987) Interrenal secretion of corticosteroids and plasma cortisol and cortisone concentrations after acute stress and during seawater acclimation in juvenile coho salmon (*Oncorhynchus kisutch*). *Gen. comp. Endocrinol.*, **68**, 431–9.

Payan, P., Mayer-Gostan, N. and Pang, P.K.T. (1981) Site of calcium uptake in the freshwater trout gill. *J. exp. Zool.*, **216**, 345–7.

Perry, S.F. and Flik, G. (1988) Characterization of branchial transepithelial calcium fluxes in freshwater trout, *Salmo gairdneri*. *Am. J. Physiol.*, **254**, R491–8.

Perry, S.F. and Laurent, P. (1989) Adaptational responses of rainbow trout to lowered external NaCl concentration: contribution of the branchial chloride cell. *J. exp. Biol.*, **147**, 147–68.

Perry, S.F. and Laurent, P. (1990) The role of carbonic anhydrase in carbon dioxide excretion, acid–base balance and ionic regulation in aquatic gill breathers, in *Animal Nutrition and Transport Processes. 2. Transport, Respiration and Excretion: Comparative and Environmental Aspects* (eds J.P. Truchot and B. Lahlou), Karger, Basel, pp. 39–57.

Perry, S.F. and Randall, D.J. (1981) Effects of amiloride and SITS on branchial ion fluxes in rainbow trout, *Salmo gairdneri*. *J. exp. Zool.*, **215**, 225–8.

Perry, S.F. and Walsh, P.J. (1989) Metabolism of isolated fish gill cells: contribution of epithelial chloride cells. *J. exp. Biol.*, **144**, 507–20.

Perry, S.F. and Wood, C.M. (1985) Kinetics of branchial calcium uptake in the rainbow trout: effects of acclimation to various external calcium levels. *J. exp. Biol.*, **116**, 411–33.

Perry, S.F. and Wood, C.M. (1989) Control and coordination of gas transfer in fishes. *Can. J. Zool.*, **67**, 2961–70.

Perry, S.F., Haswell, M.S., Randall, D.J. and Farrell, A.P. (1981) Branchial ionic uptake and acid–base regulation in the rainbow trout, *Salmo gairdneri*. *J. exp. Biol.*, **92**, 289–303.

Perry, S.F., Payan, P. and Girard, J.P. (1984) Adrenergic control of branchial chloride transport in the isolated perfused head of the freshwater trout (*Salmo gairdneri*). *J. comp. Physiol.*, **154B**, 269–74.

Perry, S.F., Kinkead, R., Gallaugher, P. and Randall, D.J. (1989) Evidence that hypoxemia promotes catecholamine release during hypercapnic acidosis in rainbow trout (*Salmo gairdneri*). *Respir. Physiol.*, **77**, 351–64.

Pickering, A.D. (1974) The distribution of mucous cells in the epidermis of the brown trout (*Salmo trutta* L.) and the char (*Salvelinus alpinus* L.). *J. Fish Biol.*, **6**, 111–18.

Pisam, M., Caroff, A. and Rambourg, A. (1987) Two types of chloride cells in the gill epithelium of a freshwater adapted euryhaline fish: *Lebistes reticulatus*; their modification during adaptation to seawater. *Am. J. Anat.*, **179**, 40–50.

Rahim, S.M., Delaunoy, J.P. and Laurent, P. (1988) Identification and immunocytochemical localization of two different carbonic anhydrase isozymes in teleostean fish erythrocytes and gill epithelia. *Histochemistry*, **89**, 451–9.

Rosen, M.W. and Cornford, N.E. (1971) Fluid friction of fish slimes. *Nature, Lond.*, **234**, 49–51.

Sardet, C. (1977) Ordered arrays of intramembrane particles on the surface of fish gills. *Cell Biol. Int. Rep.*, **1**, 409–17.

Sardet, C. (1980) Freeze fracture of the gill epithelium of euryhaline teleost fish. *Am. J. Physiol.*, **238**, R207–12.

Sardet, C., Pisam, M. and Maetz, J. (1979) The surface epithelium of teleostean fish gills. Cellular and junctional adaptations of the chloride cell in relation to salt adaptation. *J. Cell Biol.*, **80**, 96–117.

Shephard, K.L. (1981) The influence of mucus on the diffusion of water across fish epidermis. *Physiol. Zool.*, **54**, 224–9.

Soivio, A. and Hughes, G.M. (1978) Circulatory changes in secondary lamellae of *Salmo gairdneri* gills in hypoxia and anaesthesia. *Annls zool. fenn.*, **15**, 221–5.

Soivio, A. and Tuurala, H. (1981) Structural and circulatory responses to hypoxia in the secondary lamellae of *Salmo gairdneri* gills at two temperatures. *J. comp. Physiol.*, **145**, 37–43.

Solanki, T.G. and Benjamin, M. (1982) Changes in the mucus cells of the gills, buccal cavity and epidermis of the nine-spined stickleback, *Pungitius pungitius* L., induced by transferring the fish to sea water. *J. Fish Biol.*, **21**, 563–75.

Spry, D.J. and Wood, C.M. (1988) Zinc influx across the isolated, perfused head preparation of the rainbow trout (*Salmo gairdneri*) in hard and soft water. *Can. J. Fish. aquat. Sci.*, **45**, 2206–15.

Suzuki, R. and Hirano, T. (1991) Development of a homologous radioimmunoassay for eel prolactin. *Gen. comp. Endocrinol.*, **81**, 403–9.

Tang, Y., McDonald, D.G. and Boutilier, R.G. (1989) Adrenergic regulation of blood acid–base status following exhaustive exercise in seawater-adapted rainbow trout (*Salmo gairdneri*). *Physiol. Zool.*, **62**, 950–63.

Tang, Y., Nolan, S. and Boutilier, R.G. (1988) Acid–base regulation following acute acidosis in seawater-adapted rainbow trout, *Salmo gairdneri*: a possible role for catecholamines. *J. exp. Biol.*, **134**, 297–312.

Thomas, S., Fievet, B. and Motais, R. (1988) Adaptive respiratory responses of trout to acute hypoxia. I. Effects of water ionic composition on blood acid–base response and gill morphology. *Respir. Physiol.*, **74**, 77–90.

Thomson, A.J. and Sargent, J.R. (1977) Changes in levels of chloride cells and (Na^+ + K^+)-dependent ATPase in the gills of yellow and silver eels adapting to seawater. *J. exp. Zool.*, **200**, 33–40.

Tuurala, H., Part, P., Nikinmaa, M. and Soivio, A. (1984) The basal channels of secondary lamellae in *Salmo gairdneri* gills – a non-respiratory shunt. *Comp. Biochem. Physiol.*, **79A**, 35–9.

Ultsch, G.R. and Gros, G. (1979) Mucus as a diffusion barrier to oxygen: possible role in O_2 uptake at low pH in carp (*Cyprinus carpio*) gills. *Comp. Biochem. Physiol.*, **62A**, 685–9.

Vermette, M.G. and Perry, S.F. (1987) The effects of prolonged epinephrine infusion on the physiology of the rainbow trout, *Salmo gairdneri*. *J. exp. Biol.*, **128**, 255–67.

Vermette, M.G. and Perry, S.F. (1988) Adrenergic involvement in blood oxygen transport and acid–base balance during hypercapnic acidosis in the rainbow trout, *Salmo gairdneri*. *J. comp. Physiol.*, **158B**, 107–15.

Wade, J.B. (1976) Membrane structural specialization of the toad urinary bladder revealed by the freeze–fracture technique. II. The mitochondria-rich cell. *J. Membr. Biol.*, **29**, 111–26.

Weisbart, M., Chakraborti, P.K., Gallivan, G. and Eales, J.G. (1987) Dynamics of cortisol receptor activity in the gills of the brook trout, *Salvelinus fontinalis*, during seawater adaptation. *Gen. comp. Endocrinol.*, **68**, 440–48.

Wendelaar Bonga, S.E. and van der Meij, C.J.M. (1989) Degeneration and death, by apoptosis and necrosis, of the pavement and chloride cells in the gills of the teleost *Oreochromis mossambicus. Cell Tissue Res.*, **255**, 235–43.

Wendelaar Bonga, S.E. and Van der Meij, L.C.A. (1981) Effect of ambient osmolarity and calcium on prolactin cell activity and osmotic water permeability of the gills of the teleost, *Sarotherodon mossambicus. Gen. comp. Endocrinol.*, **43**, 432–42.

Wendelaar Bonga, S.E., Flik, G., Balm, P.H.M. and van der Meij, J.C.A. (1990) The ultrastructure of chloride cells in the gills of the teleost *Oreochromis mossambicus* during exposure to acidified water. *Cell Tissue Res.*, **259**, 575–85.

Wood, C.M. (1989) The physiological problems of fish in acid waters, in *Acid Toxicity and Aquatic Animals* (eds R. Morris, E.W. Taylor, D.J.A. Brown and J.A. Brown), Cambridge Univ. Press, Cambridge, pp. 125–52.

Wood, C.M. (1991) Branchial ion and acid–base transfer in freshwater teleost fish–environmental hyperoxia as a probe. *Physiol. Zool.*, **64**, 68–102.

Wood, C.M. and Goss, G.G. (1990) Kinetic analysis of the relationship between ion exchange and acid–base regulation at the gills of freshwater fish, in *Animal Nutrition and Transport Processes. 2. Transport, Respiration and Excretion: Comparative and Environmental Aspects* (eds J.P. Truchot and B. Lahlou), Karger, Basel, pp. 119–36.

Wright, P.A., Heming, T. and Randall, D.J. (1986) Downstream pH changes in water flowing over the gills of rainbow trout. *J. exp. Biol.*, **126**, 499–512.

Zadunaisky, J.A. (1984) The chloride cell: the active transport of chloride and the paracellular pathways, in *Fish Physiology*, Vol. XB (eds W.S. Hoar and D.J. Randall), Academic Press, New York, pp. 129–76.

Chapter ten

Effects of water pH on gas and ion transfer across fish gills

David J. Randall and Hong Lin

10.1 INTRODUCTION

This chapter will be restricted to a discussion of the effects of changes in freshwater pH on fish, as relatively few studies have been carried out on the effects of changes in seawater pH on fish.

Changes in freshwater pH do not have any direct effect on either oxygen uptake (Ye *et al.*, 1991) or carbon dioxide excretion (Lin and Randall, 1990) in resting fish, but reductions in water pH inhibit proton excretion (Lin and Randall, 1991) and sodium chloride influx (McDonald *et al.*, 1983; Wright and Wood, 1985). Exposure of fish to acid water results in a decrease in blood NaCl levels, and eventually a blood acidosis develops (McDonald *et al.*, 1983; Ye *et al.*, 1991). Low calcium levels exacerbate ion loss across the gills of fish in acid water and there is a marked reduction in blood NaCl. High calcium levels in acid water, however, are associated with only an acidosis and little or no reduction in blood NaCl because ion regulatory mechanisms are not impaired (McDonald, 1983).

There is a transient decrease in ammonia excretion which leads to an increase in blood ammonia levels in fish exposed to acid waters (Wright and Wood, 1985). The elevated blood ammonia levels result in the restoration of ammonia excretion to normal after a few hours (Ye *et al.*, 1991). There is no immediate increase in ventilation in fish exposed to water acidified by the addition of hydrochloric acid (Lin and Randall, 1990). A decrease in water pH due to a rise in carbon dioxide, however, will cause a marked rise in ventilation (Janssen and Randall, 1975). In this case the decrease in water

Fish Ecophysiology. Edited by J. Cliff Rankin and Frank B. Jensen. Published in 1993 by Chapman & Hall, London. ISBN 0 412 45920 5.

pH is associated with an immediate and marked blood acidosis due to an increase in Pa_{CO_2}. This acidosis will decrease blood oxygen content via the Root effect and be associated with an increase in circulating catecholamine levels, both of which can cause an increase in gill ventilation.

Exposure of fish to alkaline water is less well studied. Short-term exposure of rainbow trout, *Oncorhynchus mykiss*, to alkaline water causes a marked blood alkalosis and a reduction in both sodium influx and ammonia excretion (Wright and Wood, 1985).

10.2 PROTON EXCRETION

We estimated the degree of acidification of water passing over the gills by measuring inspired and expired water pH and the buffer capacity of the water, which was then added to ammonia excretion to give proton excretion. Proton excretion is not equivalent to net acid excretion (McDonald and Wood, 1981) because the measurement includes the effect of carbon dioxide excretion on acidification of the water.

Not surprisingly, exposure to acid or alkaline water affects proton excretion by the fish (Lin and Randall, 1991). Proton excretion increases with water pH and is completely inhibited if water pH falls below 5.5 (Fig. 10.1).

It has long been hypothesized that an $Na^+/H^+(NH_4^+)$ antiport (a counter-ion exchange mechanism) occurring in the apical membrane of the fish gill epithelia was the major pathway for sodium uptake and proton excretion and an optional excretory pathway for ammonia. The most recent evidence for

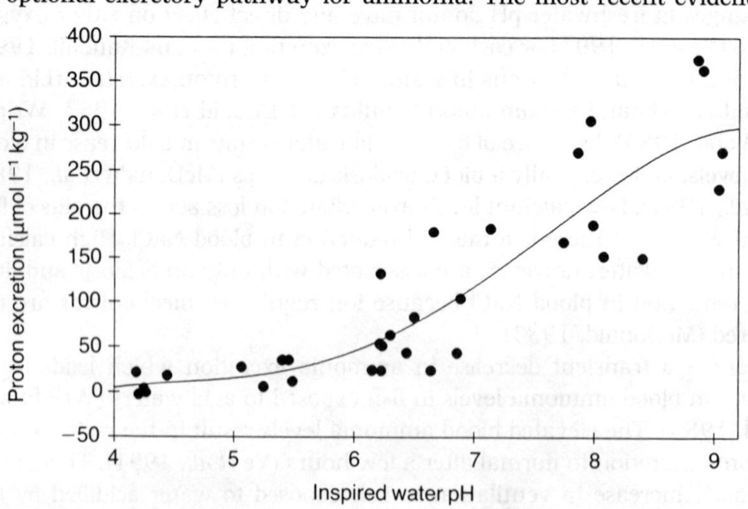

Fig. 10.1 The relationship between the proton excretion across the gills of freshwater rainbow trout and the external water pH was expressed by a fifth-order polynomial regression curve ($r^2 = 0.896$).

this conclusion was based on the fact that sodium influx and ammonia efflux in trout was coupled with an approximately 1:1 stoichiometry over an external water pH range from 4 to 9, and 94% of the sodium uptake by the gills was inhibited by 0.1 mM amiloride in water (Wright and Wood, 1985). The degree of coupling between sodium absorption and ammonia excretion, however, was loose in goldfish, *Carrasius auratus*, carp, *Cyprinus carpio*, and rainbow trout *in vivo* (Payan, 1978) and rainbow trout *in vitro* (Avella and Bornancin, 1989), and amiloride also blocks the sodium conductive channel, which is indirectly coupled with a proton-ATPase in other epithelia such as frog skin and turtle bladder (Steinmetz and Andersen, 1982; Ehrenfeld *et al.*, 1985). An indirect coupling of an electrogenic proton pump and a passive sodium entry, instead of an $Na^+/H^+(NH_4^+)$ exchanger, was proposed by Avella and Bornancin (1989) as the mechanism of sodium and proton transport across the fish gill epithelium.

In vivo studies of proton excretion in freshwater rainbow trout have provided evidence for the existence of an active proton pump in the gill epithelium (Lin and Randall, 1991). Neither proton nor ammonia excretion across the gills was found to be affected by 0.1 mM amiloride in external water (Figs. 10.2 and 10.3), a treatment reported to cause a 94% reduction of gill sodium uptake in the same species under similar experimental conditions (Wright and Wood, 1985). This indicates that sodium uptake and proton or ammonia excretion are not coupled by an $Na^+/H^+(NH_4^+)$ exchanger. Higher concentrations of amiloride, however, reduced proton and ammonia excretion, probably because application of amiloride above 0.1 mM from the

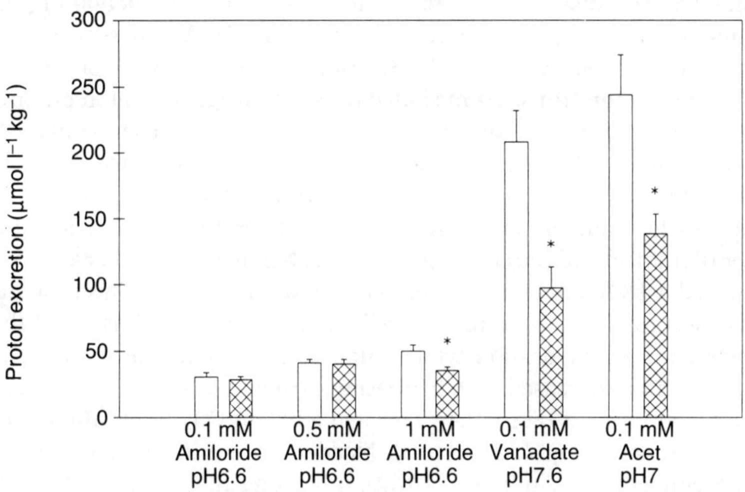

Fig. 10.2 Proton excretion across the gill epithelium of rainbow trout under control, amiloride, vanadate and acetazolamide treatments. Open columns, control; hatched columns, treatments; * indicates a significant difference between the control and treatment values under Student's two-tailed *t*-test ($P < 0.05$); bars show standard errors; $N = 6$.

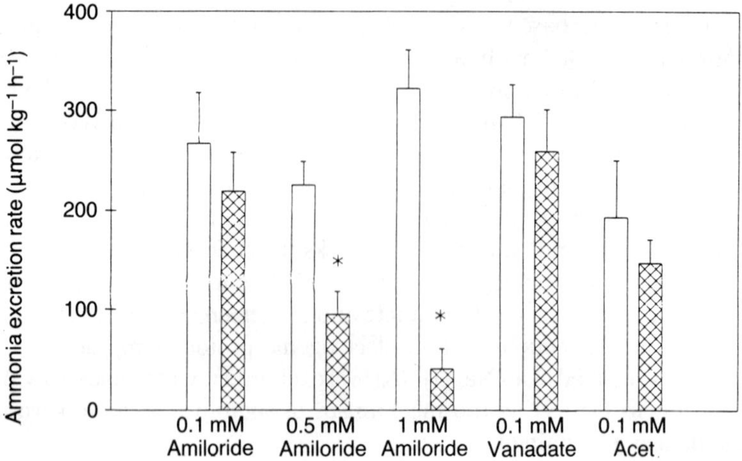

Fig. 10.3 Ammonia excretion rates of rainbow trout under control, amiloride, vanadate and acetazolamide treatments. Open columns, control; hatched columns, treatments; * indicates a significant difference between control and treatment values under student's two-tailed *t*-test ($P < 0.05$); bars show standard errors; $N = 6$. Reproduced with permission from Lin and Randall (1991).

mucosal side of the epithelium can interfere with the $Na^+/K^+(NH_4^+)$-ATPase in the basolateral membrane (Knauf *et al.*, 1976; Kleyman and Cragoe, 1988), which transfers potassium, and perhaps NH_4^+ from the blood into the epithelial cell.

Vanadate (0.1 mM) added to the water resulted in a reduction of proton excretion across the gills by more than 50% (Fig. 10.2) whereas ammonia excretion was unaffected (Fig. 10.3). Since a similar inhibitory effect of vanadate on proton transport mediated by an electrogenic proton pump was demonstrated in frog skin and turtle urinary bladder (Arruda *et al.*, 1981; Ehrenfeld *et al.*, 1985), we suggest that proton excretion across the gills is mediated by an active proton pump in trout. Treatment with 0.1 mM acetazolamide, another proton transport inhibitor in frog skin and turtle bladder (Ehrenfeld and Garcia-Romeu, 1977; Steinmetz and Andersen, 1982), also caused a 48% reduction in proton excretion (Fig. 10.2). This is probably because acetazolamide inhibits intracellular carbonic anhydrase, and therefore bicarbonate dehydration which supplies protons to the proton pump. Proton excretion increased with increasing carbon dioxide levels in a neutral external environment (Fig. 10.4). Elevated ambient CO_2 levels also enhance proton transport in frog skin and turtle bladder (Ehrenfeld and Garcia-Romeu, 1977; Steinmetz and Andersen, 1982), indicating that intracellular CO_2 is one of the sources of protons for the electrogenic pump.

We have no direct evidence that the effect of these drugs is to act only on proton excretion; there could also be effects on Cl^-/HCO_3^- exchange or any other mechanism involved with the transfer of acid equivalents across the

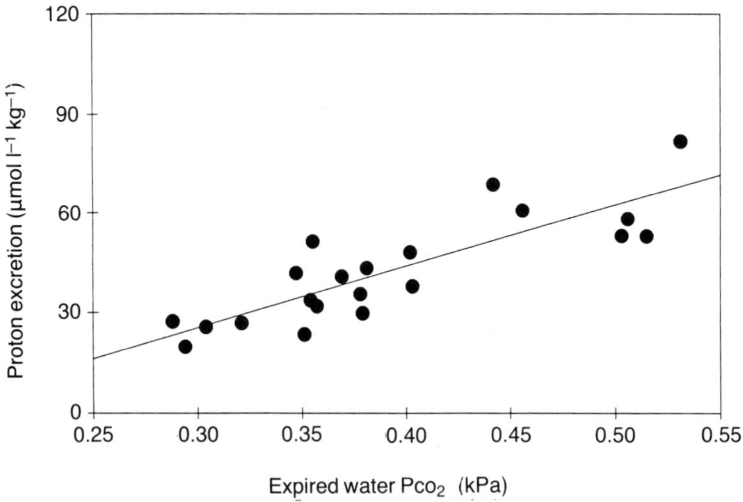

Fig. 10.4 The relationship between proton excretion across the gills of freshwater rainbow trout and expired water carbon dioxide levels. The linear regression line, $Y = 184.72X - 29.92$, has an r^2 value of 0.844 and its slope is significantly different from zero ($P < 0.05$).

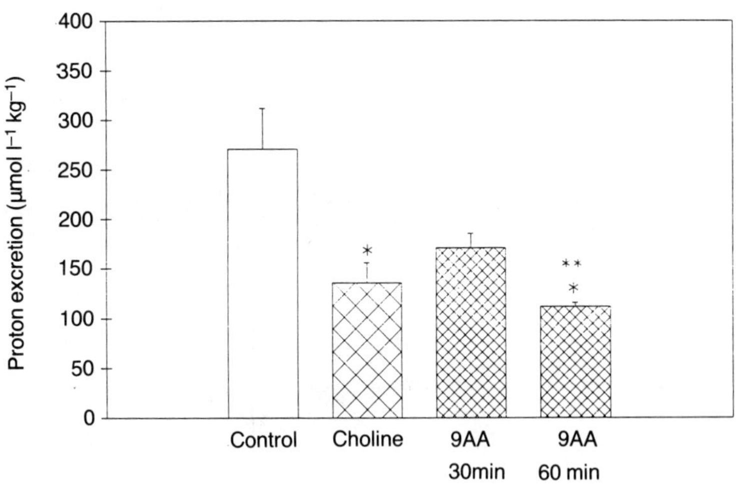

Fig. 10.5 Proton excretion across the gill epithelium of rainbow trout in control water ([NaCl] = 40 mmol l^{-1}, [$CaCl_2$] = 0.5 mmol l^{-1}), sodium-free water ([choline chloride] = 40 mmol l^{-1}, [$CaCl_2$] = 0.5 mmol l^{-1}) and 30 min and 60 min in sodium-free water with 0.2 mM 9 anthroic acid (9AA). * Indicates a significant difference between the control and treatment values, ** indicates a significant difference between the 30 min and 60 min 9AA treatment values under Student's two-tailed *t*-test ($P < 0.05$); bars show standard errors.

gills. Our conclusion that proton ATPases exist on fish gills and are affected by these drugs is by comparison with results from frog skin and turtle bladder, where isolated preparations were used and experimental conditions could be more defined than in our experiments on intact animals.

Although sodium influx and proton efflux are not linked by a $Na^+/H^+(NH_4^+)$ antiport, they are electrically coupled. The negative potential in the inner side of the apical membrane generated by the primary proton pump is the driving force for sodium uptake. This is why freshwater fish can maintain sodium uptake in a very dilute medium. Acid excretion was reduced when external sodium was replaced by choline (Fig. 10.5) and restored by the addition of sodium to sodium-free water. Theoretically, when sodium transport is inhibited, either by using sodium-free external media or by blocking the sodium channel with 0.1 mM amiloride, the negative potential in the inner side of the apical membrane should build up and eventually stop the operation of the active proton pump. Since removal of sodium from water or the addition of amiloride did not completely block acid excretion, another counter ion may also be involved.

In conclusion, the results of these studies indicate that proton excretion across the gills of freshwater trout is mainly mediated by an active proton

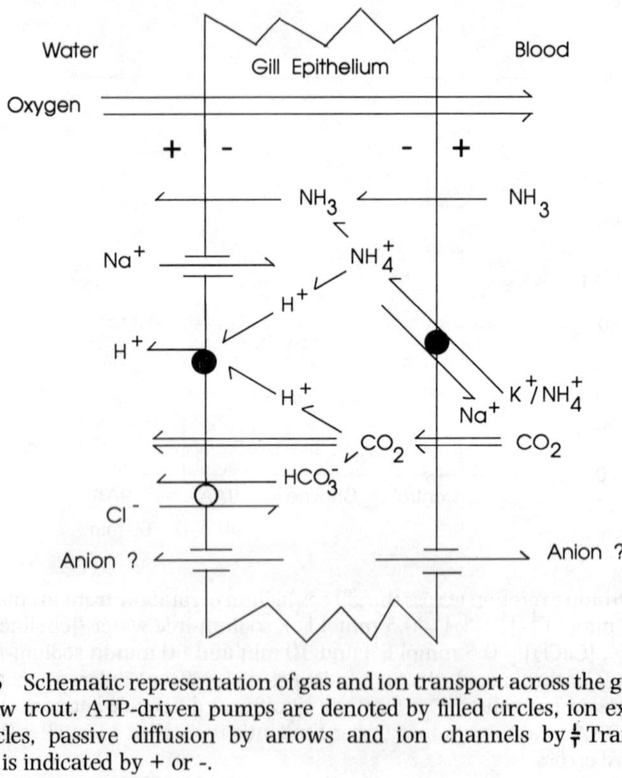

Fig. 10.6 Schematic representation of gas and ion transport across the gill epithelium of rainbow trout. ATP-driven pumps are denoted by filled circles, ion exchangers by open circles, passive diffusion by arrows and ion channels by ⫩ Transmembrane potential is indicated by + or -.

pump in the gill epithelium, which resembles the proton pump in turtle bladder and frog skin in several major characteristics. Fig. 10.6 illustrates an ion-transport model for the freshwater trout gill epithelium which summarizes the relationships between the electrogenic proton pump and other ion transport pathways.

10.3 EFFECT OF WATER PH ON PROTON TRANSPORT

Proton excretion across the gills of freshwater rainbow trout was examined under a wide range of external water pH (Fig. 10.1). When external water pH was below 5.5, proton excretion was completely inhibited, presumably due to the large hydrogen ion electrochemical gradient operating against the proton pump. Sodium uptake was also found to be inhibited by low water pH (Wright and Wood, 1985). Above pH 5.5, proton excretion increased with pH (Fig. 10.1). A linear relationship between proton transport and mucosal pH was demonstrated in both frog skin and turtle bladder (Steinmetz and Andersen, 1982; Ehrenfeld *et al.*, 1985). This indicates that the hydrogen ion electrochemical gradient across the membrane is a fundamental regulator of the proton pump activity.

10.4 INTERACTIONS BETWEEN PROTON, CARBON DIOXIDE AND AMMONIA EXCRETION

The substantial proton excretion across the gill epithelium of freshwater rainbow trout in neutral and alkaline environments causes an acidification of expired water (Lin and Randall, 1990). Besides proton excretion, the hydration of excreted CO_2 might also acidify the gill water boundary layer. This CO_2 hydration reaction, if it occurs, is thought to facilitate CO_2 and ammonia excretion across fish gills (Wright *et al.*, 1989). Measurement of bicarbonate levels in inspired and expired water, however, showed that bicarbonate concentration in inspired water was higher than that of expired water (Fig. 10.7). The reduction in bicarbonate level in expired water decreased when proton excretion was inhibited by vanadate, by acetazolamide or by 0.5, 1 mM amiloride (Fig. 10.7). This indicates that, over the external water pH range tested, CO_2 hydration did not occur as water passed over the gills but bicarbonate in the water was dehydrated due to proton excretion. Although proton excretion might change the CO_2/HCO_3^- level in the boundary water layer, the CO_2 excretion rate was unaffected by the proton pump inhibitors except acetazolamide (Lin and Randall, 1991).

Acidification of the boundary layer, whether by carbon dioxide hydration or proton excretion, will augment ammonia excretion by trapping ammonium ions and enhancing NH_3 gradients (Wright *et al.*, 1989; Randall *et al.*, 1991).

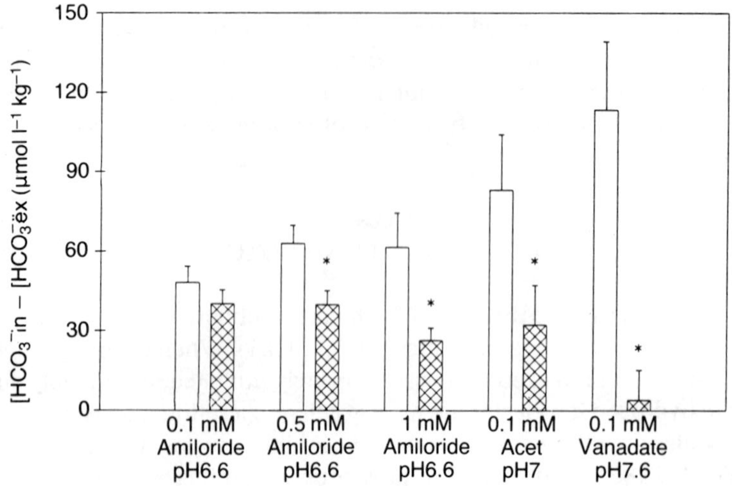

Fig. 10.7 Bicarbonate concentration differences between inspired and expired water of rainbow trout under control, amiloride, acetazolamide and vanadate treatments. Open columns, control; hatched columns, treatments; * indicates a significant difference between the control and treatment values under Student's two-tailed *t*-test ($P < 0.05$); bars show standard errors; $N = 6$. Reproduced with permission from Lin and Randall (1991).

10.5 SODIUM UPTAKE

Sodium uptake by fish is blocked by amiloride (Perry and Randall, 1981; Wright and Wood, 1985), SITS (4-acetamido-4′-isothiocyanatostilbene-2-2′-disulphonic acid) (Perry and Randall, 1981) and exposure to acid or alkaline water (Wright and Wood, 1985). Exposure of trout to 0.1 mM amiloride in the water resulted in an 84% (Perry and Randall, 1981) and 94% (Wright and Wood, 1985) inhibition of sodium uptake, indicating that most if not all sodium is taken up through an amiloride-sensitive sodium channel. The effects of water pH on sodium uptake cannot be simply related to a membrane potential generated by activity of the proton pump, as seen in frog skin, because membrane potential is increased under alkaline conditions, and this should enhance, not inhibit, sodium uptake. Alkaline conditions may result in a down regulation of sodium channels, and hence a reduction in sodium uptake despite an increased sodium electrochemical gradient.

10.6 CHLORIDE UPTAKE

Acid conditions cause a marked increase in chloride efflux but a decrease in chloride influx across trout gills (McDonald *et al.*, 1983). This is associated with a decrease in the chloride electrochemical gradient, so the increased efflux must be due to a large increase in epithelial permeability (Ye *et al.*, 1991). Chloride uptake is considered to be associated with bicarbonate excretion, because it is inhibited by SITS (Perry and Randall, 1981). In addition, Cl^-/HCO_3^- exchange may be inhibited by acid water, presumably because of a reduction in epithelial bicarbonate levels associated with an acidification of the gill epithelium. The addition to the water of the drug 9-anthroic acid (9AA), which is thought to inhibit chloride channels in turtle bladder (Stetson *et al.*, 1985), had no effect on proton excretion in fish in sodium-free water (Fig. 10.5). Thus, it would appear that 9AA-inhibited chloride channels are not present in the fish gill apical membrane. They could be present, however, in the serosal membrane (Fig. 10.6).

10.7 CARBON DIOXIDE TRANSFER

Carbon dioxide transfer across fish gills is unaffected by water pH in resting fish (Lin and Randall, 1990). The rate-limiting step determining carbon dioxide excretion is HCO_3^-/Cl^- exchange across red blood cell membrane (Perry *et al.*, 1982). Alkaline pH reduces Pa_{CO_2} whereas acid conditions have the reverse effect, but both have little effect on the rate of CO_2 excretion.

10.8 OXYGEN TRANSFER

Exposure for 24h to acid conditions has no effect on oxygen consumption in resting fish (Ye *et al.*, 1991). The development of a blood acidosis, however, may reduce blood oxygen content and impair oxygen delivery and swimming capacity. The acidosis that follows exposure to acid water does not result in the release of catecholamines, as seen following exhaustive exercise, such that oxygen transport capacity is not maintained (Ye *et al.*, 1991).

10.9 AMMONIA EXCRETION

Ammonia excretion is reduced during the initial stages of exposure to both acid and alkaline water (Wright and Wood, 1985) but then returns to normal, with elevated plasma ammonia levels (Randall and Wright, 1989) that can be toxic. Tilapia (*Oreochromis alcalicus grahami*) from Lake Magadi, Kenya, however, survive in water of pH 9–10.5 probably because they can detoxify ammonia by producing urea via the ornithine–urea cycle (Randall *et al.*, 1989). Most teleosts cannot produce urea via the ornithine–urea cycle but exposure of the tilapia, *Oreochromis mossambica*, to water of high pH and bicarbonate concentrations results in a marked increase in urea excretion (McKenzie, unpublished observations).

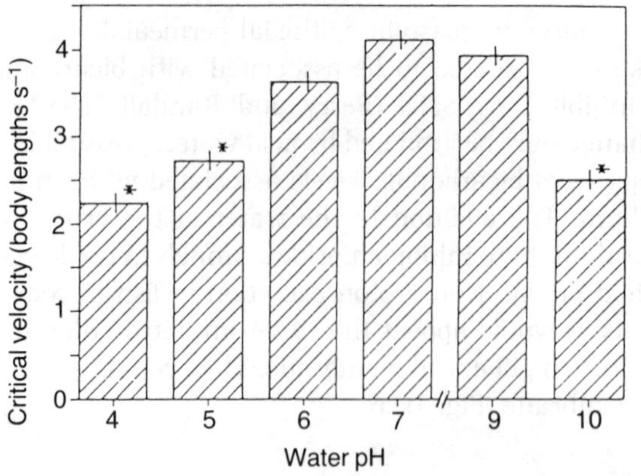

Fig. 10.8 The effect of water pH on the critical swimming velocity of rainbow trout; * indicates a significant difference from neutral control under Student's two-tailed *t*-test ($P < 0.05$); bars show standard errors. Reproduced with permission from Randall and Brauner (1991).

10.10 SWIMMING PERFORMANCE

Exercise involves the interaction of many systems, and therefore can be used to assess an animal's fitness for a particular environment. Fish exposed to acid or alkaline water have a reduced swimming ability (Fig. 10.8), an increase in H^+ ions causing a greater reduction in swimming performance than an equivalent increase in OH^- ion concentration (Ye and Randall, 1991). This reduction in performance could be related to acid–base disturbances and associated effects on oxygen transport, ionic and osmotic changes and/or ammonia accumulation in body tissues (Randall and Brauner, 1991).

REFERENCES

Arruda, J.A.L., Sabatini, S. and Westenfelder, C. (1981) Vanadate inhibits urinary acidification by the turtle bladder. *Kidney Int.*, **20**, 772–9.

Avella, M. and Bornancin, M. (1989) A new analysis of ammonia and sodium transport through the gills of the freshwater rainbow trout (*Salmo gairdneri*). *J. exp. Biol.*, **142**, 155–75.

Ehrenfeld, J. and Garcia-Romeu, F. (1977) Active hydrogen excretion and sodium absorption through isolated frog skin. *Am. J. Physiol.*, **233**, F46–54.

Ehrenfeld, J., Garcia-Romeu, F. and Harvey, B.J. (1985) Electrogenic active proton pump in *Rana esculenta* skin and its role in sodium ion transport. *J. Physiol., Lond.*, **359**, 331–55.

Janssen, R.G. and Randall, D.J. (1975) The effects of changes in pH and P_{CO_2} in blood and water on breathing in rainbow trout, *Salmo gairdneri*. *Respir. Physiol.*, **25**, 235–45.

Kleyman, T.R. and Cragoe, E.J.jun. (1988) Amiloride and its analogs as tools in the study of ion transport. *J. Membr. Biol.*, **105**, 1–21.

Knauf, H., Simon, B. and Wais, U. (1976) Non-specific inhibition of membrane-ATPase by amiloride. *Naunyn-Schmiedebergs Arch. Pharmakol.*, **292**, 189–92.

Lin, H. and Randall, D.J. (1990) The effect of varying water pH on the acidification of expired water in rainbow trout. *J. exp. Biol.*, **149**, 149–60.

Lin, H. and Randall, D.J. (1991) Evidence for the presence of an electrogenic proton pump on the trout gill epithelium. *J. exp. Biol.*, **161**, 119–34.

McDonald, D.G. (1983) The interaction of environmental calcium and low pH on the physiology of the rainbow trout, *Salmo gairdneri*. I. Branchial and renal net ion and H^+ fluxes. *J. exp. Biol.*, **102**, 123–40.

McDonald, D.G. and Wood, C.M. (1981) Branchial and renal acid and ion fluxes in the rainbow trout, *Salmo gairdneri* at low environmental pH. *J. exp. Biol.*, **93**, 101–18.

McDonald, D.G., Walker, R.L. and Wilkes, P.R.H. (1983) The interaction of environmental calcium and low pH on the physiology of the rainbow trout, *Salmo gairdneri*. II. Branchial ionoregulatory mechanisms. *J. exp. Biol.*, **102**, 141–55.

Payan, P. (1978) A studies of the Na^+/NH_4^+ exchange across the gill of the perfused head of the trout (*Salmo gairdneri*). *J. comp. Physiol.*, **124**, 181–8.

Perry, S.F. and Randall, D.J. (1981) Effects of amiloride and SITS on branchial ion fluxes in rainbow trout (*Salmo gairdneri*). *J. exp. Zool.*, **215**, 225–8.

Perry, S.F., Davie, P.S., Daxboeck, C. and Randall, D.J. (1982) A comparison of CO_2 excretion in a spontaneously ventilating blood-perfused trout preparation and saline-perfused gill preparation: contribution of the branchial epithelium and red blood cell. *J. exp. Biol.*, **101**, 47–60.

Randall, D.J. and Brauner, C. (1991) Effects of environmental factors on exercise in fish. *J. exp. Biol.*, **160**, 113–26.

Randall, D.J. and Wright, P.A. (1989) The interaction between carbon dioxide and ammonia excretion and water pH in fish. *Can. J. Zool.*, **67**, 2936–42.

Randall, D.J., Wood, C.M., Perry, S.F., Bergman, H., Maloiy, G.M.O., Mommsen, T.P. and Wright, P.A. (1989) Urea excretions a strategy for survival in a fish living in a very alkaline environment. *Nature, Lond.*, **337**(12), 165–6.

Randall, D.J., Lin, H. and Wright, P.A. (1991) Gill water flow and the chemistry of the boundary layer. *Physiol. Zool.*, **64**(1), 26–38.

Steinmetz, P.R. and Andersen, O.S. (1982) Electrogenic proton transport in epithelial membrane. *J. Membr. Biol.*, **65**, 155–74.

Stetson, D.L., Beauwens, R., Palmisano, J., Mitchell, P.P. and Steinmetz, P.R. (1985) A double-membrane model for urinary bicarbonate secretion. *Am. J. Physiol.*, **249**, F546–52.

Wright, P.A. and Wood, C.W. (1985) An analysis of branchial ammonia excretion in the freshwater rainbow trout: effects of environmental pH change and sodium uptake blockage. *J. exp. Biol.*, **114**, 329–53.

Wright, P.A., Randall, D.J. and Perry, S.F. (1989) Fish gill water boundary layer: a site of linkage between carbon dioxide and ammonia excretion. *J. Comp. Physiol.*, **158B**, 627–35.

Ye, X. and Randall, D.J. (1991) The effect of water pH on swimming performance in rainbow trout (*Salmo gairdneri*, Richardson). *Fish Physiol. Biochem.*, **9**(1), 15–21.

Ye, X., Randall, D.J. and He, X. (1991) The effect of acid water on oxygen consumption, circulating catecholamines and blood ionic and acid–base status in rainbow trout (*Salmo gairdneri*, Richardson). *Fish Physiol. Biochem.*, **9**(1), 23–30.

Chapter eleven

Endocrine responses to environmental pollutants

J. Anne Brown

11.1 INTRODUCTION

Generalized endocrine responses to stress, common to a wide variety of stressors, involve activation of (i) the hypothalamo–pituitary–interrenal axis (HPI), resulting in release of corticosteroids, and (ii) activation of the adrenergic system resulting in release of catecholamines (Mazeaud *et al.*, 1977; Donaldson, 1981; Mazeaud and Mazeaud, 1981). The primary release of hormones will inevitably lead to secondary physiological and biochemical events. Some of these, such as alterations in energy metabolism, may be immediately beneficial to the fish, and as such are in line with the general adaptation syndrome described by Selye (1950), but it is equally clear that this is not always the case, and maladaptive effects such as disturbed ionoregulation, immunosuppression, and in the longer term, depressed growth and reproductive function, have more recently become apparent.

The stress responses to crowding, handling, netting, transport and anaesthesia of fish have been extensively investigated and recently reviewed (e.g. Wedemeyer *et al.*, 1990; Barton and Iwama, 1991). The aim of the present chapter is to review the more limited information on the endocrine responses to water-borne environmental pollutants and poor water quality, and to provide a focus for future investigations. Both laboratory and field investigations are essential to enable estimation of the threshold level of pollutants where endocrine responses ensue. Such information is important for the setting of appropriate water quality criteria to protect fisheries. Endocrine studies of feral fish are particularly difficult as blood samples must generally be obtained very rapidly (pp. 277–84) and it is important to ensure that comparisons are made with fish from 'unpolluted' sites.

Fish Ecophysiology. Edited by J. Cliff Rankin and Frank B. Jensen. Published in 1993 by Chapman & Hall, London. ISBN 0 412 45920 5.

This review will consider in some detail the endocrine responses to environmental acidification, one form of water pollution which has received much attention in recent years. The effects of declining pH in rivers and lakes, with resultant mobilization of toxic species of aluminium and trace metals, is now well known (Henrikssen *et al.*, 1984; Rosseland *et al.*, 1990). The monitoring of the endocrine status of fish exposed to these condition should enable comparison of relatively resistant with more sensitive species, as well as pinpointing the role of hormones in acclimation.

11.2 ADRENOCORTICAL RESPONSE

The teleost interrenal tissue (D adrenocortical tissue of mammals) is generally found embedded in the anterior portion of the kidney (head kidney) in association with the posterior cardinal veins (van der Boon *et al.*, 1991). Histopathological assessment of the activity of the interrenal tissue by measurement of average nuclear diameter, cell size or cellular RNA can provide an index of stress (Donaldson, 1981), but the assay of circulating levels of cortisol, the major hormone secreted by the adrenocortical tissue (Balment and Henderson, 1987), has been more widely used. Plasma concentrations of cortisol generally rise fairly rapidly as stress ensues (e.g. 5–8 min after handling and confinement; Sumpter *et al.*, 1986).

Stress frequently causes osmoregulatory disturbance – a fall in the plasma osmolarity of freshwater fish or a rise in the plasma osmolarity of seawater fish Wedemeyer *et al.*, 1990; Barton and Iwama, 1991). Release of cortisol as a response to stress may have beneficial effects on ionoregulatory balance. Cortisol stimulates proliferation of chloride cells (Fu *et al.*, 1990; Laurent and Perry, 1990) and hence increases Na^+–K^+–ATPase and Ca^{2+}–ATPase activity (Flik and Perry, 1989; Madsen, 1990). In freshwater fish this stimulates active influx of sodium chloride and calcium (Flik and Perry, 1989; Laurent and Perry, 1990). Cortisol-stimulated elevation in ATPase activities could therefore potentially contribute to the ionoregulatory recovery of stressed fish.

Cortisol may in the short term have beneficial effects on carbohydrate metabolism, but the relative contributions of corticosteroid and adrenergic responses (p. 285) to the increased plasma glucose concentrations usually observed in stressed fish are still unclear. In chronic stress, the metabolic effects of prolonged high concentrations of cortisol may depress growth as nutrients normally used for growth are diverted to maintain a high blood sugar level (van der Boon *et al.*, 1991). However, recent administration of physiological levels of cortisol to rainbow trout, *Oncorhynchus mykiss*, to mimic plasma concentrations typical of chronic stress, did not result in hyperglycaemia (Andersen *et al.*, 1991).

There is considerable evidence that cortisol has a suppressive effect on the immune system, reducing the number of antibody-producing cells (Maule *et al.*, 1989), circulating lymphocytes (Pickering, 1984), and interfering with lymphokine production (Kaattari and Tripp, 1987; Tripp *et al.*, 1987), which results in a predisposition to infection by common bacteria, fungi and protozoans (Pickering, 1984; Pickering and Pottinger, 1987, 1989; Maule *et al.*, 1989).

Plasma cortisol is usually elevated at some stage after exposure of fish to water-borne pollutants (Tables 11.1, 11.2). The few exceptions to this pattern (Schreck and Lorz, 1978; Lee *et al.*, 1983) may have resulted from failure to pick up transitory increases in plasma cortisol because of limited blood sampling programmes (see below) and/or high basal levels. Although plasma

Table 11.1 Studies showing elevated plasma cortisol in fish exposed to water-borne environmental pollutants

Species	Stressor	Concentration of Stressor	Source
Oncorhynchus nerka	Cu	63.5, 635 μg l^{-1}	Donaldson and Dye (1975)
Oncorhynchus kisutch	Cu	140, 210 μg l^{-1}	Schreck and Lorz (1978)
Oncorhynchus mykiss	Cd	50 μg l^{-1}	James and Wigham (1986)
Oreochromis mossambicus	Cd	10 μg l^{-1}	Fu *et al.* (1990)
Oncorhynchus mykiss	Cr	0.02, 0.2 mg l^{-1}	Hill and Fromm (1968)
Oncorhynchus kisutch	Suspended solids	2–3 g l^{-1}	Redding *et al.* (1987)
Fundulus heteroclitus	Ammonia	Total NH_3-N, 47.1 mg l^{-1}	Spotte and Anderson (1989)
Oncorhynchus mykiss	Ammonia	Unionized NH_3-N, 240, 330 μg l^{-1}	Swift (1981)
Oncorhynchus kisutch	Ammonia	Unionized 80 μg l^{-1}	Donaldson (1981)
Ictalurus punctatus	Nitrite	2–5 mg l^{-1}	Tomasso *et al.* (1981)
	Ammonia	Total NH_3-N, 30–110 mg l^{-1}	Tomasso *et al.* (1981)
Fundulus heteroclitus	Naphthalene	200 μg l^{-1}	Di Michele and Taylor (1978)
Oncorhynchus kisutch	WSF crude oil	–	Thomas and Rice (1987)
Anguilla anguilla	Crude oil spill	–	Lopez *et al.* (1981)
Mugil cephalus	WSF No. 2 fuel oil		Thomas *et al.* (1980)
Mugil cephalus	Dibenzofuran	750 μg l^{-1}	Thomas *et al.* (1981a)
	Carbazole	10 μg l^{-1}	Thomas *et al.* (1981a)
	Dibenzo-thiophene	10 μg l^{-1}	Thomas *et al.* (1981a)
Mugil cephalus	Pento-chlorophenol	100, 200 μg l^{-1}	Thomas *et al.* (1981b)
Oncorhynchus mykiss	Phenol	4.7, 7.3 mg l^{-1}	Swift (1981)

Table 11.2 Experimental studies of cortisol responses of acid-stressed and aluminium + acid-stressed fish

Species	pH	Al* (μg l⁻¹)	Ca (mM)	Time	Response	Source
Oncorhynchus mykiss	4.7–5.1	–	0.05	5 days	Increased plasma cortisol at 2 days	Barton *et al.* (1985)
Oncorhynchus mykiss	4.7	–	0.5	7 days	Increased plasma cortisol	Brown *et al.* (1990a)
	4.7	584 T	0.5	7 days	Increased plasma cortisol Increased clearance rate	Brown *et al.* (1990a)
Oncorhynchus mykiss	4.7		0.5	21 days	Interrenal hyperplasia Increase plasma cortisol at 8 days	Brown *et al.* (1984)
Oncorhynchus mykiss	5.0		0.5	7–10 days	Increased degradation rate	Brown *et al.* (1986b)
Coregonus clupeaformis	4.1,4.4	–	0.6	14 days	Increased plasma cortisol	Scherer *et al.* (1986)
	4.7,5.0	–	0.6	14 days	NS effect plasma cortisol	Scherer et al. (1986)
Salvelinus fontinalis	4.5	–	1.5	1 year	Increased plasma cortisol at 3 months	Tam *et al.* (1987)
Salvelinus fontinalis	4.5	–	1.5	15 days	Increased at 1.5 h, recover at 5h Increased at 4 days, NS at 15 days Interrenal hyperplasia day 15	Tam *et al.* (1988)
Oncorhynchus mykiss	4.5	–	1.0	6 h – 2 days	Rapid increase for 12 h, maintained 48 h	Marshall Adams *et al.* (1985)
Salvelinus fontinalis	4.0	–	NP	24 h	Increased at 1–3 h, recovering at 4–6h, basal at 12 h	Mudge *et al.* (1977)
Oncorhynchus mykiss	4.8	–	0.1		Increased at 4 and 7 h, recovery by 15 h	Goss and Wood (1988)
	4.8	112 T	0.1		Increased at 1.5 h and increased further at 15 h	Goss and Wood (1988)
Salvelinus fontinalis	4.8	333 T	0.1	66 h	Increased at 4 h and 66 h	Wood *et al.* (1988)
Salvelinus alpinus	4.5	–	0.5	15 days	Increased within 1 day fluctuating, recovery by 15 days	Jones *et al.* (1987)
Salmo trutta	4.7–5.1	278 F	0.04	1.5 days	Increased plasma cortisol	Whitehead and Brown (1989)
	4.9	617 F		3.5 h	Increased plasma cortisol	Whitehead and Brown (1989)

* F, filterable aluminium; T, total aluminium; NP, concentration of Ca not provided; NS, not a statistically significant effect.

cortisol can undoubtedly provide a useful index of endocrine status, interpretation of responses based on a single sampling point will be difficult, as different degrees of stress will affect both the degree of response and its time course. Attempts to estimate the threshold level of pollutants causing sublethal stress by reference to the cortisol responses (e.g. Swift, 1981) must therefore be undertaken with considerable care. Low levels of stress will result in only a transitory increase in plasma cortisol, followed by acclimation with decline of cortisol to basal levels as the hormone is metabolized by the liver and its metabolites are excreted in the bile. For example, exposure of channel catfish, *Ictalurus punctatus*, to elevated levels of ammonia resulted in a rise of plasma corticosteroids peaking after 8 h exposure, followed by a marked recovery over the remaining 16 h (Tomasso *et al.*, 1981). Similarly, large amounts of suspended solids only induced moderate increases in plasma cortisol in the first 24 h of exposure, and this was followed by a gradual return to basal levels (Redding *et al.*, 1987). The use of cortisol as an indicator of mild, sublethal stress may therefore be limited unless the responses can be monitored soon after exposure to the environmental pollutant. Moderate levels of stress may produce a more rapid and sustained increase in plasma cortisol; for example, exposure of tilapia, *Oreochromis mossambicus*, to cadmium (10 $\mu g\, l^{-1}$) increased plasma cortisol within 1 h, and the increase was sustained over the first 14 days of exposure, with a recovery after 35 days exposure (Fu *et al.*, 1990). Higher levels of stress typically induce larger perturbations in plasma cortisol which may be followed by an initial decline, but often rise again later, indicating a failure to adapt. At very high levels of stress, a rapid rise to a maintained high concentration of plasma cortisol is likely to be followed by death.

An additional complication in interpretation of plasma cortisol responses is the general use of radioimmunoassays measuring total, i.e. bound and free, cortisol. It is usually assumed that any increase in total cortisol will reflect an increase in free cortisol, the physiologically active component, but this is not necessarily true. The relative proportions of free and bound cortisol may change without altering total cortisol, or an increase in total cortisol may merely reflect an increase in the protein-bound cortisol. Also, measurement of plasma cortisol provides only a static assessment of corticosteroid status. An increase in plasma cortisol for 5 days or more increases the metabolic clearance rate (Redding *et al.*, 1984; Brown *et al.*, 1990a). Increased clearance rates will mask the increased secretion. Clearance studies provide a fuller picture than measurement of plasma cortisol, but their increased complexity means that relatively few have been undertaken to date.

Hydrocarbons

Fish can be exposed to various types of oil and hydrocarbons in various habitats. Exposure to hydrocarbons generally leads to increased plasma cortisol (Table 11.1). The study of eels *Anguilla anguilla*, after the *Amoco Cadiz* oil spill provides an excellent example of the use of both interrenal activity and plasma cortisol as a means of investigating the stress response to pollutants in the field. Interrenal hypertrophy and elevated plasma cortisol persisted for 8 months after the spill, after which time the glandular tissue showed signs of exhaustion (Lopez *et al.*, 1981).

Acidic water and aluminium

Changes in plasma cortisol in fish exposed to acidic water with and without elevated aluminium have been reported in a variety of teleosts (Table 11.2). In studies of the effects of acidic water alone, interspecific variations in response, differing experimental protocols, and differing pH and other water quality parameters (particularly Ca, which is not always controlled despite its established importance in determining acid toxicity) make exact comparisons difficult. In the rainbow trout, *Oncorhynchus mykiss*, the most widely studied teleost, the degree of cortisol response has been quite variable, with recovery of basal cortisol levels within 1–3 days, despite continued exposure to water of around pH 4.7–5.7 (Barton *et al.*, 1985; Goss and Wood, 1988), while other studies noted elevated cortisol after 1 or 3 weeks of exposure to acidic water (Brown *et al.*, 1984, 1986a, 1990a). This may reflect a biphasic response, implying a lack of acclimation. However, it seems highly likely that different strains of rainbow trout differ in their acid sensitivity. Brook trout exposed to pH 4.5 for 1 year had elevated plasma cortisol for as long as 3 months (Tam *et al.*, 1987).

Our own studies have investigated the endocrine responses of the brown trout, *Salmo trutta*, to acid stress (Brown *et al.*, 1989). Trout exposed to water of pH 4 showed similar increases in plasma cortisol after 3 h in water with calcium concentrations of 0.05 mM or 2.8 mM (Fig. 11.1). When in low-calcium water, these trout did not survive to a 2 days sampling point; the higher calcium regime was clearly less stressful, and after 48 h, although plasma cortisol was still significantly elevated, there were signs of some possible recovery (Fig. 11.1). In a less stressful acid regime, where water pH was allowed to drift upwards overnight and so ranged between 4 and 4.6, after 2 days there was a significant increase in plasma cortisol concentration only in the group of fish held in the high-calcium regime, but over the following 5 days these trout recovered. In contrast, in the low-calcium acid regime there was a dramatic increase in plasma cortisol at the 7 day sampling point (Fig. 11.2). These results demonstrate the importance of water calcium as a crucial factor determining the endocrine responses.

Although many studies have suggested an increase in plasma cortisol in acid-stressed fish, cytological studies of the interrenal tissue suggested an initial decrease in the activity of the interrenal cells (at pH 4, Mudge *et al.*, 1977). Decreased cortisol clearance and increased plasma binding of cortisol (after 2 days at pH 4.2, Nichols and Huntly, 1986) may largely account for elevated plasma cortisol. In longer-term exposure, there appears to be hypertrophy of the interrenal tissue, suggesting an increased secretion rate (Tam *et al.*, 1988).

In the wild, fish are likely to be exposed to a combination of acid soft water and mobilized metals – e.g. aluminium and trace metals (Exley and Phillips,

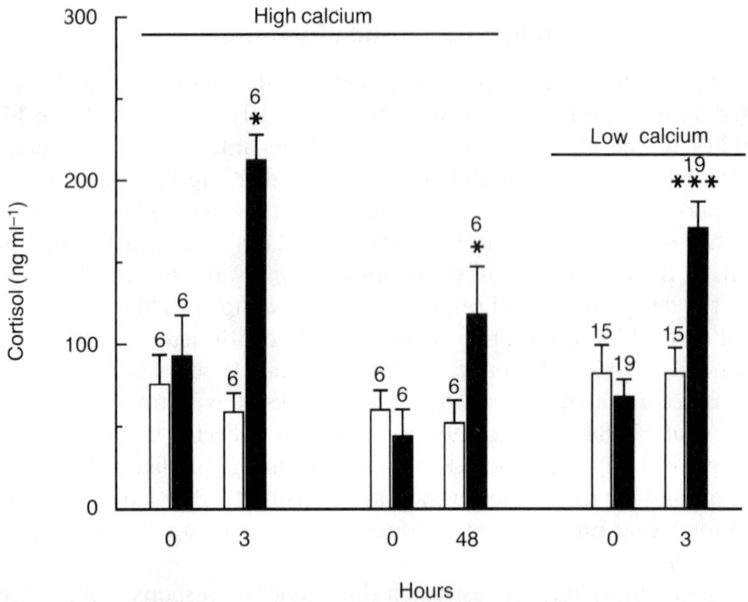

Fig. 11.1 Plasma cortisol concentrations in cannulated brown trout, *Salmo trutta*, exposed to control (open columns, pH 7) and acidic water (shaded columns, pH 4) of low (0.05 mM) or high (2.8 mM) calcium concentration. Blood samples were collected from individual fish at time 0, before any manipulation of water quality and again at 3 h and 48 h (in high Ca water only), after reduction of water pH. Values are means ± SE, number of fish given above relevant columns. Mann–Whitney *U* test of percentage changes relative to initial concentrations of individual fish: *, $P < 0.05$; ***, $P < 0.001$. Drawn from data in Brown *et al.* (1989).

1988) – and fish losses occur at higher pHs than those used in most laboratory experiments on the endocrine effects of acid alone. Investigations of the adrenocortical responses to environmentally realistic combinations of acidic soft water (pH 5–5.5) and aluminium are therefore important. We were fortunate to have the opportunity to examine the responses of brown trout caged in an acidic soft-water stream running through a close canopy forest (yearly mean pH 5.3, Ca 0.05 mM, total Al 320 $\mu g\ l^{-1}$; Whitehead and Brown, 1989)–a stream which had no indigenous trout. An adjacent moorland stream (annual mean pH 6.9, Ca 0.07 mM, Al 70 $\mu g\ l^{-1}$) acted as a circumneutral reference stream. Brown trout were held in these two streams for 1.5 days, during which time rainfall exaggerated the differences in water chemistry of the two streams so that the pH of the acidic stream declined to 5.1 (filterable Al = 278 $\mu g\ l^{-1}$) compared with the neutral stream (pH 7.3, filterable Al 53 $\mu g\ l^{-1}$). Plasma cortisols of the acid + aluminium-exposed fish (382 ± 69 ng ml^{-1}, $n = 5$) were seven times as high as those of fish from the neutral stream (53 ± 19 ng ml^{-1}, $n = 5$). These measurements may be the first cortisol measurements obtained from fish during a natural episode of acid rain, and clearly demonstrate an endocrine stress response.

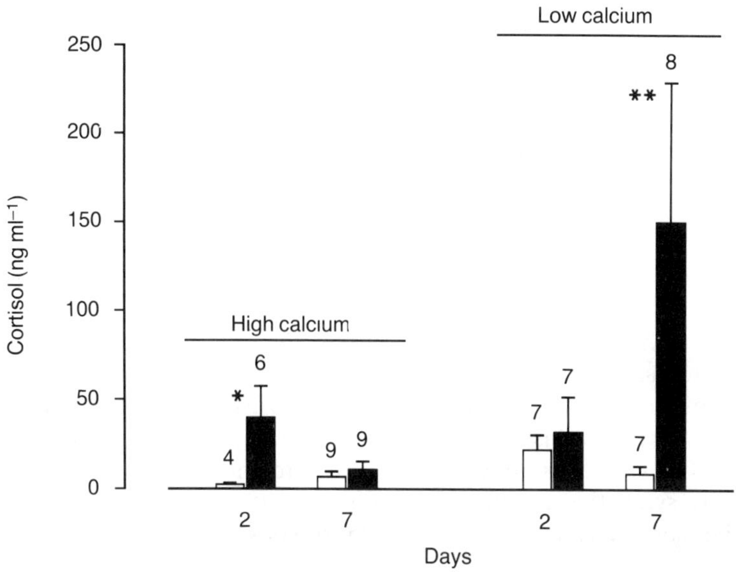

Fig. 11.2 Plasma cortisol concentrations of groups of brown trout, *Salmo trutta*, exposed to neutral water (open columns, control, pH 7) or acidic water (shaded columns, pH 4–4.6) with low (0.05 mM) or high (2.8 mM) calcium content for 2 or 7 days. Values are means ± SE, number of animals given above relevant column. Mann–Whitney *U* test. *, $P < 0.05$, **, $P < 0.01$. Redrawn from Brown *et al.* (1989).

Experimental manipulations of the stream water chemistry (dosing in sequence along its length with additional acid, acid + aluminium, and lime slurry for a period of 3.5 h) provided further information on the cortisol responses. The highest cortisol levels occurred in fish from the acid-dosed and acid + aluminium-dosed zones (Fig. 11.3). Both of these regions had high levels of filterable aluminium. Liming (to reduce water pH and aluminium) reduced plasma cortisol. This may, however, be only a short-term response. Atlantic salmon, *Salmo salar*, caged for 3 months in a soft-water acidic river in Nova Scotia showed increased corticosteroid production by the head kidney which could not be alleviated by liming (Sangalang *et al.*, 1990).

Although the study of caged fish in the wild can be very informative, the manipulation of water chemistry is easier to control in the laboratory. Also, there is considerable spread in the responses of individual fish, and therefore the serial collection of blood samples before and after manipulation of water chemistry is usually a more sensitive approach as each animal can act as its own control. In studies of laboratory-cannulated brown trout sampled at 6 h intervals, there was no increase in plasma cortisol when the fish were held in water of pH 5 (Ca 0.05 mM) (Brown, unpublished). However, addition of aluminium to the acidic water, progressively increasing filterable aluminium from 5 μg l^{-1} at 0 h to 80 μg l^{-1} at 6 h and 280 μg l^{-1} at 36 h, caused a rapid and progressive increase in plasma cortisol. These conditions were lethal for

62% of the fish within the 36 h study period. Similar cortisol responses occurred in rainbow trout (Goss and Wood, 1988), where pH 4.8 alone induced a transitory effect on plasma cortisol, whereas pH 4.8 plus Al (112 $\mu g\ l^{-1}$ total aluminium) produced a rapid, pronounced and prolonged rise in plasma cortisol with a final further rise as fish approached death (Goss and Wood, 1988).

11.3 ADRENERGIC RESPONSES

In teleost fish, the richly innervated chromaffin cells, which secrete the bulk of the circulating catecholamines, adrenaline and noradrenaline, are located in the anterior part of the kidney (head kidney), usually lining the walls of the posterior cardinal veins running through the kidney (Mazeaud and Mazeaud, 1981). The catecholamines are released into the circulation immediately following sympathetic stimulation (Nilsson, 1983); plasma catecholamines are therefore elevated very rapidly in response to disturbance, capture,

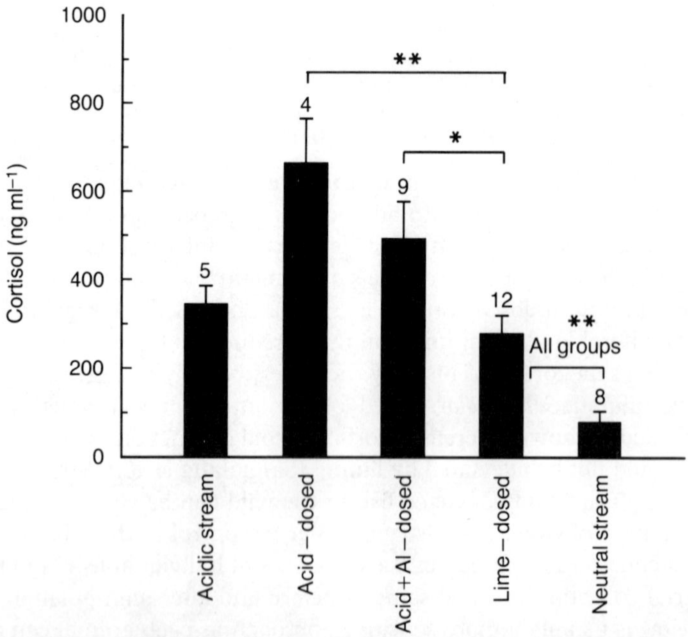

Fig. 11.3 Plasma cortisol concentrations in brown trout, *Salmo trutta*, held in an acidic soft-water stream divided into four zones: natural stream conditions (pH 4.9, Ca 0.05 mM, filterable Al 302 $\mu g\ l^{-1}$), acid-dosed (pH 4.7, Ca 0.06 mM, filterable Al 413 $\mu g\ l^{-1}$), acid plus aluminium-dosed (pH 4.8, Ca 0.06 mM, filterable Al 617 $\mu g\ l^{-1}$), and lime-dosed (pH 7.3, Ca 0.15 mM, filterable Al 216 $\mu g\ l^{-1}$) compared with values for fish held in an adjacent circumneutral stream. Values are means ± SE, numbers of fish are given above relevant column. Student–Newman–Keuls multiple comparison test: *, $P < 0.05$; **, $P < 0.01$. Redrawn from Whitehead and Brown (1989).

handling and surgical stress (Mazeaud *et al.*, 1977; Mazeaud and Mazeaud, 1981; Epple *et al.*, 1982; Ling and Wells, 1985).

Release of catecholamines during stress has several beneficial effects on oxygen transport.

1. Catecholamines maintain the oxygen-carrying capacity of the blood despite the plasma acidosis which occurs in many forms of acute stress (e.g. Primmett *et al.*, 1986). This is achieved by stimulating a Na^+/H^+ exchange mechanism that protects the intracellular pH of erythrocytes (see Jensen *et al.*, Chapter 6, this volume, for details).
2. Catecholamines result in increased blood flow through the gas-exchanging lamellae of the gills (Booth, 1979) and increased branchial permeability to oxygen, resulting in increased oxygen uptake (Pettersson & Johansen, 1982; Perry *et al.*, 1985).
3. Catecholamines stimulate splenic contraction (Nilsson, 1983), releasing additional erythrocytes into circulation, increasing both the buffering capacity and the oxygen-carrying capacity of the blood.
   The latter effect also inevitably increases blood viscosity, which will contribute to increased blood pressure. Direct cardiovascular effects of catecholamines, an increase in stroke volume and systemic vasoconstriction, may have greater effect on blood pressure and be responsible for the ensuing hypertension.

Catecholamines have been suggested to make an important contribution to acid–base regulation after stress, as 'stress levels' of catecholamines elevate branchial H^+ excretion in correlation with an increased difference between branchial Na^+ and Cl^- net fluxes (Vermette and Perry, 1987). An increase in the branchial permeability by catecholamines (Rankin and Bolis, 1984) may result in water gain in freshwater fish and dehydration in marine fish. Hence, catecholamines are unlikely to have any beneficial effect on osmotic and ionic regulation. The benefits on oxygen uptake and acid–base regulation, along with metabolic changes, may, however, outweigh the deleterious effects on osmoregulation.

Stress in fish is typically associated with an increase in plasma glucose. Both corticosteroid and catecholaminergic responses may participate in the hyperglycaemic response to stress. Particularly persuasive evidence that catecholamines are involved in the regulation of plasma glucose has been provided by injection of the eel with physiological doses of catecholamines (Epple *et al.*, 1989). Moderate elevation of catecholamines increases the active form of hepatic glycogen phosphorylase, which stimulates glycogenolysis and gluconeogenesis (Wright *et al.*, 1989).

Whereas many studies have shown elevated circulating catecholamines after severe exercise or physical disturbance, there is very little information on the adrenergic responses to environmental pollutants. Of the vast array of potential environmental pollutants, two of the few for which the adrenergic responses have been investigated are the effects of acidic pH and aluminium in acidic water. Here the effects are of particular interest, given the potential beneficial effects of catecholamines on oxygen carriage and acid–base regulation during acidosis.

Exposure of brown trout to soft water (Ca 0.05 mM) of pH 4 for 6 h resulted in a transitory increase in plasma catecholamines (Fig. 11.4). This water regime was ultimately lethal, and a final rise in catecholamines before death may be expected, as occurred in rainbow trout exposed to pH 4 (Ye *et al.*, 1991). In rainbow trout, plasma catecholamines were little affected for up to 48 h of exposure to pH 4, but after 72 h, noradrenaline and adrenaline were elevated to 7 and 50 times their respective control values (Ye *et al.*, 1991). Exposure of brown trout to pH 5, an environmentally realistic pH, resulted in a transitory increase in noradrenaline after 6 h exposure (mean 75 ± 14 nM compared with a concentration of 39 ± 8 nM in neutral water immediately prior to the switch-over to acidic water, $n = 5$ fish). Fish survived well at this pH, but when aluminium was added, mortalities occurred (pp. 281–2) and all fish showed a profound release of noradrenaline and adrenaline – adrenaline increased from 5 ± 1 nM to 25 ± 7 nM and noradrenaline from 34 ± 3 nM to 82 ± 18 nM after 24 h exposure (Brown, unpublished observations). Both catecholamines increased progressively thereafter. Increased plasma concentrations of both adrenaline and noradrenaline have also been reported in the rainbow trout exposed for 46 h to pH 5 containing aluminium (60 $\mu g\ l^{-1}$ total aluminium), but the time course of response was not investigated in this species (Witters *et al.*, 1991).

Although release of catecholamines has now been shown in acid + aluminium-stressed fish, it is as yet unclear whether any of the beneficial physiological actions mentioned above actually occur. Further studies integrating adrenergic, osmoregulatory and respiratory effects are essential.

In addition to the generalized endocrine responses described above, more specific endocrine responses, such as release of plasma prolactin or stimulation of the thyroid gland, may be elicited by particular environmental pollutants.

11.4 PROLACTIN

Prolactin has been implicated in the maintenance of plasma electrolytes of freshwater fish by decreasing the gill permeability to ions and osmotic permeability to water (Clarke and Bern, 1980). There is also strong evidence that prolactin has an important role in stimulating calcium uptake via the branchial Ca^{2+} pump (Flik *et al.*, 1989a). Prolactin can also stimulate proliferation of epidermal mucocytes.

Prolactin has not been suggested to be part of the generalized endocrine response to stress, but may be part of a homeostatic mechanism to assist in

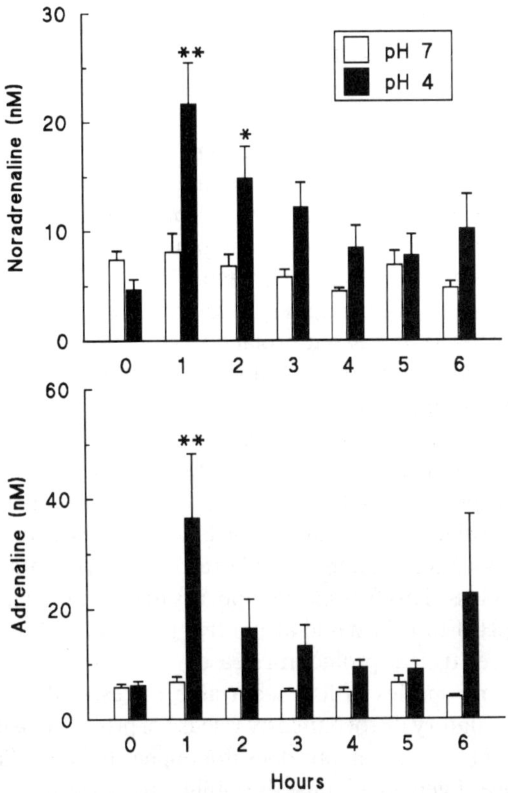

Fig. 11.4 Plasma catecholamine concentrations of brown trout, *Salmo trutta*, exposed to pH 4 (shaded columns, $n = 13$) compared with those of 'control' fish (open columns, $n = 7$) held at pH 7. Student–Newman–Keuls multiple comparison test: *, $P < 0.05$; **, $P < 0.01$ relative to time 0h. Note differing vertical scales. (Brown and Whitehead, unpublished data).

osmotic and ionic balance in response to environmental stressors causing ionoregulatory disturbance. For example, addition of cadmium to fresh water depresses calcium uptake at the gills and results in disturbed calcium balance. In tilapia, *Oreochromis mossambicus*, exposure to sublethal levels of cadmium (10 μg l^{-1}) initially reduced plasma calcium, but after 2 weeks of continued exposure, plasma calcium was restored (Fu *et al.*, 1989). Morphometric studies of the pituitary cells secreting prolactin in these fish, during the first 4 days of exposure, indicated increased endoplasmic reticulum and Golgi bodies and a decreased density of prolactin granules, implying increased secretion, less storage and greater release. However, decreased activity of the prolactin cells thereafter suggested the involvement of additional mechanisms of resistance to Cd. This may involve a cortisol response, as cortisol was elevated in these conditions from 1 h after exposure to at least 14 days later (Fu *et al.*, 1990).

Release of prolactin appears to form part of the homeostatic mechanism of some acid-exposed teleosts. This appears to be particularly pronounced in acid-resistant species such as tilapia and mudminnow, *Umbra pygmaea*. Exposure of tilapia to pH 3.5–4 resulted in an initial decline in plasma osmolarity, sodium and calcium, but this was followed by a recovery (Wendelaar Bonga *et al.*, 1984a,b). In parallel to the recovery in ionic balance, the volume of pituitary prolactin cell increased, with an increase in Golgi bodies but a decreased density of storage granules. Also, increased incorporation of ^{3}H lysine by prolactin-secreting lobes of the pituitary *in vitro* suggests increased prolactin synthesis and release after exposure to acidic water (Wendelaar Bonga *et al.*, 1984a,b). After long-term acclimation of tilapia to acid conditions (3 months at pH 4.5), increased prolactin cell activity and synthesis *in vitro* occurred alongside a substantial reduction in the acid-increased branchial permeability to sodium (Flik *et al.*, 1989b).

The endocrine status of the mudminnow, *U. pygmaea*, is also of interest as this species survives in the wild at pH < 3.0. Examination of wild populations has shown high prolactin activity in fish in neutral waters or at pH < 3, but at pH 3 to 6.5, prolactin cell activity was low (Wendelaar Bonga *et al.*, 1990). This difference was reproducible in laboratory experiments and appeared to be related to ion loss. This fish species appears to be osmotically extremely well adapted from pH 3 to 6.5, while at neutral pH or at < 3 it is suggested that ionic disturbances trigger prolactin release.

Although some species of fish seem able to respond to specific stressors causing ionoregulatory disturbance by release of prolactin, a decline in plasma osmolarity and/or individual ions does not appear to be a reliable stimulus for prolactin release. Even in tilapia, the ability to respond to acidic water by synthesis and release of prolactin, and to restore plasma electrolytes, appears to be related to the rate of pH decline. Rapid acidification suppresses the prolactin response, while a gradual reduction of water pH may allow compensatory responses as there is no noticeable decline in plasma osmolarity (Wendelaar Bonga *et al.*, 1988). This may lie behind the apparent lack of a stimulation of prolactin release in the more acid-sensitive brook trout, *Salvelinus fontinalis* (Fryer *et al.*, 1988).

11.5 THYROID RESPONSE

The thyroid gland of teleosts generally consists of isolated follicles found throughout the subpharyngeal and parapharyngeal region, around the ventral aorta and afferent branchial vessels, and in the head kidney. There is strong evidence that in fish the thyroid gland releases thyroxine (T_4), and that tri-iodothyronine (T_3), the physiologically active hormone, is primarily formed by peripheral conversion of T_4 to T_3, by monodeiodinase in the liver and possibly kidney (Eales, 1985; Leatherland *et al.*, 1990). In circulation, most of the thyroid hormones are bound to specific binding plasma proteins, with < 0.3% free T4 and < 0.2% free T3 (Eales and Shostak, 1985). The free concentrations of T3 and T4, however, correlate well with total hormone levels (Eales and Shostak, 1985), so that measurement of total amounts may provide a reasonable index of peripheral thyroid status.

The actions of thyroid hormones are still unclear. Thyroid hormones may potentiate the effect of cortisol on gill Na^+–K^+-ATPase in tilapia (Dange, 1986), but in the rainbow trout no synergism between T4 and cortisol on gill Na^+–K^+–ATPase or chloride cell number was apparent, while cortisol elevated both parameters (Madsen, 1990).

The work on thyroid responses in relation to pollutants is very limited, although reports of thyroid hyperplasia (goitres) in teleosts are not uncommon (Leatherland, 1982). Dietary or water-borne environmental goitrogens are suspected to inhibit iodide uptake or iodination by the gland, resulting in lowered synthesis of thyroid hormones and thus increased activity of the hypothalamo–pituitary–thyroid axis. Polychlorinated biphenyls (PCBs) do not appear to be responsible (Leatherland, 1982) and as yet the cause of these goitres remains enigmatic.

Exposure of catfish, *Heteropneustes fossilis* and *Clarias batrachus*, to malathion (organophosphorus), endosulfan (organochlorine), and carbaryl (carbamate) pesticides caused changes in circulating thyroid hormones (Yadav and Singh, 1986; Sinha *et al.*, 1991), but as yet no consistent picture has emerged.

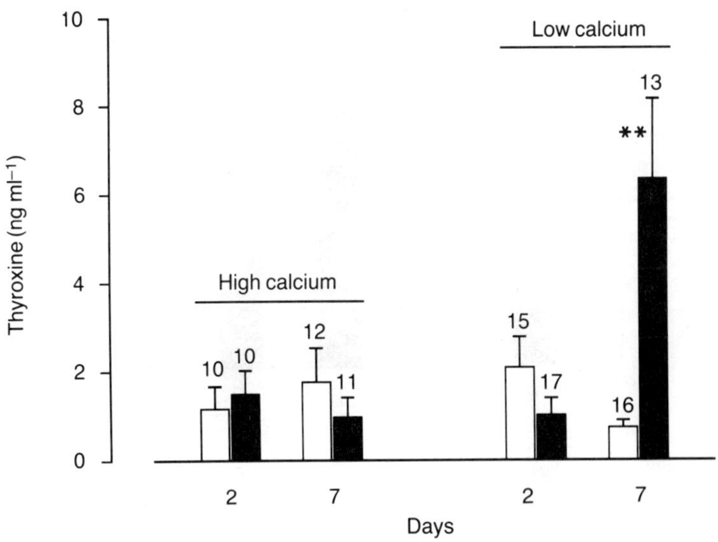

Fig. 11.5 Plasma thyroxine concentrations in groups of brown trout, *Salmo trutta*, exposed to an acidic pH (shaded columns, pH 4–4.6) for 2 or 7 days with low (0.05 mM) or high (2.8 mM) calcium concentrations compared with those of control fish (open columns, pH 7). Values are means ± SE, numbers of fish are given above relevant column. Mann–Whitney *U* test: ** $P < 0.01$. Redrawn from Brown *et al.* (1989).

Aside from these studies, work on thyroid responses to pollutants has been largely restricted to the responses to acid or acid + aluminium exposure.

Studies on the rainbow trout do not present an entirely consistent picture, but taken together they give the impression of relatively little thyroid response to an acidic pH (Brown *et al.*, 1984, 1986a,b). However, all of the laboratory studies have employed water with a high calcium content, which would alleviate the deleterious effects of acidic waters on branchial function. A recent field study of rainbow trout comparing populations in two acidic rivers in Nova Scotia suggested depressed plasma T_3 in the more acidic river (pH 4.7–5.2; Brown *et al.*, 1990b). This is in line with recent work on thyroid hormone kinetics of the rainbow trout exposed to acidic water containing aluminium (541 $\mu g\ l^{-1}$; Ca 0.5 mM) which depressed T_3 production (Brown *et al.*, 1990).

The observations in the rainbow trout contrast markedly with the conclusions reached in our own studies of the brown trout. In the brown trout, thyroxine (T_4) increased within 3 h of exposure to pH 4 (at both Ca 0.05 mM and Ca 2.8 mM; Brown *et al.*, 1989) but recovered within 48 h in the high-calcium regime. Fish did not survive 2 days' exposure to the more stressful low-calcium regime. With a less stressful acid regime (pH 4–4.6 varying on a diurnal cycle), there was no increase in T_4 after 2 days in soft

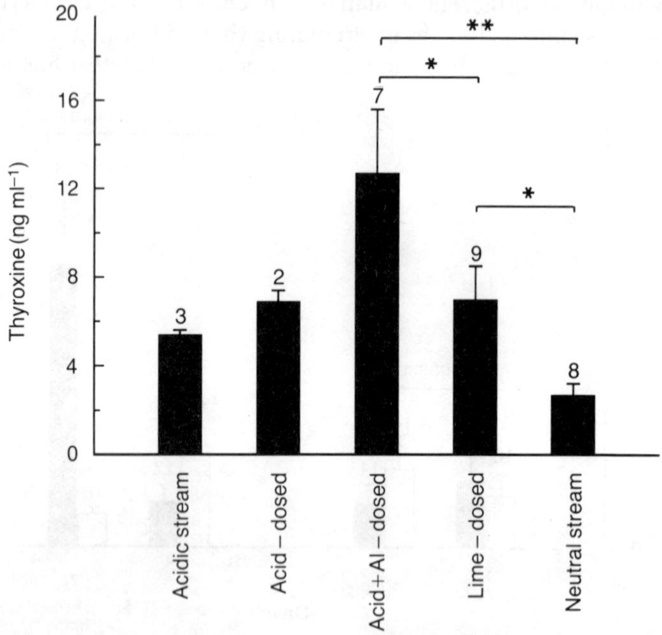

Fig. 11.6 Plasma thyroxine concentrations in *Salmo trutta* held in an acidic soft-water stream. For explanation of zones and water chemistry see Fig. 11.3 Values are means ± SE, numbers of fish above error bars. Student–Neuman–Keuls multiple comparison test: *, $P < 0.05$; **, $P < 0.01$. Redrawn from Whitehead and Brown (1989).

water, but after 7 days T4 was dramatically elevated, although there were large variations in individual responses (Fig. 11.5). T3 was unchanged at any of the time intervals, suggesting good regulation of liver monodeiodinase. The elevation in circulating thyroxine in acid-stressed brown trout appears to be a consistent response, as it was also noted in the studies of fish held in an acidic soft-water stream (p. 281) and liming downstream alleviated this response (Fig. 11.6; Whitehead and Brown, 1989).

11.6 FUTURE STUDIES

Monitoring of the endocrine responses of fish to environmental pollutants permits a more exact understanding of their sublethal stress responses than the investigation of secondary effects, especially where secondary effects may arise by more than one route. The ultimate challenge for the future is to investigate the influence of combinations of pollutants which are in reality what fish experience in the wild. For example, it is only in recent years that the effects of aluminium and low pH have been investigated, but exposure to these stressors in combination with sublethal levels of a variety of metals which will be mobilized by acidic water, and other likely freshwater pollutants such as nitrates, would be more realistic. In the marine environment, complex cocktails of pollutants such as 'production waters' from crude oil extraction similarly require attention.

REFERENCES

Andersen, D.E., Reid, S.D., Moon, T.W. and Perry, S.F. (1991) Metabolic effects associated with chronically elevated cortisol in rainbow trout (*Oncorhynchus mykiss*). *Can. J. Fish. aquat. Sci.*, **48**, 1811–17.

Balment, R.J. and Henderson, I.W. (1987) Secretion of endocrine glands and their relationship to osmoregulation, in *Fundamentals of Comparative Endocrinology* (eds I. Chester Jones, P.M. Ingleton and J.G. Phillips), Plenum Press, New York, pp. 413–508.

Barton, B.A. and Iwama, G.K. (1991) Physiological changes in fish from stress in aquaculture with emphasis on the responses and effects of corticosteroids. *A. Rev. Fish Diseases*, **1**, 3–26.

Barton, B.A., Weiner, G.S. and Schreck, C.B. (1985) Effect of prior acid exposure on physiological responses of juvenile rainbow trout (*Salmo gairdneri*) to acute handling stress. *Can. J. Fish. aquat. Sci.*, **42**, 710–17.

Booth, J.H. (1979) The effect of oxygen supply – epinephrine and acetylcholine on the distribution of blood flow in trout gills. *J. exp. Biol.*, **83**, 31–9.

Brown, S.B., Eales, J.G., Evans, R.E. and Hara, T.J. (1984) Interrenal, thyroidal and carbohydrate responses of rainbow trout (*Salmo gairdneri*) to environmental acidification. *Can. J. Fish. aquat. Sci.*, **41**, 36–45.

Brown, S.B., Evans, R.E. and Hara, T.J. (1986a) Interrenal, thyroidal, carbohydrate and electrolyte responses in rainbow trout (*Salmo gairdneri*) during recovery from the effects of acidification. *Can. J. Fish. aquat. Sci.*, **43**, 714–18.

Brown, S.B. Eales, J.G. and Hara, T.J. (1986b) A protocol for estimation of cortisol plasma clearance in acid-exposed rainbow trout (*Salmo gairdneri*). *Gen. comp. Endocrinol.*, **62**, 493–502.

Brown, J.A., Edwards, D. and Whitehead, C. (1989) Cortisol and thyroid hormone responses to acid stress in the brown trout, *Salmo trutta* L. *J. Fish Biol.*, **35**, 73–84.

Brown, S.B., MacLatchy, D.L., Hara, T.J. and Eales, J.G. (1990a) Effects of low ambient pH and aluminium on plasma kinetics of cortisol T_3 and T_4 in rainbow trout, *Oncorhynchus mykiss Can. J. Zool.*, **68**, 1537–43.

Brown, S.B., Evans, R.E., Majewski, H.S., Sangalang, G.B. and Klaverkamp, J.F. (1990b) Responses of plasma electrolytes, thyroid hormones, and gill histology in Atlantic salmon (*Salmo salar*) to acid and limed river waters. *Can. J. Fish. aquat. Sci.*, **47**, 2431–40.

Clarke, W.C. and Bern, H.A. (1980) Comparative endocrinology of prolactin, in *Hormonal Proteins and Peptides*, Vol. VIII (ed. C.H. Li), Academic Press, New York, pp. 105–97.

Dange, A.D. (1986) Branchial Na^+/K^+-ATPase activity in freshwater and seawater acclimated tilapia *Oreochromis (Sarotherodon) mossambicus*: effects of cortisol and thyroxine. *Gen. comp. Endocrinol.*, **62**, 341–3.

DiMichele, L. and Taylor, M.H. (1978) Histopathological and physiological responses of *Fundulus heteroclitus* to naphthalene exposure. *J. Fish. Res. Bd Can.*, **35**, 1060–66.

Donaldson, E.M. (1981) The pituitary–interrenal axis as an indicator of stress in fish, in *Stress and Fish* (Ed. A.D. Pickering), Academic Press, New York, pp. 11–47.

Donaldson, E.M. and Dye, H.M. (1975) Corticosteroid concentrations in sockeye salmon (*Oncorhynchus nerka*) exposed to low concentrations of copper. *J. Fish. Res. Bd Can.*, **32**, 533–9.

Eales, J.G. (1985) The peripheral metabolism of thyroid hormones and regulation of thyroidal status in poikilotherms. *Can. J. Zool.*, **63**, 1217–31.

Eales, J.G. and Shostak, K. (1985) Free T_4 and T_3 in relation to total hormone free hormone indices and protein in plasma of rainbow trout and Arctic charr. *Gen. comp. Endocrinol.*, **58**, 291–302.

Epple, A., Vogel, W.H. and Nibbio, A.J. (1982) Catecholamines in head kidney and body blood of eels and rats. *Comp. Biochem. Physiol.*, **71C**, 115–18.

Epple, A. Hathaway, C.B. and Nibbio, B. (1989) Circulatory catecholamines in the eel: origins and functions. *Fish Physiol. Biochem.*, **7**, 273–8.

Exley, C. and Phillips, M.J. (1988) Acid rain: implications for the farming of salmonids, in *Recent Advances in Aquaculture* (eds, J.M. Muir and R.R. Roberts), Croom Helm, London pp. 225–341.

Flik, G. and Perry, S.F. (1989) Cortisol stimulates whole body calcium uptake and the branchial calcium pump in freshwater trout. *J Endocr.*, **120**, 83–8.

Flik, G., Fenwick, J.C. and Wendelaar Bonga, S.E. (1989a) Calcitropic actions of prolactin in North American eel (*Anguilla rostrata* Le Sueur). *Am. J. Physiol.*, **257**, 74–9.

Flik, G, Van der Velden, J.A., Seegers, H.C.M., Kolar, Z. and Wendelaar Bonga, S.E. (1989b) Prolactin cell activity and sodium fluxes in tilapia *Oreochromis mossambicus* after long-term acclimation to acid water. *Gen. comp. Endocrinol.*, **75**, 39–45.

Fryer, J.N., Tam, W.H., Valentine, B. and Tikkala, R.E. (1988) Prolactin cell cytology, plasma electrolytes and whole body sodium efflux in acid stressed brook trout, *Salvelinus fontinalis*. *Can. J. Fish. aquat. Sci.*, **45**, 1212–21.

Fu, H., Lock, R.A.C. and Wendelaar Bonga, S.E. (1989) Effect of cadmium on prolactin cell activity and plasma electrolytes in the freshwater teleost *Oreochromis mossambicus*. *Aquat. Toxicol.*, **14**, 295–306.

Fu, H., Steinebach, O.M., van den Hamer, C.J.A., Balm, P.H.M. and Lock, R.A.C. (1990) Involvement of cortisol and metallothionein-like proteins in the physiological responses of tilapia (*Oreochromis mossambicus*) to sublethal cadmium stress. *Aquat. Toxicol.*, **16**, 257–70.

Goss, G.G. and Wood, C.M. (1988) The effects of acid and acid/aluminium exposure on circulating plasma cortisol levels and other blood parameters in the rainbow trout, *Salmo gairdneri*. *J. Fish Biol.*, **32**, 63–76.

Henrikssen, A., Skogheim, O.K. and Rosseland, B.O. (1984) Episodic changes in pH and aluminium speciation kill fish in a Norwegian salmon river. *Vatten*, **40**, 255–60.

Hill, C.W. and Fromm, P.O. (1968) Response of interrenal gland of rainbow trout (*Salmo gairdneri*) to stress. *Gen. comp. Endocrinol.*, **11**, 69–77.

James, V.A. and Wigham, T. (1986) The effects of cadmium on prolactin cell activity and plasma cortisol levels in the rainbow trout (*Salmo gairdneri*). *Aquat. Toxicol.*, **8**, 273–80.

Jones, K.A., Brown, S.B. and Hara, T.J. (1987) Behavioural and biochemical studies of onset and recovery from acid stress in Arctic char (*Salvelinus alpinus*). *Can. J. Fish. aquat. Sci.*, **44**, 373–81.

Kaattari, S.L. and Tripp, R.A. (1987) Cellular mechanisms of glucocorticoid immunosuppression in salmon. *J. Fish Biol.*, **31** (Supp-A), 129–32.

Laurent, P. and Perry, S.F. (1990) Effects of cortisol on gill chloride cell morphology and ionic uptake in the freshwater trout, *Salmo gairdneri*. *Cell Tissue Res.*, **259**, 429–42.

Leatherland, J.F. (1982) Environmental physiology of the teleost thyroid gland: a review. *Env. Biol. Fishes*, **7**, 83–110.

Leatherland, J.F., Redding, P.K., Yong, A.N., Leatherland, A. and Lam, T.J. (1990) Hepatic 5′-monodeiodinase activity in teleosts *in vitro*: a survey of thirty-three species. *Fish Physiol. Biochem.*, **8**, 1–10.

Lee, R.M., Gerking, S.D. and Jezierska, B. (1983) Electrolyte balance and energy mobilization in acid-stressed rainbow trout *Salmo gairdneri*, and their relation to reproductive success. *Env. Biol. Fishes*, **8**, 115–23.

Ling, N. and Wells, R.M.G. (1985) Plasma catecholamines and erythrocyte swelling following capture stress in a marine teleost fish. *Comp. Biochem. Physiol.*, **82C**, 231–4.

Lopez, E., Leloup-Hatey, J., Hardy, A., Lallier, F., Martelly, E., Oudot, J., Peignoux-Deville, J. and Fontaine, Y.A. (1981) Modifications histopathologiques et stress chez des Anguilles soumises a une exposition prolongée aux hydrocarbures, in *Amoco Cadiz. Conséquences d'une Pollution accidentelle par les Hydrocarbures. Fates and Effects of the Oil Spill*, Centre National pour l'Exploitation des Océans, Paris, pp 645–53.

Madsen, S.S. (1990) Effect of repetitive cortisol and thyroxine injections on chloride cell number and Na^+/K^+-ATPase activity in gills of freshwater acclimated rainbow trout *Salmo gairdneri*. *Comp. Biochem. Physiol.*, **95A**, 171–5.

Marshall Adams, S., Burtis, C.A. and Beauchamp, J.J. (1985) Integrated and individual biochemical responses of rainbow trout (*Salmo gairdneri*) to varying durations of acidification stress. *Comp. Biochem. Physiol.*, **82C**, 301–10.

Maule, A.G., Tripp, R.A., Kaattari, S.L. and Schreck, C.B. (1989) Stress alters immune function and disease resistance in chinook salmon (*Oncorhynchus tshawytscha*). *J. Endocr.*, **120**, 135–42.

Mazeaud, M.M. and Mazeaud, F. (1981) Adrenergic responses to stress in fish, in *Stress and Fish* (ed. A.D. Pickering), Academic Press, London, pp. 49–75.

Mazeaud, M.M., Mazeaud, F. and Donaldson, E.M. (1977) Primary and secondary effects of stress in fish: some new data with a general review. *Trans. Am. Fish. Soc.*, **105**, 201–12.

Mudge, J.E., Dively, J.L., Neff, W.H. and Anthony, A. (1977) Interrenal histochemistry of acid-exposed brook trout (*Salvelinus fontinalis* Mitchill). *Gen. Comp. Endocrinol.*, **31**, 208–15.

Nichols, D.J. and Huntly, P. (1986) Cortisol kinetics and plasma protein binding in acid-stressed trout. *J. Endocr.*, **111** (Supp.), Abs. 113.

Nilsson, S. (1983) *Autonomic Nerve Function in the Vertebrates*, Springer Verlag, Berlin, 248 pp.

Perry, S.F., Daxboek, C. and Dobson, G.P. (1985) The effect of perfusion flow rate and adrenergic stimulation on oxygen transfer across the isolated saline-perfused head of rainbow trout. (*Salmo gairdneri*). *J. exp. Biol.*, **116**, 251–69.

Pettersson, K. and Johansen, K. (1982) Hypoxic vasoconstriction and the effects of adrenaline on gas exchange efficiency in fish gills. *J. exp. Biol.*, **97**, 263–73.

Pickering A.D. (1984) Cortisol-induced lymphocytopenia in brown trout, *Salmo trutta* L. *Gen. comp. Endocrinol.*, **53**, 252–9.

Pickering, A.D. and Pottinger, T.G. (1987) Crowding causes prolonged leucopenia in salmonid fish, despite interrenal acclimation. *J. Fish Biol.*, **30**, 701–2.

Pickering, A.D. and Pottinger, T.G. (1989) Stress responses and disease resistance in salmonids: effects of chronic elevation of plasma cortisol. *Fish Physiol. Biochem.*, **7**, 253–8.

Primmett, D.R.N., Randall, D.J., Mazeaud, M. and Boutilier, R.G. (1986) The role of catecholamines in erythrocyte pH regulation and oxygen transport in rainbow trout (*Salmo gairdneri*) during exercise. *J. exp. Biol.*, **122**, 139–48.

Rankin, J.C. and Bolis, L. (1984) Hormonal control of water movement across the gills, in *Fish Physiology*, Vol. XB, *Gills*, Part B, *Ion and Water Transfer* (eds W.S. Hoar and D.J. Randall), Academic Press, London, pp. 117–201.

Redding, J.M. and Schreck, C.B. (1987) Physiological effects on coho salmon and steelhead of exposure to suspended solids. *Trans. Am. Fish. Soc.*, **116**, 737–44.

Redding, J.M., Patiño, R. and Schreck, C.B. (1984) Clearance of corticosteroids in yearling coho salmon, *Oncorhynchus kisutch*, in freshwater and seawater after stress. *Gen. comp. Endocrinol.*, **54**, 433–43.

Redding, J.M., Schreck, C.B. and Everest, F.H. (1987) Physiological effect on coho salmon and steelhead of exposure to suspended solids. *Trans. Am. Fish. Soc.*, **116**, 737–44.

Rosseland, B.O., Eldhurst, T.D. and Staurnes, M. (1990) Environmental effects of aluminium. *Env. Geochem. Health*, **12**, 17–27.

Sangalang, G.B., Freeman, H.C., Uthe, J.F. and Sperry, L.S. (1990) Effects of diet or liming on steroid hormone metabolism and reproduction in Atlantic salmon (*Salmo salar*) held in an acidic river. *Can. J. Fish. aquat. Sci.*, **47**, 2422–30.

Scherer, E., Harrison, S.E. and Brown, S.B. (1986) Locomotor activity and blood plasma parameters of acid-exposed lake whitefish, *Coreogonus clupeaformis*. *Can. J. Fish. aquat. Sci.*, **43**, 1556–61.

Schreck, C.B. and Lorz, H.W. (1978) Stress responses of coho salmon (*Oncorhynchus kisutch*) elicited by cadmium and copper and potential use of cortisol as an indicator of stress. *J. Fish. Res. Bd Can.*, **35**, 1124–9.

Selye, H. (1950) Stress and the general adaptation syndrome. *Br. Med. J.*, **1**, 1383–92.

Sinha, N., Lal, B. and Singh, T.P. (1991) Pesticide-induced changes in circulating thyroid hormones in the freshwater catfish *Clarius batrachus*. *Comp. Biochem. Physiol.*, **100C**, 107–10.

Spotte, S. and Anderson, G. (1989) Plasma cortisol changes in seawater-adapted mummichogs (*Fundulus heteroclitus*) exposed to ammonia. *Can. J. Fish. aquat. Sci.*, **46**, 2065–9.

Sumpter. J.P., Dye, H.M. and Benfey, T.J. (1986) The effects of stress on plasma ACTH, αMSH and cortisol levels in salmonid fish. *Gen. comp. Endocrinol.*, **62**, 377–85.

Swift, D.J. (1981) Changes in selected blood component concentrations of rainbow trout, *Salmo gairdneri* Richardson, exposed to hypoxia in sublethal concentrations of phenol or ammonia. *J. Fish Biol.*, **19**, 45–61.

Tam, W.H., Birknett, L. Makaran, R., Payson, P.D., Whitney, D.K. and Yu, C.K.C.

(1987) Modification of carbohydrate metabolism and liver vitellogenin function in brook trout (*Salvelinus fontinalis*) by exposure to low pH. *Can. J. Fish. aquat. Sci.*, **44**, 630–35.

Tam, W.H., Fryer, J.N., Ali, I., Dallaire, M.R. and Valentine, B. (1988) Growth inhibition, gluconeogenesis and morphometric studies of the pituitary and interrenal cells of acid-stressed brook trout *Salvelinus fontinalis*. *Can. J. Fish. aquat. Sci.*, **45**, 1197–1211.

Thomas, P., Woodin, B.R. and Neff, J.M. (1980) Biochemical responses of the striped mullet *Mugil cephalus* to oil exposure. 1. Acute responses interrenal activations and secondary stress responses. *Mar. Biol.*, **59**, 141–9.

Thomas, P. Wofford, H.W. and Neff, J.M. (1981a) Biochemical stress responses of striped mullet (*Mugil cephalus* L) to fluorine analogs. *Aquat. Toxicol.*, **1**, 329–42.

Thomas, P., Carr, R.S. and Neff, J.M. (1981b) Biochemical responses of mullet *Mugil cephalus* and polychaete worms *Nereis virens* to pentachlorophenol, in *Biological Monitoring of Marine Pollutants* (eds F.J. Vernberg, A. Calabrese, F.P. Thurberg and W.B. Vernberg), Academic Press, New York, pp. 73–104.

Thomas, R.E. and Rice, S.D. (1987) Effect of water-soluble fraction of Cook Inlet crude oil on swimming performance and plasma cortisol in juvenile coho salmon (*Oncorhynchus kisutch*). *Comp. Biochem. Physiol.*, **87C**, 177–80.

Tomasso, J.R., Davis, K.B. and Simco, B.A. (1981) Plasma corticosteroid dynamics in channel catfish (*Ictalurus punctatus*) exposed to ammonia and nitrite. *Can. J. Fish. aquat. Sci.*, **38**, 1106–12.

Tripp, R.A., Maule, A.G., Schreck, C.B. and Kaattari, S.L. (1987) Cortisol mediated suppression of salmonid lymphocyte responses *in vitro*. *Develop. comp. Immunol.*, **11**, 565–76.

Van der Boon, J. van der Thillart, G.E.E.J.M and Addink, A.D.F. (1991) The effects of cortisol administration on intermediary metabolism in teleost fish. *Comp. Biochem. Physiol.*, **100A**, 47–53.

Vermette, M.G. and Perry, S.F. (1987) The effects of prolonged epinephrine infusion on the physiology of the rainbow trout, *Salmo gairdneri*. II. Branchial solute fluxes. *J. exp. Biol.*, **128**, 255–67.

Wedemeyer, G.A., Barton, B.A. and McLeay, D.J. (1990) Stress and acclimation, in *Methods for Fish Biology* (eds C.B. Schreck and P.B. Moyle), American Fisheries Soc., Bethesda, MD, pp. 451–89.

Wendelaar Bonga, S.E., van der Meij, J.C.A. and Flik, G. (1984a) Prolactin and acid stress in the teleost *Oreochromis* (formerly *Sarotherodon) mossambicus. Gen. comp. Endocrinol.*, **55**, 323–32.

Wendelaar Bonga, S.E., van der Meij, J.C.A. van der Krabben, W.A.W.A. and Flik, G. (1984b) The effect of water acidification on prolactin cells and pars intermedia PAS-positive cells in the teleost fish *Oreochromis* (formerly *Sarotherodon) mossambicus* and *Carassius auratus. Cell Tissue Res.*, **238**, 601–9.

Wendelaar Bonga, S.E., Balm, P.H.M. and Flik, G. (1988) Control of prolactin secretion in the teleost *Oreochromis mossambicus*: effects of water acidification. *Gen. comp. Endocrinol.*, **72**, 1–12.

Wendelaar Bonga, S.E., Flik, G. van der Velden, J.A. and Kolar, Z. (1990) Prolactin cell activity and sodium balance in the acid tolerant mudminnow *Umbra pygmaea* in acid and neutral water. *Gen. comp. Endocrinol.*, **78**, 421–32.

Whitehead, C. and Brown, J.A. (1989) Endocrine responses of brown trout, *Salmo trutta* L to acid, aluminium and lime dosing in a Welsh hill stream. *J. Fish Biol.*, **35**, 59–71.

Witters, H.E., Van Puymbroeck, S. and Vanderborght, O.L.J. (1991) Adrenergic responses to physiological disturbances in rainbow trout, *Oncorhynchus mykiss*, exposed to aluminium at acid pH. *Can. J. Fish. aquat. Sci.*, **48**, 414–20.

Wood, C.M., Simons, B.P., Mount, D.R. and Bergman, H.L. (1988) Physiological evidence of acclimation to acid/aluminium stress in brook trout (*Salvelinus fontinalis*). 2. Blood parameters by cannulation. *Can. J. Fish. aquat. Sci.*, **45**, 1597–1605.

Wright, P.A., Perry, S.F. and Moon, T.W. (1989) Regulation of hepatic gluconeogenesis and glycogenolysis by catecholamines in rainbow trout during environmental hypoxia. *J. exp. Biol.*, **147**, 169–88.

Yadav, A.K. and Singh, T.P. (1986) Effect of pesticide on circulating thyroid hormone levels in the freshwater catfish *Heteropneustes fossilis* (Bloch). *Env. Res.*, **39**, 136–42.

Ye, X., Randall, D.J. and He, X. (1991) The effect of acid water on oxygen consumption, circulating catecholamines and blood ionic and acid base status in rainbow trout (*Salmo gairdneri*, Richardson). *Fish Physiol. Biochem.*, **9**, 23–3.

Chapter twelve

Branchial mechanisms of acclimation to metals in freshwater fish

D.G. McDonald and C.M. Wood

12.1 INTRODUCTION

In most aquatic habitats the concentrations of trace metals are below threshold for detectable effects upon fish species. Threshold levels range from < 1 to > 1000 μg l^{-1} depending upon the toxicity of the metal (Spry *et al.*, 1981). However, through a variety of human activities, such as combustion of fossil fuels, pulp and paper production, cement production, and extractive metallurgy, levels of metals have increased in aquatic ecosystems in a number of industrialized areas of the world (Stumm and Morgan, 1981). Metals increase either directly, as a result of atmospheric deposition and waste water discharge and runoff (e.g. Pb, Hg, Cd, Cu, and Zn), or indirectly, through increased solubilization and mobilization from sediments owing to acidification (e.g. Al and Fe). While both marine and freshwater ecosystems are threatened, soft fresh waters that are poorly buffered, (i.e. low in $Ca(HCO_3)_2$) are particularly sensitive because they are the only aquatic habitats prone to acidification.

Metal toxicity to freshwater organisms, particularly in the context of environmental acidification, has previously been reviewed in depth (e.g. Spry *et al.*, 1981; Campbell and Stokes, 1985; McDonald *et al.*, 1989). The metals which are of greatest concern in fresh waters are Hg, Pb, Cd, Cu, Zn, Al, Mn and Cr (approximately in order of decreasing toxicity). Although the toxicity of each of these has been studied quite thoroughly, sublethal physiological effects have only been studied in detail for Al (at environmentally relevant pHs, 4.0–5.5), Cd, Cu, and Zn (reviews: McDonald *et al.*, 1989; Wood, 1992).

Fish Ecophysiology. Edited by J. Cliff Rankin and Frank B. Jensen. Published in 1993 by Chapman & Hall, London. ISBN 0 412 45920 5.

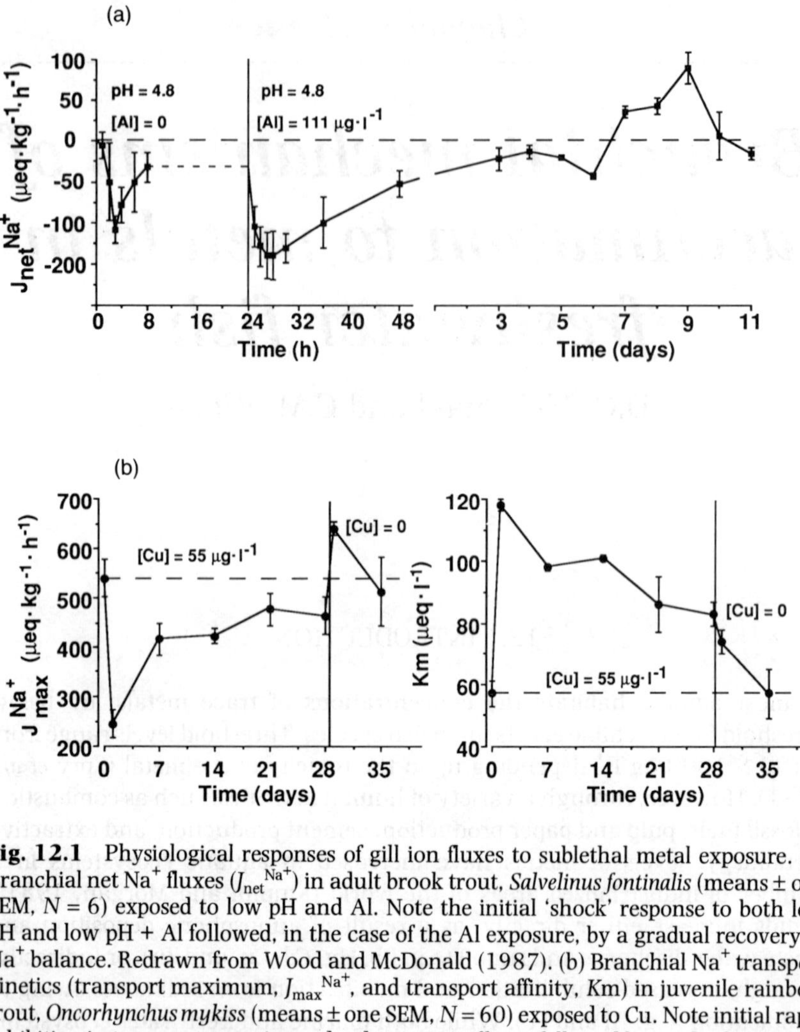

Fig. 12.1 Physiological responses of gill ion fluxes to sublethal metal exposure. (a) Branchial net Na^+ fluxes ($J_{net}^{Na^+}$) in adult brook trout, *Salvelinus fontinalis* (means ± one SEM, $N = 6$) exposed to low pH and Al. Note the initial 'shock' response to both low pH and low pH + Al followed, in the case of the Al exposure, by a gradual recovery in Na^+ balance. Redrawn from Wood and McDonald (1987). (b) Branchial Na^+ transport kinetics (transport maximum, $J_{max}^{Na^+}$, and transport affinity, *Km*) in juvenile rainbow trout, *Oncorhynchus mykiss* (means ± one SEM, $N = 60$) exposed to Cu. Note initial rapid inhibition in transport capacity and reduction in affinity followed by gradual recovery. Redrawn from Laurén and McDonald (1987a).

Less extensive data have been reported for Cr (Van der Putte *et al.*, 1982) and Mn (Reader *et al.*, 1988).

In spite of some limitations in the available data, one common theme that emerges is the central importance of ionoregulatory disturbances to the toxic mechanism of the metals. The main, or at least initial, target of the metals is the gills, and the disturbances are such to suggest that both gill transport and barrier functions have become impaired. Furthermore, the disturbances are usually characterized by an initial 'shock' phase of fairly short duration (usually less than 2 days for sublethal exposures) where the disturbances

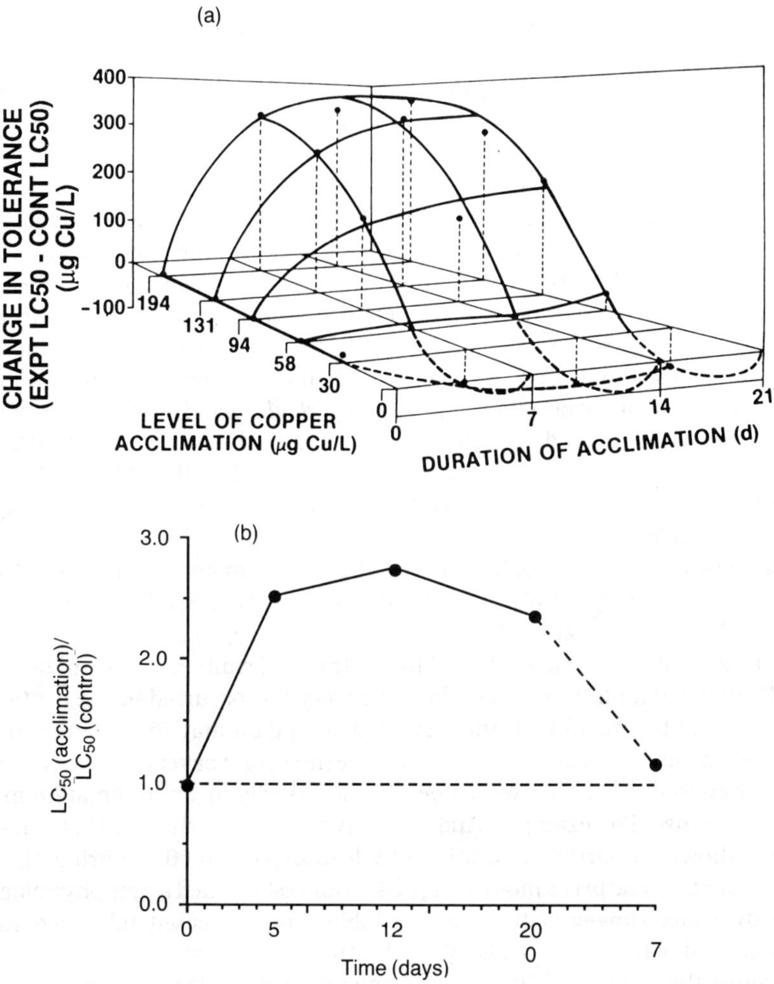

Fig. 12.2 Acclimation to metals. (a) Change in lethal level of copper for juvenile rainbow trout acclimated to any combination of copper from 0 to 194 $\mu g\ l^{-1}$ and time from 0 to 21 days. Note that copper tolerance reached a plateau in about a week and did not increase appreciably with increasing [Cu]. Reproduced with permission from Dixon and Sprague (1981). (b) Change in lethal level of zinc in juvenile rainbow trout acclimated to Zn (~ 2 mg l^{-1}) for 20 days. Note that zinc tolerance reached a plateau (~ 2.5 × LC_{50} of control fish) in about 1 week and with return to clean water for 7 days the tolerance is lost. Redrawn from Bradley *et al.* (1985).

develop fairly rapidly, and a longer-term 'recovery' phase where the disturbances diminish gradually. Examples of this phenomenon for sublethal Al and Cu are shown in Fig. 12.1. The animal may, in some instances, fully recover in the continued presence of the metal, but more usually it comes into a new physiological steady state.

Applying a broad definition of acclimation (Fry, 1971), this recovery would indicate that the animal had 'acclimated' to the metal. However, in the toxicology literature, **acclimation** has typically been used in a more restricted sense to mean an increased tolerance of an elevated, usually lethal, concentration of a toxicant arising from chronic exposure to a sublethal concentration of that toxicant. The criteria for this type of acclimation are either a reduction in the magnitude of physiological disturbances when challenged with a higher concentration of the toxicant (e.g. Audet and Wood, 1988; McDonald and Milligan, 1988; Wood *et al.*, 1988b,c; Walker *et al.*, 1991), an increase in the LT_{50} (time to 50% mortality at a single lethal concentration, e.g. Kito *et al.*, 1982b; McDonald *et al.*, 1991b) or, more rigorously, an increase in the LC_{50} (estimated concentration of the metal required to kill 50% of the population in a given time, e.g. Bradley *et al.*, 1985). Metals for which acclimation has been demonstrated according to these criteria in freshwater fish are aluminium (Orr *et al.*, 1986; Wood *et al.*, 1988b,c; McDonald *et al.*, 1991b; Reid *et al.*, 1991; Walker *et al.*, 1991), copper (Dixon and Sprague, 1981; Buckley *et al.*, 1982; McCarter and Roch, 1983, 1984), cadmium (Pascoe and Beatty, 1979; Kito *et al.*, 1982b; Duncan and Klaverkamp, 1983; Benson and Birge, 1985) and zinc (Sprague, 1970; Bradley *et al.*, 1985). Examples of Cu and Zn acclimation are shown in Fig. 12.2.

In this review we have adopted the restricted definition of acclimation, but will employ data from studies where recovery has occurred in the continued presence of the metal with the view that such data may indicate the source of the increased tolerance to the metal. One important caveat, however, is that acclimation in the broad sense does not necessarily mean acclimation in the narrow sense. For example, Audet and Wood (1988) showed that rainbow trout showed a partial restoration of gill transport function during chronic exposure to acidic pH (3 months at pH 4.8) and established a new physiological steady state. However, these fish exhibited no increased tolerance when challenged with a more acidic pH (pH 4.0).

Since the initial and, in many instances, primary target of metals is the gills, it follows that increased tolerance may originate from specific biochemical, physiological and/or morphological adaptations centred on the gills. Effects of metals on gill structure and function are now widely reported, but it is not always clear whether such effects represent damage or the repair and compensatory alterations from which increased tolerance arises. Since the distinction is not always clear, it is worthwhile reviewing both phenomena.

12.2 EFFECTS OF METALS ON GILL MORPHOLOGY

The branchial epithelium is an extremely sensitive, highly dynamic and metabolically active tissue, responsible not only for gas exchange, but also ion regulation, acid–base regulation and nitrogenous waste excretion. The mitochondrion-rich chloride cells, which make up about 10% of the volume of gill cells (Mueller *et al.*, 1991), play a pre-eminent role in ion transport; their apical exposure and ion transport intensity are closely correlated (Perry and Laurent, 1989; Laurent and Perry, 1990; Goss *et al.*, 1991). The mucous cells are less numerous than the chloride cells (~ 1% of volume of gill cells; Mueller *et al.*, 1991), but are capable of producing substantial quantities of mucus when animals are stressed.

The morphological effects of metal exposure have been comprehensively reviewed by Mallatt (1985) and fall into two broad categories: accumulated damage and compensatory responses. At moderate levels of exposure, the damage consists of separation of epithelial layers, tissue oedema, and clubbing of lamellae, while at more severe levels, tissue necrosis and rupture and fusion of secondary lamellae become more prominent. As Mallatt concluded, the necrosis and cell rupture "are amongst the most dose-dependent of branchial lesion types, as they are clearly more often reported under lethal than sub-lethal conditions". This physical damage can clearly explain the increases in ionic permeability and reductions in transport function commonly seen during the early phases of metal exposures (Section 12.3), as well as some of the short-term defence mechanisms which quickly ensue (Section 12.4). The later compensatory responses include hypertrophy and hyperplasia of mucous cells and chloride cells, and a general thickening of the filamental and lamellar epithelia. These responses, which appear to be associated with the repair of gill damage, probably play important roles in recovery and acclimation (Section 12.5).

A particularly dramatic example of the nature and the time course of these damage/repair and compensatory processes is that for Al acclimation shown in Fig. 12.3 (Mueller *et al.*, 1991). Here, brook trout were exposed to 75 μg l^{-1} Al at pH 5.2 for up to 24 days. The damage reached a peak at day 4 (Fig. 12.3(b)) and consisted of necrosis, hyperplastic epithelial cells separating from the basal lamina, and extensive lamellar fusion. The fusion was so severe that lamellar surface area was reduced by as much as 75%. Nonetheless, fish survived and subsequently showed progressive repair of the gills. By day 13 (Fig. 12.3(c)), discrete lamellae were once again visible, mucous cells remained hyperplastic and hypertrophic, and there was still some residual lifting of the epithelium. Identifiable chloride cells were reduced to about 4% compared with 10% in controls (Fig. 12.3(a)), but undifferentiated cells, which are believed to be immature chloride cells, made up 24% of the epithelium, compared with only 8% in controls. By day 13, aluminium tolerance had increased, for when these fish were exposed to lethal Al challenge they survived about twice as long as controls (McDonald *et al.*, 1991b). By day 24 of continued Al exposure (Fig. 12.3(d)), the gills appeared more normal, with the exception of hypertrophic mucous cells and chloride cells, and there were indications of chloride cell hyperplasia. Observations over a much longer period (147 d) in a similar exposure confirmed this chronic proliferation of chloride cells (Tietge *et al.*, 1988).

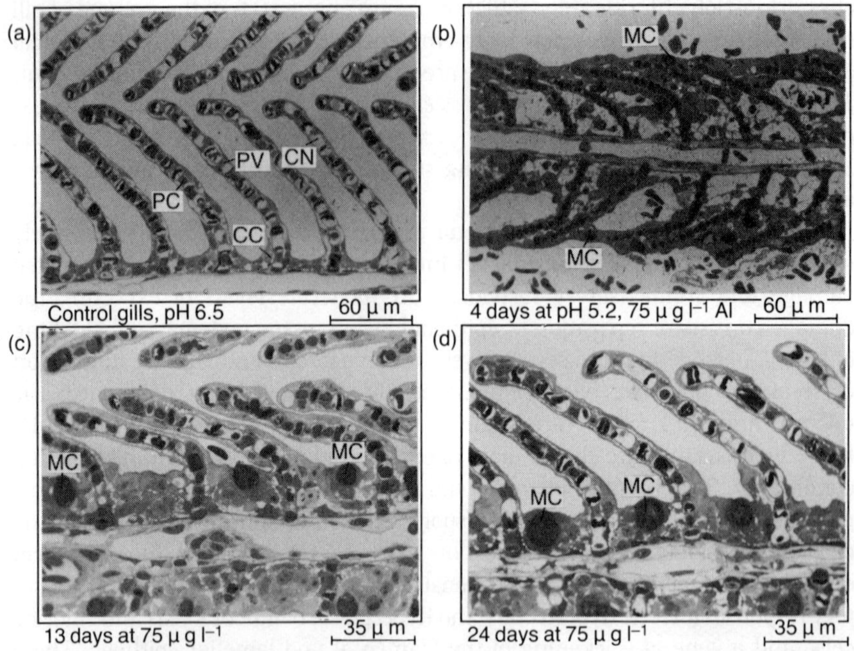

Fig. 12.3 Morphological responses of brook trout gills to aluminium exposure (micrographs from Mueller *et al.*, 1991). (a) Normal gills from control trout, maintained at pH 6.5. (b) Gills from trout exposed to Al (pH 5.2, 75 μg l^{-1} nominal) for 4 days. Note detachment of tissue from the basal lamina and the extensive lamellar fusion. Necrotic and degenerating cells fill the interlamellar regions. (c) Gills from trout exposed to Al for 13 days. Note discrete lamellae, hyperplasia on primary filament and hypertrophic mucous cells (dark staining cells on primary filament). (d) Gills from trout exposed to Al for 24 days. Note the extensive repair of secondary lamellae, and hypertrophic mucous cells. CC, chloride cell; PC, pillar cell; PV, pavement cell; CN, capillary network; MC, mucous cell.

Whether for other metals, the damage/repair process is as dramatic is not known. While there is an extensive literature on the effects of metals on gill morphology, observations have generally been taken at a single point during the exposure. There is a clear need for studies that document the changes with time and attempt to relate the changes to the development of metal tolerance and the restoration of physiological processes. Our impression, however, is that sublethal aluminium has rather more severe morphological effects than those occurring during acclimation to other metals. Nonetheless, it is our view that a necessary part of the acclimation process is the initial cellular damage. Such damage then initiates a repair process characterized by a considerable increase in mitotic activity and a rapid change-over in cell components. It is well established in vertebrate epithelial cell populations that "under unusual circumstances, such as during a period of cell damage, these populations are capable of recovering by developing an abnormally high mitotic rate. Therefore, cellular homeostasis is very much based on the control of mitosis" (Conte, 1980). In other words, without the damage to act as a signal, cellular turnover and renewal would be very slow (estimated half-time is 16 days for freshwater fish gills; Conte and Lin, 1967). Evidence to support the conclusion that damage is necessary for acclimation is the finding that the absence of acclimation to low pH (Audet *et al.*, 1988; Audet and Wood, 1988) was accompanied by very little in the way of damage to the gills (Audet and Wood, unpublished data) despite the development of a physiological disturbance.

12.3 'SHOCK' PHASE

The morphological damage phase described above roughly corresponds to the physiological 'shock' phase. The latter is characterized by inhibition of ion uptake at low metal levels, with stimulation of ionic efflux at higher metal levels thereby leading to net ion losses (Al, Wood and McDonald, 1987 (Fig. 12.1(a)); Cu, Laurén and McDonald, 1987a (Fig. 12.1(b)); Zn, Spry and Wood, 1985; Cd, Reid and McDonald, 1988). A direct poisoning of branchial ion-transport ATPases (Stagg and Shuttleworth, 1982; Staurnes *et al.*, 1984; Laurén and McDonald, 1987b; Verbost *et al.*, 1988) is implicated in the inhibition of ionic uptake, whereas disruption and widening of tight junctions (which normally seal the intercellular diffusion pathways) are implicated in the increased efflux (e.g. Freda *et al.*, 1991). The effects of Cu, Al and Cr are directed toward disturbing Na^+ and Cl^- balance, whereas Cd, Zn and Mn are directed mainly towards disrupting Ca^{2+} balance.

Respiratory effects have also been reported, but in most circumstances appear to require a much higher toxicant threshold than for effects on ionoregulation. In fact, for most metals, the levels needed to induce marked disturbances in respiratory gas exchange, at least in resting fish, are well above those likely to occur in field situations. Al is an important exception, and has been shown to inhibit gas exchange at environmentally realistic levels (Rosseland, 1980; Neville, 1985; Walker *et al.*, 1988; Wood *et al.*, 1988a,c; Playle *et al.*, 1989; Witters *et al.*, 1991). This respiratory toxicity of Al is dependent upon water pH, and is seen clearly only under moderately acidic conditions (pH > 5.0). At lower pH, ionoregulatory effects of Al predominate, as with other metals. Both the chemical speciation and physical solubility of Al are critically dependent on water pH. Playle and Wood (1991) have recently presented evidence that respiratory toxicity may be associated with precipitation of neutral $Al(OH)_3$ in the more alkaline gill microenvironment, whereas ionoregulatory toxicity is likely to be associated in the traditional way with the binding of charged Al species to gill ligands.

It is important to note that most studies have been performed on resting fish. Relatively small inhibitory effects on respiratory capacity at rest (e.g. due to mucus and/or increased diffusion distance) may become much more pronounced when the fish are forced to exercise. For example, while routine O_2 consumption and body lactate levels of juvenile rainbow trout were only marginally altered on the first few days of sublethal Al exposure, there was a 25–30% reduction in maximum O_2 uptake and a 10–15% reduction in maximum swimming speed (R.W. Wilson, H.L. Bergman, and C.M. Wood, unpublished observations). Similar effects were seen during sublethal exposure to relatively low levels of Cu (Waiwood and Beamish, 1978a).

12.4 BRANCHIAL DEFENCES AGAINST ACUTE METAL TOXICITY

One of the first lines of defence against metal exposure is secretion of mucus by the gills. Although there is now some doubt as to whether there is a mucus coating on the secondary lamellae under normal (i.e. non-stress) conditions (Handy and Eddy, 1991), there is general agreement (e.g. Varanasi and Markey, 1978; Lock and van Overbeeke, 1981; Eddy and Fraser, 1982; Miller and Mackay, 1982; Handy and Eddy, 1989a) that production of mucus is an important mechanism for protecting gill tissues from toxic metals. One of the few studies to examine mucus secretion directly in relation to metal exposure (in addition to histological examination of the gills) is that of Lock and van Overbeeke (1981), who showed that Hg, particularly in the ionic form, was a strong stimulus to mucus secretion in rainbow trout. Indeed, mucus release was a more sensitive indicator of Hg exposure than mucus content (via analysis of sialic acid residues) of gills, and this in turn was a more sensitive indicator than the number of mucocytes in gills. The importance of this secretion was illustrated recently in studies by Playle and Wood (1989, 1991) on Al-exposed rainbow trout. These authors showed that aluminium binding by gill tissue could account for only 10–15% of that extracted from the water passing over the gills. Although mucus secretion was not measured in this study, it seems likely that the bulk of the Al extracted from the water was bound by mucus secreted by the gills and quickly sloughed off. Studies with body mucus have confirmed this ability of mucus to selectively bind a number of different metals (McKone *et al.*, 1971; Varanasi and Markey, 1978; Miller and MacKay, 1982; Part and Lock, 1983; Handy *et al.*, 1989). Indeed Part and Lock (1983) were able to show that mucus not only bound Cd and Hg, but also significantly retarded their rate of diffusion.

The protective action of mucus in keeping metals away from the epithelial surface appears incontrovertible. However, the other function commonly attributed to mucus secretion, that of retarding electrolyte loss, is less clear cut. Mucus certainly inhibits O_2 diffusion (Ultsch and Gros, 1979), thereby contributing to the respiratory effects outlined above, but neither Marshall (1978) nor Part and Lock (1983) could find any evidence that mucus lowers the diffusion rates of Na^+, Cl^-, or Ca^{2+} *in vitro*. A more likely scenario, investigated by Shephard (1982) and supported by the recent measurements of Handy and Eddy, (1989b), is that the presence of mucus creates an unstirred layer with much higher concentrations of major electrolytes in the matrix next to the epithelial surface than next to the dilute external water. The presence of this thick standing gradient, in contrast to the precipitous gradient which would occur in the absence of mucus, would be expected to both retard ion loss and facilitate active ion uptake across the epithelium.

Reduction in the rate of ion loss across the gills, whether or not due to mucus secretion, certainly occurs and is an important second line of defence against metal exposure. At least some of this reduction can be attributed to a reduction in the outward diffusion gradient because of depression of plasma ion levels, but there is also evidence of a direct reduction in the ionic permeability of the gills (Laurén and McDonald, 1986). This is quite a rapid response, and as such probably occurs while the gill damage is still accumulating. Its precise physiological and ultrastructural basis is unknown. In addition to mucus secretion, the lamellar fusion and clubbing characteristic of severe gill damage may serve as a protective response to diminish the amount of vulnerable gill surface area and increase the diffusion distance, as suggested by Mallatt (1985). Osmotic swelling of the epithelial cells may also increase the blood-to-water diffusion distance, and has been suggested to mechanically reduce the permeability of paracellular diffusion channels (McDonald and Rogano, 1986). Finally, hormonal mechanisms may act specifically to stabilize the tight junctions on the paracellular channels, to stabilize the binding of protective divalent cations (Ca^{2+}, Mg^{2+}) on the external surface, and to promote mucus secretion. Prolactin, which is known to be mobilized in response to ionic losses in freshwater fish, is a prime candidate for all these functions, though a variety of other hormones may also be involved (Wendelaar Bonga and Balm, 1989).

12.5 RECOVERY AND THE ORIGIN OF METAL TOLERANCE

If the fish survives the initial shock phase, there then follows a phase of compensation and repair which occurs in the continued presence of the toxicant and appears to be intimately associated with the process of acclimation. At a morphological level, the gills return to a more normal appearance, though with generally persistent hypertrophy and hyperplasia of mucous and chloride cells, and a general thickening of lamellar and filamental epithelia (Mallatt, 1985). At a physiological level, ionic uptake rates tend to recover, ionic loss rates remain reduced (Fig. 12.1), and there is a partial or complete recovery of internal physiology. At the same time, the animal becomes more resistant to acute challenge with a more severe level of the toxicant. The precise time course may vary with the metal, the species, the water chemistry

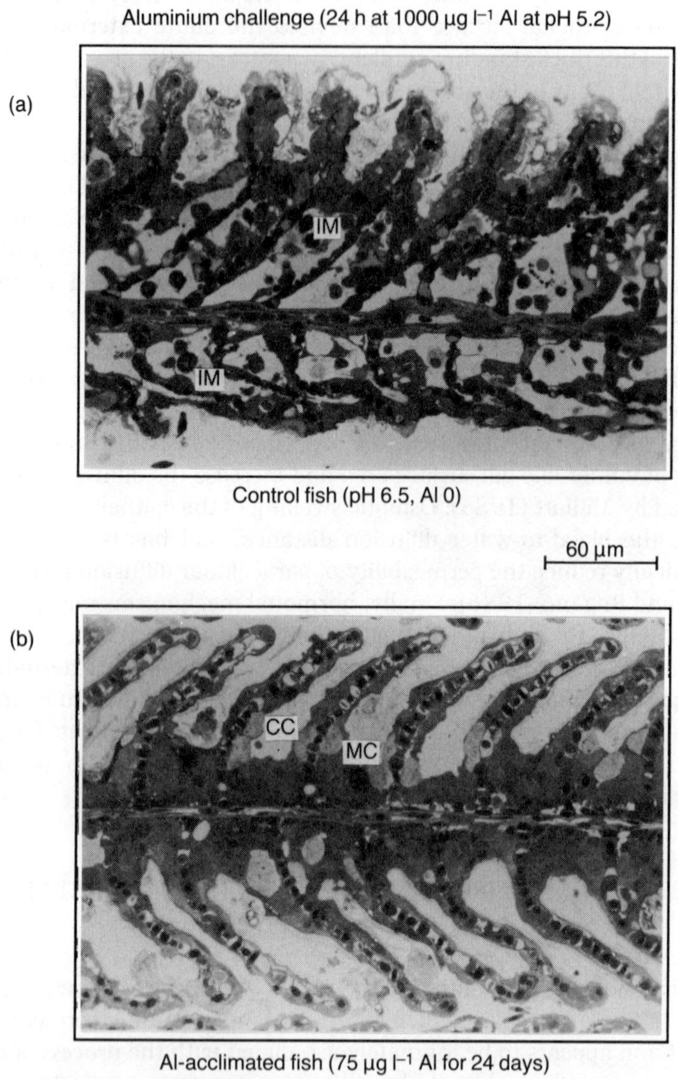

Fig. 12.4 Morphological responses of brook trout gills to aluminium challenge (24 h at pH 5.2 and 1000 μg l^{-1} Al; micrographs from Mueller *et al.*, 1991). (a) Gills from control fish (pH 6.5, 0 Al). Note extensive lifting of filament epithelium and detachment of lamellar epithelium from basal lamina. White blood cells, immature mucous cells (IM) and necrotic debris fill the fluid-filled interlamellar regions. (b) Gills from Al-acclimated fish (24 days at pH 5.2 and 75 μg l^{-1} Al). Note that gill damage is much less than in controls. There is hyperplasia of the primary filament, and proliferation and hypertrophy of chloride cells (CC) and mucous cells (MC).

and the temperature, but in general the development of acclimation is surprisingly rapid. In most studies, there was typically a twofold increase in LC_{50} relative to controls in about 5–7 days of exposure, followed by a plateau (Cu, Dixon and Sprague, 1981 (Fig. 12.2(a)); Zn, Bradley *et al.*, 1985 (Fig. 12.2(b)); Cd, Duncan and Klaverkamp, 1983, Klaverkamp and Duncan, 1987; Al, Orr *et al.*, 1986), though McDonald *et al.*, (1991b) found that acclimation of brook trout to Al was not significant until day 13, and continued to increase to day 24. Benson and Birge (1985), working on fathead minnow *Pimephales promelas*, also found a longer period for acclimation to Cd (14 days), but interestingly, loss of acclimation upon return to clean water occurred within 7 days. Bradley *et al.* (1985) also reported that acclimation (to Zn) was lost within 7 days (Fig. 12.2(b)).

The development of metal tolerance, for it to be centred at the gills, must involve adaptations that reduce the toxic impact of the metal on branchial morphology and function. This phenomenon is dramatically illustrated in Fig. 12.4, which compares gills from Al-acclimated and non-acclimated brook trout 24 h into a lethal Al challenge (~ 1000 $\mu g\ l^{-1}$ at pH 5.2 for 48 h). The extent of branchial damage is much less in the acclimated animals (and confined to some lifting of the lamellar epithelium and dilation of the lymphoid spaces) than in the non-acclimated fish, even though both fish had survived up to the point of sampling. Rates of ion loss were correspondingly lower in the acclimated animals, indicating a reduced physiological impact on the gills. Based, in part, on earlier discussions of metal tolerance (Bryan, 1976; Coombs, 1980; Simkiss, 1984), at least three mechanisms can be proposed to explain this phenomenon: (1) alterations to the barrier properties of the tissue that act to decrease the net rate of metal entry, (2) an increase in the storage and detoxification of the metal once it has entered the gill tissue, and (3) an increase in resistance of metal-sensitive branchial processes to metal poisoning.

12.6 INCREASED TOLERANCE VIA DECREASED METAL ACCUMULATION BY THE GILLS

As Bryan (1976) has noted, metal-tolerant marine invertebrates are less permeable to Zn, Cd and Mn than intolerant populations, but more permeable to Cu. Interestingly, the limited data available on fish suggest a similar pattern.

In rainbow trout the rate of accumulation of zinc by gill tissue was slower in acclimated fish (12 days at 2.0 $mg\ l^{-1}$ Zn) than in control fish (Bradley *et al.*, 1985) when both groups were exposed to lethal concentrations of Zn (Zn $\geq$ 5.0 $mg\ l^{-1}$). This conclusion is based on the observation that acclimated fish lived longer in lethal Zn but at death had similar or lower gill Zn levels than control fish.

Similarly, Al-acclimated rainbow trout (3 weeks at pH 5.2 and 27 $mg\ l^{-1}$ Al) showed a 50% reduction in Al uptake by the gills when challenged (118 mg l^{-1} at pH 5.2) compared with non-acclimated fish (Reid *et al.*, 1991). McDonald *et al.* (1991b) found similar results with brook trout acclimated to Al for 3–4 weeks (75 μg l^{-1} at pH 5.2) and then challenged with higher levels of Al. In this study, mortality rate was much higher in the non-acclimated fish, and dead fish had much higher gill Al levels than survivors. For Al, at least, it is very clear that mortality is closely related to the amount of Al accumulated on the gills.

For other metals, the effects of acclimation on uptake rates have not been assessed under the circumstances where the animals were challenged with elevated metal levels. In terms of routine branchial uptake rates (i.e. uptake of metal at the acclimation level), there was a significant reduction in branchial Cd uptake in minnows, *Phoxinus phoxinus*, (assessed by exposure to ^{109}Cd for 24 h at the same concentration as the acclimation level) after 1 month at 0.3 μg l^{-1} Cd (Wicklund-Glynn and Olsson, 1991). In contrast, after 2 weeks of acclimation to 100 μg l^{-1} Cu, gill Cu uptake in coho salmon, *Oncorhynchus kisutch*, was, in fact, double that of controls (McCarter and Roch, 1984).

Thus, at present, there is reasonably good evidence that increased tolerance of Zn, Al and Cd (but not of Cu) is accompanied by a reduction in branchial accumulation of the metals. It is less clear what changes in the gills are responsible for this reduction, but two possibilities can be suggested. The first, and perhaps most obvious, possibility is an increase in the rate of mucus secretion by the gills (Section 12.3). In Al-acclimated brook trout, there was a marked hypertrophy of mucous cells, a doubling of their volume density with a 75% increase in their surface area, and a reduction in the gill Al burden over time (McDonald *et al.*, 1991b; Mueller *et al.*, 1991). However, this was not accompanied by an increase in gill sialic acid content (McDonald *et al.*, 1991b), which is probably the major cation-binding component of fish mucus (Harris *et al.*, 1973). The most detailed study of mucus production, that of Lock and van Overbeeke (1981), found no evidence that branchial mucus secretion increased over time during continued Hg exposure. Mucus secretion was extremely sensitive to Hg concentration (particularly ionic Hg), but the gill mucus secretion/production did not apparently adapt to continued Hg exposure. Thus, while mucus secretion is probably the first line of defence against metal poisoning as discussed above, it is still an open question whether acclimation enhances this secretion.

The other possibility is that acclimation induces a qualitative change in the metal-binding properties of the gill surfaces. This possibility was investigated by Reid *et al.* (1991), using a rapid radio-ligand binding assay to measure gill metal surface binding. These authors showed that after 27 days at sublethal Al, rainbow trout gills exhibited an increased binding affinity for Ca^{2+}, a decreased binding affinity for Al and a decline in the potency of Al at displacing Ca^{2+}. These changes would contribute to reducing the reactivity of aluminium at the gill surface. They could, in fact, reflect changes in the composition of the mucus, since the method of Reid *et al.* (1991) does not distinguish between the apical epithelial surface and its external mucus layer. Indeed it would be interesting to apply the mucus dialysis/competitive binding approach developed by Handy *et al.* (1989) to this question. Zuchelkowski *et al.* (1985), for example, reported increased synthesis of neutral sulfomucins (which do not contain sialic acid) in epidermal mucus in response to acid stress in catfish. Such a change would lower the metal-binding capacity of the mucus. Similarly, Varanasi *et al.* (1975) reported that both Pb and Hg induced fundamental changes in the structure of polar sites of epidermal mucus in rainbow trout *Oncorhynchus mykiss*, which persisted even after a period of depuration. If, however, the gill damage prompts an acceleration of cellular turnover (as suggested above), then alterations to membrane composition (charge characteristics of integral membrane proteins etc.) could also be expected. It is worth noting in this regard the report by Bolis *et al.* (1984) of major changes in both the phospholipid and fatty acid composition of brown trout gill membranes during acid stress (pH 4–4.5 for 100–120 h).

12.7 INCREASED TOLERANCE VIA INCREASED METAL STORAGE AND DETOXIFICATION IN GILL TISSUE

In reviewing metal dynamics of marine organisms, mainly invertebrates, from contaminated environments, Bouquegneau *et al.* (1984) concluded that heavy metals can be stored and detoxified by a compartmentation process either within membrane-limited vesicles or by binding to specific proteins (metallothioneins and other metal-binding proteins).

In fact, the most often stated reason for increased metal tolerance in fishes is the enhanced synthesis of metallothioneins (MT) (e.g. Klaverkamp *et al.*, 1984; Chapman, 1985; Hamilton and Mehrle, 1986; Hodson, 1988; Fu *et al.*, 1990; Hogstrand and Haux, 1991). MTs are exceptionally heat stable, cytosolic proteins, MW 6000–7000, which have a high sulphur (thiol) content as a result of cysteine residues. The latter are responsible for the metal binding: 6–7 metal atoms per mole of protein. Since radioimmunoassays for fish MTs have been developed only recently (Hogstrand *et al.*, 1989; Hogstrand and Haux, 1990), most measurements to date have been indirect – via analysis of heat-stable fractions, either for sulphur content (via pulse polarography or by measurement of acid soluble thiols (AST) which also include free cysteine and glutathione) or for their ability to bind certain cations (e.g. ^{109}Cd). The sites of greatest induction of MT are the liver and kidneys, but this protein (or others like it) is also found in the gills of fish, and increased synthesis has been reported in response to metal exposure (review: Hamilton and Mehrle, 1986). The hypothesis is that metal entering the gills would be sequestered and rendered non-toxic by binding to MTs or other metal-binding proteins. At least two studies have shown a strong quantitative correlation between the time course of induction of gill MT and the development of metal tolerance (Zn, Bradley *et al.*, 1985; Cd, Benson and Birge, 1985). However, many others have found rather different responses, or no MT response at all. No common or clear-cut response pattern has yet emerged. Consequently each metal is considered separately below.

Copper

Although copper in the water is a potent stimulant of hepatic MT in both a concentration- and time-dependent fashion (McCarter and Roch, 1983), there appears to be little, if any, change in metal-binding ligands in the gills. After 1 month of Cu exposure (55 μg l^{-1}), gills of rainbow trout showed no increase in either AST, glutathione, cysteine, or ^{109}Cd-binding proteins (Laurén and McDonald, 1987b). Furthermore, although the total amount of copper in the gills increased very little with copper exposure, the exchangeable copper pool increased from 25% after 24 h of exposure to 52% of the total gill copper after 7 days and remained stable thereafter. If there were any increase in the concentration of any Cu-binding ligand, then the exchangeable pool should decrease rather than increase.

Zinc

Zinc, in contrast to Cu, does apparently stimulate gill metal-binding ligands, although responses vary. In rainbow trout, exposure to Zn > 0.7 mg l^{-1} for 5 days led to a doubling of heat-stable, thiol-rich protein in gill (Bradley *et al.*, 1985). Exposure of white suckers *Catostomus commersoni*, to a similar Zn level for 7 days also increased apparent MT levels in the gills (Klaverkamp and Duncan, 1987). The studies of Thomas *et al.* (1985) suggest the induction of gill MT by environmental exposure to Zn at a much lower level (0.1 mg l^{-1}). In contrast, in a later study on rainbow trout, Spry and Wood (1989) showed that exposure to 0.5 mg l^{-1} Zn caused a slight decline in gill AST, but led to a 1.7-to-5-fold increase in an HSP (heat-stable protein) fraction binding ^{109}Cd. This response was also stimulated by elevated Zn in the diet. Furthermore, the ^{109}Cd binding was not to the MT fraction but to a fraction of an apparent MW of < 3000, indicating that a non-MT binding protein was induced.

Cadmium

Again, responses vary. In an early report, Nöel-Lambot *et al.* (1978) showed that Cd exposure in eels, *Anguilla anguilla* (180 days at 13 mg l^{-1} in seawater) induced an MT-like protein in the gills which bound Cd, but did not quantify the increase. Similarly, in freshwater carp *Cyprinus carpio*, (31 days at 5 mg l^{-1} Cd), Kito *et al.* (1982a) reported a progressive increase in the Cd-binding MT fraction in gill extracts. In the fathead minnow, increases in Cd-binding protein (apparent

MT) paralleled the development of tolerance during exposure to a low level of Cd (10 μg l^{-1}; Benson and Birge, 1985). This was later corroborated in tilapia, *Oreochromis mossambicus*, exposed to 10 μg l^{-1} Cd by Fu *et al.* (1990), who also found induction of a low MW (< 3000) metal-binding ligand in the gills, similar to the observation of Spry and Wood (1989) in zinc-exposed trout. The possibility of induction of non-MT Cd-binding ligands was previously suggested by Stone and Overnell (1985). Further complicating the story are earlier observations of Thomas *et al.* (1983, 1985), Klaverkamp and Duncan (1987), and very recent data of Wicklund-Glynn and Olsson (1991). Thomas and co-workers concluded that trout chronically exposed to low levels of Cd (9–54 μg l^{-1}) manifested an increase in low MW non-MT metal-binding ligands only, whereas those exposed to low-level Zn (0.1 mg l^{-1}) exhibited MT induction. Klaverkamp and Duncan (1987) found a virtually identical pattern in the gills of white sucker exposed to much higher levels of Cd (0.7 mg l^{-1}) or Zn (0.9 mg l^{-1}) for 7 days. Wicklund-Glynn and Olsson (1991) showed that chronic exposure to extremely low Cd (up to 91 days at only 0.3 μg l^{-1}) had no effect on MT levels (assessed by pulse polarography) in the gills of minnows, even though gill Cd increased 20-fold compared with controls. The MT content of the gills (20 nmol g^{-1}) was enough to sequester all Cd taken up from the water (80 pmol g^{-1}). Nevertheless, significant levels of Cd were detected in the non-MT fraction.

Aluminium

As far as we are aware, the gills of Al-acclimated fish have not been examined for MT induction. There is very good evidence, however, that Al is sequestered in electron-dense particles, both on the apical surfaces of the gills and within the cells (Karlsson-Norrgren *et al.*, 1986a,b; Youson and Neville, 1987; Evans *et al.*, 1988; Goossenaerts *et al.*, 1988). According to Youson and Neville, the intracellular particles are membrane-bound Al precipitates (Al was identified in the particles by X-ray microanalysis) and are therefore in a detoxified form. These authors suggest that the particles resemble lysosomes which have condensed the metal after endocytosis of surface-bound Al and, in turn, function to isolate the metal prior to exocytosis. Thus a mechanism appears to exist for the branchial storage and excretion of Al. However, it is still an open question whether Al acclimation potentiates this mechanism so as to protect gill function.

12.8 INCREASED TOLERANCE VIA INCREASED RESISTANCE OF METAL-SENSITIVE PROCESSES

One of the most sensitive processes in the gills to metal poisoning is ion transport, and one of the most common responses to chronic sublethal metal exposure is an increased synthesis of ATPases. This has been demonstrated for both Na^{+}+ K^{+} ATPase (Cu, Stagg and Shuttleworth, 1982; Laurén and

McDonald, 1987b) and Ca^{2+} ATPase (Cu, Zn, Cd, Shephard and Simkiss, 1978; Zn, Watson and Beamish, 1981). In the case of Al, there is at least indirect evidence of increased synthesis; Na^+ transport activity *in vivo* measured at pH 6.5 in Al-acclimated fish (≥ 10 weeks at 150 μg l^{-1}) was over three times higher than that in fish held under the same acid soft-water conditions (pH 5.2, Ca^{2+} = 0.25 mM) but without Al (McDonald and Milligan, 1988). The cellular basis for these changes is most likely the observed hypertrophy and hyperplasia of chloride cells in the gill epithelium. This phenomenon has now been clearly observed in fish chronically exposed to sublethal Zn (Mathiessen and Brafield, 1973), Cd (Oronsaye and Brafield, 1984; Fu *et al.*, 1990), and Al (Karlsson-Norggren *et al.*, 1986a,b; Evans *et al.*, 1988; Tietge *et al.*, 1988; Mueller *et al.*, 1991). That the cellular proliferation is responsible for the increased ATPase activity is supported by observations that soft-water exposure (1–2 weeks) of rainbow trout, a treatment which leads typically to hyperplasia and hypertrophy of chloride cells with at least a doubling of their exposed surface area, is accompanied by an increase in both NaCl and Ca^{2+} transport activity (Perry and Wood, 1985; McDonald and Rogano, 1986; Perry and Laurent, 1989). However, for both Cu and Al, the adjustments were such to simply return ion transport rates more or less to normal in the continued presence of the metal (Laurén and McDonald, 1987a (Fig. 12.1(b)); McDonald *et al.*, 1991b). It is therefore unlikely that this synthesis of additional transport enzyme units is the source of the increased tolerance of fish to elevated levels of the metal.

One cannot rule out, however, that the tolerance arises through changes in the amounts of conformation of certain target proteins within the gills. Since disruption of cellular integrity through displacement of membrane Ca^{2+} is thought to be the basis of much of the initial toxic effect of metals (McDonald, 1983; McDonald *et al.*, 1989), more precise regulation of tight junction dimensions, and a higher binding affinity of cell surface ligands for Ca^{2+} would be highly adaptive changes. While direct evidence is lacking except for Al (Reid *et al.*, 1991), recent studies of the physiological basis of low pH tolerance indicate that species that resist low pH exposure are characterized by gills which have precisely these features (Freda and McDonald, 1988; Gonzalez and McDonald, 1991; McDonald *et al.*, 1991a).

12.9 CONCLUSIONS

Although there are considerable gaps in our knowledge (Section 12.10), it is now possible to propose a general model for acclimation to metals as summarized in Fig. 12.5. First, we reiterate the view that the metal level in the environment must exceed the threshold to cause branchial damage, a threshold which in turn is higher than that needed to cause a detectable physiological disturbance. The cellular damage then initiates a repair process which includes

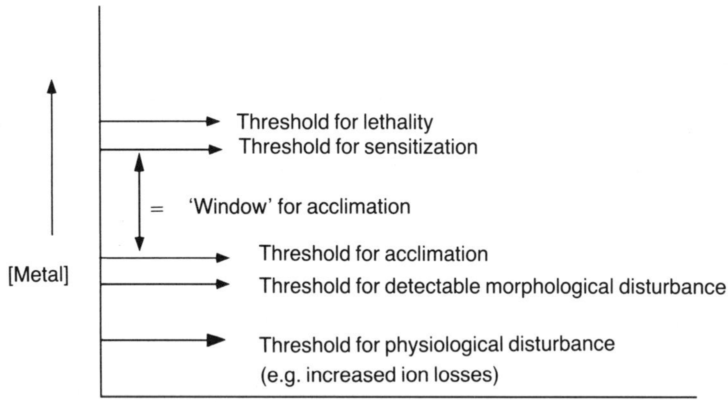

Fig. 12.5 General model for metal acclimation. See text for further details.

the expression of tolerance mechanism(s). The present evidence (Fig. 12.2) suggests that the tolerant state develops within a week or so, but once developed, tolerance does not increase further with time, nor with any increase in the metal level above the threshold concentration. Indeed, as metal levels increase, one can assume that the threshold for lethality will be approached. Before that point is reached, it is probable that the accumulated damage to the gills could be such that metal tolerance begins to decline (i.e. threshold for sensitization). The range between the thresholds for acclimation and sensitization is the window over which acclimation will occur. The probability of acclimation occurring is then a function of the width of this window. A narrow window for a particular metal means that in this case, acclimation would be an elusive phenomenon. One would then have to question its environmental relevance. Although this window has not been well defined for any metal, it is probable that each metal is characterized by a different window that is influenced to a variable extent by water chemistry variables known to affect metal toxicity, such as pH, hardness (Ca^{2+}, Mg^{2+}) and alkalinity.

Although the precise nature of the tolerance mechanism(s) has yet to be worked out, the weight of evidence points to the conclusion that the mechanism is probably different for each metal, despite the fact that the nature of the gill damage is broadly similar for all metals studied so far. Nonetheless, the tolerance mechanisms may well be related to one another for those metals that operate by similar toxic mechanisms (e.g. Zn and Cd; Cu and Al). For example, Cd acclimation has been shown to increase Zn tolerance and vice versa (Kito *et al.*, 1982b; Duncan and Klaverkamp, 1983; Thomas *et al.*, 1985; Klaverkamp and Duncan, 1987). However, fish acclimated to Cu showed increased sensitivity to Zn (Dixon and Sprague, 1981), which probably reflects the fact that these two metals have very different toxic mechanisms (Cu: Na^+ and Cl^- balance disruption; Zn: Ca^{2+} balance disruption).

12.10 FUTURE RESEARCH DIRECTIONS

Our main objective in this review has been to illustrate the complexity of the branchial mechanisms of metal tolerance in freshwater fish, as we currently understand them, and to emphasize that further research in this area is very necessary. Such research is likely to be both interesting and productive. The following are some suggestions for future research directions, although this is by no means an all-inclusive list.

There is a clear need to examine a wider range of metals which have been shown to have a toxic impact on the gills (e.g. Hg, Cr, Ni, Pb, Mn) as well as to examine in greater detail the morphological, physiological and biochemical impact on the gills of the commonly studied metals we have reviewed here (Cu, Zn, Cd, Al).

In terms of morphological impact, there is a need for more studies on the nature and time course of metal damage and repair in the gills, as well as specific studies of cellular turnover rates during acclimation, using such techniques as ^{3}H-thymidine autoradiography (e.g. Conte and Lin, 1967; Chretien and Pisam, 1986) or more simply by analysis of DNA/RNA ratios in gill tissues during repair (e.g. Houlihan *et al.*, 1989). Such studies will test the hypothesis that gill damage is required to initiate the increased rate of cellular turnover that is thought to be the basis for increased metal tolerance.

There is also a need for more studies which precisely define the window of acclimation for each metal and examine how this window is affected by water quality variables (e.g. pH, hardness, alkalinity).

An important question which has received only slight attention (Section 12.9) is the extent to which acclimation to one metal may increase tolerance to other metals, i.e. crossover. Such studies may help clarify which mechanisms are general phenomena, and which are metal-specific, as well as casting further light on which metals share common mechanisms of toxicity and tolerance.

Finally, there is a need to quantify the metabolic cost of acclimation, by evaluating effects on feeding, growth rate, conversion efficiency, swimming performance, and routine and active metabolic rates (e.g. Lett *et al.*, 1976; Waiwood and Beamish, 1978a,b). Such studies will reveal whether a fish protected against metal toxicity by acclimation has really benefited in an ecological context.

However, a detailed understanding of the nature of metal tolerance will more likely come from specific and detailed studies on gill tissues of fish prior to, during, and after completion of acclimation to metal. Such studies should include the following.

1. An examination of the impact of metal challenge (i.e. at lethal or near-lethal concentrations of the acclimation metal) on mucus secretion, rate of metal accumulation and gill morphology in acclimated and

non-acclimated fish. Such studies will test the notion that acclimation, in effect, reduces the 'reactivity' of the gills to the metal.

2. An analysis of the metal-binding properties of the gill surfaces in relation to metal acclimation. The radio-ligand binding assay developed by Reid (Reid and McDonald, 1991; Reid *et al.*, 1991) will be useful here in evaluating quantitative and qualitative changes in gill binding characteristics (i.e. affinity and capacity) for the acclimation metal as well as for other metals, and Ca^{2+} and H^{+}.
3. An analysis of qualitative changes in biochemical composition of gill tissues, including membrane lipids, metallothioneins and other metal-binding ligands, transport ATPases, and epidermal mucus. Such studies will eventually be necessary to determine precisely whether the metal is less reactive because less is accumulated, or accumulated metal is more readily stored, detoxified or excreted or gill functions are more resistant to metal poisoning.

ACKNOWLEDGEMENTS

Supported by Strategic Grants in Environmental Quality to the authors from the Natural Sciences and Engineering Research Council of Canada. We thank Dr Mary Ellen Mueller for generously supplying the photomicrographs used in Figures 12.3 and 12.4.

REFERENCES

Audet, C. and Wood, C.M. (1988) Do rainbow trout acclimate to low pH? *Can. J. Fish. aquat. Sci.*, **45**, 1399–1405.

Audet, C. and Wood, C.M. (1992). Branchial morphological and endocrine responses of rainbow trout (*Oncorhynchus mykiss*) during a long-term sublethal acid exposure in which acclimation did not occur. *Can. J. Fish. aquat. Sci.* (in press).

Audet, C., Munger, R.S. and Wood, C.M. (1988) Long-term sublethal acid exposure in rainbow trout (*Salmo gairdneri*) in soft water: effects on ion exchanges and blood chemistry. *Can. J. Fish. aquat. Sci.*, **45**, 1387–98.

Benson, W.H. and Birge, W.J. (1985) Heavy metal tolerance and metallothionein induction in fathead minnows: results from field and laboratory investigations. *Environ. Tox. Chem.*, **4**, 209–12.

Bolis, C.L., Cambria, A. and Fama, M. (1984) Effects of acid stress on fish gills, in *Toxins, Drugs, and Pollutants in Marine Animals* (eds C.L. Bolis, A. Cambria and M. Fama), Springer-Verlag, Berlin, pp. 122–9.

Bouquegneau, J.M., Martoja, M. and Truchet, M. (1984) Heavy metal storage in marine animals under various environmental conditions, in *Toxins, Drugs, and Pollutants in Marine Animals* (eds C.L. Bolis, A. Cambria and M. Fama), Springer-Verlag, Berlin, pp. 147–60.

Bradley, R.W., DuQuesnay, C. and Sprague, J.B. (1985) Acclimation of rainbow trout, *Salmo gairdneri* Richardson, to zinc: kinetics and mechanism of enhanced tolerance induction. *J. Fish Biol.*, **27**, 367–79.

Bryan, G.W. (1976) Some aspects of heavy metal tolerance in aquatic organisms, in *Effects of Pollutants on Aquatic Organisms* (ed. A.P.M. Lockwood), Cambridge Univ. Press, Cambridge, pp. 7–34.

Buckley, J.T., Roch, M., McCarter, J.A., Rendell, C.A. and Matheson, A.T. (1982) Chronic exposure of coho salmon to sublethal concentrations of copper. I. Effect on growth, on accumulation and distribution of copper, and on copper tolerance. *Comp. Biochem. Physiol.*, **72C**, 15–19.

Campbell, P.G.C. and Stokes, P.M. (1985) Acidification and toxicity of metals to aquatic biota. *Can. J. Fish. aquat. Sci.*, **42**, 2034–49.

Chapman, G.A. (1985) Acclimation as a factor influencing metal criteria, in *Aquatic Toxicology and Hazard Assessment: 8th Symp. ASTM STP 891* (ed. D.J. Hansen), American Society for Testing and Materials, Philadelphia, PA, pp. 119–36.

Chretien, M. and Pisam, M. (1986) Cell renewal and differentiation in the gill epithelium of freshwater and seawater adapted euryhaline fish as revealed by ^{3}H-thymidine autoradiography. *Biol. Cell.*, **56**, 137–50.

Conte, F.P. (1980) The ionocyte chalone: a chemical regulator of chloride cell proliferation and transformation, in *Epithelial Transport in the Lower Vertebrates* (ed. B. Lahlou), Cambridge Univ. Press, Cambridge, pp. 287–96.

Conte, F.P. and Lin, D.H.Y. (1967) Kinetics of cellular morphogenesis in gill epithelium during seawater adaptation of *Oncorhynchus* (Walbaum). *Comp. Biochem. Physiol.*, **23**, 945–57.

Coombs, T.L. (1980) Heavy metal pollutants in the aquatic environment, in *Animals and Environmental Fitness*. Vol. 1 (ed. R. Gilles), Pergamon Press, New York, pp. 283–302.

Dixon, D.G. and Sprague, J.B. (1981) Acclimation to copper by rainbow trout (*Salmo gairdneri*) – a modifying factor in toxicity. *Can. J. Fish. aquat. Sci.*, **38**, 880–88.

Duncan, D.A. and Klaverkamp, J.F. (1983) Tolerance and resistance to cadmium in white suckers (*Catostomus commersoni*) previously exposed to cadmium, mercury, zinc, or selenium. *Can. J. Fish. aquat. Sci.*, **40**, 128–38.

Eddy, F.B. and Fraser, J.E. (1982) Sialic acid and mucus production in rainbow trout (*Salmo gairdneri* Richardson) in response to zinc and seawater. *Comp. Biochem Physiol.*, **73C**, 357–9.

Evans, R.E., Brown, S.B. and Hara, T.J. (1988) The effects of aluminium and acid on gill morphology in rainbow trout, *Salmo gairdneri*. *Env. Biol. Fishes*, **22**, 299–311.

Freda, J. and McDonald, D.G. (1988) Physiological correlates of interspecific variation in acid tolerance in fish. *J. exp. Biol.*, **136**, 243–58.

Freda, J., Sanchez, D.A. and Bergman, H.L. (1991) Shortening of branchial tight junctions in acid exposed rainbow trout. *Can. J. Fish. aquat. Sci.*, **48**, 2028–33.

Fry, F.E.J. (1971) The effects of environmental factors on the physiology of fish, in *Fish Physiology*, Vol. VI (eds W.S. Hoar and D.J. Randall), Academic Press, New York, pp. 1–98.

Fu, H., Steinebach, O.M., van den Hamer, C.J.A., Balm, P.H.M. and Lock, R.A.C. (1990) Involvement of cortisol and metallothionein-like proteins in the physiological responses of tilapia (*Oreochromis mossambicus*) to sublethal cadmium stress. *Aquat. Toxicol.*, **16**, 257–70.

Gonzalez, R.J. and McDonald, D.G. (1991) The relationship between oxygen consumption and ion loss in a freshwater fish. *J. Exp. Biol.*, **163**, 317–32.

Goossenaerts, C., van Grieken, R., Jacob, W., Witters, H. and Vanderborght, O. (1988) A microanalytical study of the gills of aluminium-exposed rainbow trout (*Salmo gairdneri*). *Int. J. environ. anal. Chem.*, **34**, 227–37.

Goss, G.G., Perry, S.F., Wood, C.M. and Laurent, P. (1992) Relationships between ion and acid–base regulation at the gills of freshwater fish. *J. exp. Zool.*, in press.

Hamilton, S.J. and Mehrle, P.M. (1986) Metallothionein in fish: review of its importance in assessing stress from metal contaminants. *Trans. Am. Fish. Soc.*, **115**, 596–609.
Handy, R.D. and Eddy, F.B. (1989a) Surface absorption of aluminium by gill tissue and body mucus of rainbow trout, *Salmo gairdneri*, at the onset of episodic exposure. *J. Fish Biol.*, **34**, 865–74.
Handy, R.D. and Eddy, F.B. (1989b) The ionic composition of rainbow trout body mucus. *Comp. Biochem. Physiol.*, **93A**, 571–5.
Handy, R.D. and Eddy, F.B. (1991) The absence of mucus on the secondary lamellae of unstressed rainbow trout, *Oncorhynchus mykiss* (Walbaum). *J. Fish Biol.*, **38**, 153–5.
Handy, R.D., Eddy, F.B. and Romain, G. (1989) *In vitro* evidence for the ionoregulatory role of rainbow trout mucus in acid, acid/aluminium, and zinc toxicity. *J. Fish Biol.*, **35**, 737–47.
Harris, J.E., Watson, A. and Hunt, S. (1973) Histochemical analysis of mucous cells in the epidermis of brown trout, *Salmo trutta* L. *J. Fish Biol.*, **5**, 345–51.
Hodson, P.V. (1988) The effect of metal metabolism on uptake, disposition, and toxicity in fish. *Aquat. Toxicol.*, **11**, 3–18.
Hogstrand, C. and Haux, C. (1990) Metallothionein as an indicator of heavy-metal exposure in two subtropical fish species. *J. exp. mar. Biol. Ecol.*, **138**, 69–84.
Hogstrand, C. and Haux, C. (1991) Binding and detoxification of heavy metals in lower vertebrates with reference to metallothionein. *Comp. Biochem. Physiol.*, **100C**, 137–41.
Hogstrand, C., Olsson, P.-E. and Haux, C. (1989) A radioimmunoassay for metallothionein in fish. *Mar. Env. Res.*, **28**, 191–4.
Houlihan, D.F., Hall, S.J. and Gray, C. (1989) Effects of ration on protein turnover in cod. *Aquaculture*, **79**, 103–9.
Karlsson-Norggren, L., Dickson, W., Ljungberg, O. and Runn, P. (1986a) Acid water and aluminium exposure: gill lesions and aluminium accumulation in farmed, brown trout, *Salmo trutta*. L. *J. Fish Dis.*, **9**, 1–8.
Karlsson-Norrgren, L., Björklund, I., Ljungberg, O. and Runn, P. (1986b) Acid water and aluminium exposure: experimentally induced gill lesions in brown trout, *Salmo trutta* L. *J. Fish. Dis.*, **9**, 11–25.
Kito, H., Kazawa, T., Ose, Y., Sato, T. and Ishikawa, T. (1982a) Formation of metallothionein in fish. *Comp. Biochem. Physiol.*, **73C**, 129–34.
Kito, H., Kazawa, T., Ose, Y., Sato, T. and Ishikawa, T. (1982b) Protection by metallothionein against cadmium toxicity. *Comp. Biochem. Physiol.*, **73**, 135–9.
Klaverkamp, J.F. and Duncan, D.A. (1987) Acclimation to cadmium toxicity by white suckers: cadmium binding capacity and metal distribution in gill and liver cytosol. *Environ. Tox. Chem.*, **6**, 275–89.
Klaverkamp, J.F., Macdonald, W.A., Duncan, D.A. and Wagemann, R. (1984) Metallothionein and acclimation to heavy metals in fish: a review, in *Contaminant Effects on Fisheries* (eds W. Cairns, P.V. Hodson and J.O. Nriagu), Wiley, New York, pp. 99–113.
Laurén, D.J. and McDonald, D.G. (1986) Influence of water hardness, pH, and alkalinity on the mechanisms of copper toxicity in juvenile rainbow trout, *Salmo gairdneri*. *Can. J. Fish. aquat. Sci.*, **43**, 1488–96.
Laurén, D.J. and McDonald, D.G. (1987a) Acclimation to copper by rainbow trout: physiology. *Can. J. Fish. aquat. Sci.*, **44**, 99–104.
Laurén, D.J. and McDonald, D.G. (1987b) Acclimation to copper by rainbow trout: biochemistry. *Can. J. Fish. aquat. Sci.*, **44**, 105–11.
Laurent, P. and Perry, S.F. (1990) Effects of cortisol on gill chloride cell morphology and ionic uptake in the freshwater rainbow trout, *Salmo gairdneri*. *Cell Tissue Res.*, **259**, 429–42.

Lett, P.F., Farmer, G.J. and Beamish, F.W.H. (1976) Effect of copper on some aspects of the bioenergetics of rainbow trout (*Salmo gairdneri*). *J. Fish. Res. Bd Can.*, **33**, 1335–42.

Lock, R.A.C. and van Overbeeke, A.P. (1981) Effects of mercuric chloride and methylmercuric chloride on mucus secretion in rainbow trout. *Comp. Biochem. Physiol.*, **69C**, 67–73.

McCarter, J.A. and Roch, M. (1983) Hepatic metallothionein and resistance to copper in juvenile coho salmon. *Comp. Biochem. Physiol.*, **74C**, 133–7.

McCarter, J.A. and Roch, M. (1984) Chronic exposure of coho salmon to sublethal concentrations of copper – III. Kinetics of metabolism of metallothionein. *Comp. Biochem. Physiol.*, **77C**, 83–7.

McDonald, D.G. (1983) The effects of H^+ upon the gills of freshwater fish. *Can. J. Zool.*, **61**, 691–703.

McDonald, D.G. and Milligan, C.L. (1988) Sodium transport in the brook trout, *Salvelinus fontinalis*: effects of prolonged low pH exposure in the presence and absence of aluminum. *Can. J. Fish. aquat. Sci.*, **45**, 1606–16.

McDonald, D.G. and Rogano, M.S. (1986) Ion regulation by the rainbow trout in ion-poor water. *Physiol. Zool.*, **59**, 318–31.

McDonald, D.G., Reader, J.P. and Dalziel, T.R.K. (1989) The combined effects of pH and trace metals on fish ionoregulation, in *Acid Toxicity and Aquatic Animals* (eds R. Morris, E.W. Taylor, D.J.A. Brown and J.A. Brown), Cambridge Univ. Press, Cambridge, pp. 221–42.

McDonald, D.G., Freda, J., Cavdek, V., Gonzalez, R., and Zia, S. (1991a) Interspecific differences in gill morphology of freshwater fish in relation to tolerance of low pH environments. *Physiol. Zool.*, **64**, 124–44.

McDonald, D.G., Wood, C.M., Rhem, R.G., Mueller, M.E., Mount, D.R. and Bergman, H.L. (1991b) Nature and time course of acclimation to aluminum in juvenile brook trout (*Salvelinus fontinalis*). I. Physiology. *Can. J. Fish. aquat. Sci.*, **48**, 2006–15.

McKone, C.E., Young, R.G., Bache, C.A. and Lisk, D.J. (1971) Rapid uptake of mercuric ion by goldfish. *Env. Sci. Tech.*, **5**, 1138–9.

Mallatt, J. (1985) Fish gill structural changes induced by toxicants and other irritants: a statistical review. *Can. J. Fish. aquat. Sci.*, **42**, 630–48.

Marshall, W.S. (1978) On the involvement of mucous secretion in teleost osmoregulation. *Can. J. Zool.*, **56**, 1088–91.

Mathiessen, P. and Brafield, A.E. (1973) The effects of dissolved zinc on the gills of the stickleback, *Gasterosteus aculeatus* (L.). *J. Fish Biol.*, **5**, 607–13.

Miller, T.G. and Mackay, W.C. (1982) Relationship of secreted mucus to copper and acid toxicity in rainbow trout. *Bull. Environ. Contam. Toxicol.*, **28**, 68–74.

Mueller, M.E., Sanchez, D.A., Bergman, H.L., McDonald, D.G., Rhem, R.G. and Wood, C.M. (1991) Nature and time course of acclimation to aluminum in juvenile brook trout (*Salvelinus fontinalis*). II. Histology. *Can. J. Fish aquat. Sci.*, **48**, 2016–27.

Neville, C.M. (1985) Physiological response of juvenile rainbow trout, *Salmo gairdneri*, to acid and aluminum – prediction of field responses from laboratory data. *Can. J. Fish. aquat. Sci.*, **42**, 2004–19.

Nöel-Lambot, F., Gerday, Ch. and Distèche, A. (1978) Distribution of Cd, Zn and Cu in liver and gills of the eel, *Anguilla anguilla*, with special reference to metallothioneins. *Comp. Biochem. Physiol.*, **61C**, 177–87.

Oronsaye, J.A.O. and Brafield, A.E. (1984) The effect of dissolved cadmium on the chloride cells of the gills of the stickleback, *Gasterosteus aculeatus* L. *J. Fish Biol.*, **25**, 253–8.

Orr, P.L., Bradley, R.W., Sprague, J.B. and Hutchinson, N.J. (1986) Acclimation-induced change in toxicity of aluminum to rainbow trout (*Salmo gairdneri*). *Can. J. Fish. aquat. Sci.*, **43**, 243–6.

Part, P. and Lock, R.A.C. (1983) Diffusion of calcium, cadmium and mercury in a mucous solution from rainbow trout. *Comp. Biochem. Physiol.*, **76C**, 259–63.

Pascoe, D. and Beattie, J.H. (1979) Resistance to cadmium by pre-treated rainbow trout alevins. *J. Fish Biol.*, **14**, 303–8.

Perry, S.F. and Laurent, P. (1989) Adaptational responses of rainbow trout to lowered external NaCl: contribution of the branchial chloride cell. *J. exp. Biol.*, **147**, 147–68.

Perry, S.F. and Wood, C.M. (1985) Kinetics of branchial calcium uptake in the rainbow trout: effects of acclimation to various external calcium levels. *J. exp. Biol.*, **116**, 411–33.

Playle, R.C. and Wood, C.M. (1989) Water pH and aluminum chemistry in the gill micro-environment of rainbow trout during acid and aluminum exposure. *J. comp. Physiol.*, **159B**, 539–50.

Playle, R.C. and Wood, C.M. (1991) Mechanisms of aluminium extraction and accumulation at the gills of rainbow trout, *Oncorhynchus mykiss* (Walbaum), in acid soft water. *J. Fish Biol.*, **38**, 791–805.

Playle, R.C., Goss, G.G. and Wood, C.M. (1989) Physiological disturbance in rainbow trout (*Salmo gairdneri*) during acid and aluminum exposures in soft water of two calcium concentrations. *Can. J. Zool.*, **67**, 314–24.

Reader, J.P., Dalziel, T.K.R. and Morris, R. (1988) Growth, mineral uptake and skeletal calcium deposition in brown trout, *Salmo trutta* L., yolk-sac fry exposed to aluminium and manganese in soft acid water. *J. Fish Biol.*, **32**, 607–24.

Reid, S.D. and McDonald, D.G. (1988) Effects of cadmium, copper, and low pH on ion fluxes in the rainbow trout, *Salmo gairdneri*. *Can. J. Fish. aquat. Sci.*, **45**, 244–53.

Reid, S.D. and McDonald, D.G. (1991) Metal binding activity of the gills of rainbow trout (*Oncorhynchus mykiss*). *Can. J. Fish. aquat. Sci.*, **48**, 1061–8.

Reid, S.D., Rhem, R.G. and McDonald, D.G. (1991) Acclimation to sublethal aluminum: modifications of metal–gill surface interactions of juvenile rainbow trout (*Oncorhynchus mykiss*). *Can. J. Fish. aquat. Sci.*, **48**, 1996–2005.

Rosseland, B.O. (1980) Physiological responses to acid water in fish. 2. Effects of acid water on metabolism and gill ventilation in brown trout, *Salmo trutta* L., and brook trout, *Salvelinus fontinalis* Mitchell, in *Proc. Int. Conf. Ecol. Impact Acid Precip. SNSF project, SNSF, Oslo, Norway*, pp. 348–9.

Shephard, K. (1982) The influence of mucus on the diffusion of ions across the esophagus of fish. *Physiol. Zool.*, **55**, 23–34.

Simkiss, K. (1984) Effects of metal ions on respiratory structures, in *Toxins, Drugs, and Pollutants in Marine Animals* (eds C.L. Bolis, A. Cambria and M. Fama), Springer-Verlag, Berlin, pp. 137–45.

Shephard, K. and Simkiss, K. (1978) The effects of heavy metal ions on Ca^{2+} ATPase extracted from fish gills. *Comp. Biochem. Physiol.*, **61B**, 69–72.

Sprague, J.B. (1970) Measurement of pollutant toxicity to fish. II. Utilizing and applying bioassay results. *Water Res.*, **4**, 3–32.

Spry, D.J. and Wood, C.M. (1985) Ion flux rates, acid–base status, and blood gases in rainbow trout, *Salmo gairdneri*, exposed to toxic zinc in natural soft water. *Can. J. Fish. aquat. Sci.*, **42**, 1332–41.

Spry, D.J. and Wood, C.M. (1989) The influence of dietary and waterborne zinc on heat-stable metal ligands in rainbow trout: quantification by ^{109}Cd radioassay and evaluation of the assay. *J. Fish Biol.*, **35**, 557–76.

Spry, D.J., Wood, C.M. and Hodson, P.V. (1981) The effects of environmental acid on freshwater fish with particular reference to the softwater lakes in Ontario and the modifying effects of heavy metals. A literature review. *Can. tech. Rep. Fish. Aquat. Sci.*, **999**, 144 pp.

Stagg, and Shuttleworth, T.J. (1982) The effects of copper on ionic regulation by the gills of the seawater-adapted flounder (*Platichthys flesus* L.). *J. comp. Physiol.*, **149**, 83–90.

Staurnes, M., Sigholt, T. and Reite, O.B. (1984) Reduced carbonic anhydrase and Na^+–K^+-ATPase activity in gills of salmonids exposed to aluminium-containing acid water. *Experientia*, **40**, 226–7.

Stone, H. and Overnell, J. (1985) Non-metallothionein cadmium-binding proteins. *Comp. Biochem. Physiol.*, **80C**, 9–14.

Stumm, W. and Morgan, J.J. (1981) *Aquatic Chemistry. An Introduction Emphasizing Chemical Equilibria in Natural Waters*, 2nd edn., J. Wiley & Sons, New York, 780 pp.

Thomas, D.G., Cryer, A., del G. Solbe, J.F. and Kay, J. (1983) A comparison of the accumulation and protein binding of environmental cadmium in the gills, kidney and liver of rainbow trout (*Salmo gairdneri*). *Comp. Biochem. Physiol.*, **76C**, 241–6.

Thomas, D.G., Brown, M.W., Shurben, D., del G. Solbe, J.F., Cryer, A. and Kay, J. (1985) A comparison of the sequestration of cadmium and zinc in the tissue of rainbow trout (*Salmo gairdneri*) following exposure to the metals singly or in combination. *Comp. Biochem, Physiol.*, **82C**, 55–62.

Tietge, J., Johnson, R. and Bergman, H.L. (1988) Morphometric changes in gill secondary lamellae of brook trout (*Salvelinus fontinalis*) after long-term exposure to acid and aluminum. *Can. J. Fish. Aquat. Sci.*, **45**, 1643–8.

Ultsch, G.R. and Gros, G. (1979) Mucus as a diffusion barrier to oxygen: possible role in O_2 uptake at low pH in carp (*Cyprinus carpio*) gills. *Comp. Biochem. Physiol.*, **62A**, 685–9.

Van der Putte, I.M., Laurier, B.H.M. and van Eijk, G.J.M. (1982) Respiration and osmoregulation in rainbow trout (*Salmo gairdneri*) in response to hexavalent chromium at different pH values. *Aquat. Toxicol.*, **2**, 99–112.

Varanasi, U. and Markey, D. (1978) Uptake and release of lead and cadmium in skin and mucus of coho salmon. *Comp. Biochem. Physiol.*, **60C**, 187–91.

Varanasi, U., Robisch, P.A. and Malins, D.C. (1975) Structural alterations in fish epidermal mucus produced by water-borne lead and mercury. *Nature, Lond.*, **258**, 431–2.

Verbost, P.M., Flik, G., Lock, R.A.C. and Wendelaar Bonga, S.E. (1988) Cadmium inhibits plasma membrane calcium transport. *J. Membr. Biol.*, **102**, 97–104.

Waiwood, K.G. and Beamish, F.W.H. (1978a) Effects of copper, pH, and hardness on the critical swimming performance of rainbow trout (*Salmo gairdneri* Richardson). *Water Res.*, **12**, 611–19.

Waiwood, K.G. and Beamish, F.W.H. (1978b) The effect of copper, hardness and pH on the growth of rainbow trout (*Salmo gairdneri* Richardson). *J. Fish Biol.*, **13**, 591–8.

Walker, R.L., Wood, C.M. and Bergman, H.L. (1988) Effects of low pH and aluminum on ventilation in the brook trout, *Salvelinus fontinalis. Can. J. Fish. aquat. Sci.*, **45**, 1614–22.

Walker, R.L., Wood, C.M. and Bergman, H.L. (1991) The effects of long term pre-exposure to sublethal concentrations of acid and aluminum on the ventilatory response to aluminum challenge in brook trout (*Salvelinus fontinalis*). *Can. J. Fish. aquat. Sci.*, **48**, 1989–95.

Watson, T.A. and Beamish, F.W.H. (1981) Effects of zinc on branchial ATPase activity *in vivo* in rainbow trout, *Salmo gairdneri. Comp. Biochem. Physiol.*, **66C**, 77–82.

Wendelaar Bonga, S.E. and Balm, P.H.M. (1989) Endocrine responses to acid stress in fish, in *Acid Toxicity and Aquatic Animals* (eds. R. Morris, E.W. Taylor, D.J.A. Brown and J.A. Brown), Cambridge Univ. Press, Cambridge, pp. 243–63.

Wicklund-Glynn, A. and Olsson, P.-E. (1991) Cadmium turnover in minnows, *Phoxinus phoxinus*, pre-exposed to water-borne cadmium. *Environ. Toxicol. Chem.*, **10**, 383–94.

Witters, H.E., Van Puymbroeck, S. and Vanderborght, O.L.J. (1991) Adrenergic response to physiological disturbances in rainbow trout, *Oncorhynchus mykiss*, exposed to aluminum at acid pH. *Can. J. Fish. aquat. Sci.*, **48**, 414–20.

Wood, C.M. (1992) Flux measurements as indices of H^+ and metal effects on freshwater fish. *Aquat. Toxicol.*, in press.

Wood, C.M. and McDonald, D.G. (1987) The physiology of acid/aluminum stress in trout. *Annis. Soc. r. zool. Belg.*, **117**, 399–410.

Wood, C.M., Playle, R.C., Simons, B.P., Goss, G.G. and McDonald, D.G. (1988a) Blood gases, acid–base status, ions, and hematology in adult brook trout (*Salvelinus fontinalis*) under acid/aluminum exposure. *Can. J. Fish. aquat. Sci.*, **45**, 1575–86.

Wood, C.M., McDonald, D.G., Booth, C.E., Simons, B.P., Ingersoll, C.G. and Bergman, H.L. (1988b) Physiological evidence of acclimation to acid/aluminum stress in adult brook trout (*Salvelinus fontinalis*). 1. Blood composition and net sodium fluxes. *Can. J. Fish. aquat. Sci.*, **45**, 1587–96.

Wood, C.M., Simons, B.P., Mount, D.R. and Bergman, H.L. (1988c) Physiological evidence of acclimation to acid/aluminum stress in adult brook trout (*Salvelinus fontinalis*). 2. Blood parameters by chronic cannulation. *Can. J. Fish. aquat. Sci.*, **45**, 1597–1605.

Youson, J.H. and Neville, C.M. (1987) Deposition of aluminum in the gill epithelium of rainbow trout (*Salmo gairdneri* Richardson) subjected to sublethal concentrations of the metal. *Can. J. Zool.*, **65**, 647–56.

Zuchelkowski, E.M., Pinkstaff, C.A. and Hinton, D.E. (1985) Mucosubstance histochemistry in control and acid-stressed epidermis of brown bullhead catfish, *Ictalurus nebulosus*. *Anat. Rec.*, **212**, 327–35.

Chapter thirteen

Phenotypic plasticity of fish muscle to temperature change

Ian A. Johnston

13.1 INTRODUCTION

The swimming performance of teleost fish is highly dependent on temperature. Natural selection has acted to adjust locomotory performance to particular thermal environments involving selective changes at the level of the genome. Johnston and co-workers have investigated muscle contraction associated with escape behaviour in teleosts from polar, temperate and tropical environments (Johnston, 1990). Force production shows perfect temperature compensation, whereas rate parameters, including force development, cross bridge cycle times and relaxation are significantly slower in Antarctic than tropical species (Johnston and Altringham, 1985; Johnson and Johnston, 1991a). It would appear that constraints to the adaptation of muscle proteins in low-temperature environments limit the maximum swimming speeds that can be attained (Johnston *et al.*, 1991). Specialized biochemical adaptations to particular temperature regimes restrict the geographical ranges of polar and tropical species. Some, mostly temperate species, experience large seasonal changes in water temperature. In these species, phenotypic adaptations to temperature enable swimming performance to be adjusted to local conditions within certain limits (Johnston and Dunn, 1987). A fall in water temperature during the winter months may result in a dramatic reduction in the availability of prey species. Fish with seasonal food supplies often enter a dormant state at low temperature, which is associated with a range of behavioural and metabolic responses serving to maintain energy reserves until the return of more favourable conditions (Lemons and Crawshaw, 1985). In contrast, some

Fish Ecophysiology. Edited by J. Cliff Rankin and Frank B. Jensen. Published in 1993 by Chapman & Hall, London. ISBN 0 412 45920 5.

species which continue to forage during the winter show improvements in physiological performance at low temperature (Hazel and Prosser, 1974). Temperature acclimation responses usually require several weeks to complete and are modified by other environmental factors, particularly daylength (Kleckner and Sidell, 1985). Specialist cold-water fishes may also show phenotypic adaptations which improve locomotory performance at the upper end of their thermal range during the warm summer months (Johnson and Johnston, 1991b).

13.2 MUSCLE FUNCTION AND TEMPERATURE

Most fish swim by the sequential activation of myotomes on alternate sides of the body. This results in a wave of lateral bending from the head to the tail, producing a reactive thrust from the water. The myotomes contain a spectrum of muscle fibre types which are specialized for function at different speeds (Bone, 1978; Johnston and Altringham, 1991). In most species, red fibres form a superficial strip in the region of the lateral line nerve, and the remaining bulk of the myotome is composed of white muscle. Slow red muscle fibres are used for low swimming speeds, whereas the faster-contracting white muscle fibres are required for high cruising and maximum speeds. Forces produced by the muscle fibres are transmitted to the water via skeletal and connective tissue elements and the skin, although the details are not fully understood (Wainwright, 1983; Johnston and Altringham, 1991). The geometric arrangements of muscle fibres and myotomes would appear to be critical for the generation of the locomotory wave. The majority of superficial muscle fibres run parallel to the longitudinal axis of the fish, whereas deeper fibres make angles of up to 40° (Alexander, 1969). The complex orientation of the deeper muscle fibres is thought to allow them to shorten by a similar amount when the body bends (Alexander, 1969; Rome and Sosnicki, 1991).

Two main methods have been developed for estimating changes in sarcomere length during swimming. In the first method, the longitudinal axis of the fish is calculated from its digitized outline in successive frames of a high-speed ciné or video recording of swimming. On certain assumptions, the fractional shortening (strain) of superficial fibres can be calculated from the local curvature of the longitudinal axis (van Leeuwen *et al.*, 1990). The second method involves constructing a calibration curve relating the local curvature of the backbone to measured values of sarcomere length (Rome and Sosnicki, 1991). Briefly, fish are bent into shapes similar to those found in swimming and the muscle fibres are allowed to go into rigor so that they generate force, shorten and become taut. Rigor forces are 30–40% of the maximum isometric tension and this is similar to the peak forces generated during swimming, which suggests that the fibre–myosepta connections are stretched to a similar extent. The rigor fish are rapidly frozen in liquid nitrogen, photographed in silhouette, and the sarcomere lengths are measured from longitudinal sections of steaks taken from various points along the body. Rome and Sosnicki (1991) found that sarcomere length for the superficial fibres was a function of both radius of curvature of the body and the distance from the backbone. For a given body curvature, the strain of the deeper fast fibres was independent of the distance from the backbone and only a quarter of that for the superficial slow fibres.

Rome *et al.* (1990) used high-speed cinematography to calculate changes in backbone curvature of carp *Cyprinus carpio*, swimming at different speeds in a water tunnel. This enabled the strain fluctuations and shortening speed of the red muscle fibres to be calculated (Fig. 13.1 (b–d)). Sarcomere length excursions for the slow muscle fibres were typically from 1.91 to 2.22 mm for all but the most anterior myotomes, corresponding to a strain of ± 7.5% resting fibre length. Sarcomere length excursions were found to be largely independent of swimming speed and temperature. Since the frequency of tailbeats increases with speed, the sarcomeres must undergo the same length changes in a shorter time as speed

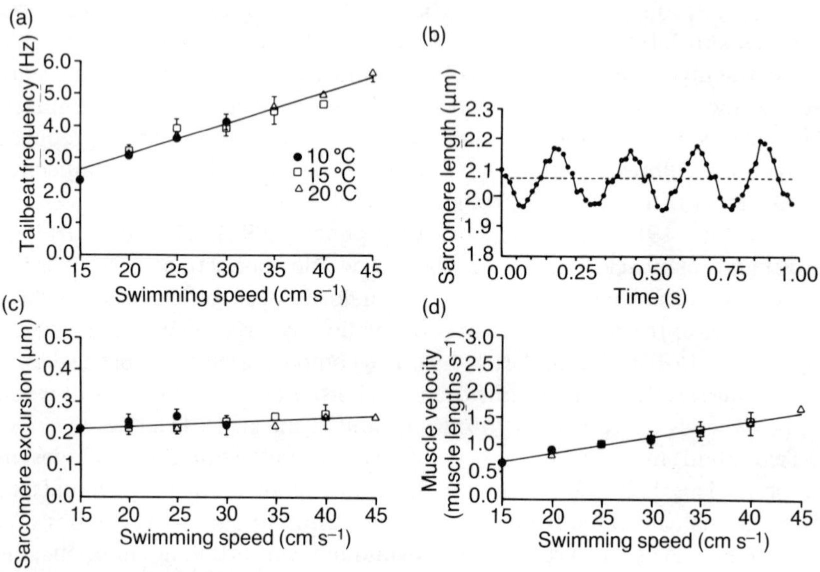

Fig. 13.1 The influence of temperature on muscle action during swimming in common carp, *Cyprinus carpio*. (a) Acute effects of temperature on tailbeat frequency in fish swimming at steady speeds measured from high-speed motion pictures. (b) Record showing changes in the sarcomere length of red muscle fibres in the middle myotomes of a fish swimming at 40 cm s^{-1}. Resting sarcomere length (dotted line) was 2.06 μm. (c) Relationship between swimming speed and sarcomere length excursions at three experimental temperatures. (d) Relationship between swimming speed and the maximum strain rate of muscle fibres at three temperatures. Note that muscle velocities increase linearly with swimming speed at all temperatures. Points are means; error bars represent SE. Adapted from Rome *et al.* (1990) with permission of the Company of Biologists Ltd.

increases (Fig. 13.1 (a)). However, at any given swimming speed, the muscle shortening velocity *in vivo* (*V*) is the same at all temperatures (Fig. 13.1 (d)). Rome and Sosnicki (1991) found that the maximum shortening speed (*V*max) of the red muscle fibres *in vitro* decreased with a Q_{10} of 1.63 over the range 20 to 10 °C, whereas the maximum isometric stress decreased with a Q_{10} of 1.13. They estimated that a 12–14 cm carp swimming at 2.4 body lengths s^{-1} would require a 50% greater cross-sectional area of red muscle fibres to generate the same power at 10 °C as at 20 °C.

Electromyographical studies in the common carp have shown that there is a sequential recruitment of slow red > fast red > fast white fibres with increasing speed (Johnston *et al.*, 1977). As the temperature is reduced, carp recruit their anaerobic fast muscle fibres at lower swimming speeds, in order to compensate for the reduced power output of the aerobic fibre types (Rome *et al.*, 1984). Fast and slow muscle fibres are used over a similar range of *V*/*V*max (0.17 to 0.36) at a given temperature (Rome *et al.*, 1988). The useful range of strain rates is probably explained by decreases in muscle efficiency at low *V*/*V*max (Curtin and Woledge, 1988), and by decreases in power output at high values of *V*/*V*max (Rome *et al.*, 1990). Following several weeks' acclimation to low temperatures, the speed at which fast muscle fibres are first recruited increases relative to that in acutely exposed fish (Rome *et al.*, 1985). Changes in recruitment patterns with temperature acclimation are thought to reflect phenotypic adaptations in the swimming muscles (Johnston and Dunn, 1987).

Not all species show the same pattern of swimming behaviour in response to acute changes in temperature. For example, in striped bass, *Morone saxatilis*, the slope of the relationship between tailbeat frequency and swimming velocity was found to decrease as the test temperature was reduced (Sisson and Sidell, 1987). Thus, at the same tailbeat frequency, striped bass can swim at a higher speed at cold than warm temperatures. The increased stride length (distance moved during one complete tailbeat), at low temperatures implies either an increased efficiency of power transmission into propulsion or a lower cost of swimming. At sustainable speeds, striped bass swim using only their red muscle. The swimming speed at which the fast muscle fibres were first recruited in 16–18 cm fish increased from around 2.0 body lengths s^{-1}, at 9 °C, to 2.8 body lengths s^{-1} at 25 °C (Sisson and Sidell, 1987). Thus both carp and striped bass show a compression of the recruitment order of muscle fibre types at low temperatures.

13.3 TEMPERATURE ACCLIMATION OF SWIMMING PERFORMANCE

Fry and Hart (1948) used a rotating annular chamber with an outer glass wall to determine the effects of temperature on the maximum speeds that goldfish, *Carassius auratus* L., could maintain for 2 minutes (Fig. 13.2). When the chamber was rotated, the fish reacted by swimming to maintain a constant

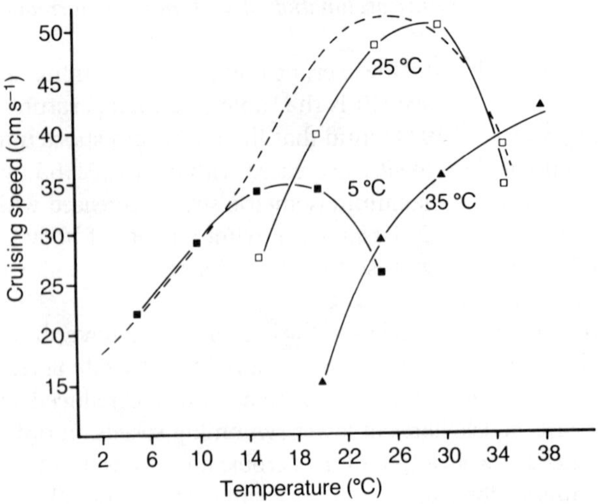

Fig. 13.2 The effects of temperature on the maximum sustainable swimming speeds of goldfish acclimated to 5 °C (■), 25 °C (❐) and 35 °C (▲). The dotted line represents the speed attained by the fish when tested at its acclimation temperature. Adapted from Fry and Hart (1948).

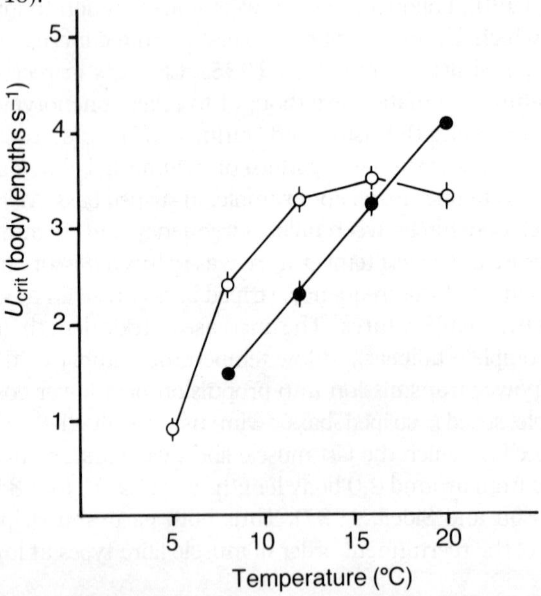

Fig. 13.3 Effects of temperature on the critical swimming speed (*U*crit) in body lengths s^{-1} of common carp, *Cyprinus carpio* L., acclimated to either 8 °C (open symbols) or 20 °C (closed symbols) for a minimum of 2 months (photoperiodic regime 12h light:12h dark). The fish were forced to swim against a current in a Brett respirometer for periods of 60 min. $U\text{crit} = U_i + (t_1/t_2\ U_{ii})$ where U_i is the highest speed maintained for the 60 min period (cm s^{-1}), t_1 is the time taken for the fish to fatigue on the final velocity increment (min) at the fatigue time, t_2 is the prescribed period of swimming (60 min), and U_{ii} is the velocity increment between trials. Points are means; error bars represent ± SE (n = 8) (unpublished data of Johnston and Johnston).

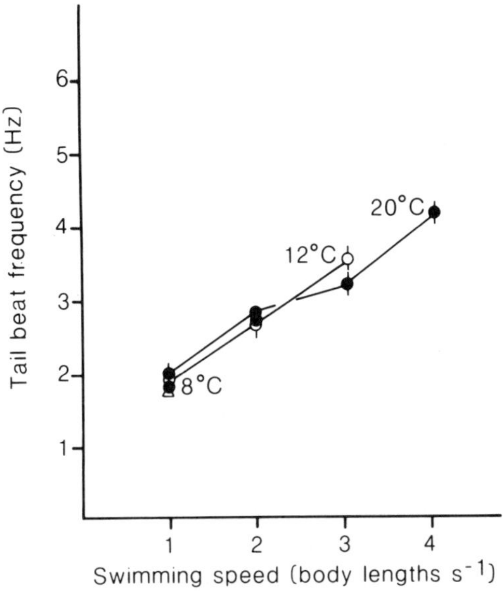

Fig. 13.4 The relationship between swimming speed and tailbeat frequency for common carp acclimated to 8 (open symbols) and 20 °C. Fish acclimated to 8 °C were tested at both 8 °C (open triangles) and 12 °C (open circles). Fish acclimated at 20 °C were tested at 20 °C. Error bars represent mean ± SE. (n = 8) (unpublished data of Johnston and Johnston, 1992.)

position with reference to the laboratory. The relatively high speeds that can be maintained for such short periods probably require the recruitment of both fast and slow muscle fibre types. Fry and Hart found that cold acclimation increased the maximum speeds that could be sustained at low temperatures, whereas performance at high temperatures was reduced (Fig. 13.2). Cold acclimation has similarly been shown to increase maximum cruising speed (Ucrit) at low temperature in the common carp *Cyprinus carpio L.* (Fig. 13.3). However, only the range of swimming speeds was altered by temperature acclimation. The tailbeat frequency (Fig. 13.4) and amplitude required to swim at a given steady speed were independent of acclimation temperature.

13.4 PLASTICITY OF MUSCLE PHENOTYPES

Proportions of fast and slow muscle fibres

Improvements in sustained swimming performance with cold acclimation are partly explained by an increase in the volume of red muscle fibres. Johnston and Lucking (1978) used histochemical techniques to characterize the myotomal muscle fibre types of goldfish acclimated for 3 months to either 3, 18 or

31 °C. They found that, at a point 8–10 myotomes from the tail, aerobic muscle fibre types occupied 8.3% of the total muscle cross-sectional area at 31 °C compared with 12.5% at 3 °C. This was due to significant increases in the numbers and diameters of slow red fibres with cold acclimation, as has been reported for several other species, including the striped bass (Jones and Sidell, 1982).

Muscle protein composition

Fast and slow muscle fibres contain various isoforms of the contractile proteins which differ with respect to their amino acid sequences and functional characteristics. Contractile protein isoforms can represent multi-gene families of proteins or the products of RNA processing. In some species, the expression of contractile protein isoforms varies according to the acclimation temperature and correlates with adaptive changes in locomotory performance (Johnston *et al.*, 1990). All vertebrate myosins have a similar structure, consisting of two heavy chains (MHC) and two light chains (LC). The two heavy chains form a coiled coil over their C-terminal halves and then separate to form a globular S1 head at the NH_2 terminus end. The S1 head contains the actin binding site and the ATPase active site. ATPase activity and muscle shortening speed are largely determined by myosin heavy chain composition, but are modulated by other proteins, particularly the myosin light chains (Greaser *et al.*, 1988; Bottinelli *et al.*, 1991).

Johnston *et al.*, (1975) found that myofibrillar ATPase activity from fast muscle of goldfish was almost three times higher in 1°C- than 31°C-acclimated fish at the cold acclimation temperature. This was associated with an increase in the susceptibility of the ATPase from cold-acclimated fish to thermal denaturation, which suggests some structural change in the myosin. Similar increases in myofibrillar ATPase at low temperature have been reported for other cyprinid species including common carp (Crockford and Johnston, 1990; Hwang *et al.*, 1990), roach *Rutilus rutilus L.*, and tench *Tinca tinca L.*, (Heap *et al.*, 1985). These changes in ATPase activity are apparent within 2 weeks and are complete within about 4 weeks. This time course and the absence of acclimatory responses in starved individuals is consistent with the involvement of protein synthesis (Heap *et al.*, 1986). Hwang *et al.* (1990) have found small differences in the peptide maps of α-chymotrypsin-digested fast muscle myosin heavy chains (MHC) between carp acclimated to either 10 °C or 30 °C. Support for changes in the expression of fast muscle myosin heavy chains genes with temperature acclimation also comes from molecular biological studies. Gerlach *et al.* (1990) constructed a genomic library from carp acclimated to 18 °C and isolated and partially characterized 28 non-overlapping clones containing MHC sequences, indicating a minimum of 28 MHC genes. Plasmids containing a 7-kb probe specific to carp fast muscle MHC hybridized more strongly to mRNA from 28 °C than 18 °C fish, indicating

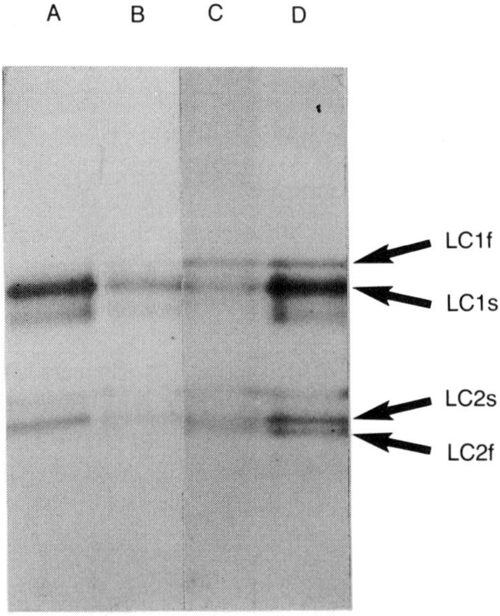

Fig. 13.5 A silver-stained SDS-PAGE gel of electrophoretically purified myosin prepared form slow muscle fibre bundles isolated from the abductor superficialis muscle of the common carp. Lanes A and B were from 20 °C-acclimated carp; lanes C and D were from 8 °C-acclimated carp. LC_{1s}, myosin light chain 1 slow; LC_{1f}, myosin light chain 1 fast; LC_{2s}, myosin light chain 2 slow; LC_{2f}, myosin light chain 2 fast. Reproduced with permission (from Langfeld *et al.* (1991).

changes in the expression of MHCs. Crockford and Johnston (1990) also found evidence for the expression of an additional myosin light chain in single fast muscle fibres from cold-acclimated carp. The LC_{1f}:LC_{3f} ratio (molar ratio of light chains 1 and 3), was also higher in 8 °C- than 20 °C-acclimated fish (Crockford and Johnston, 1990). In mammals and birds, LC_{1f} and LC_{3f} are produced from a single gene by alternate transcription and differential splicing of the mRNA (Nabeshima *et al.*, 1984), although this may not be the case in fish (Libera *et al.*, 1991). Langfeld *et al.* (1991) purified native myosin from slow fibre bundles isolated from the pectoral fin superficial abductor muscle of the common carp. Fibre bundles from 20 °C-acclimated carp contained almost exclusively slow myosin light chains (LC_{1s} and LC_{2s}), whereas preparations from 8 °C-acclimated carp contained a significant proportion of fast myosin light chains (LC_{1f} and LC_{2f}) in addition to LC_{1s} and LC_{2s} (Fig. 13.5). Histochemical studies revealed that preparations from both cold- and warm-acclimated fish contained an average of 3% fast oxidative fibres in addition to slow fibres. Thus cold acclimation results in the expression of myosin light chain isoforms in slow muscles which are normally associated with faster-contracting fibre types (Langfeld *et al.*, 1991).

Goldfish actomyosin preparations from which the tropomyosin–troponin complex had been extracted at low ionic strength were found to have the same

ATPase activity irrespective of the acclimation temperature (Johnston, 1979). This suggests that some components(s) of thin filament Ca^{2+}-regulatory proteins may also be involved in the acclimatory response, although low-ionic-strength washing may have also altered the proportions of myosin light chains. Crockford and Johnston (1990) found that fast muscle fibres from 20 °C-acclimated carp contained an additional minor isoform of troponin I relative to 8 °C–acclimated fish, whereas the electrophoretic mobilities of isoforms of tropomyosin, troponin T and troponin C were identical (Crockford and Johnston, 1990).

Changes in myofibrillar ATPase activity with temperature acclimation may not be particularly widespread in teleost fish. Species which do not show adaptations in myofibrillar ATPase activity with temperature acclimation include mummichogs, *Fundulus heteroclitus* (Sidell, 1983), striped bass (Moerland and Sidell, 1986) and chain pickerel, *Esox niger* (Sidell and Johnston, 1985).

13.5 MUSCLE CONTRACTILE PROPERTIES

Muscle physiologists commonly study the mechanical properties of isolated muscle fibres at constant length (isometric contractions) or during shortening against a constant load (isotonic contractions). The influence of temperature acclimation on the isometric and isotonic properties of carp muscle fibres has been investigated using both skinned and live preparations. *In vivo* reductions in temperature result in a relative alkalinization of muscle pH (Walsh and Moon, 1982). Using skinned preparations, it was found that the pCa–force relationship of slow muscle fibres was shifted to the right by decreasing the temperature and to the left by alkalinization (Johnston *et al.*, 1990). Studies on acid–base changes with temperature *in vivo* suggest that the direct effects of cooling on the pCa–force curve are probably larger than those produced by relative alkalinization (Walsh and Moon, 1982). This would result in a net shift of the pCa–force curve to the right at low temperatures, increasing the calcium concentration required to maintain a given force, and thus promoting relaxation. The Ca^{2+}-concentration required for half-maximal force does not appear to change with acclimation temperature (Johnston *et al.*, 1990). In contrast, the maximum Ca^{2+}-activated tension generated by carp skinned fibres does increase significantly at low temperatures following several weeks below 10 °C. For example, at 7 °C, maximum tension (P_0) was found to be 38% higher in slow fibres and 57% higher in fast muscle fibres in 7 °C-acclimated relative to 23 °C-acclimated carp (Johnston *et al.*, 1985). Similar results have been obtained with intact electrically excitable preparations. Langfeld *et al.* (1991) found that the maximum tetanic tension (*P*max) of slow fibres from the pectoral fin abductor superficialis muscle was 32% higher in 8 °C- than 20 °C-acclimated fish at 8 ° C (Fig. 13.6(a)). *P*max was similar when measured

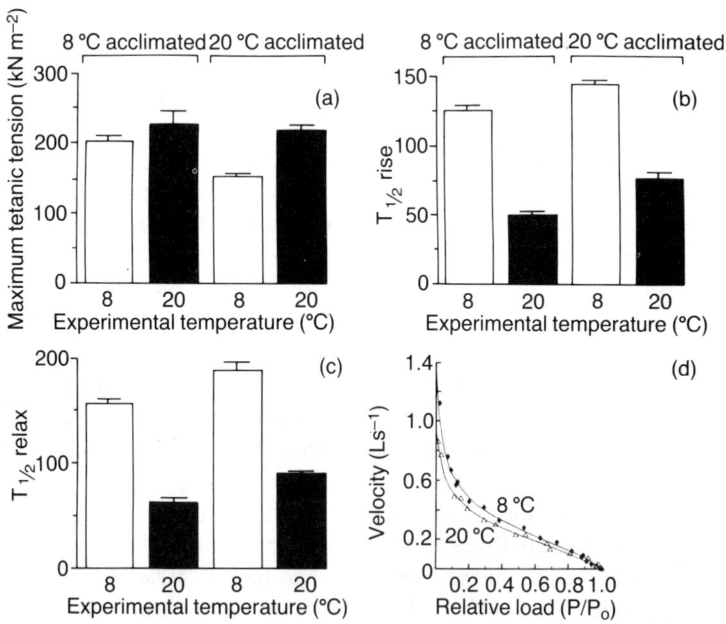

Fig. 13.6 Contractile properties of slow muscle fibres isolated from the superficial region of the pectoral fin abductor superficialis muscle of common carp. Fish were acclimated for 6–12 weeks to either 8 °C or 20 °C. (a) Maximum tetanic tension (kN m^{-2}); (b) $T_{1/2}$ rise; half-time (ms) from the first stimulus to 50% of maximum isometric tension. Muscle fibre bundles were stimulated at 25 Hz to produce maximal fused tetani. (c) $T_{1/2}$ relax; half-time (ms) from the last stimulus to 50% fall in peak tension. (d) Typical force–velocity curves of fibres from cold- and warm-acclimated fish (cold, filled symbols; warm, open symbols). The data were fitted to a hyperbolic linear equation:

$$V = [B(1 - P/P_0) / (A+P/P_0)] + C(1 - P/P_0)$$

(from Langfeld *et al.*, 1991). Error bars represent SE of the mean (n = 6).

at the acclimation temperature of each group, indicating near perfect temperature compensation of tension generation. The fraction of muscle fibre cross-sectional area occupied by myofibrils shows only small and variable changes with acclimation temperature, which indicates that altered force-generating capacities are due to changes in myosin composition (Johnston and Maitland, 1980). Cold acclimation also results in a large increase in the maximum contraction velocity (*V*max) of fast muscle fibres at low temperatures (Johnston *et al.*, 1985), and a more modest increase in *V*max for slow muscle fibres (Fig. 13.6 (d)). The curvature of the force–velocity (*P–V*) relationship of live fibres had been shown of be independent of acclimation temperature (Fig. 13.6(d)). Values for the maximum power output of slow fibres, measured at the cold acclimation temperature, were 47% higher in 8 ° C-acclimated (26.5 W kg^{-1} wet muscle mass) than in 20 °C-acclimated carp (18.0 W kg^{-1} wet muscle mass) (Langfeld *et al.*, 1991).

Skinned fibres isolated from cold-acclimated carp fail to relax completely following maximal activations at 23 °C. The resulting Ca-insensitive force component is associated with abnormal crossbridge formation between actin and myosin filaments and very slow contraction velocities (Johnston *et al.*, 1985). Thus, the improved low-temperature contractile performance of the myosin isoforms expressed during the winter is at the expense of a loss of function at higher temperatures. Crockford and Johnston (1990) investigated the limits to the plasticity of contractile properties in common carp. They found that at an experimental temperature of 0 °C, values for *P*o and *V*max in fast muscle fibres were similar in carp acclimated to 2, 5, 8 or 11 °C. However, both parameters were at least twofold higher in 2 °C- than 23 °C-acclimated fish, and were intermediate in 15 °C-acclimated fish. Measured at the acclimation temperature, *V*max was 1.8 times, and Po was 1.4 times higher in 23 °C than 2 °C carp. This corresponds to a partial capacity adaptation to temperature across the entire thermal range, consistent with observations of swimming performance.

Cold acclimation also modifies the twitch contraction kinetics of muscle fibres in common carp. Rates of force development and relaxation for isometric tetani in slow muscle fibres were found to be 15–20% faster in 8 °C- than 20 ° C-acclimated fish at the cold acclimation temperature (Fig. 13.6 (b and c)). Much larger changes in twitch duration occur in fast myotomal muscle fibres. For example, the half-times for twitch activation and relaxation for a fast myotomal nerve muscle preparation are only around half as long at low temperatures, in 8 °C compared with 20 °C-acclimated carp (Fig. 13.7). The mechanisms underlying changes in twitch duration are likely to be complex and may also vary between species. In both common carp and goldfish, cold acclimation results in a decrease in the average diameter of myofibrils which would shorten diffusion path lengths from the sarcoplasmic reticulum (SR) to the Ca^{2+}-binding sites on troponin C. This, together with increased crossbridge cycle times at low temperature, would be expected to increase the rate of muscle force development in cold-acclimated individuals. Penney and Goldspink (1980) reported that fast myotomal muscles had a higher surface density of SR in goldfish acclimated to 5 °C than to 30 °C. In contrast, Fleming *et al.*, (1990) found that the surface and volume densities of SR in fast and slow muscles of the common carp showed no consistent differences with acclimation temperature. In this case, the faster rates of relaxation in the muscles of cold-acclimated individuals were related to changes in the density and/or kinetics of SR Ca^{2+}-ATPase. At the cold acclimation temperature, Ca^{2+}-ATPase activity was 60% higher for SR vesicles isolated from 8 °C-compared with 20 °C-acclimated fish. The fast muscles of fish contain high concentrations of soluble low molecular weight Ca^{2+}-binding proteins (parvalbumins) which are thought to have a role in promoting relaxation. However, the concentrations (0.61–0.68 mmol kg^{-1} wet weight) and the relative proportions of parvalbumin isoforms are unaffected by acclimation temperature (Fleming *et al.*, 1990).

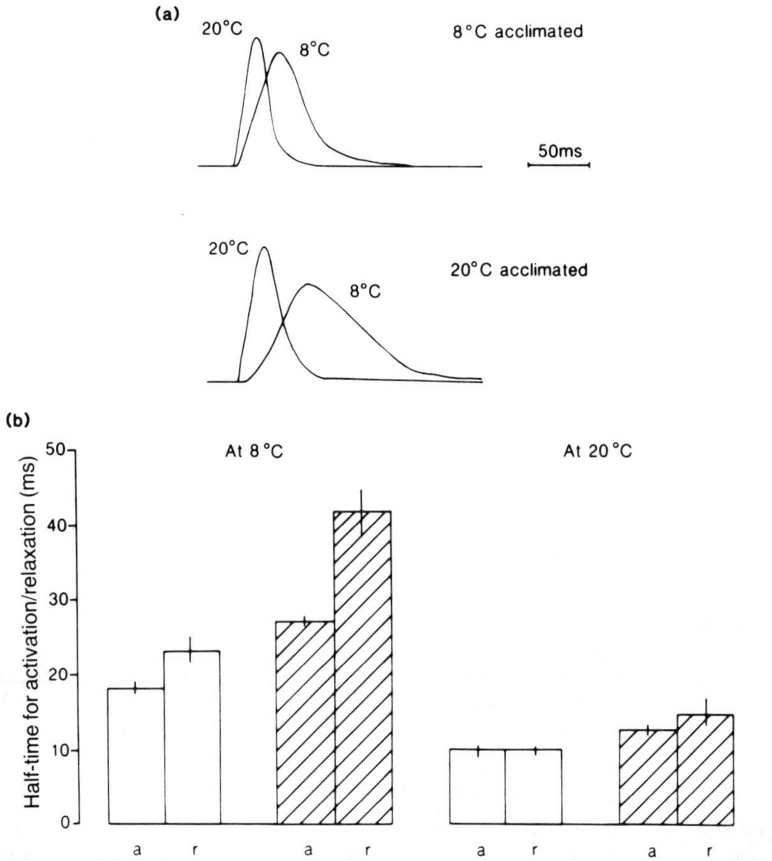

Fig. 13.7 Isometric twitch contraction times recorded form a fast nerve–muscle preparation from abdominal myotomes of common carp, *Cyprinus carpio.* Fish were acclimated for a minimum of 2 months to either 8 °C (open bars) or 20 °C (shaded bars), and experiments were performed at both temperatures for each group. (a) Typical force records. (b) Half-times for twitch activation (a) and relaxation (r) in ms. Values represent mean ± SE (n = 6). Reproduced with permission from Fleming *et al.* (1990).

During locomotion, muscles rarely undergo isometric and/or isotonic contractions (Johnston and Altringham, 1988). Muscle fibres *in vivo* alternately shorten and lengthen, and their state of activation continuously varies during the strain cycle. Computer models using experimental data on the timing of the electromyogram (e.m.g.) wave in relation to changes in sarcomere length show that muscle fibres in anterior myotomes develop their maximum force at the start of the shortening phase of the cycle (van Leeuwen, 1991). This results in net positive work over the whole of the tailbeat cycle (van Leeuwen *et al.*, 1990). During continuous swimming, the changes in length of myotomal muscle fibres are essentially sinusoidal (Hess and Videler, 1984; van Leeuwen *et al.*, 1990). Altringham and Johnston (1990) have

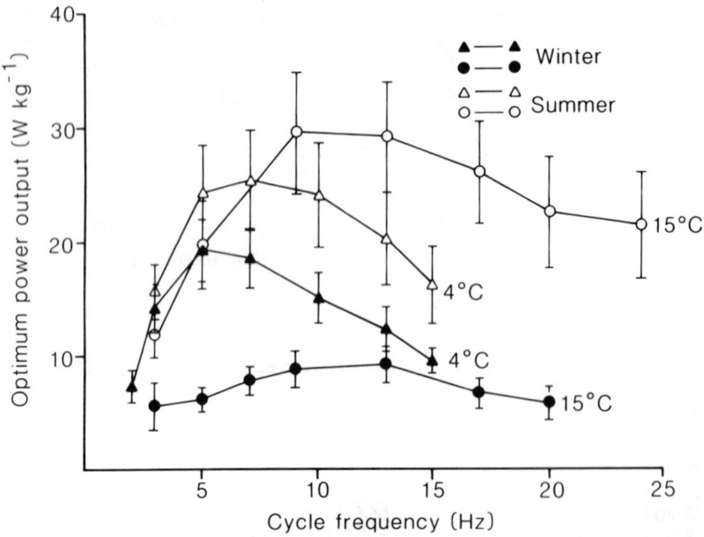

Fig. 13.8 The relationship between cycle frequency and power output (W kg^{-1}) for fast muscle fibres performing oscillatory work. Fibres were isolated form the anterior myotomes of the marine teleost the short-horned sculpin, *Myoxocephalus scorpius*. Strain and stimulus parameters were adjusted at each cycle frequency to maximize power output. Filled symbols, winter-acclimatized fish; open symbols, summer-acclimated fish; values represent mean ± SE (from Johnson and Johnston, 1991b).

measured the power output of isolated fast and slow muscle fibres subject to imposed sinusoidal length changes about *in situ* resting length, and stimulated during each cycle. Values of strain, and the number and timing of stimuli, were adjusted to maximize power at the range of frequencies found during swimming. In short-horned sculpin, *Myoxocephalus scorpius* L., 23–29 cm total length, maximum power at 4 °C was produced at a cycle frequency of 2 Hz for slow fibres (5–8 W kg^{-1} wet muscle mass) and 5–7 Hz for fast muscle fibres (25–35 W kg^{-1} wet muscle mass). The short-horned sculpin is a cold-tolerant fish widely distributed in shallow seas around northern Europe and into the Arctic circle. In order to lower the freezing point of the plasma during cold winters, the liver secretes a polypeptide antifreeze (Fletcher *et al.*, 1989). Around St Andrews (Scotland) this species experiences sea temperatures ranging from 3–5 °C in winter to 12–16 °C during the summer. Johnson and Johnston (1991b) found that the strain required to produce maximum power output was independent of temperature. However, the cycle frequency required for maximum power output increased from around 5–7 Hz at 4 °C to 9–13 Hz at 15 °C (Fig. 13.8). They found that the peak tension under conditions for maximum power output was around two times higher at all cycle frequencies in summer- than winter-caught fish. Although the maximum power output of muscle fibres at 4 °C showed little seasonal variation, it was three times higher at 15 °C in summer- (30 W kg^{-1}) than in winter- (9 W kg^{-1}) acclimatized fish (Fig. 13.8). Thus, in cold-tolerant marine species, the major seasonal adjustments in mechanical properties occur at high temperatures.

13.6 MUSCLE METABOLISM

The metabolic responses of fish muscle to temperature vary with the time course of adaptation, and with the acclimation temperature (Johnston and Dunn, 1987). Ultrastructural studies have shown that in many species, cold acclimation is associated with an increase in the volume density of muscle mitochondria (Jankowsky and Korn, 1965; Johnston and Maitland, 1980; Tyler and Sidell, 1984). For example, the volume densities of mitochondria in the myotomal muscles of striped bass were found to increase by 50–60% following 8 weeks' acclimation from 25 °C to 5 °C, reaching 45% and 4% in red and white fibres respectively (Fig. 13.9). This largely reflected an increase in the number of mitochondrial profiles in cross-sections. Since the relative mass of aerobic fibres also increased with cold acclimation, the total volume of mitochondria per 100 g body mass was almost three times higher in 5 °C- (7.6 cm^3) than 25 °C-acclimated (2.6 cm^3) fish (Egginton and Sidell, 1989). It is thought that mitochondria in the subsarcolemmal zone have a role in supplying ATP for the active transport of substrates across the sarcolemma, whereas intermyofibrillar mitochondria support muscle

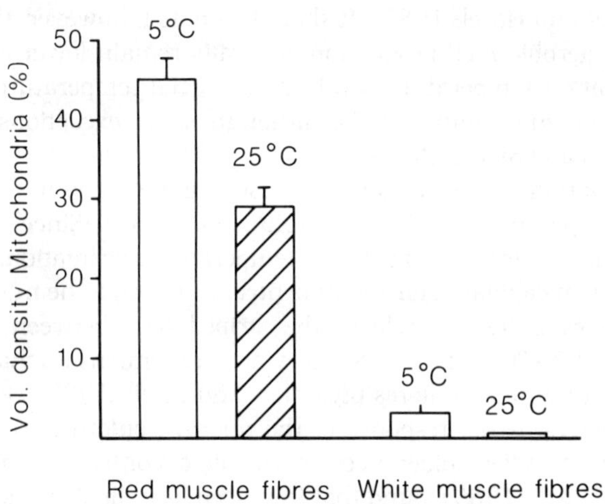

Fig. 13.9 The volume density (V_V) of mitochondria in fibres from red and white muscles of striped bass, *Morone saxatilis*, acclimated to either 25 °C or 5 °C. Values represent mean ± SE. *V*v represents Σ organelle cross-sectional areas/muscle fibre cross-sectional area. Data from Egginton and Sidell (1989).

contraction. Temperature acclimation produces changes in the relative distributions of mitochondria within muscle fibres, although the results are variable between species. For example, cold acclimation increased the proportion of mitochondria found in the subsarcolemmal zone in crucian carp *Carassius carassius* L. (Johnston and Maitland, 1980), but not in striped bass (Egginton and Sidell, 1989). In the striped bass, mitochondria were found to occur in clustered arrays in the red muscle of 5 °C-acclimated individuals such that the surface area of mitochondria exposed to the muscle cytoplasm was independent of acclimation temperature (Egginton and Sidell, 1989). Van den Thillart and Modderkolk (1978) isolated mitochondria from the red and white muscles of goldfish acclimated to 30, 20 or 5 °C. They found a significant increase in oxygen consumption and cytochrome *c* oxidase activity per mg mitochondrial protein in the cold-acclimated fish. This may reflect an increase in the concentration of respiratory enzymes within the mitochondria and/or changes in membrane composition which enhance enzyme activity (Hazel and Prosser, 1974). Stereological analyses have shown that the surface density of cristae in muscle mitochondria is unaffected by temperature acclimation in both goldfish (Tyler and Sidell, 1984) and striped bass (Egginton and Sidell, 1989).

The higher volume densities of mitochondria in the muscles of cold-acclimated fish boost the rate of ATP synthesis at low temperatures, reduce the average diffusion path lengths between the mitochondria and myofibrils, and increase the area of exchange surface between mitochondria and cytoplasm. These last two effects will help to compensate for a reduction in the diffusion coefficients of metabolites at low temperature, which have Q_{10}s of up to 1.5–2.9 (Sidell and Hazel, 1987). It should be noted, however, that absolute capacities for aerobic ATP production generally remain lower at cold- than warm-acclimation temperatures (indicating partial temperature compensation) (Johnston and Dunn, 1987), although some exceptions have been reported (Jones and Sidell, 1982).

Cold acclimation in crucian carp also results in increases in the number of capillaries (C) per muscle fibre (F) (Johnston, 1982). Since muscle fibre cross-sectional area is unchanged with temperature acclimation, this reflects a true increase in capillary density. In contrast, although the numerical ratio of C:F increases in red muscle of the striped bass between acclimation temperatures of 25 °C and 5 °C, capillary density remains unchanged due to hypertrophy of the muscle fibres (Egginton and Sidell, 1989). Some of these species differences in the response of the microvasculature to temperature acclimation may reflect differences in the lipid content of muscle fibres. Egginton and Sidell (1989) found that in striped bass, lipid droplets increased 13-fold from 0.6 to 8% in red muscle fibres following acclimation from 25 to 5 °C. They suggested that enhanced diffusion of oxygen in lipid may obviate the need for increases in capillary density in cold-acclimated individuals. In support of this hypothesis, it was found that the white muscle of striped bass, which does not accumulate lipid, showed an increase in both C:F and numerical capillary density at low temperatures.

Temperature acclimation also influences the relative importance of different metabolic fuels in muscle. For example, in striped bass, cold acclimation has been shown to result in an increased capacity for ^{14}C-1-palmitate oxidation, particularly in the red muscle fibres (Jones and Sidell, 1982). In contrast, maximal activities of enzymes associated with the main anaerobic metabolic pathways (phosphogen hydrolysis and glycolysis) usually decrease or remain unchanged with cold acclimation (review: Guderley and Blier, 1987). Guderley and Foley (1990) found that phosphofructokinase and pyruvate kinase activities were significantly higher in the swimming muscles of winter- than summer-acclimatized ninespine sticklebacks, *Pungitius pungitius*, but only in the male, not the female fish. Unfortunately, no information is available on the actual flux of metabolites through anaerobic pathways at different acclimation temperatures, although rates of ATP production would be expected to decrease at low temperatures. Temperature is known to affect the concentration of metabolic substrates and products as well as the microenvironment of enzymes via alterations to membrane lipids and intracellular pH (Hazel and Prosser, 1974).

13.7 FUTURE PROSPECTS

Almost all studies on temperature acclimation in fish have been concerned with adult stages. For many species, a temperature range of up to 15 °C can be covered by a single metabolic profile (Guderley, 1990). Thus phenotypic adaptations in muscle properties usually require adult fish to be acclimated to temperatures substantially below their thermal optimum for locomotion. Recent research suggests that the larval and juvenile stages of the life cycle may be much more sensitive to temperature change (Vieira and Johnston, 1992; Calvo and Johnston, 1992). For example, only a 5 °C difference in acclimation temperature was found to significantly alter the proportion and distribution of tonic and red muscle fibres in the myotomes of turbot larvae, *Scophthalmus maximus* L. (Calvo and Johnston, 1992). The influence of development on the phenotypic plasticity of fish muscle is therefore a particularly exciting area for future research.

ACKNOWLEDGEMENT

The author is grateful for research support from the Natural Environmental Research Council of the UK.

REFERENCES

Alexander, R. McN. (1969) The orientation of muscle fibres in the myomeres of fishes. *J. mar. biol. Ass. U.K.*, **49**, 263–90.

Altringham, J.D. and Johnston, I.A. (1990) Modelling muscle power output in a swimming fish. *J. exp. Biol.*, **148**, 395–402.

Bone, Q. (1978) Locomotor muscle, in *Fish Physiology*, Vol. VII (eds W.S. Hoar and D.J. Randall), Academic Press, New York, pp. 361–424.

Bottinelli, R., Schiaffino, S. and Reggiani, C. (1991) Force–velocity relations and myosin heavy chain isoform compositions of skinned fibres from rat skeletal muscle. *J. Physiol., Lond.*, **437**, 655–72.

Calvo, J. and Johnston, I. (1992) Influence of rearing temperature on the distribution of muscle fibre types in the turbot (*Scopthalmus maximus*) at metamorphosis. *J. exp. Mar. Biol. Ecol.* (in press).

Crockford, T. and Johnston, I.A. (1990) Temperature acclimation and the expression of contractile protein isoforms in the skeletal muscles of the common carp (*Cyprinus carpio* L.). *J. comp. Physiol.*, **160B**, 23–30.

Curtin, N.A. and Woledge, R.C. (1988) Energetic cost of power output by isolated fibre bundles from dogfish white muscle. *J. Physiol., Lond.*, **407**, 74P.

Egginton, S. and Sidell, B.D. (1989) Thermal acclimation induces adaptive changes in subcellular structure of fish skeletal muscle. *Am. J. Physiol.*, **256**, R1–9.

Fleming, J.R., Altringham, J.D. and Johnston, I.A. (1990) The effects of temperature acclimation on muscle relaxation in the carp: a mechanical, biochemical and ultrastructural study. *J. exp. Zool.*, **255**, 286–95.

Fletcher, G.L., King, M.H., Kao, M.H. and Shears, M.A. (1989) Anti-freeze proteins in the urine of marine fish. *Fish Physiol. Biochem.*, **6**, 121–7.

Fry, F.E.J. and Hart, J.S. (1948) Cruising speed of goldfish in relation to water temperature. *J. Fish. Res. Bd Can.*, **7**, 175–99.

Gerlach, G.-F., Turay, L., Malik, K.T.A., Lida, J., Scutt A. and Goldspink, G. (1990) Mechanisms of temperature acclimation in the carp: a molecular biology approach. *Am. J. Physiol.*, **259**, R237–44.

Greaser, M.L., Moss, R.L. and Reiser, P.J. (1988) Variations in contractile properties of rabbit single muscle fibres in relation to troponin T isoforms and myosin light chains. *J. Physiol., Lond.*, **406**, 85–98.

Guderley, H. (1990) Functional significance of metabolic responses to thermal acclimation in fish muscle. *Am. J. Physiol.*, **259**, R245–52.

Guderley, H. and Blier, P. (1987) Thermal acclimation in fish: conservative and labile properties of swimming muscle. *Can. J. Zool.*, **66**, 1105–15.

Guderley, H. and Foley, L. (1990) Anatomic and metabolic responses to thermal acclimation in the ninespine stickleback, *Pungitius pungitius*. *Fish Physiol. Biochem.*, **8**, 465–74.

Hazel, J.R. and Prosser, C.L. (1974) Molecular mechanisms of temperature compensation in poikilotherms. *Physiol. Rev.*, **54**, 620–77.

Heap, S.P., Watt, P.W. and Goldspink, G. (1985) Consequences of thermal change on the myofibrillar ATPase of 5 freshwater teleosts. *J. Fish Biol.*, **26**, 733–8.

Heap, S.P., Watt, P.W. and Goldspink, G. (1986) Myofibrillar ATPase activity in the carp (*Cyprinus carpio*): interactions between starvation and environmental temperature. *J. exp. Biol.*, **123**, 373–82.

Hess, F. and Videler, J.J. (1984) Fast continuous swimming of saithe (*Pollachius virens*): a dynamic analysis of bending movements and muscle power. *J. exp. Biol.*, **109**, 229–51.

Hwang, G.C., Watabe, S. and Hashimoto, K. (1990) Changes in carp myosin ATPase induced by temperature acclimation. *J. comp. Physiol.*, **160B**, 233–9.

Jankowsky, H.D. and Korn, H. (1965) The influence of acclimation temperature on the mitochondrial content of fish muscle. *Naturwissenschaften*, **52**, 642.

Johnson, T.P. and Johnston, I.A. (1991a) Temperature adaptation and the contractile properties of live muscle fibres from teleost fish. *J. comp. Physiol.*, **161**, 27–36.

Johnson, T.P. and Johnston, I.A. (1991b) Power output of fish muscle fibres performing oscillatory work: effects of acute and seasonal temperature change. *J. exp. Biol.*, **157**, 409–23.

Johnston, I.A. (1979) Calcium regulatory proteins and temperature acclimation of actomyosin from a eurythermal fish (*Carassius auratus* L.). *J. comp. Physiol.*, **129**, 163–7.

Johnston, I.A. (1982) Capillarisation, oxygen diffusion distances and mitochondrial content of carp muscles following acclimation from summer to winter temperatures. *Cell Tissue Res.*, **222**, 579–96.

Johnston, I.A. (1990) Cold adaptation in marine organisms. *Phil. Trans. R. Soc.*, **326B**, 655–67.

Johnston, I.A. and Altringham, J.D. (1985) Evolutionary adaptation of muscle power output to environmental temperature: force–velocity characteristics of skinned fibres isolated from Antarctic, temperate and tropical marine fish. *Pflügers Arch. ges. Physiol.*, **405**, 136–40.

Johnston, I.A. and Altringham, J.D. (1988) Muscle function in locomotion. *Nature, Lond.*, **335**(6193), 767–8.

Johnston, I.A. and Altringham, J.D. (1991) Movement in water: constraints and adaptations, in *Biochemistry and Molecular Biology of Fishes*, Vol. 1 (eds P.W. Hochachka and T. Mommsen), Elsevier, Amsterdam, pp. 249–68.

Johnston, I.A. and Dunn, J. (1987) Temperature acclimation and metabolism in ectotherms with particular reference to teleost fish, in *Temperature and Animal Cells* (ed. K. Bowler), *Soc. Exp. Biol. Symp.*, **XXXI**, 67–93.

Johnston, I.A. and Lucking, M. (1978) Temperature induced variation in the distribution of different types of muscle fibres in the goldfish (*Carassius auratus*). *J. comp. Physiol.*, **124**, 111–16.

Johnston, I.A. and Maitland, B. (1980) Temperature acclimation in crucian carp: a morphometric study of muscle fibre ultra-structure. *J. Fish Biol.*, **17**, 113–25.

Johnston, I.A., Davison, W. and Goldspink, G. (1975) Adaptations in myofibrillar ATPase activity induced by temperature acclimation. *FEBS Lett.*, **50**, 293–5.

Johnston, I.A., Davison, W. and Goldspink, G. (1977) Energy metabolism of carp swimming muscles. *J. comp. Physiol.*, **114**, 203–16.

Johnston, I.A., Fleming, J.R. and Crockford, T. (1990) Thermal acclimation and muscle contractile properties in cyprinid fish. *Am. J. Physiol.*, **259**, R231–6.

Johnston, I.A., Johnson, T.P. and Battram, J.C. (1991) Low temperature limit burst swimming performance in Antarctic fish, in *The Biology of Antarctic Fish* (eds B. di Prisco, B. Maresca and B. Tota), Springer-Verlag, Berlin, Heidelberg, 179–90.

Johnston, I.A., Sidell, B.D. and Driedzic, W.R. (1985) Force–velocity characteristics and metabolism of carp muscle fibres following temperature acclimation. *J. exp. Biol.*, **119**, 239–49.

Jones, P.L. and Sidell, B.D. (1982) Metabolic responses of striped bass (*Morone saxatilis*) to temperature acclimation II. Alterations in metabolic carbon sources and distributions of fibre types in locomotory muscle. *J. exp. Zool.*, **219**, 163–71.

Kleckner, N.W. and Sidell, B.D. (1985) Comparisons of maximal activities of enzymes from tissues of thermally-acclimated and naturally-acclimatized chain pickerel (*Esox niger*). *Physiol. Zool.*, **58**, 18–28.

Langfeld, K.S., Crockford, T.C. and Johnston, I.A. (1991) Temperature acclimation in the common carp: force–velocity characteristics and myosin subunit composition of slow muscle fibres. *J. exp. Biol.*, **155**, 291–304.

Leeuwen, J.L. van (1991) Optimum power output and structural design of sarcomeres. *J. theor. Biol.*, **149**, 229–56.

Leeuwen, J.L. van, Lankheet, M.J.M., Akster, H.A. and Osse, J.W.M. (1990) Function of red axial muscles of carp (*Cyprinus carpio*): recruitment and normalized power output during swimming in different modes. *J. Zool., Lond.*, **220**, 123–45.

Lemons, D.E. and Crawshaw, L.I. (1985) Behavioural and metabolic adjustments to low temperatures in the largemouth bass (*Micropterus salmoides*). *Physiol. Zool.*, **58**, 175–80.

Libera, L.D., Carpene, E., Theibert, J. and Collins, J.H. (1991) Fish myosin alkali light chains originate from two different genes. *J. Muscle Res. Cell Motil.*, **12**, 366–71.

Moerland, T.S. and Sidell, B.D. (1986) Biochemical responses to temperature in the contractile protein complex of striped bass (*Morone saxatilis*). *J. exp. Biol.*, **238**, 287–95.

Nabeshima, Y., Fujii-kuriyama, Y., Muramatsu, M. and Ogata, K. (1984) Alternative transcription and two modes of splicing result in two myosin light chains from one gene. *Nature, Lond.*, **308**, 333–8.

Penney, R.K. and Goldspink, G. (1980) Temperature adaptation of sarcoplasmic reticulum of fish muscle. *J. therm. Biol.*, **5**, 63–8.

Rome, L.C. and Sosnicki, A.A. (1991) The influence of temperature on mechanics of red muscle in carp. *J. Physiol., Lond.*, **427**, 151–69.

Rome, L.C., Loughna, P.T. and Goldspink, G. (1984) Muscle fibre recruitment as a function of swim speed and muscle temperature in carp. *Am. J. Physiol.*, **247**, R272–9.

Rome, L.C., Loughna, P.T. and Goldspink, G. (1985) Temperature acclimation: improved sustained swimming performance at low temperature. *Science, Wash. D.C.*, **228**, 194–6.

Rome, L.C., Funke, R.P., Alexander, R. McN., Lutz G., Aldridge, H.D.J.N., Scott, F. and Freadman, M. (1988) Why animals have different muscle fibre types. *Nature, Lond.*, **355**, 824–7.

Rome, L.C., Funke, R.P. and Alexander, R. McN. (1990) The influence of temperature on muscle velocity and sustained swimming performance in swimming carp. *J. exp. Biol.*, **154**, 163–78.

Sidell, B.D. and Hazel, J.R. (1987) Temperature affects the diffusion of small molecules through cytosol of fish muscle. *J. exp. Biol.* **129**, 191–203.

Sidell, B.D. and Johnston, I.A. (1985) Thermal sensitivity of contractile function in chain pickerel, (*Esox niger*). *Can. J. Zool.*, **63**, 811–16.

Sidell, B.D., Johnston, I.A., Moerland, T.S. and Goldspink, G. (1983) The eurythermal myofibrillar protein complex of the mummichog (*Fundulus heteroclitus*): adaptation to a fluctuating thermal environment. *J. comp. Physiol.*, **153B**, 167–73.

Sisson, J.E. and Sidell, B.D. (1987) Effect of thermal acclimation on muscle fibre recruitment of swimming striped bass (*Morone saxatilis*). *Physiol. Zool.*, **60**, 310–20.

Thillart, G. van den and Modderkolk, J. (1978) The effect of acclimation temperature on the activation energies of state III respiration and on the unsaturation of membrane lipids of goldfish mitochondria. *Biochem. biophys. Acta*, **510**, 38–51.

Tyler, S. and Sidell, B.D. (1984) Changes in mitochondrial distribution and diffusion distances in muscle of goldfish (*Carassius auratus*) upon acclimation to warm and cold temperatures. *J. exp. Zool.*, **232**, 1–10.

Vieira, V.L.A. and Johnston, I.A. (1992) Influence of temperature on muscle fibre development in larvae of the herring (*Clupea harengus*) L. *Mar. Biol.*, **112**, 331–41.

Wainwright, S.A. (1983) To bend a fish, in *Fish Biomechanics* (eds P.W. Webb and D. Weihs), Praeger, New York, pp. 69–91.

Walsh, P.J. and Moon, T.W. (1982) The influence of temperature on extracellular and intracellular pH in the American eel, *Anguilla rostrata* (Leseueur). *Respir. Physiol.*, **50**, 129–40.

Chapter fourteen

Recent advances in the ecophysiology of Antarctic notothenioid fishes: metabolic capacity and sensory performance

J.C. Montgomery and R.M.G. Wells

14.1 INTRODUCTION

The Antarctic fishes are a fascinating group, consisting of a high proportion of endemic species in the suborder Notothenioidei living in the continental shelf seas at temperatures close to freezing. A number of recent reviews cover many aspects of their biology: their general biology (Andriashev, 1987; Eastman, 1990), evolution (Eastman and Grande, 1989; Eastman, 1991), reproduction (North and White, 1987), ecology (Hubold, 1992), and comparative physiology (Macdonald *et al.*, 1987).

Ecological factors impose selection pressures on a population, and stable environments are the exception rather than the rule. The Antarctic seas are an example of exceptional stability with regard to temperature, oxygen, and salinity, but there are also extreme changes in seasonal illumination under the sea ice cover, and in food supply, which are expected to influence the metabolic capacity and sensory performance of the fauna. The responses of the fish to these environmental challenges provide the ecophysiologist with some powerful adaptationist arguments; a consequence of the apparent 'goodness of fit', or interactions between organism and environment, has been to view their traits as optimized by natural selection. It is reasonable to suppose

Fish Ecophysiology. Edited by J. Cliff Rankin and Frank B. Jensen. Published in 1993 by Chapman & Hall, London. ISBN 0 412 45920 5.

that if adaptation is a process fitting an organism to its environment (Gans, 1988), then most structures in Antarctic fishes arose in other habitats. How then, were the present phenotypes and their roles generated and maintained?

The Antarctic provides a special opportunity for studying the physiological consequences of ecological differences between fishes, without the attendant difficulty of comparing species from diverse phylogenetic backgrounds. The present review has two themes. First, the monophyletic radiation of notothenioid fishes has occurred over a comparatively short evolutionary time scale and in a low-competition environment (Eastman, 1991). Therefore, we should not expect to find perfect matching between structural design and function. Second, some of what we observe in Antarctic fishes may be epiphenomenal and reflect current utility, or simply the plasticity of the phenotype, and it cannot be assumed that all the observed traits arose by natural selection.

14.2 RESPIRATION AND METABOLISM

Oxygen uptake rates

The depressing effect of low temperature on metabolic rate led 19th Century biologists to surmise that life was barely sustainable in the frigid waters of Antarctica. The first estimates of metabolic rates from resting Antarctic fish were based on oxygen uptake measurements ($\dot{M}O_2$) made by Wohlschlag (1960), who concluded that the rates for nototheniids were higher than expected from downward extrapolation of rates from temperate and tropical fish, thus leading to the concept of 'metabolic cold adaptation' (MCA), viewed as an adaptation to counter the rate-depressing effect of a permanently cold (-1.9 °C) marine environment. The concept, however, has attracted criticism because: (i) measured rates might be elevated artefactually by capture and handling stresses; (ii) since Antarctic fish become moribund just a few degrees above zero, and temperate fish cannot survive freezing conditions, how should data be extrapolated? (data from acute, or acclimatory experiments, or interspecific comparisons of $\dot{M}O_2$ from fish living at their natural temperature are possibilities); (iii) comparisons with appropriate ecotypes of marine perciform fish are more appropriate than with freshwater species, but from shallow or deep water? and (iv) what is the evolutionary advantage of MCA if high resting rates leave little energy remaining for growth and reproduction?

These questions were reviewed by Macdonald *et al.* (1987), who concluded that metabolic rates were probably elevated, but not to the extent earlier suggested. Further, elevated rates should be seen not as adaptations to low temperature per se, but as consequences of an increased energy flux associated with other adaptive changes, such as antifreeze formation and renal secretory function, and temperature adaptation in other tissues.

Table 14.1 Metabolic rates of Antarctic fishes calculated for 100 g fish at temperatures close to freezing

Species	Resting $\dot{M}O_2$* (mg kg^{-1} h^{-1})	Routine $\dot{M}O_2$ (mg kg^{-1} h^{-1})
A. Mobile, semi-pelagic predators and zooplanktivores		
Nototheniidae		
Pagothenia borchgrevinki	48.5[a], 42.8[b]	70.0[c]
Notothenia gibberifrons	26.7[d]	68.8[e]
N. neglecta	36.1[d]	87.3[f]
N. rossii	63.9[d]	–
N. nudifrons	–	67.4[g]
N. coriiceps	–	94.6[e]
N. angustifrons	–	91.1[g]
N. magellanica	–	102.0[h]
Channichthyidae		
Pseudochaenichthys georgianus	31.9[i]	–
Myctophidae		
Electrona antarctica	25.7[j]	49.0[j]
Gymnoscopelus braueri	24.1[j]	39.1[j]
Gymnoscopelus opisthopterus	16.2[j]	27.3[j]
B. Benthic, intermittently active		
Nototheniidae		
Trematomus bernacchii	55.7[a]	–
T. pennellii	26.7[a]	–
T. hansoni	36.1[a]	–
Bathydraconidae		
Gymnodraco acuticeps	45.3[a]	–
C. Benthic, sluggish		
Nototheniidae		
Trematomus lönnbergi	34.8[a]	–
Artedidraconidae		
Pogonophryne scotti	17.4[k]	29.0[k]
Harpagifer georgianus	23.3[g]	–
Channichthyidae		
Chaenocephalus aceratus	30.3[d]	56.2[f]
Pagetopsis macropterus	35.0[l]	60.0[l]
Zoarcidae		
Lycodichthys dearborni	10.2[m]	19.4[n]
Bathylagidae		
Bathylagus antarcticus	7.0[j]	18.0[j]

* Sources: a, Wells (1987); b, Forster *et al.* (1987); c, Davison *et al.* (1990); d, Holeton (1970); e, Hemmingsen *et al.* (1969); f, Ralph and Everson (1968); g, Morris and North (1984); h, Hureau *et al.* (1977); i, Hemmingsen and Douglas (1977); j, Torres and Somero (1988a); k, Saint-Paul *et al.* (1988); l, Robilliard and Dayton (1969); m, Wells (1986); n, Wohlschlag (1964).

Torres and Somero (1988a,b) found higher than expected rates of $\dot{M}O_2$ for Antarctic species, and for temperate fish living at great depth. By comparing the metabolism of congeneric species with similar ways of living, these authors have been able to separate temperature from depth as factors influencing metabolism, and they advanced the novel idea that both factors contribute to MCA.

Recent work on energy metabolism in the Antarctic has focused on the ecological differences between species. Measurements have confirmed the higher-than-expected rates of metabolism in notothenioids (Forster *et al.*, 1987; Wells, 1987; Saint-Paul *et al.*, 1988; Torres and Somero, 1988a,b). Within the notothenioids, and among other Antarctic fishes, there is an evident relationship between resting $\dot{M}O_2$ and activity (Table 14.1). From these data, we may generalize that it is the most active species which most clearly show MCA, and that these species tend to be notothenioids from the high Antarctic continental shelf seas. The oceanic mesopelagic fishes include dominant myctophid and bathylagid faunas (Hubold, 1992), which are less active.

Moreover, $\dot{M}O_2$ measured at various swimming speeds indicates that Antarctic fish have aerobic scopes for activity similar to those of fish from warmer waters (Forster *et al.*, 1987; Johnston *et al.*, 1991). It may be concluded from metabolic studies on *Notothenia neglecta* (Johnston *et al.*, 1991) and the pelagic *Pleuragramma antarcticum* (Hubold *et al.*, 1990) that less energy is available for sustained locomotory activity and for growth.

A number of questions remain unanswered in relation to metabolism at the organismic level. Are there subtle seasonal changes in $\dot{M}O_2$; is MCA a summer phenomenon; is MCA underestimated by expression of wet fish weight, which includes proportionately high tissue water content and lipid stores?

Biochemical approaches to metabolism

Questions about cold adaptation are now focusing in a new direction. Sidell (1991) asks the question, have Antarctic fish become adapted to allow raised levels of metabolic activity? If biochemical indices from the activities of cytochrome oxidase and citrate synthase enzymes are used, the answer appears affirmative. By estimating maximal activities of enzymes from central pathways of aerobic metabolism in ecotypically similar Antarctic and Temperate Zone marine fishes, Crockett and Sidell (1990) proposed a selective expansion of aerobic metabolism to support activity at low temperature. Specifically, enzymes of fatty acid oxidation showed up to 27-fold increases in activity in the nototheniids family, but MCA could not be demonstrated in either aerobic or anaerobic carbohydrate metabolism. These comparisons were made at 1°C, although it might be pointed out that ventricular oxidative capacity in *Trematomus newnesi* is still higher than that in the tautog, *Tautoga onitis*, when compared at 1 °C and 10 °C respectively.

Torres and Somero (1988a,b) have compared lactate dehydrogenase (LDH) and citrate synthase activities in Antarctic and temperate fish from depths of 0–1000 m with a view to evaluating the potential for expenditure of muscular energy by respective anaerobic and aerobic pathways. Both enzymes decline in their activity with depth of their occurrence, but citrate synthase activity declines less than expected (and less than LDH), again providing an enzymatic basis for fatty acyl MCA. The somatic tissues of icefish, *Chaenocephalus aceratus*, however, appear well endowed with glycolytic enzymes for the anaerobic generation of ATP (Bacila *et al.*, 1989). These biochemical studies have provided compelling reasons to examine not basal metabolic rates, but active rates.

Further evidence for cold adaptation is provided by facilitated information transfer from DNA to RNA and protein transcription (Haschemeyer, 1985). Rates of protein synthesis in Antarctic fishes are greater than predicted for subzero temperatures, indicating cold adaptation at the level of molecular synthesis (Smith and Haschemeyer, 1980).

Surfaces for gas exchange

Gill surface areas in notothenioids are similar to, or lower than, those in other teleosts, and although oxygen extraction from the branchial water current may be low, high ventilation flow ensures adequate blood oxygenation (review: Macdonald *et al.*, 1987). Within the notothenioids there are clear differences in gill morphology relating to activity levels. The gill lamellae are more closely packed in *Pleuragramma antarcticum* (one of the very few pelagic nototheniids) than those in the more sedentary *Notothenia gibberifrons* (Kunzmann, 1990). *P. antarcticum* has a relatively small gill area by comparison with temperate species, and despite its pelagic description, appears to hover in the water column at neutral density for prolonged periods (Hubold *et al.*, 1990). The absence of haemoglobin in the icefish, *C. aceratus*, is not compensated by an increase in gill area, but by exceptionally high ventilation flow (Hughes, 1972). The energy cost of ventilation as a function of mass and increased viscosity of seawater at low temperature has not been determined.

The fine structure of the gills in notothenioid fish is similar to that of other teleosts, except for a high density of mucous cells which may limit gas diffusion (Boyd *et al.*, 1980). Low numbers of chloride cells are also observed, which correlate with the relatively high saline composition of nototheniid plasma. The rate of gas diffusion decreases with a reduction in temperature (Q_{10} = 1.22), but there is no evidence for a compensatory reduction in the thickness of exchange surfaces (Franklin and Davison, 1988).

In view of the requirement to maximize oxygen uptake and minimize water flux across the gill lamellae, gill perfusion studies in the haemoglobinless icefish are of particular importance. Rankin (1989) has addressed this problem in *C. aceratus*, and found that while catecholamines produce similar vascular effects to those seen in other teleosts, they have no effect on water flux, thus supporting the idea of lamellar recruitment during exercise.

Scales are reduced or absent among Antarctic fish, and it might be supposed that the dermis complements the branchial system as an accessory site for gas exchange. At rest, nearly half of the total gas flux is dermal in both the icefish, *C. aceratus* (Hemmingsen and Douglas, 1970), and in the zoarcid *Lycodichthys* (= *Rhigophila*) *dearborni* (Wells, 1986). The bathydraconid, *Gymnodraco acuticeps* (lit. 'naked dragon with needles' [teeth]) has no scales, and though a moderately well-developed cutaneous vasculature is evident, particularly in the head region, the skin is not otherwise modified for respiration, and remains secondary to the gills as a respiratory surface (Eastman and Hikida, 1991). The importance of transcutaneous exchange will, of course, diminish as a fish increases its demand for oxygen through activity. Nototheniids have only modest transcutaneous oxygen exchange at rest (Wells, 1987).

Growth rates

Growth rates are central to discussion of MCA, because fish with faster growth rates require matching high metabolic rates to supply the necessary energy for growth (De Vries and Eastman, 1981; Hubold, 1991). There is evidence that *Notothenia rossii* has a faster growth than do other notothenioids (Shust and Pinskaya, 1978), but this may be influenced by the intensive fishing pressure on the species, and the fact that it is absent from the colder Ross Sea. Slow growth rates are reported from fragmentary information on nototheniids (Wohlschlag, 1960), and from recapture data of the largest fish in the Ross Sea, *Dissostichus mawsoni* (DeVries and Eastman, 1981). Recent work confirms that low growth rates are characteristic of notothenioids from the Antarctic continental shelf seas and are slower than those from the sub-Antarctic (Kock, 1985; Hubold *et al.*, 1990; Hubold 1991). These observations lead us to the view that metabolic constraints influence the growth rates of Antarctic fish.

14.3 THE CARDIOVASCULAR SYSTEM

Anatomy of the heart

The heart is central to the functional requirements of a circulatory system, and interspecific differences in myocardial architecture may be expected to reflect ecophysiological considerations. Weight-specific cardiac mass is greater in *Notothenia neglecta* and *N. rossii* than that in *N. gibberifrons*, which is less active (Holeton, 1970; Harrison *et al.*, 1991). Similarly, it is greater in the cryopelagic *Pagothenia borchgrevinki* than in the benthic *Trematomus bernacchii*, and both appear larger than those of comparable temperate benthic teleosts (Axelsson *et al.*, 1992). The haemoglobinless icefishes are most unusual in that the relative cardiac mass is about three times that of red-blooded nototheniids, and in similar proportion to the hearts of tuna and mammals (Johnston *et al.*, 1983; Harrison *et al.*, 1991; Tota *et al.*, 1991a). The icefish, however, is not cardiomegalic in the sense of having hypertrophied ventricular walls.

The gross anatomy of the generalized fish myocardium shows differentiation into a compact outer layer bearing a coronary arterial supply, and an inner spongy layer formed of radially enclosed spaces separated by trabeculae. The development of the compact layer is highly correlated with the degree of athleticism displayed by the fish. It may be absent in sedentary benthic fish and yet constitutes up to 40% of the myocardium in extremely active species (Santer, 1985; Davie, 1990; Satchell, 1991). This poses an evolutionary problem for notothenioids because they have radiated from a sedentary benthic habitat into a wide range of ecotypes, including the pelagos. Thus the compact layer and associated coronary system is entirely missing in the semi-pelagic *P. borchgrevinki* (C. Franklin, 1991, pers. comm.), and as in other notothenioids, the ventricular myocardium appears entirely spongy and highly trabeculate (Tota *et al.*, 1991a).

The icefishes are again unusual among teleosts in having a subepicardium supplied with a well-developed arterial system (Feller *et al.*, 1983, 1985), which in the case of *C. aceratus* penetrates the lacunae of the ventricular lumen (Harrison *et al.*, 1991). This cannot yet be termed a coronary supply because the origin of the vessels has not been established. The lacunar network in the spongy layer appears more complex than that of other fish, including *N. rossii*, and is served by Thebesian vessels that provide for shunting blood between the lacunar spaces (Feller *et al.*, 1985; Tota *et al.*, 1988).

Interspecific differences in ventricular myoglobin content correlate well with aerobic activity (review: Satchell, 1991). *N. neglecta* is an active notothenioid containing 20 nM myoglobin g^{-1} ventricle (Johnston and Harrison, 1987), a value comparable to that in rainbow trout (Douglas *et al.*, 1985). Less active notothenioids have visibly paler ventricles. Icefish are again unusual in apparently lacking myoglobin (Hamoir, 1988), but Douglas *et al.* (1985) have detected up to 1 nM g^{-1} in some icefish by using sensitive difference-spectroscopy.

Driedzic *et al.* (1982) have evaluated myocardial performance in relation to oxygen availability for hearts with differing myoglobin content, and recorded a drastic fall in the performance of hypoxic hearts with myoglobin-poor myocardia. This observation accords with the poor performance of the icefish (*C. aceratus*) heart under hypoxia (Hemmingsen *et al.*, 1972).

One might expect that myocytes lacking myoglobin would have smaller dimensions to facilitate oxygen diffusion. However, not only is there no difference in myocyte diameter between icefish and nototheniids (Feller *et al.*, 1985), but morphometric measurements in *C. aceratus* and *N. neglecta* show that myocytes are significantly larger in both length and diameter in comparison with those of temperate species, and are assembled into very loosely packed myofibrils (Harrison *et al.*, 1991). A further peculiarity of icefish myocytes is that mitochondrial densities are some six times higher than those in either *Notothenia* spp. or temperate species (Johnston and Harrison, 1987; Hamoir, 1988). High mitochondrial densities are not a usual feature of myoglobin-poor hearts. In addition, icefish myocytes contain exceptionally high levels of glycolytic enzymes, particularly lactate dehydrogenase (LDH), and hexokinase (Johnston *et al.*, 1983; Hamoir, 1988; Bacila *et al.*, 1989). Feller *et al.* (1991) assert that cardiac LDH in the icefish is an adaptation to hypoxia tolerance. These observations are perplexing, because the high mitochondrial density is not accompanied by elevated cytochrome oxidase activity, and thus cannot be related to high oxidative capacity (Johnston and Harrison, 1987). Most likely, the rate-depressing effect of low temperature on enzymes of oxidative phosphorylation accounts for high mitochondrial density in icefish myocytes. Increased mitochondrial density following low-temperature acclimation of temperate teleosts tends to support this hypothesis (Sidell, 1983), though the phenomenon is not universal (Bowler and Tirri, 1990). Cytochrome is a porphyrin-based haem pigment like myoglobin, and its oxygen affinity constant in icefish might well be low enough to provide an effective oxygen gradient. Alternatively, the mitochondria may be associated with enzymes of greater capacity for ATP yield through glycolytic or fatty acid pathways, and activities of key enzymes in the latter pathway are significantly higher in ventricular tissue from Antarctic notothenioids compared with temperate fishes measured at the same temperature (Crockett and Sidell, 1990; Sidell, 1991). Because oxygen is four to five times more soluble in lipid than in water, the high intracellular fat content of aerobic tissues in Antarctic fishes may significantly enhance oxygen delivery to mitochondria (Sidell, 1991). Here, we must be cautious in seeking adaptive explanations. A glycolytic role for high LDH activity has not been demonstrated, and it may be that the enzyme plays a structural role in the myocardium, as does its cardiac isozyme in the case of vertebrate eye lens crystallin (de Jong *et al.*, 1989).

Cardiac performance

The suite of complementary features characterizing the cardiovascular system of the icefish has been reviewed by Tota *et al.* (1991a) and may be summarized:

1. proportionately large heart volume and blood vessels of comparatively large diameter;
2. blood volume two- to four-fold higher than that in other teleosts, and absence of red blood cells;

3. spectacular cardiac output (150–300 ml min^{-1} kg^{-1}), exceeding that of the most athletic teleosts, achieved through large stroke volume coupled with low beat frequency;
4. low output pressure in the ventral aorta and minimal resistance in the gill lamellae.

The mechanical performance of the icefish heart, isolated from neurohumoral influences, has been investigated by Tota *et al.* (1991), who concluded that the organ is a specialized volume pump whose output is regulated by the Starling mechanism in response to preload. (The Starling mechanism describes the increased force of ventricular contraction when ventricular blood volume is increased.) Constant stroke work is possible only over a narrow range of afterload which corresponds to the low ventral aortic pressures measured in icefish by Hemmingsen *et al.*, (1972). Both the ventricle and the bulbus are highly compliant, with the latter functioning to dampen systolic pulses, and possibly delaying the time at which valves must close to prevent aortic backflow.

The complex and unusual microanatomy of the icefish heart reveals components that appear to increase the elastic modulus of the cardiac wall, and is entirely consistent with its exceptional physiological performance in adjusting cardiac output to a range of preloads without raising heartbeat frequency (Harrison *et al.*, 1991; Tota *et al.*, 1991a). Stroke volume is therefore the major determinant of systolic function in the icefish, and thus responses

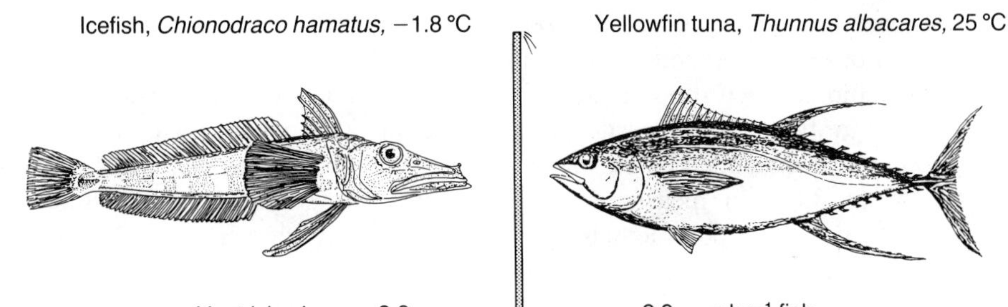

Fig. 14.1 Comparison of cardiac performance in icefish and tuna. Performance statistics derive from perfused hearts with peak loading. The height of fluid in the tubes connected to the hearts represents afterload equivalent to pressures in the ventral aorta. Data normalized to 1 kg body weight and based on Tota *et al.* (1991b) and Farrell *et al.* (1992).

Table 14.2 Relative ventricular mass (RVM), maximum cardiac (CO) and power outputs (PO) of isolated perfused teleost fish hearts induced by preloading

Species	RVM	CO	PO	Reference
	(% body wt)	(ml min^{-1} kg^{-1})	(mW kg^{-1} fish)	
Eel, *Anguilla dieffenbachii*	0.05	22.3	1.8	Franklin and Davie (1991)
Rainbow trout, *Oncorhynchus mykiss*	0.07	42.4	3.9	Graham and Farrell (1989)
Yellowfin tuna, *Thunnus albacares*	0.29	108.0	21.9	Farrell *et al.* (1992)
Icefish, *Chaenocephalus aceratus*	0.39	119.0	6.0	Tota *et al.* (1991b)

to preload during exercise will improve cardiac output. Intrapericardial pressures appear to be important for filling the heart, since puncturing the pericardium *in situ* results in ventricular collapse (Feller *et al.*, 1985). The spectacular performance of the icefish and tuna hearts are compared in Fig. 14.1 where the tuna heart is seen to operate as a pressure pump and achieves a high chronotropic cardiac output through homeometric regulation. A comparison of peak mechanical performance of the icefish heart against those from other teleosts with much higher metabolic rates is given in Table 14.2.

Recent functional studies by Axelsson *et al.* (1992) on the cardiovascular systems of two nototheniids are an important advance in understanding the relationship between the red-blooded nototheniids and the icefish. They have recorded resting ventral aortic pressures of 3.6 kPa in the cryopelagic *P. borchgrevinki*, and 3.1 kPa in the sedentary benthic *T. bernacchii*, which are similar to 3.0 kPa measured in *C. hamatus*, but despite the respectable haematocrits in the nototheniids, the values are low by comparison with those from temperate teleosts. Further, resting cardiac output is approximately 68% greater in the active nototheniid (29.6 ml min^{-1} kg^{-1}) compared with the benthic species (17.6 ml min^{-1} kg^{-1}), which correlates cardiac performance with life style. Regulation of cardiac output is unusual in the cryopelagic *P. borchgrevinki*, in that heart rate doubles to raise output by 75% during sustained swimming. Further, heart rate in the nototheniids at rest is comparatively low (10–12 beats min^{-1}) by comparison with icefish (*C. aceratus*: 17.5 beats min^{-1}), but the rate more than doubles when cholinergic inhibitions are relieved (Axelsson *et al.*, 1992). It is necessary to emphasize that these are among the first *in vivo* measurements of cardiac output measured directly from miniature calibrated Doppler probes, and are thus free from errors inherent in using the Fick principle. The Fick principle states that oxygen consumption is equal to cardiac outflow multiplied by the difference between arterial and mixed venous oxygen contents, and assumes that oxygen uptake takes place exclusively across the gills and that the gills themselves do not consume oxygen.

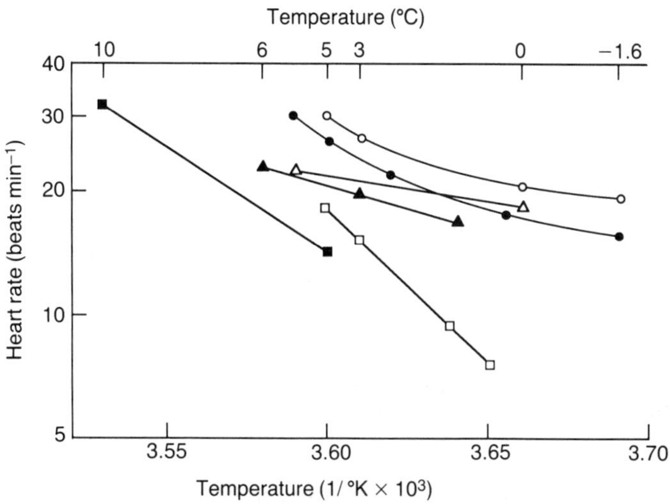

Fig. 14.2 Arrhenius plots for *Pagothenia borchgrevinki* heart (○), atrium (●) (Wells, unpublished), icefish, *Chionodraco hamatus*, heart (Δ) (Tota *et al.*, 1991b), sole, *Solea vulgaris*, heart (▲) (Sureau *et al.*, 1989), perch, *Perca fluviatilis*, atrium (❑) (Bowler and Tirri, 1990), and sea raven, *Hemitripterus americana*, heart (■) (Bailey and Driedzic, 1989). Axes are frequency in beats per minute, temperature and the reciprocal of absolute temperature.

Temperature

Increasing the temperature of the icefish, *C. hamatus*, heart by 5 °C raises heart rate and cardiac output by one-fifth, but unlike temperate teleosts, stroke volume remains unaffected (Tota *et al.*, 1991). Similarly, for the red-blooded nototheniid, *Pagothenia borchgrevinki*, there is no change in ventral aortic pressure for a 5 °C rise (Grigg and Wells, unpublished), compared with a doubling in pressure for the sea raven, *Hemitripterus americanus* (Bailey and Driedzic, 1989). The chronotropic response of Antarctic fish hearts to temperature shows higher intrinsic rates when compared with hearts from temperate fish acclimated to low temperature, and sedentary fish seem chronotropically less sensitive to temperature (Fig. 14.2).

Blood viscosity

Viscosity approximately doubles when a liquid is cooled from 20 °C to 0 °C, and for blood, this will directly affect vascular resistance and therefore the work that the heart must perform to maintain circulation. Nearly 25% of resting energy expenditure is directed towards this task in the icefish (Hemmingsen *et al.*, 1972). In addition to temperature, the major contribution to blood viscosity is red cell mass, and it might be expected that reducing the number of erythrocytes would improve blood flow and result in energy savings. The viscosity of blood in the icefish, *Chionodraco hamatus*, is indeed lower (3.99 cP) than that in the red-blooded nototheniid *Pagothenia bernacchii* (4.91 cP) at -1.8 °C and exhibits non-Newtonian flow behaviour. Thus at higher shear rates (equivalent to flow), the viscosity decreases and red cells contribute less to resistance (Wells *et al.*, 1990). The effective delivery of oxygen to tissues depends on both blood flow and oxygen capacity (Wells and Weber, 1991), and so it is apparent that the high solubility of oxygen and high flow rate in channichthyiid blood are factors tending to optimize delivery.

Interestingly, there appears to be little difference in the viscosity of blood between *T. bernacchii* and blood from the New Zealand blue cod, *Parapercis colias*, cooled to the same temperature over a range of shear rates (Macdonald and Wells, 1991). This suggests that there are no special adaptations for increasing erythrocyte deformability at Antarctic temperatures.

Future directions in comparative cardiology

Further measurements of cardiac output that are independent of the Fick principle are required because inherent errors due to gill shunts and metabolism, and transcutaneous gas exchange, may have resulted in gross over-estimations (Satchell, 1991; Tota *et al.*, 1991; Axelsson *et al.*, 1992). Questions concerning the distribution of blood in the heart throughout the cardiac cycle, blood flow from the bulbus during diastole, and the proportion of ventricular blood volume expelled during systolic contraction remain largely unaddressed. New advances in technology using principles of echotomographic imaging with ultrasound probes offer considerable promise in this direction (P.S. Davie; B. Tota, 1991, pers. comms)

Neurohumoral control mechanisms in Antarctic fish hearts also require elucidation. Our unpublished data from *Pagothenia borchgrevinki* indicate a high vagal tonus, and the isolated atrium appears chronotropically insensitive to catecholamines. Furthermore, the consequences of low ventral aortic pressure should be investigated, because an important function of the circulation is to transmit pressure for functions such as erectile tissue, fluid exchange in capillary beds, and in ultrafiltration. With respect to the last, in the absence of glomerular filtration in the notothenioid kidney (Eastman and DeVries, 1986), it may be that high pressures are not required and that this in turn may result in some energetic savings.

Aside from these intriguing conjectures, studies in comparative cardiac design from the pelagic species *Pleuragramma*, the giant midwater predator *Dissostichus*, and from representatives of the Harpagiferidae and bathydraconid notothenioids which include species with extremely low haemoglobin values (Wells *et al.*, 1980), as well as nototheniids from lower latitudes, would be invaluable in further elucidating ecophysiological trends. The origin and distribution of arterial vessels to the myocardium requires clarification and may be expected to show correlation with swimming performances.

14.4 THE BLOOD OXYGEN TRANSPORT SYSTEM

The solubility of gas increases with a reduction in temperature. The volumes of oxygen and carbon dioxide that are transported in physical solution in the body fluids of Antarctic fishes are therefore much higher than those of fish from warmer waters and, in conjunction with low metabolic rates, it is both expected and observed that the quantity of haemoglobin (Hb) and numbers of red blood cells might be low in Antarctic species (Wells *et al.*, 1980).

The Channichthyidae, or icefishes, are unique among vertebrates in that all members of the family lack red blood cells, and the oxygen-carrying capacity of the blood is only about one-tenth that of other notothenioids. It is often assumed that the icefishes pay a price for their lack of Hb, so that they are easily exhausted and have poor responses to hypoxic insults (Hemmingsen and Douglas, 1972). But far from being relatively sedentary species, several are highly active predators (Macdonald *et al.*, 1987). Moreover, during a failure of the aquarium system at Terra Nova Bay Station in 1988, the fall in oxygen levels resulted in heavy mortality of captive red-blooded nototheniids, but the icefishes, *Chionodraco hamatus* and *Cryodraco antarcticus*, survived.

Among the red-blooded notothenioids, there is a clear relationship between blood Hb content or haematocrit, and the activity levels of the fish; active species have higher oxygen-carrying capacities (Wells *et al.*, 1980). The concentration of Hb in the cells (MCHC), however, does not appear to be related to activity, and is lower in polar species than that in temperate and tropical fishes. Low MCHC does not confer an advantage in reducing blood viscosity (Wells *et al.*, 1990), and its significance for Antarctic fishes remains unknown.

The role of Hb in the red-blooded nototheniids has been questioned following the demonstration that *Trematomus bernacchii*, like most temperate inactive benthic fish, tolerates exposure to carbon monoxide (di Prisco *et al.*, 1991). The species also survives exsanguination through serial venesection down to haematocrit values $< 0.5\%$, but in resting, control fish there is always an arterio–venous oxygen content difference, demonstrating that the Hb is functional (Wells and Macdonald, unpublished). The extent to which other components of the respiratory system may be regulated in anaemia may provide a fruitful avenue for further research.

Molecular ecology of haemoglobin

Haemoglobin multiplicity is common in many fish and may be categorized according to the functional scheme of Weber and Wells (1989): (i) possession of either a single Hb component, or multiple haemoglobins with similar pH- dependence of oxygen affinity (e.g. inactive species such as carp, flatfish); (ii) multiple Hbs with significant functional differences; anodal components sensitive to pH and allosteric modulation, and cathodal components with reduced functional plasticity (e.g. trout); (iii) apparently temperature-insensitive Hbs (e.g. tuna).

A number of Antarctic notothenioids lack multiple Hb components (Wells *et al.*, 1980). D'Avino and di Prisco (1988), di Prisco (1988), and Kunzmann (1991) have examined additional nototheniids and identified a major anodal Hb component comprising ≈ 90% of total Hb, with a second minor Hb; harpagiferid and bathydraconid notothenioids may posses only one Hb component.

The major and minor Hb components have similar oxygen-binding properties (di Prisco *et al.*, 1988, 1990) and thus fall into category (i), along with other sedentary benthic species. One exception is the active cryopelagic species, *Trematomus newnesi*, which has two major components that appear functionally distinct (di Prisco *et al.*, 1991). Structural studies on the major and minor Hb components have revealed a common β-chain in *Notothenia neglecta* (di Prisco *et al.*, 1990), but in the bathydraconid *Cygnodraco mawsoni*, it is the α-globin chain that is shared (Caruso *et al.*, 1991).

The zoarcid fishes of Antarctica are not related to notothenioids, but derive from a cosmopolitan family. They have four or five Hb components which are functionally distinct with respect to pH and organic phosphate modulation (di Prisco *et al.*, 1990) and fall into category (ii), despite their having metabolic rates among the lowest for any vertebrate (Wells, 1986).

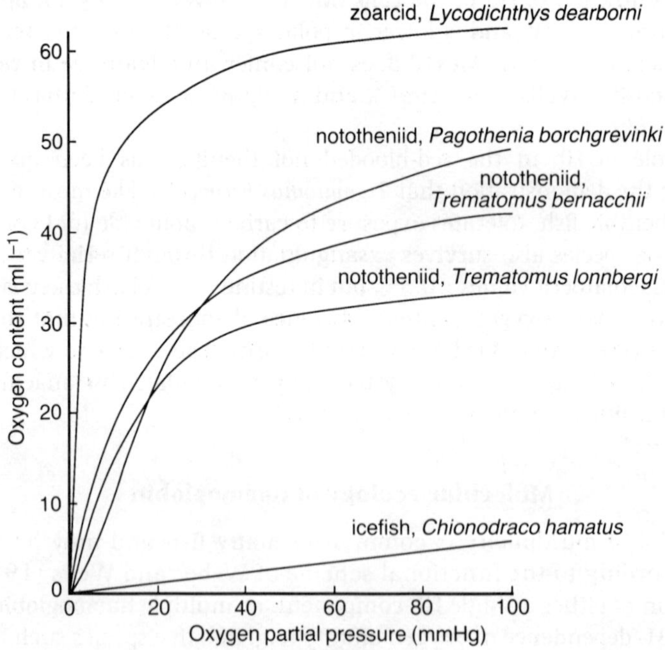

Fig. 14.3 Blood oxygen-binding curves of five Antarctic fish (pH 8.1, -1.5 °C) expressed in terms of oxygen content as a function of partial pressure.

Notothenioid Hbs have intrinsically low affinities for oxygen, and there is a strong correlation between the activity of a species and P_{50}; active species have lower affinity for oxygen than do sedentary species, which enables oxygen to be transferred to tissues at relatively high oxygen tensions. The oxygen-transport capacity of blood for five Antarctic fish with contrasting ecotypes is indicated in Fig. 14.3. The advantage of haemoglobin over dissolved oxygen in plasma is obvious from comparison with the icefish. Within the Nototheniidae, both oxygen-carrying capacity and affinity for oxygen are related to activity. *Trematomus lonnbergi* is the least active nototheniid and has the highest affinity for oxygen and a low oxygen-carrying capacity. The higher activity of the cryopelagic *P. borchgrevinki* is matched by its low blood oxygen affinity ($P_{50} = 20.7$ mmHg), sigmoid oxygen dissociation curve (Hill's coefficient, *n*, = 2), and high carrying capacity. The borch also has a marked Bohr effect ($\Delta \log P_{50} \Delta$ pH), enabling efficient gas transfer when blood pH falls. The closely related *Trematomus pennellii* (= *centronotus*) is a sedentary benthic scavenger, typically seen underwater curled up inside cup-like sponges, and has a higher affinity for oxygen ($P_{50} = 13.5$ mmHg), with modest Bohr effect, and a more hyperbolic shaped oxygen-binding curve ($n = 1.4$) (Tetens *et al.*, 1984). The zoarcid *Lycodichthys* (= *Rhigophila*) *dearborni* is a sluggish species not of the endemic notothenioids, and has one of the lowest recorded metabolic rates for fish (Wells, 1986). It has an extremely high blood oxygen affinity (Tetens *et al.*, 1984) and high carrying capacity. The latter feature may reflect aspects of its behaviour (such as burrowing) that are as yet unknown. A comparison of blood parameters with muscle buffering capacity, a further indicator of aerobic scope (Wells *et al.*, 1988), is given in Table 14.3.

The observation that these activity-related properties of whole blood are also manifested in purified haemolysates (Wells and Jokumsen, 1982) suggests that they are due to intrinsic molecular properties, rather than occurring through modulation of protein function by allosteric modifiers. The presence

Table 14.3 Physiological indicators of activity in Antarctic fishes. Hb and blood oxygen affinity constants, P_{50} at pH 8.1 and -1.5 °C, from Tetens *et al.* (1984), and muscle buffering capacities from Wells *et al.* (1988)

Species	Hb (mM)	P_{50} (mmHg)	Buffer capacity (slykes[1])
Zoarcidae			
Lycodichthys dearborni	0.63	4.3	3.5 ± 0.3
Nototheniidae			
Trematomus lönnbergi	0.29	11.9	–
T. bernacchii	0.38	13.5	13.6 ± 0.8
Pagothenia borchgrevinki	0.46	20.7	16.1 ± 1.3
Channichthyidae			
Pagetopsis macropterus	0.00	–	3.4 ± 0.2

[1] Slyke = μmol base required to titrate 1g tissue by one pH unit.

of a strong Bohr effect in the bathydraconid, *Parachaenichthys charcoti*, and in the nototheniid, *Notothenia gibberifrons*, occurs despite their relatively inactive habits (di Prisco, 1988).

The presence of single major Hb components within a group of closely related fishes, each with characteristic behaviours and energetic requirements, is unusual, and contrasts with the situation seen, for example, in catastomid fishes where adaptive explanations for Hb multiplicity are offered (Powers, 1980). Single or predominant Hbs in notothenioids need not find adaptive explanations in terms of the constancy of the physico-chemical environment; the single component in tench, *Tinca tinca*, enables it to thrive in a highly variable environment (Jensen and Weber, 1982). It is tempting, however, to speculate that the reduction in the number of genes encoding globin synthesis represents a stage in evolution on the path to total gene deletion seen in the icefishes.

Functional studies on both Hb (Qvist *et al.*, 1977; Wells and Jokumsen, 1982) and on whole blood (Tetens *et al.*, 1984) have shown that oxygen binding in nototheniids is an exothermic process comparable to that of Hbs from temperate fishes (e.g. $\Delta H° = -62$ kJ mol^{-1} for *T. bernacchii*). These results are at variance with di Prisco *et al.* (1991), who record that temperature has no effect on the binding constant of *T. bernacchii* Hb.

Another unexpected finding has been the marked Root effect (equivalent to an exaggerated Bohr effect), in the major nototheniid Hb component (di Prisco *et al.*, 1988, 1992; Kunzmann, 1991). This poses a puzzle, because Root effect Hbs are typically associated with active teleosts having a swim bladder or a well-developed choroid rete, where the Root Hbs assist in transferring oxygen into the bladder or the eye. But all notothenioids lack swim bladders, and several have no choroid body (Eastman, 1988), including *P. borchgrevinki* which has a Root effect; di Prisco *et al.* (1988) have postulated a role for these Hbs in buoyancy regulation, whereby oxygen may be secreted into hollow interstitial spaces. Alternatively, there has simply been no selection pressure for deletion of the Root effect.

In physiological terms, the Root effect cannot be distinguished from a large Bohr effect, but at the molecular level, it has been repeatedly correlated with a serine substitution on position (F9)93 of the β-globin chain, which stabilizes the COOH-terminal salt bridges in the T-structure, thus reducing the oxygen association constant (Weber and Wells, 1989). The key F9 serine residue has recently been identified in the β-chain sequence of the major component in some Antarctic nototheniid Hbs having a Root effect, but it is absent from the minor component of the bathydraconid, *Cygnodraco mawsoni*, despite its Root effect (Caruso *et al.*, 1991; di Prisco *et al.*, 1991). Uniquely, *G. acuticeps* and *T. newnesi* have neither a choroid rete, nor a Root Hb, yet both have serine residues at F9 (di Prisco *et al.*, 1991). One must caution, however, that though the Root effect is most marked at about pH 7.5 in Hb solution at 20 °C (di Prisco *et al.*, 1991), the lowest pH value recorded in fully stressed and lactate-loaded *Dissostichus mawsoni* was 7.71 (Qvist *et al.*, 1977). It has not been reported for whole blood from Antarctic fishes under physiological conditions (Tetens *et al.*, 1984). This paradox may be resolved by analogy with the controversial Root Hb from the *Xenopus* toad, which possesses the structural feature of an SER(F9) residue, and a Root effect at the molecular level, but this cannot be demonstrated under physiological conditions that might be adaptive (Bridges *et al.*, 1985; Brunori *et al.*, 1987). These matters are, of course, largely hypothetical, and demand further functional studies of a most interesting group of haemoglobins.

Molecular adaptation v. current utility

Many species respond to environmental change by acclimatory processes in the oxygen transport system (Weber and Wells, 1989). In contrast with fishes from habitats that periodically become hypoxic, which show expected adaptive adjustments in blood oxygen-binding characteristics, it might be supposed that Antarctic fishes, with their low absolute requirement for oxygen, and generous and constant environmental oxygen supply, might not possess similar mechanisms for dealing with environmentally induced hypoxia. When acclimated for 14 days to an oxygen tension approximately one-third normal value, *Pagothenia borchgrevinki* revealed a 40% increase in blood oxygen-carrying capacity, and a red cell ATP decrease of 27% with concomitant shift in P_{50} from 31.1 to 20.6 mmHg (pH 8.0 and -1.5 °C) (Wells *et al.*, 1989). Wells (1990) cautions against the dangers of interpreting such epiphenomena as adaptive in the sense of having been selected in evolution. These unexpected observations might best be interpreted as non-adaptive features, rather than those 'designed' for current utility to meet specific environmental challenges, and they may simply indicate considerable phenotypic flexibility.

14.5 RESPONSES TO STRESS

The way in which Antarctic fish respond to stress can provide important clues as to the mechanisms of control, and the robustness of the systems. Several Antarctic fish make little attempt to escape when captured. For example, the icefish, *Chionodraco hamatus*, adopts a tonically immobile posture with the mouth in extended gape; the zoarcid, *Lycodichthys dearborni*, may appear quite flaccid when handled from the aquarium.

Haematological responses to capture and exercise stress are very rapid in Antarctic fish. Haematocrit rises dramatically, partly from erythrocyte swelling, but also representing a real increase in oxygen-carrying capacity (Qvist *et al.*, 1977; Wells *et al.*, 1984; Davison *et al.*, 1988; Egginton *et al.*, 1991). The spleen appears to be the major source of erythrocytes during hypoxic stress (Wells *et al.*, 1989).

Compensation for agitational stress in *D. mawsoni* also occurs through ATP regulation of blood oxygen affinity which compensates for acidosis. After 24 h on a set line, arterial pH fell from 8.2 to 7.7, and the highest lactate levels were a modest 2 mmol l^{-1} (Qvist *et al.*, 1977). *P. borchgrevinki* that had swum to exaustion did not attain these levels of lactate, possibly due to limitation on ATP synthesis by glycolytic pathways (Davison *et al.*, 1988). Recent work with *Notothenia neglecta*, an 'ambush feeder', has shown that the major effect of strenuous burst activity is a respiratory, rather than metabolic, acidosis (Egginton *et al.*, 1991). The accumulated fixed-acid load during exercise stress is therefore lower in Antarctic fish compared with active temperate teleosts (Wells *et al.*, 1986), a result which agrees with their low estimated metabolic rates.

When stressed, marine fishes tend to gain ions and their plasma osmolarity rises, and both osmotic and ionic disturbances are expected from acute heat stress. A 10 min exposure of *P. borchgrevinki* to 10 °C (which is lethal if extended) resulted in a haematocrit rise of up to 40%, and a rapid rise in plasma chloride and osmolarity (Franklin *et al.*, 1991). Disturbed values were rapidly restored within 12 h when returned to normal temperature. Regulation of blood oxygen delivery through increased ATP-modulation of haemoglobin function has also been observed in heat-stressed nototheniids (Qvist *et al.*, 1977; Tetens *et al.*, 1984).

Recovery to control values of haematocrit and resting oxygen uptake is at least as rapid as in active temperate species, and is normally complete within 8–24 h (Wells *et al.*, 1984; Forster *et al.*, 1987; Davison *et al.*, 1988; Saint-Paul *et al.*, 1988; Franklin *et al.*, 1991). These data show that red-blooded nototheniid fish possess robust responses to stress.

Whereas many teleosts show cardiovascular responses to stress when inspired oxygen is decreased, nototheniids appear unusual in having no hypoxic bradycardia (Axelsson *et al.*, 1992). Yet both *P. borchgrevinki* and *T. pennellii* (= *centronotus*) regulate M_{O_2} down to critical oxygen tensions that correlate with their activity (Wells, 1987).

One of the more peculiar observations in Antarctic fish from McMurdo Sound has been the widespread occurrence of a deformity giving the gills a fluffy, white appearance. Fusion of the lamellae, and tissue hyperplasia, has been observed in 16% of the cryopelagic species *P. borchgrevinki* (Franklin and Davison, 1988). Up to 30% of some fish populations appear affected in more recent years, and the observations have been extended to *T. bernacchii* and *T. hansoni* (Macdonald and Wells, unpublished observations). The aetiology of these so-called X-cell tumours is unknown, but transformation of neoplastic cells as a consequence of toxic substances in the water cannot be ruled out (Franklin and Davison, 1988). The disease clearly affects aerobic potential through decreasing the effective gill surface area and increasing diffusion distances; chloride secretion is also impaired (Davison *et al.*, 1990). The consequence is a reduced condition factor, perhaps through poor performance in catching prey. It is an intriguing idea from Davison *et al.* (1990) that the low metabolism and high oxygen solubility in Antarctic fish enables these fish to survive the disability, whereas more active temperate species might not.

14.6 SENSORY ECOPHYSIOLOGY OF NOTOTHENIOID FISHES

Sensory systems form the interface between the animal and the environment and are among the key attributes which determine survival and success. There is a well-documented association between sensory capability and behavioural ecology (Ali and Klyne, 1985; Montgomery, 1991). Antarctic fishes, like virtually all marine teleosts, have patent olfactory, visual, auditory, mechanosensory lateral line and tactile senses, but in general terms they lack any extreme sensory specializations comparable to the visual or lateral line adaptations of deep-water fishes, or to the electrosensory systems of some teleosts. The general tendency is towards the development of non-visual senses, and the purpose of this section is to examine the relationship between sensory physiology and ecology of notothenioid fishes.

Sensory biology of *Trematomus bernacchii*

We know more about the sensory biology of *T. bernacchii* than that of any other notothenioid, and this species may serve as a benchmark against which we can compare the sensory capabilities of other species (Macdonald and Montgomery, 1991). *T. bernacchii* feeds predominantly on benthic invertebrates (Oliver and Slattery, 1985; Foster and Montgomery, 1992) though some fish are also taken. Observations of behaviour from a sub-ice chamber indicate that *T. bernacchii* undertakes a stereotyped prey search (Janssen, Montgomery and Sideleva, unpublished). Fish rest on the bottom for a short while, then move forward 1–2 body lengths with one or two flaps of the pectoral fins. The interpretation of this behaviour is that during the resting phase, the fish is listening and feeling for prey located on or in the substratum. On several instances, fish were observed to burrow into the sponge mat, presumably in search of prey. Observations of fish in aquaria (Janssen, unpublished) show that animals surgically deprived of vision will adopt a similar search pattern. During the stationary phase, a light touch to the area of the mandible, or a disturbance in the area of the mouth, will result in the fish biting at the source of the stimulus. A touch of the pelvic fin will induce a repositioning to bring the mouth over the place where the fin was when touched; a light touch to the anal fin, which normally lies in contact with the substratum – or a disturbance adjacent to the trunk lateral line – results in a repositioning to bring the pelvic fin into contact with the appropriate area

(Fig. 14.4). These observations indicate the importance of the somatosensory and mechanosensory lateral line systems for the location of benthic prey, and demonstrate that vision is not necessary for feeding in these fish. However, visual cues may elicit feeding in *T. bernacchii* since they can be caught on unbaited lures, and an olfactory component to feeding is indicated by the attraction of fish to baits and to baited traps. One way of assessing the relative importance of different senses for normal behaviour is to produce conflicting cues and observe which of the cues dominates. In experiments of this sort (Janssen, unpublished) where the fish are approaching a visual target, a lateral line stimulus displaced to one side of the target will routinely provoke a misplaced strike, indicating that lateral line is probably the dominant sensory modality for these fish. A similar effect is observed when jigging for the cryopelagic species, *Pagothenia borchgrevinki*. Individuals are initially attracted visually to the lure, and will often sprint in from some distance and then stop at a distance of about 3 cm from the target as if awaiting an additional cue. Subsequent strikes can often be infuriatingly inaccurate and may well be the result of attacks directed at water disturbance generated by the lure, rather than at the lure itself. Thus vision and olfaction appear to give way to mechanoreception as target distances close.

The visual system of *T. bernacchii* has been studied anatomically and physiologically (Eastman, 1988; Pankhurst and Montgomery, 1989). This species has a pronounced corneal iridescence that is thought to act like a venetian blind, reducing light scatter into the eye from relatively bright downwelling light, while preserving lateral vision. The corneal iridescence and the side and forward location of the eyes on the head (Fig. 14.4) indicate that the lateral and forward visual axes are the directions of most visual interest

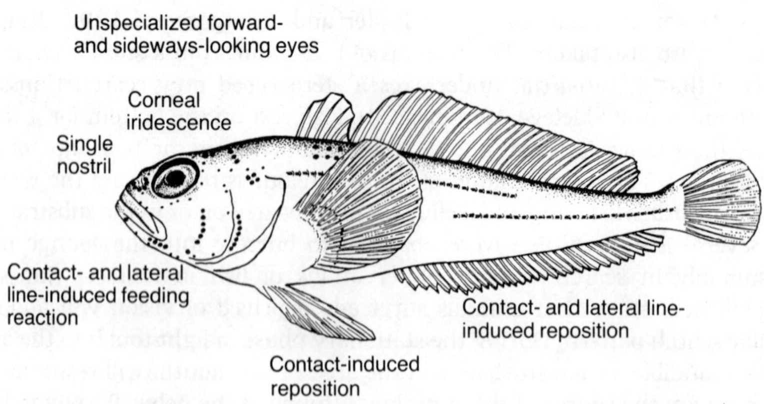

Fig. 14.4 Sensory biology of *Trematomus bernacchii*. The prominent lateral-facing eyes, single nostril and lateral line system, and the importance of skin contact and lateral line stimulation for feeding responses, are indicated. Large dots represent free-standing neuromasts, smaller dots the pores of lateral line canals. Fish length approximately 200 mm.

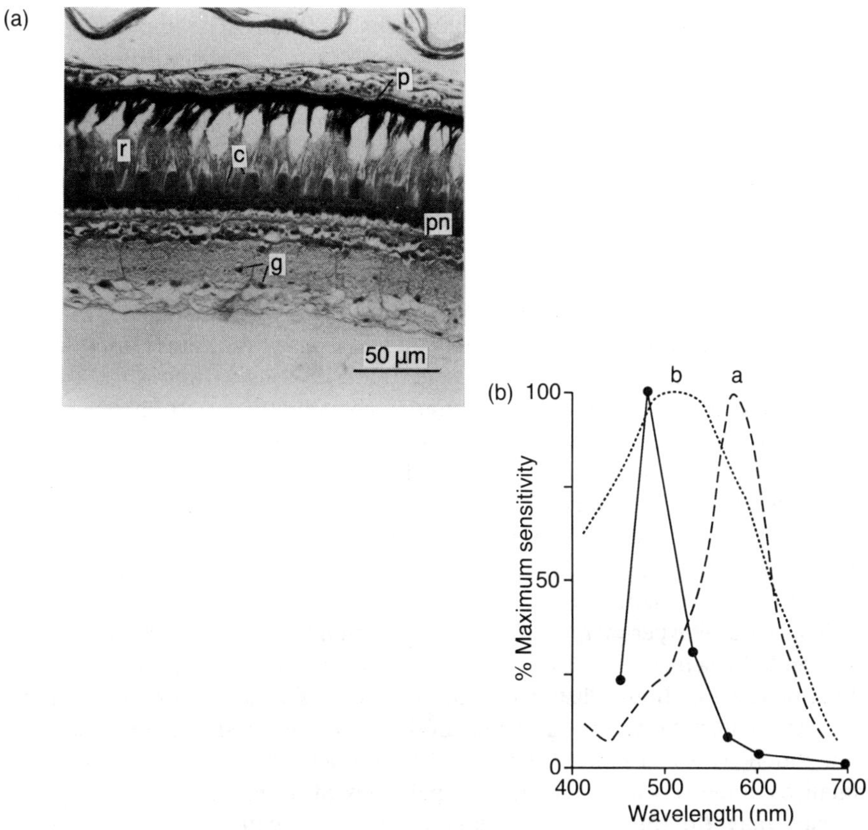

Fig. 14.5 Visual anatomy and physiology of *T. bernacchii*. (a) Cross-section through the retina showing single and double cones (c), rods (r), pigmented epithelium (p), photoreceptor nuclei (pn), and ganglion cells (g), and the relative convergence between photoreceptors and ganglion cells. (b) Spectral sensitivity of the visual system (solid curve) compared with the spectral irradiance of ambient light (broken curves) with (a) and without (b) ice algae present.

for this species. The retina of *T. bernacchii* is dominated by twin and single cones (Fig. 14.5) with cell counts of 13 cones, 27 rods, and 5 ganglion cells per μm^2, giving a low-sensitivity convergence ratio (cones and rods: ganglion cells) of 8:1. *T. bernacchii* has a choroid body which assists with oxygen supply to the retina, and the presence of which is considered a primitive character for notothenioids (Eastman, 1988). These visual characteristics are similar to those of coastal fishes elsewhere in the world and do not appear especially adapted to the Antarctic photic conditions.

The conclusions based on anatomical observations are supported by physiological studies using electroretinographic measures of visual sensitivity and spectral properties (Pankhurst and Montgomery, 1989). This study shows that visual sensitivity is comparable to that in other shallow-water teleosts and that the maximum sensitivity is to blue light (480 nm). The implications of these findings are that the discrimination of small visual targets will be severely limited by the low light conditions that exist under snow and ice cover, at depth, or during the Antarctic winter. Visual capabilities will also be impaired by the growth in summer of ice algae, which absorb much of the blue light penetrating the ice, and produce a spectral shift in the ambient light away from the region of greatest sensitivity (Fig. 14.5(b)). There are situations when this species would experience bright light levels and vision may be important, such as shallow-water habitats which become ice-free in summer, and particularly in the pelagic juvenile phase where they live just below the ice and are presumed to feed visually on small copepods (Pankhurst and Montgomery, 1990). It is possible to envisage that a juvenile feeding stage, timed to coincide with high levels of primary productivity and suitable photic conditions, followed by an ontogenetic shift to non-visual feeding, could account for the lack of visual specialization in adult *T. bernacchii* and other notothenioids. In summary, *T. bernacchii* has a laterally directed visual axis, anatomically and functionally vision shows little evidence of specialization, and for adult fish there are many circumstances where visual function must be limited by low light levels.

Behavioural experiments with *T. bernacchii* indicate that lateral line stimuli are treated essentially as tactile stimuli. The lateral line has in the past been described as 'touch-at-a-distance', and this may be the closest description that we have of its action. One advantage of the lateral line system is that the fish can 'touch' a moving object before physical contact is made with the fish. Even small asymmetries in the sensory capabilities of a predator over the prey species may tip the balance in favour of the predator. Thus the major advantage and use of the lateral line may be for the fish to 'feel' its prey without itself being detected.

The lateral line system consists of a number of hair cell receptors clumped together in groups called neuromasts which may occur either free standing or as canal organs (Fig. 14.4). Anatomical studies of the lateral line canals in *T. bernacchii* indicate extensive variability in morphology on different parts of the body. Canals of the head tend to be longer, and have a greater diameter than the trunk lateral line canal. For example, on a fish of about 160 mm standard length, the interpore distance on the mandibular canal is around 4 mm, whereas the equivalent measure for the trunk canal is less than 2 mm. The diameter of mandibular canals for this fish would be about 500 μm, whereas the trunk canal is less than 200 μm (Coombs and Montgomery, 1991). Extensive modelling studies of the mechanical factors involved in the excitation of lateral lines (Denton and Gray, 1983, 1988) have led to the expectation that canal morphology would have a direct influence on lateral line function. The first direct test of this expectation has been made on the Antarctic fish *T. bernacchii* by comparing the physiology of head and trunk

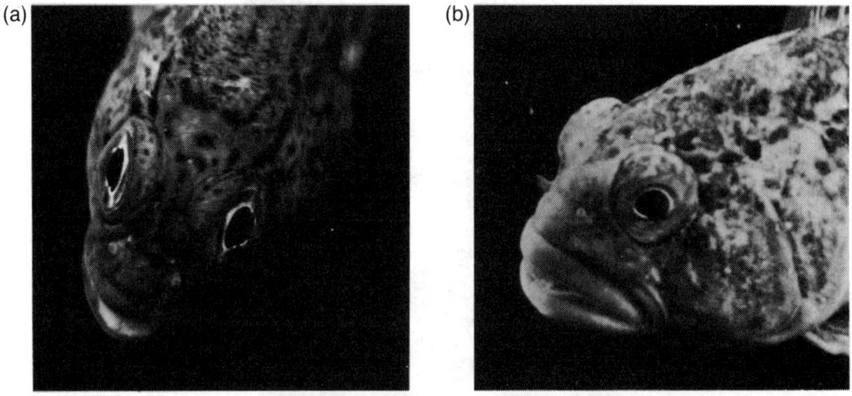

Fig. 14.6 (a) Dorsal view of *Trematomus hansoni*, showing the upward visual axis, and (b) lateral view of *T. bernacchii*, showing the lateral visual axis.

lateral line systems (Coombs and Montgomery, 1991). Electrophysiological recordings were made of the responses of single lateral line sensory afferents to a sinusoidally vibrating sphere over a range of stimulus frequencies. In accord with theoretical expectations, the frequency–response curves show a relatively flat response shape at low frequencies when plotted as a function of stimulus acceleration. However, the decline in responsiveness that begins at around 45 Hz and continues at about 18 dB per octave occurs at a lower frequency, and is steeper than the responses predicted by modelling studies. Furthermore, the difference expected between the responses of head and trunk canals due to the differences in morphology are not evident in the physiology. These results suggest that (i) there may be considerable variability in the morphology with little consequence for function, and (ii) there is additional processing between the canal mechanics and the response of afferent fibres.

The physiology of free-standing neuromasts shows that they respond to the velocity component of a periodic stimulus and have a flat frequency response up to around 40 Hz. Free-standing neuromasts are more sensitive to water currents and low-frequency stimuli than are canal neuromasts, and are presumably more susceptible to noise interference produced by the fishes' own movements. The exact contributions of these two submodalities (i.e. free-standing and canal neuromasts) in fish behaviour are as yet unknown.

Comparative ecophysiology of vision

The basic visual ecophysiology exhibited by *T. bernacchii* shows relatively little change throughout the other notothenioids. A change of visual axis is shown in several species; for example *T. hansoni* has no corneal iridescence and the eyes are placed more dorsally on the head (Fig. 14.6). This change to a more dorsal visual axis is correlated with an increase in prey items taken from the water column (Foster and Montgomery, 1992). Dorsally directed vision is

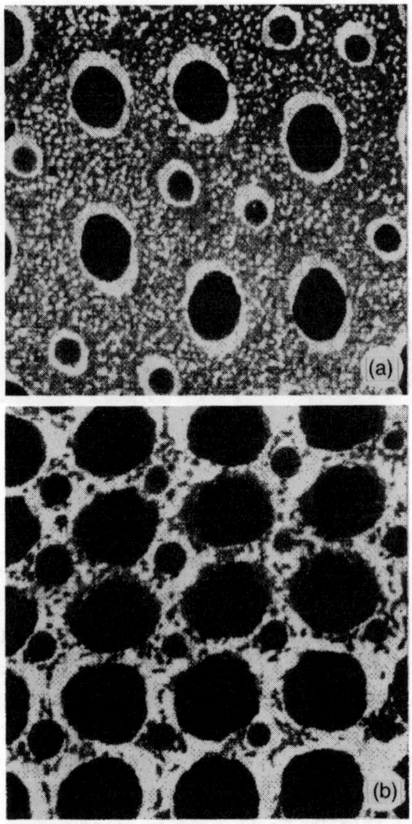

Fig. 14.7 Scan of tangential sections through the photo-mosaic of (a) *D. mawsoni* and (b) *T. bernacchii*, showing the packing of the photoreceptive cells. *D. mawsoni* exhibits a hexagonal mosaic comprising six twin cones and two centrally-placed single cones, with rods densely interspersed. *T. bernacchii* shows a square mosaic with twin cones forming the corners, single cones between them, and fewer rods scattered between the cones. The diameter of a double cone is approximately 15 μm.

suited to the detection of opaque objects by silhouetting them against the downwelling light. However, direct observations from the sub-ice chamber show that transparent prey such as the pteropod *Limacina* are more easily seen by looking laterally, essentially using downwelling light, and a dark background as dark-field illumination. The pelagic planktivore *P. borchgrevinki* has laterally directed eyes. As in *T. bernacchii*, the visual feeding is likely to be often impaired by low light levels which occur beneath ice and snow cover (Montgomery *et al.*, 1989).

If anything, the visual characteristics of *T. bernacchii* are probably less sensitive than those of the non-Antarctic notothenioids, including the primitive *Bovichthys*. Despite living in shallow-water, brightly lit habitats, non-Antarctic notothenioids have higher densities of rods and greater convergence ratios than *T. bernacchii*. The only extensively modified retina is that of the relatively deep-water *Dissostichus mawsoni*, which has a rod-dominated retina (Fig. 14.7) and a very high convergence ratio (Eastman, 1988).

The lack of visual specialization in some deep-water species such as *T. lönnbergi*, including retention of cones, may possibly be related to the detection of bioluminescence. Feeding on bioluminescent organisms may be indicated by the darkened lining of the oral cavity and gut in this species, which may shield the fish from being illuminated by recently consumed prey species, thereby becoming less evident to prey and larger predators (Eastman, 1981).

Comparative ecophysiology of touch

In *T. bernacchii*, the somatosensory system is important for the detection of benthic prey. It is difficult to imagine that the sense of touch has the same importance for pelagic species taking planktonic prey. However, for the plunderfishes (Harpagiferidae), prey capture is closely related to somatosensory stimuli. These fishes have a range of species-specific mental (chin) barbels which are used both as a fishing lure and as a prey sensor. Under dark conditions, touching the barbel of a *Histiodraco velifer* results in a strike, indicating that the somatosensory stimulus alone is sufficient to initiate prey capture (Janssen, unpublished). One imagines that most fish utilizing lures of this kind depend on the visibility of the lure to attract prey, and in turn use visual cues to initiate a strike. A visually mediated interaction would certainly have the advantage for the predator of not getting the lure bitten by the prey species. For *Histiodraco*, the deployment of the lure in dark conditions, and the interaction of visual and somatosensory cues to initiate and control the strike, remain to be investigated.

Comparative ecophysiology of the lateral line

The lateral line of the cryopelagic species *P. borchgrevinki* has been physiologically investigated (Montgomery and Macdonald, 1987; Montgomery *et al.*, 1988; Montgomery, 1989) and has been shown to be suited to the detection of planktonic prey (Montgomery and Macdonald, 1990). The frequency–response properties of lateral line afferents in *P. borchgrevinki* are comparable with those of *T. bernacchii*.

Across the whole suborder Notothenioidei, there is considerable variation in lateral line morphology. The most structurally elaborated lateral lines occur as membranous canals in some of the channichthyids (Montgomery, Coombs, and Janssen, unpublished), and large and open cephalic lateral lines in genera such as *Aethotaxis* (deWitt, 1962), *Bathydraco* (Fig. 14.8), *Pleurogramma* (DeWitt and Hopkins, 1977), and *Cryothenia* (Daniels, 1981). In fact, the notothenioid radiation is a particularly favourable system in which to examine the form-and-function relationship in lateral line systems. Montgomery, Coombs, and Janssen (unpublished) have initiated investigations into lateral line morphology and function in six species. Despite dramatic differences in

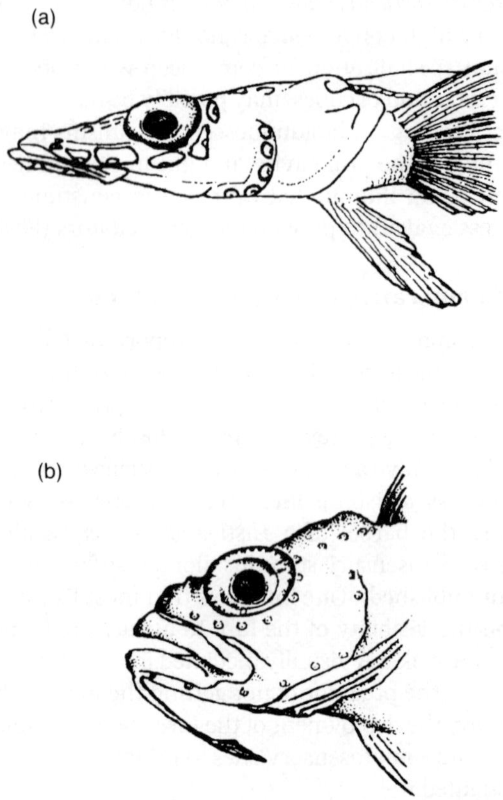

Fig. 14.8 (a) Lateral profile of the head of *Bathydraco macrolepis*, showing the enlarged lateral line canals and pores, and (b) lateral profile of the head of *Histiodraco velifer*, showing the mental barbel.

anatomy, particularly between the smaller species and the giant *D. mawsoni*, the physiology of lateral line afferents does not show any significant differences. It seems unlikely, however, that the diversity of lateral line structures in Antarctic fish bears no relation to niche specialization. Examination of further species and a better understanding of the sensory modality may be required before we can interpret form and functional relationships in lateral line systems within an adaptive framework (Coombs *et al.*, 1991).

14.7 ECOPHYSIOLOGY AND EVOLUTION OF NOTOTHENIOID FISHES

The Antarctic coastal fish fauna is dominated by perciform fishes of the suborder Notothenioidei. Only two other groups are represented by significant species numbers: the relatively deep-water zoarcids and liparids. Given the age and size of the Antarctic marine ecosystem, this represents a surprising lack of diversity (Andriashev, 1987). Why other fish failed to survive during the evolution of the Antarctic ichthyofauna, and the notothenioids thrived to form an extensive species radiation, is open to speculation. Eastman and Grande (1989) can point to no single characteristic that is unique to this group.

In other ecosystems, maximal species diversity occurs in shallow water, whereas in Antartica it occurs at greater depths (300–600 m). Eastman and Grande (1989) believe that glacial "habitat and faunal destruction may have caused the local extinction of most of the early fish fauna, leaving an ecological void which was filled by a notothenioid fauna tolerant of deep water conditions."

Deep water tolerance is not, however, a simple characteristic. While our physiological understanding of the direct effects of pressure is rudimentary, it does appear that in some respects pressure parallels the effects of low temperature. Increased pressure leads to decreased membrane viscosity as does decreased temperature, and the biochemical adaptations of deep-water fishes parallel those of cold-water fishes through increased unsaturated lipids maintaining membrane fluidity. Torres and Somero (1988a) have shown similarities in enzyme activities between cold-adapted and deep-sea-dwelling fishes.

As a group, the notothenioids are not as reliant on vision as are most other groups of fishes. Teleosts are generally visual and diurnal. There are many exceptions to this, including the notothenioids which survive and feed through the winter darkness, and in the low light levels under ice and snow that occur during summer. Recent behavioural and physiological work indicates that non-visual senses such as touch and mechanosensory lateral lines play an important part in feeding behaviour. The ability to feed under low light is a characteristic which is useful at depth, during high-latitude winter darkness, and under ice and snow cover, and suggests that a common suite of characters is required for both depth and cold tolerance.

One character which may also have played a role in the success of notothenioids is their reproductive strategy of benthic spawning; large yolky eggs are typical of benthic fishes (North and White, 1987). The larvae hatch at a large size, and typically occupy a niche in the top of the water column, using ice cover for shelter, where they feed on immature copepods and small pelagic molluscs. The majority of temperate marine teleosts are open-water spawners producing floating eggs – a feature that could well be a liability in ice-covered seas.

Even without the problem of freezing, most fish simply cannot survive at -2 °C. Success of Antarctic notothenioids is no doubt in part due to what Dudley and Gans (1991) call "excessive construction". Thus the phenotype (form) is not optimized for function under a particular set of environmental circumstances, but provides considerable plasticity. Examples are the allosteric behaviour of erythrocyte factors in the hypoxic responses of *Pagothenia* (Wells *et al.*, 1989), and the notothenioid Root effect (di Prisco *et al.*, 1992).

Without doubt, the most spectacular feature among the notothenioids is the loss of haemoglobin in the icefishes. Yet the fundamental organization of several aspects of the respiratory and cardiovascular systems seems to indicate that the loss of haemoglobin may not, in itself, have been an adaptive change. As pointed out by Tota *et al.* (1991), many of the cardiocirculatory features that permit life without haemoglobin are related to the intrinsic properties of the icefish heart, i.e. a spongy myocardium working as a high-volume pump. The notothenioid phenotypes are therefore a subset of those carried by the original population, and the icefish represents just one extreme in teleost heart design. These aspects illustrate well the constraints under which natural selection must operate.

Whatever the complex of traits was that allowed the notothenioids to thrive where other fish groups failed, it is apparent that their dominance is one of the main features of Antarctic ichthyology. Although their radiation appears to provide excellent examples of ecophysiological correlation and adaptation to environment, the general impression is not one of extreme specialization of either form or function.

Temperature compensation of physiological processes is invariably partial (Macdonald *et al.*, 1987), and sensory and metabolic specializations are generally modest relative to other groups such as deep-sea fishes. This lack of extreme specialization could be the result of the relatively recent origin of radiation, it could be due to absolute constraints on processes imposed by temperature, or it could result from the low competition under which evolution has occurred. As Gans (1988) reminds us, there is no sense in which animals have to be optimal to survive or thrive; they have only to be adequate.

ACKNOWLEDGEMENTS

The authors are grateful to friends and colleagues who made freely available their latest research findings, and in particular to S. Coombs, P.S. Davie, J.T. Eastman, M.E. Forster, C.E. Franklin, G. Hubold, J. Janssen, I.A. Johnston, N.W. Pankhurst, G. di Prisco, and B. Tota, for providing unpublished material for discussion.

REFERENCES

Ali, M.A. and Klyne, M.A. (1985) Physiology and functional morphology of the vertebrate retina. *Fortschr. Zool.*, **30**, 633–48.

Andriashev, A.P. (1987) A general review of the Antarctic bottom fish fauna, in *Proceedings of the Fifth Congress of European Ichthyology, Stockholm 1985* (eds S.O. Kullander and B. Fernholm), Swedish Mus. Nat. Hist., Stockholm, pp. 357–72.

Axelsson, M., Davison, W., Forster, M.E. and Farrell, A.P. (1992) Cardiovascular responses of the red-blooded Antarctic fishes, *Pagothenia bernacchii* and *P. borchgrevinki*. *J. exp. Biol.*, **167**, 179–201.

Bacila, M., Rosa, R., Rodrigues, E., Lucchiari, P.H. and Rosa, C.D. (1989) Tissue metabolism of the ice-fish *Chaenocephalus aceratus* Loenberg. *Comp. Biochem. Physiol.*, **92B**, 313–18.

Bailey, J.R. and Driedzic, W.R. (1989) Effects of acute temperature change on cardiac performance and oxygen consumption of a marine fish, the sea raven (*Hemitripterus americanus*). *Physiol. Zool.*, **62**, 1089–1101.

Bowler, K. and Tirri, R. (1990) Temperature dependence of the heart isolated from cold or warm acclimated perch (*Perca fluviatilis*). *Comp. Biochem. Physiol.*, **96A**, 177–80.

Boyd, R.B., DeVries, A.L., Eastman, J.T. and Pietra, G.G. (1980) The secondary lamellae of the gills of cold water (high latitude) teleosts. *Cell Tissue Res.*, **213**, 361–7.

Bridges, C.R., Pelster, B. and Scheid, P. (1985) Oxygen binding in blood of *Xenopus laevis* (Amphibia) and evidence against Root effect. *Respir. Physiol.*, **61**, 125–36.

Brunori, M., Bellelli, A., Giardina, B., Condo, S. and Perutz, M.F. (1987) Is there a Root effect in *Xenopus* hemoglobin? *FEBS Lett.*, **221**, 161–6.

Caruso, C., Rutigliano, B., Romano, M. and di Prisco, G. (1991) The hemoglobins of the cold-adapted Antarctic teleost *Cygnodraco mawsoni*. *Biochim. biophys. Acta*, **1078**, 273–82.

Coombs, S., and Montgomery, J.C. (1991) Similar frequency responses from fibers innervating morphologically different parts of the lateral line system in an Antarctic fish. *Soc. Neurosci. Abst.*, **17**, 630.

Coombs, S., Janssen, J. and Montgomery, J.C. (1991) Functional and evolutionary implications of periopheral diversity in lateral line systems, in *Evolutionary Biology of Hearing* (eds D.B. Webster, R.R. Fay and A.N. Popper), Springer-Verlag, New York, pp. 267–94.

Crockett, E.L. and Sidell, B.D. (1990) Some pathways of energy metabolism are cold adapted in Antarctic fishes. *Physiol. Zool.*, **63**, 472–88.

Daniels, R.A. (1981) *Cryothenia peninsulae* a new genus and species of notothenioid fish from the Antarctic Peninsula. *Copeia*, **1981**, 558–62.

Davie, P.S. (1990) *Pacific Marlins: Anatomy and Physiology*, Massey Univ. Press, Palmerston North N.Z., 88 pp.

D'Avino, R. and di Prisco, G. (1988) Antarctic fish hemoglobin: an outline of the molecular structure and oxygen binding properties – I. Molecular structure. *Comp. Biochem. Physiol.*, **90B**, 579–84.

Davison, W., Forster, M.E., Franklin, C.E. and Taylor, H.H. (1988) Recovery from exhausting exercise in an Antarctic fish, *Pagothenia borchgrevinki*. *Polar Biol.*, **8**, 167–71.

Davison, W., Franklin, C.E. and Carey, P.W. (1990) Oxygen uptake in the Antarctic teleost *Pagothenia borchgrevinki*. Limitations imposed by X-cell gill disease. *Fish Physiol. Biochem.*, **8**, 69–77.

Denton, E.J. and Gray, J.A.B. (1983) Mechanical factors in the excitation of clupeid lateral lines. *Proc. R. Soc.*, **218B**, 1–26.

Denton, E.J. and Gray, J.A.B. (1988) Mechanical factors in the excitation of the lateral line of fishes, in *Sensory Biology of Aquatic Animals* (eds A. Atema, R.R. Fay, A.N. Popper and W.N. Tavolga), Springer-Verlag, New York, pp. 595–617.

DeVries, A.L. and Eastman, J.T. (1981) Physiology and ecology of notothenioid fishes of the Ross Sea. *J. R. Soc. N.Z.*, **11**, 329–40.

DeWitt, H.H. (1962) A new Antarctic nototheniid fish with notes on two recently described nototheniiforms. *Copeia*, **1962**, 826–33.

DeWitt, H.H. and Hopkins, T.L. (1977) Aspects of the diet of the Antarctic silverfish, *Pleurogramma antarcticum*, in *Adapatations within Antarctic Ecosystems* (ed. G.A. Llano), Smithsonian Institution, Washington, D.C. pp. 557–67.

di Prisco, G. (1988) A study of hemoglobin in Antarctic fishes: purification and characterization of hemoglobins from four species. *Comp. Biochem. Physiol.*, **90B**, 631–7.

di Prisco, G., D'Avino, R., Camardella, L., Caruso, C., Romano, M. and Rutigliano, B. (1990) Structure and function of hemoglobin in Antarctic fishes and evolutionary implications. *Polar Biol.*, **10**, 269–74.

di Prisco, G., D'Avino, R., Caruso, C., Tamburini, M., Camardella, L., Rutigliano, B., Carratore, V. and Romano, M. (1991) The biochemistry of oxygen transport in red-blooded Antarctic fish, in *Biology of Antarctic Fishes* (eds. G. di Prisco, B. Maresca and B. Tota), Springer-Verlag, Berlin, pp. 263–81.

di Prisco, G., Giardina, B., D'Avino, R., Condo, S.G., Bellelli, A. and Brunori, M. (1988) Antarctic fish hemoglobins: an outline of the molecular structure and oxygen binding properties – II. Oxygen binding properties. *Comp. Biochem. Physiol.*, **90B**, 585–91.

Douglas, E.L., Peterson, K.S., Gysi, J.R. and Chapman, D.J. (1985) Myoglobin in the heart tissue of fishes lacking hemoglobin. *Comp. Biochem. Physiol.*, **81A**, 855–88.

Driedzic, W.R., Stewart, J.M. and Scott, D.L. (1982) The protective effect of myoglobin during hypoxic perfusion of isolated fish hearts. *J. mol. cell. Cardiol.*, **14**, 673–7.

Dudley, R. and Gans, C. (1991) A critique of symmorphosis and optimality models in physiology. *Physiol. Zool.*, **64**, 627–37.

Eastman, J.T. (1981) Morphological specializations in Antarctic fishes. *Antarct. J. U.S.*, **16**, 146–7.

Eastman, J.T. (1988) Ocular morphology in Antarctic notothenioid fishes. *J. Morph.*, **196**, 283–306.

Eastman, J.T. (1990) The biology and physiological ecology of notothenioid fishes, in *Fishes of the Southern Ocean* (eds O. Gon and P.C. Heemstra), J.L.B. Smith Inst. Ichthyology, Grahamstown, South Africa, pp. 34–51.

Eastman, J.T. (1991) Evolution and diversification of Antarctic notothenioid fishes. *Am. Zool.*, **31**, 93–109.

Eastman, J.T. and DeVries, A.L. (1986) Renal glomerular evolution in Antarctic notothenioid fishes. *J. Fish Biol.*, **29**, 649–62.

Eastman, J.T. and Grande, L. (1989) Evolution of the Antarctic fish fauna with emphasis on recent notothenioids, in *Origins and Evolution of the Antarctic Biota* (ed. J.A. Crame), U.S. Geol. Soc. Spec. Pub. No. 47, pp. 241–52.

Eastman, J.T. and Hikida, R.S. (1991) Skin structure and vascularization in the Antarctic notothenioid fish *Gymnodraco acuticeps*. *J. Morph.*, **208**, 347–65.

Egginton, S., Taylor, E.W., Wilson, R.W., Johnston, I.A. and Moon, T.W. (1991) Stress response in the Antarctic teleosts (*Notothenia neglecta* Nybelin and *N. rossii* Richardson). *J. Fish Biol.*, **38**, 225–35.

Farrell, A.P., Davie, P.S., Franklin, C.E., Johansen, J.A. and Brill, R.W. (1992) Cardiac physiology in tunas: I. Perfused heart preparations from yellowfin and skipjack tuna. *Can. J. Zool.*, (in press).

Feller, G., Bassleer, R., Goessens, G. and Hamoir, G. (1983) Relative size and myocardial structure of the heart of an Antarctic fish devoid of haemoglobin and myoglobin, *Channichthys rhinoceratus*. *J. Zool., Lond.*, **199**, 5–7.

Feller, G., Goessens, G., Gerday, C. and Bassleer, R. (1985) Heart structure and ventricular ultrastructure of hemoglobin- and myoglobin-free icefish *Channichthys rhinoceratus*. *Cell Tissue Res.*, **242**, 669–76.

Feller, G., Pauly, J.-P., Smal, A., O'Carra, P. and Gerday, C. (1991) The lactate dehydrogenase of the icefish heart: biochemical adaptations to hypoxia tolerance. *Biochim. biophys. Acta*, **1079**, 343–7.

Forster, M.E., Franklin, C.E., Taylor, H.H. and Davison, W. (1987) The aerobic scope of an Antarctic fish, *Pagothenia borchgrevinki* and its significance for metabolic cold adaptation. *Polar Biol.*, **8**, 155–9.

Foster, B.A. and Montgomery, J.C. (1992) Planktivory in benthic nototheniid fish in McMurdo Sound, Antarctica. *Env. Biol. Fishes*, in press.

Franklin, C.E. and Davie, P.S. (1991) Effect of adrenaline on the performance of a perfused eel heart. *Comp. Biochem. Physiol.*, **16**, 173–7.

Franklin, C.E. and Davison, W. (1988) X-cells in the gills of an Antarctic teleost, *Pagothenia borchgrevinki*. *J. Fish Biol.*, **32**, 341–53.

Franklin, C.E., Davison, W. and Carey, P.W. (1991) The stress response of an Antarctic teleost to an acute increase in temperature. *J. therm. Biol.*, **16**, 173–77.

Gans, C. (1988) Adaptation and the form-function relation. *Am. Zool.*, **28**, 681–97.

Graham, M.S. and Farrell, A.P. (1989) The effect of temperature acclimation and adrenaline on the performance of a perfused trout heart. *Physiol. Zool.*, **62**, 113–34.

Hamoir, G. (1988) Biochemical adaptation of the muscles of the Channichthyidae to their lack in hemoglobin and myoglobin. *Comp. Biochem. Physiol.*, **90B**, 557–9.

Harrison, P., Zummo, G., Farina, F., Tota, B. and Johnston, I.A. (1991) Gross anatomy, myoarchitecture, and ultrastructure of the heart ventricle in the haemoglobinless icefish *Chaenocephalus aceratus*. *Can. J. Zool.*, **69**, 1339–47.

Haschemeyer, A.E.V. (1985) Cold adaptation and genome organization in an Antarctic fish species. *Comp. Biochem. Physiol.*, **81B**, 523–30.

Hemmingsen, E.A. and Douglas, E.L. (1970) Respiratory characteristics of the hemoglobin-free fish *Chaenocephalus aceratus*. *Comp. Biochem. Physiol.*, **33**, 733–44.

Hemmingsen, E.A. and Douglas, E.L. (1972) Respiratory and circulatory responses in a hemoglobin-free fish, *Chaenocephalus aceratus*, to changes in temperature and oxygen tension. *Comp. Biochem. Physiol.*, **43A**, 1031–43.

Hemmingsen, E.A. and Douglas, E.L. (1977) Respiratory and circulatory adaptations to the absence of hemoglobin in chaenichthyid fishes, in *Adaptations within Antarctic Ecosystems* (ed. G.A. Llano), Smithsonian Inst., Washington, DC, pp. 479–87.

Hemmingsen, E.A., Douglas, E.L. and Grigg, G.C. (1969) Oxygen consumption in an Antarctic hemoglobin-free fish, *Pagetopterus macropterus*, and in three species of *Notothenia*. *Comp. Biochem. Physiol.*, **43A**, 1031–43.

Hemmingsen, E.A., Douglas, E.L. Johansen, K. and Millard, R.W. (1972) Aortic blood flow and cardiac output in the hemoglobin-free fish *Chaenocephalus aceratus*. *Comp. Biochem. Physiol.*, **43A**, 1045–51.

Holeton, G.F. (1970) Oxygen uptake and circulation by a hemoglobinless Antarctic fish (*Chaenocephalus aceratus* Lonnberg) compared with three red-blooded Antarctic fish. *Comp. Biochem. Physiol.*, **34**, 457–71.

Hubold, G. (1991) Ecology of notothenioid fishes in the Weddell Sea, in *Biology of Antarctic Fish* (eds G. di Prisco, B. Maresca, and B. Tota), Springer-Verlag, Berlin, pp. 3–22.

Hubold, G., Hagen, W., Kunzmann, A., Wohrmann, A.P.A. and von Dorrien, C.F. (1990) Adaptations of *Pleuragramma antarcticum* to life in high Antarctic waters, in *Proc. 2nd Int. Conf. Biol. Antarctic Fishes* (eds. G. di Prisco and B. Maresca), IIGP Press, Naples, pp. 62–3.

Hughes, G.M. (1972) Distribution of oxygen tension in the blood and water along the secondary lamellae of the icefish gill. *J. exp. Biol.*, **56**, 481–92.

Hureau, J.C., Petit, D., Fine, J.M. and Marneux, M. (1977) New cytological, biochemical, and physiological data on the colorless blood of the Channichthyidae (Pisces, Teleostei, Perciformes), in *Adaptations within Antarctic Ecosystems* (ed. G.A. Llano), Smithsonian Institute, Washington, DC, pp. 459–77.

Jensen, F.B. and Weber, R.E. (1982) Respiratory properties of tench blood and hemoglobin. Adaptation to hypoxic–hypercapnic water. *Mol. Physiol.*, **2**, 235–50.

Johnston, I.A. and Harrison, P. (1987) Morphometrics and ultrastructure of myocardial tissue in notothenioid fishes. *Fish Physiol. Biochem.*, **3**, 1–6.

Johnston, I.A., Clarke, A. and Ward, P. (1991) Temperature and metabolic rate in sedentary fish from the Antarctic, North Sea and Indo-West Pacific Ocean. *Mar. Biol.*, **109**, 191–5.

Johnston, I.A., Fitch, N., Zummo, G., Wood, R.E., Harrison, P. and Tota, B. (1983) Morphometric and ultrastructural features of the ventricular myocardium of the hemoglobinless icefish *Chaenocephalus aceratus. Comp. Biochem. Physiol.*, **76A**, 475–80.

de Jong, W.W., Hendricks, W., Mulders, J.W.M. and Bloemendal, H. (1989) Evolution of eye lens crystallins: the stress connection. *Trends Biochem. Sci.*, **14**, 365–8.

Kock, K.-H. (1985) Marine habitats – Antarctic fish, in *Key Environments in Antarctica* (eds W.N. Bonner and D.W.H. Walton), Pergamon Press, Oxford, pp. 173–92.

Kunzmann, A. (1990) Gill morphometrics of two Antarctic fish species, *Pleuragramma antarcticum* and *Notothenia gibberifrons. Polar Biol.*, **11**, 9–18.

Kunzmann, A. (1991) Blutphysiologie hochantarktischer Fische. *Polarforschung*, **59**, 129–39.

Macdonald, J.A. and Montgomery, J.C. (1991) The sensory of notothenioid fish, in *Biology of Antarctic Fish* (eds G. di Prisco, B. Maresca and B. Tota), Springer-Verlag, Berlin, pp. 145–62.

Macdonald, J.A. and Wells, R.M.G. (1991) Viscosity of body fluids from Antarctic notothenioid fishes, in *Biology of Antarctic Fishes* (eds G. di Prisco and B. Maresca), Springer-Verlag, Berlin, pp. 163–78.

Macdonald, J.A., Montgomery, J.C. and Wells, R.M.G. (1987) Comparative physioloy of Antarctic fishes. *Adv. Mar. Biol.*, **24**, 321–88.

Meyer-Rochow, V.B. and Klyne, M.A. (1982) Retinal organisation of the eyes of three nototheniid fishes from the Ross Sea (Antarctica). *Gegenbaurs morph. Jb.*, **128**, 762–77.

Montgomery, J.C. (1989) Lateral line detection of planktonic prey, in *The Mechanosensory Lateral Line, Neurobiology and Evolution* (eds S. Coombs, P. Gorber and H. Munz), Springer-Verlag, New York, pp. 561–74.

Montgomery, J.C. (1991) “Seeing” with non visual senses: mechano-sensory and electrosensory systems of fish. *News physiol. Sci.*, **6**, 73–7.

Montgomery, J.C. and Macdonald, J.A. (1987) Sensory tuning of lateral line receptors in Antarctic fish to the movements of planktonic prey. *Science*, **235**, 195–6.

Montgomery, J.C. and Macdonald, J.A. (1990) Effects of temperature on the nervous system: implications for behavioural performance. *Am. J. Physiol.*, **259**, R191–6.

Montgomery, J.C. Macdonald, J.A. and Housley, G.D. (1988) Lateral line function in an Antarctic fish related to the signals produced by planktonic prey. *J. Comp. Physiol.*, **163**, 827–33.

Montgomery, J.C., Pankhurst, N.W. and Foster, B.A. (1989) Limitations on visual feeding in the planktivorous Antarctic fish *Pagothenia borchgrevinki. Experientia*, **45**, 395–7.

Morris, D.J. and North, A.W. (1984) Oxygen consumption of five species of fish from South Georgia. *J. exp. Mar. Biol. Ecol.*, **78**, 75–86.

North, A.W. and White, M.G. (1987) Reproductive strategies of Antarctic fish, in *Proc. Fifth Congress of European Ichthyology, Stockholm 1985* (eds S.O. Kullander and B. Fernholm), Swedish Museum of Natural History, Stockholm, pp. 381–90.

Oliver, J.S. and Slattery, P.N. (1985) Effects of crustacean predators on species composition and population structure of soft-bodied infauna from McMurdo Sound, Antarctica. *Ophelia*, **24**, 155–75.

Pankhurst, N.W. and Montgomery, J.C. (1989) Visual function in four Antarctic nototheniid fishes. *J. exp. Biol.*, **142**, 311–24.

Pankhurst, N.W. and Montgomery J.C. (1990) Ontogeny of vision in the Antarctic fish *Pagothenia borchgrevinki* (Nototheniidae). *Polar Biol.*, **10**, 419–22.

Powers, D.A. (1980) Molecular ecology of teleost fish hemoglobins: strategies for adapting to changing environments. *Am. Zool.*, **20**, 139–62.

Qvist, J., Weber, R.E., DeVries, A.L. and Zapol, W.M. (1977) pH and haemoglobin oxygen affinity in blood from the Antarctic cod *Dissostichus mawsoni*. *J. exp. Biol.*, **67**, 77–88.

Ralph, R. and Everson, I. (1968) The respiratory metabolism of some Antarctic fish. *Comp. Biochem. Physiol.*, **27**, 299–307.

Rankin, J.C. (1989) Blood circulation and gill water fluxes in *Chaenocephalus aceratus* Lönnberg. *Antarct. Spec. Topic, Br. Antarc. Surv.*, 87–91.

Robilliard, G.A. and Dayton, P.K. (1969) Notes on the biology of the chaenichthyid fish *Pagetopsis macropterus* from McMurdo Sound, Antarctica. *Antarct. J. U.S.*, **4**, 304–6.

Saint-Paul, U., Hubold, G. and Ekau, W. (1988) Acclimation effects on routine oxygen consumption of the Antarctic fish *Pogonophryne scotti* (Artedidraconidae). *Polar Biol.*, **9**, 125–8.

Santer, R.M. (1985) Morphology and innervation of the fish heart. *Adv. anat. embryol. Cell Biol.*, **89**, 1–102.

Satchell, G.H. (1991) *Physiology and Form of Fish Circulation*, Cambridge Univ. Press, Cambridge, 235 pp.

Shust, K.V. and Pinskaya, I.A. (1978) Age and rate of growth of six species of nototheniid fish (Family Nototheniidae). *J. Ichthyol.*, **18**, 743–9.

Sidell, B.D. (1983) Cellular acclimitization to environmental change by quantitative alterations in enzymes and organelles, in *Cellular Acclimatization to Environmental Change* (Soc. Exp. Biology, Seminar Ser. 17) (eds A.R. Cossins and P. Sheterline), Cambridge Univ. Press, Cambridge, pp. 103–20.

Sidell, B.D. (1991) Physiological roles of high lipid content in tissues of Antarctic fish species, in *Biology of Antarctic Fish* (eds G. di Prisco, B. Maresca and B. Tota), Springer-Verlag, Berlin, pp. 220–310.

Smith, M.A.K. and Haschemeyer, A.E.V. (1980) Protein metabolism and cold adaptation in Antarctic fishes. *Physiol. Zool.*, **53**, 373–82.

Sureau, D., Lagardere, J.P. and Pennec, J.P. (1989) Heart rate and its cholinergic control in the sole (*Solea vulgaris*) acclimatized to different temperatures. *Comp Biochem. Physiol.*, **92A**, 49–51.

Tetens, V., Wells, R.M.G. and DeVries, A.L. (1984) Antarctic fish blood: respiratory properties and the effects of thermal acclimation. *J. exp. Biol.*, **109**, 265–79.

Torres, J.J. and Somero, G.N. (1988a) Metabolism, enzyme activities and cold adaptation in Antarctic mesopelagic fishes. *Mar. Biol.*, **98**, 169–80.

Torres, J.J. and Somero, G.N. (1988b) Vertical distribution and metabolism in Antarctic mesopelagic fishes. *Comp. Biochem. Physiol.*, **90B**, 521–8.

Tota, B., Farina, F. and Zummo, G. (1988) Ultrastructural aspects of functional interest in the ventricular myocardial wall of the Antarctic fish *Chaenocephalus aceratus*. *Comp. Biochem. Physiol.*, **90B**, 561–6.

Tota, B., Agnisola, A., Schioppa, M., Acierno, R., Harrison, P. and Zummo, G. (1991a) Structural and mechanical characteristics of the heart of the icefish *Chionodraco hamatus* (Lönnberg), in *Biology of Antarctic Fish*, (eds G. di Prisco, B. Maresca and B. Tota), Springer-Verlag, Berlin, pp. 204–19.

Tota, B., Acierno, R. and Agnisola, C. (1991b) Mechanical performance of the isolated and perfused heart of the haemoglobinless Antarctic icefish *Chionodraco hamatus* (Lönnberg): effects of loading conditions and temperature. *Phil. Trans. R. Soc.*, **332B**,

191–8.

Weber, R.E. and Wells, R.M.G. (1989) Hemoglobin structure and function, in *Comparative Pulmonary Physiology Current Concepts* (ed. S.C. Wood), Marcel Dekker, New York, pp. 279–310.

Wells, R.M.G. (1986) Cutaneous oxygen uptake in the Antarctic icequab, *Rhigophila dearborni* (Pisces: Zoarcidae). *Polar Biol.*, **5**, 175–80.

Wells, R.M.G. (1987) Respiration of Antarctic fish from McMurdo Sound. *Comp. Biochem. Physiol.*, **88A**, 417–24.

Wells, R.M.G. (1990) Hemoglobin physiology in vertebrate animals: a cautionary approach to adaptationist thinking, in *Advances in Comparative and Environmental Physiology*, Vol. 6 (ed. R.G. Boutilier), Springer-Verlag, Berlin, pp. 143–61.

Wells, R.M.G. and Jokumsen, A. (1982) Oxygen binding properties of hemoglobins from Antarctic fishes. *Comp. Biochem. Physiol.*, **71B**, 469–73.

Wells, R.M.G. and Weber, R.E. (1991) Is there an optimal haematocrit for rainbow trout, *Oncorhynchus mykiss* (Walbaum)? An interpretation of recent data based on blood viscosity measurements. *J. Fish Biol.*, **38**, 53–65.

Wells, R.M.G., Ashby, M.D., Duncan, S.J. and Macdonald, J.A. (1980) Comparative study of the erythrocytes and haemoglobins in nototheniid fishes from Antarctica. *J. Fish Biol.*, **17**, 517–27.

Wells, R.M.G., Tetens, V. and DeVries, A.L. (1984) Recovery from stress following capture and anaesthesia of Antarctic fish: haematology and blood chemistry. *J. Fish Biol.*, **25**, 567–76.

Wells, R.M.G., McIntyre, R.G., Morgan, A.K. and Davie, P.S. (1986) Physiological stress responses in big gamefish after capture: observations on plasma chemistry and blood factors. *Comp. Biochem. Physiol.*, **84**, 565–71.

Wells, R.M.G., Summers, G., Beard, L.A. and Grigg, G.C. (1988) Ecological and behavioural correlates of intracellular buffering capacity in the muscles of Antarctic fishes. *Polar Biol.*, **8**, 321–5.

Wells, R.M.G., Grigg, G.C., Beard, L.A. and Summers, G. (1989) Hypoxic responses in a fish from a stable environment: blood oxygen transport in the Antarctic fish *Pagothenia borchgrevinki*. *J. exp. Biol.*, **141**, 97–111.

Wells, R.M.G., Macdonald, J.A. and di Prisco, G. (1990) Thin-blooded Antarctic fishes: a rheological comparison of the haemoglobin-free icefishes *Chionodraco kathleenae* and *Cryodraco antarcticus* with a red-blooded nototheniid, *Pagothenia bernacchii*. *J. Fish Biol.*, **36**, 595–609.

Wohlschlag, D.E. (1960) Metabolism of an Antarctic fish and the phenomenon of cold adaptation. *Ecology*, **41**, 287–92.

Wohlschlag, D.E. (1964) Respiratory metabolism and growth of some Antarctic fishes, in *Biologie Antarctique* (Scientific Committee on Antarctic Research, *Proc. 1st Symp. Antarctic Biology, Paris*) (eds R. Carrick, M.W. Holdgate and J. Prevost), Hermann, Paris, pp. 489–502.

Chapter fifteen

Ecophysiology of intertidal fish

C.R. Bridges

15.1 INTRODUCTION

The intertidal environment may be characterized by extreme changes in water temperature, pH, salinity, oxygen and carbon dioxide tensions, and these may occur on a daily basis or be influenced by season or by the prevailing weather conditions. Temperature, osmoregulatory and respiratory problems may also be exacerbated by aerial exposure in some intertidal fish species. This brief review will concentrate on temperature, osmotic, respiratory and metabolic adaptations of intertidal fish which enable them to survive and colonize this extreme environment. Gibson (1982) has defined an intertidal fish as "a species which relies on utilization of the intertidal zone for completion of all or an essential part of its life history". This definition will therefore be used throughout the chapter.

There have been a number of reviews on the biology of intertidal fish and the reader is referred to these for more details of their life history (Gibson, 1969, 1982; Zander, 1972; Horn and Gibson, 1988). The present chapter concerns itself specifically with the physiological adaptations of fish to the intertidal environment.

15.2 THE INTERTIDAL ENVIRONMENT

The intertidal environment is characterized by the diurnal changes brought about by the daily tidal cycles of ebb (emersion) and flood (immersion). Parameters such as P_{O_2}, P_{CO_2} and pH are under biological control, whereas temperature and salinity are dependent upon climatic conditions.

Fish Ecophysiology. Edited by J. Cliff Rankin and Frank B. Jensen. Published in 1993 by Chapman & Hall, London. ISBN 0 412 45920 5.

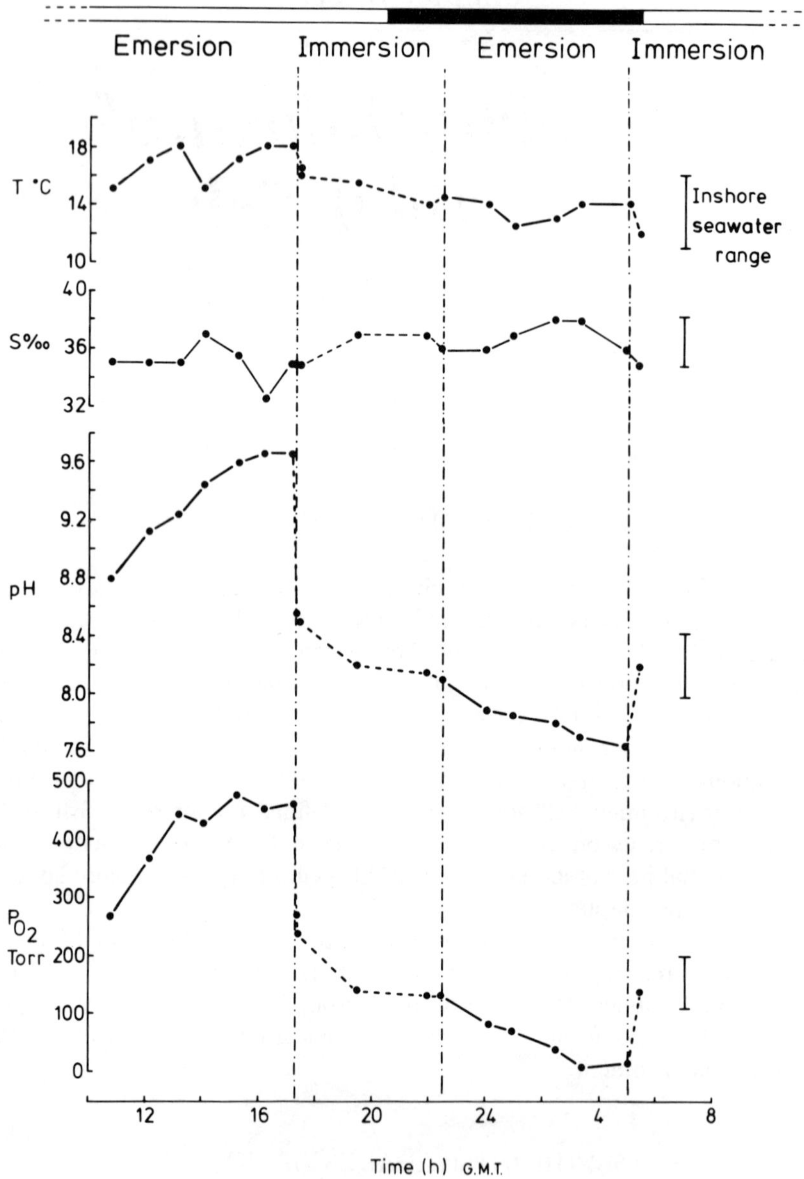

Fig. 15.1 Field measurements of temperature (*T*), salinity (*S*), pH and oxygen partial pressure (P_{O_2}) made over a 20 h period (9–10 September, 1981) in an intertidal rockpool (5.2 m above chart datum); darkness is indicated by the black horizontal bar. Modified from Bridges *et al.* (1984).

Table 15.1 Extreme values for environmental parameters measured on a diurnal basis

Study area	Env.	Temp. (ÉC)		P_{O_2} (Torr)		P_{CO_2}+ (Torr)		pH		Salinity (‰)		Source
		max.	min	max.	min.	max.	min.	max.	min.	max.	min	
E. Canada	R	28.5	15	–	–	–	–	–	–	29.0	28.1	Klugh (1924)
Cape Town	R	25.2	11.8	489	23	–	–	<9.0	<8.0	–	–	Stephenson *et al.* (1934)
N. Wales	R	25.6	12.7	423	90	–	–	9.3	8.3	40.6	33.8	Pyefinch (1943)
S. India	T	33.0	25.0	238	49	–	–	–	–	38.0	37.0	Ramachandran Nair *et al.* (1965)
Sweden	R	30.5	18.9	303	58	–	–	10.0	8.0	–	–	Ganning (1971)
S. Wales	R	24.9	13.6	442	6.2	–	–	9.4	7.2	34.1	32.3	Daniel and Boyden (1975)
W. Canada	R	24.0	12.5	–	–	–	–	–	–	–	–	Nakamura (1976)
France	R	25.5	12.5	577	3.2	2.72	0.0001	10.2	7.3	36.7	34.5	Truchot and Duhamel-Jouve (1980)
Scotland	R	22.0	14.5	435	40	0.25	0.01	8.9	8.25	34.0	29.0	Morris and Taylor (1983)
Isle of Man	R	18.0	12.5	475	9	1.07	0.004	9.6	7.7	38.0	32.5	Bridges *et al.* (1984)

Env., environment (R, Rockpool; T, Tidepool); max. values refer to daylight emersion; min. values refer to night emersion.
+ P_{CO_2} values are reversed, with a daylight minimum and a night maximum.

Diurnal (diel) changes

During high tide periods all regions show similar physico-chemical values to those in the open sea; it is only during the emersion period that values may change dramatically. Figure 15.1 shows a typical example of diurnal variations in physico-chemical parameters in an intertidal rockpool. During daylight emersion periods, the surface water temperature rises due to warming by the sun. Concomitantly pH increases from 8.7 to 9.6 due to a decrease in P_{CO_2} (not shown) from 10^{-1} to 10^{-2} Torr, resulting from algal photosynthesis. During darkness, where respiration by microorganisms, animals and plants produces CO_2, pH decreased to values around 7.6 and P_{CO_2} reached 1 Torr. In parallel marked changes are observed in oxygen tensions. During daylight emersion, P_{O_2} increased to 460 Torr and at night decreased to 16 Torr. Maximum and minimum changes recorded on a diurnal basis are summarized in Table 15.1 for mainly temperate waters. The absolute maximum or minimum values recorded for a particular pool are dependent upon four main factors: (1) the tidal level on the shore, which determines the length of the emersion period; (2) the size and the depth of the pool; (3) climatic conditions such as exposure, shade and weather; and (4) biological factors such as the biomass of the pool population.

Seasonal changes

Seasonal changes in physico-chemical parameters have been thoroughly studied by Morris and Taylor (1983) for rockpools in Millport, Scotland. There are two ways in which seasonal variations may occur: (1) an increase in the range of values in any cycle, and (2) a change in the maximum and minimum values independent of any change in range. This is true for pH and to a lesser extent for temperature. For temperature and P_{O_2} the largest variations between maximum and minimum values occurred in the summer (July). In winter the difference between extremes was much reduced. The largest difference in salinity and pH extremes occurred in December and March respectively where rainfall was extensive.

Rates of change of environmental parameters

The rate of change as a function of time for environmental variables is shown in Table 15.2 and in part in Fig. 15.2. These values represent the slower changes which occur over the emersion period. In general the rates of change of temperature, P_{O_2} and pH are greater during daylight than at night, whereas P_{CO_2} values increase more rapidly at night. The magnitude of the rate of change for temperature is remarkably constant for all areas, and the lower rate at night may be due to the buffering effect of the bedrock retaining heat (Ganning, 1971; Daniel and Boyden, 1975). Truchot and Duhamel-Jouve

Table 15.2 Rates of change as a function of time (t) for environmental variables in an intertidal rockpool. Values are taken or calculated from authors' data

Study Area	$\Delta T/\Delta t$ (°C h^{-1})		$\Delta P_{O_2}/\Delta t$ (Torr h^{-1})		$\Delta P_{CO_2}/\Delta t$ (Torr h^{-1})		Δ pH/Δt (pH unit h^{-1})		Source
	Day	Night	Day	Night	Day	Night	Day	Night	
Canada	+1.9	–	–	–	–	–	–	–	Klugh (1924)
South Africa	–	–	+63	-20	–	–	+0.23	–	Stephenson *et al.* (1934)
Wales	+0.9	-0.4	+42	-21	–	–	+0.14	–	Pyefinch (1943)
India	+0.8	-0.6	+32	-16	–	–	–	–	Ramchandram Nair *et al.* (1965)
Sweden	–	–	+35	-24	–	–	–	–	Ganning (1971)
Wales	+1.6	-0.1	+58	-26	–	–	–	–	Daniel and Boyden (1975)
France	+1.8	-0.2	+60	-40	-0.008	+0.2	+0.3	-0.2	Truchot, 1988; Truchot and Duhamel-Jouve (1980)
Scotland	+1.75	-0.7	+28	-47	-0.019	+0.164	+0.1	-0.1	Morris and Taylor (1983)
Isle of Man	+0.4	-0.3	+46	-13	-0.008	+0.1	+0.13	-0.08	Bridges *et al.* (1984)

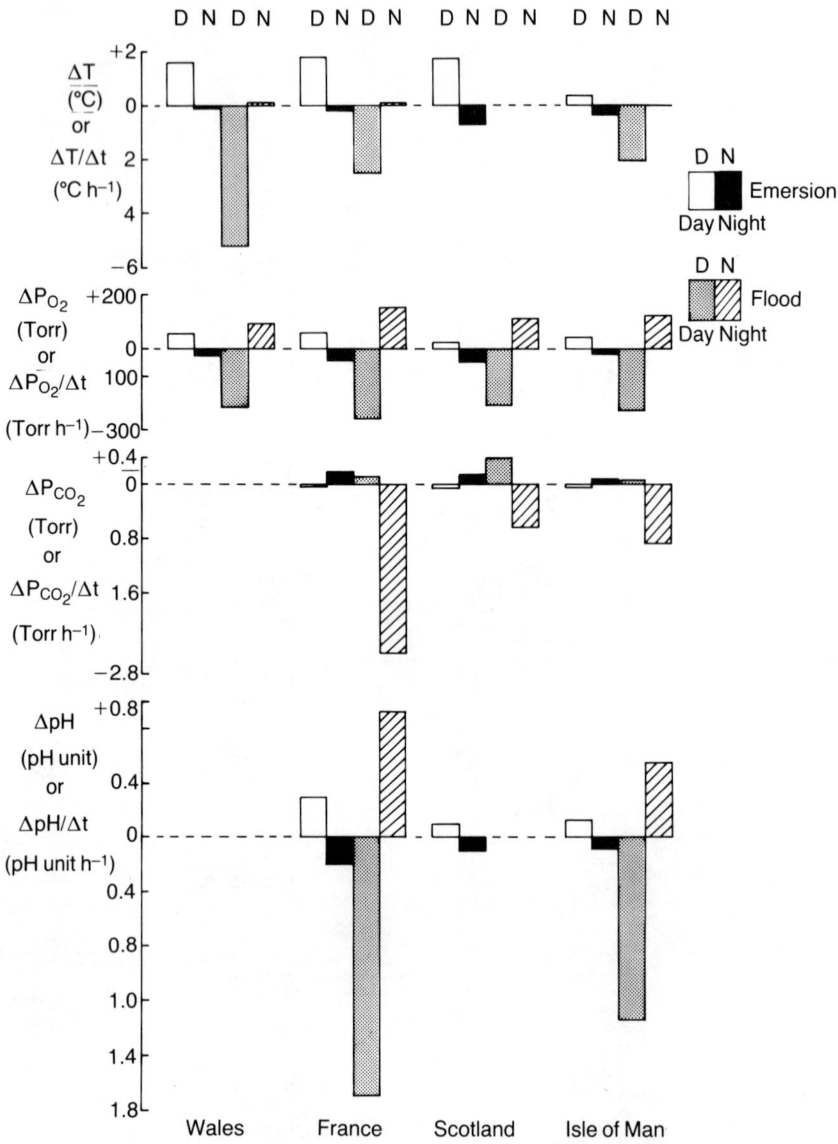

Fig. 15.2 Rates of change in environmental parameters in different coastal rockpools during the emersion period (data taken from Table 15.2 for emersion changes) and instantaneous changes recorded in the same parameters on flooding (data sources: Wales, Pyefinch, 1943; France, Truchot and Duhamel-Jouve, 1980; Scotland, Morris and Taylor, 1983; Isle of Man, Bridges *et al.*, 1984).

(1980) suggest that O_2 diffusion between water and the atmosphere may limit the oxygen increase during the day and the oxygen exhaustion at night. For carbon dioxide this is not as significant, as the diffusion gradients are larger and more important for oxygen (Morris and Taylor, 1983). Even more extreme and rapid changes are observed during the flood period (Fig. 15.2), where the change may occur over a period of only a few minutes as the pool is flooded or immersed. Immersion during daylight causes the largest change in all parameters.

Soft sediments

Environmental parameter studies in soft sediments are relatively rare. Thermal gradients in soft sediments have been recorded in mudskipper burrows in Kuwait, with burrow temperatures up to 12 °C lower than air temperatures (El-Ziady *et al.*, 1979; Tytler and Vaughan, 1983). In winter, burrow temperatures at dawn were higher than the corresponding mud or air temperatures, which gradually increased above the burrow temperature as emersion progressed (El-Ziady *et al.*, 1979). Low Po_2 values of 5 to 25 Torr have been recorded in the burrows of various mudskippers (El-Ziady *et al.*, 1979; Gordon *et al.*, 1985) at depths varying from 5 to 50 cm. A number of gobies inhabit the burrows of intertidal thalassinid shrimps where oxygen tensions of less than 3 Torr have been recorded 4 h after tidal exposure at a depth of 15–18 cm (Congleton, 1974).

15.3 TEMPERATURE

Temperature has a very varied and important role in controlling both the vertical distribution and survival of intertidal fish, since these organisms are poikilotherms and are unable to regulate body temperature actively. There remains, however, the use of thermoregulatory behaviour to avoid insolation, and evaporative water loss may also be used for cooling.

Temperature preference and temperature tolerance

Bull (1936) showed that a number of intertidal fish species were able to detect a 0.03 °C change in water temperature and could therefore react to a temperature gradient. Doudoroff (1938) rejected the term 'optimum temperature' for *Girella nigricans* since this species selected a higher temperature in the laboratory than that found in the field. Fish acclimated at higher temperatures selected a higher temperature level, but this effect was slight and only temporary. More emphasis was therefore placed on the fast adaptation to rapid changes in temperature, where heat resistance could be gained rapidly after a rise in temperature and lost slowly after cooling (Doudoroff, 1942). Norris (1963) reappraised and amplified the work of Doudoroff to show thermotaxis in the juveniles of this species towards a mean temperature of 26 °C. Prolonged constant temperature acclimation in the laboratory masked the ability of the fish to respond to temperature changes, thus explaining the different laboratory results previously published. In a naturally occurring temperature gradient of rockpools with a temperature difference of 9.5 °C, fish were found to select a temperature between 20 and 29.5 °C with a mean value similar to that found in laboratory experiments. Transforming juvenile fish selected the highest temperature, behaviour which may have directed an inshore migration to the highest pools where predators are relatively scarce and food abundant.

Acclimatization to cold was gained slowly by *G. nigricans* and fish were killed at temperatures well above 0 °C (Doudoroff, 1942). Cold adaptation may also involve the generation of 'antifreeze' in the blood of intertidal fishes. Duman and DeVries (1974) found that only specimens of *Anoplarchus pupurescens* from the Bering Sea were able to synthesize 'antifreeze' and the same species from the coast of California could not.

Temperature sensitivity can also be dependent upon ancestry, as a member of the boreal cottids was more tolerant of cold than a member of the more tropical Blenniidae but less tolerant of warm water conditions (Graham, 1970). Acclimation of these species to different temperatures changed the lower lethal temperature, but the upper lethal temperature remained relatively constant in both species. In all experiments the lethal temperature was well beyond that found in the natural environment. In *Blennius pholis* the critical thermal maxima and minima could be changed by acclimation (Campbell and Spencer-Davies, 1975), and in *Pomatoschistus microps* not only acclimation temperature but also an increase in salinity influences the lethal temperature (Tolksdorf, 1978).

All the examples so far dealt with have referred to aquatic exposure. The most extreme temperature changes may be found during emersion on aerial exposure due to stranding or through voluntary excursions on land. Here the effect of insolation is of critical importance.

Rockskippers die at a lower body temperature when exposed to aerial temperature extremes in comparison with water values, survival time being increased by evaporative cooling. They show no signs of thermoregulatory behaviour, but like their aquatic counterparts are able to detect a small difference in water temperature (Brown *et al.*, 1991). Rockskippers in general do not migrate on land for large distances and prefer to remain in contact with water, therefore avoiding desiccation (Brown *et al.*, 1991). Thermal resistance to high and low temperatures shows set patterns in the clingfish distribution, with all species being more cold tolerant. Exposure to high temperatures, as in many intertidal species, may only be for a brief period of time at low tide, thus enabling these fish to tolerate what would normally be a lethal temperature.

The mudskippers are exposed to insolation and are truly amphibious, spending longer periods of time out of water (up to 7 minutes, Gordon *et al.*, 1978) compared to rockskippers. Behavioural thermoregulation can substantially lower body temperatures compared with air temperatures in some mudskippers, and El-Ziady *et al.* (1979) and Tytler and Vaughan (1983) showed clearly that the burrow can be used to avoid higher temperatures in the daytime and also to avoid low temperatures in the morning during the winter. Basking can also be used to raise the body temperature in winter (Tytler and Vaughan, 1983).

Desiccation resistance and evaporative water loss

The vertical distribution of many intertidal fish is controlled to a large extent by desiccation tolerance. Table 15.3 and Fig. 15.3 show a comparison (if, however, limited, as they are not for similar weight ranges, temperatures or humidities) for intertidal species ranging from the mudskippers through clingfish, rockskippers and rocky shore fish. Some amphibians are also included for comparison.

Mudskippers can usually survive desiccation up to -21 to -22% of their body weight in the shade (Gordon *et al.*, 1969, 1978). At near saturation values they can survive in air for 30–37 h, and thermal stress appears to be greater than desiccation stress when exposed to the sun because evaporative water loss is low (Gordon *et al.*, 1969), survival times depending upon insolation and wind conditions.

Clingfish can withstand higher evaporative water losses and can survive for long periods in air, in the shade (Table 15.3). There appears to be no means of controlling evaporative water loss in *Sicyases sanguineus* as evaporative water loss at death is higher in windy conditions and not correlated with body size (Gordon *et al.*, 1970). In clingfish, survival in air is correlated with the vertical zonation of these species on the shore (Eger, 1971). These fish use both behavioural strategies, e.g. curling up at the sides of a rock to reduce exposed surface area, and morphological strategies, e.g. possession of large mucous cells and a thickened epidermis in the skin, to reduce water loss.

Rockskippers are not well adapted to aerial exposure for long periods and prefer to remain close to water (Table 15.3; Graham, 1973; Graham *et al.*, 1985). In *Mnierpes macrocephalus*, Graham (1973) found a negative correlation between weight loss and body size, indicating again the role of evaporative water loss.

Rocky shore fish are able to withstand long periods in air and can survive decreases in body weight up to 17 to 24% of their body weight (Table 15.3 and Fig. 15.3; Daniel, 1971; Horn and Riegle, 1981). Similarly, survival time in amphibious stichaeoid fish increases with body size, and a good correlation was obtained between the ability to avoid water loss and the vertical distribution of each species (Horn and Riegle, 1981); *Cebidichthys violaceus*, with the highest endurance of water loss and smallest surface area to weight ratio, was found highest on the shore. In three of the species studied, live and dead fish showed the same rate of water loss. In *Amoplarchus purpurescens* and *Xererpes*

Table 15.3 Survival time and weight loss under varying experimental conditions in intertidal fish and amphibians

Species	Survival time (h)	Weight loss[‡] (% body wt.)	Experimental conditions	Humidity (%)	Temp. (°C)	Source
Mudskippers						
Periophthalmus sobrinus	24	-21	Shade	70–80	23–30	Gordon *et al.* (1969)
P. subrinus	0.8	-12	Sun	70–80	43–44	Gordon *et al.* (1969)
Periophthalmus cantonensis	5.5	-22	Still, shade	40–55	23–24	Gordon *et al.* (1978)
P. cantonensis	1.0	-14	Moving, shade	51–57	27–28	Gordon *et al.* (1978)
P. cantonensis	0.25	-8	Moving, sun	55–56	30–31	Gordon *et al.* (1978)
Clingfish						
Sicyases sanguineus	41	-9	Still, shade	60–76	12–16	Gordon *et al.* (1970)
S. sanguineus	19	-22	Wind, shade	60–76	12–16	Gordon *et al.* (1970)
S. sanguineus	2.5	-12	Wind, sun	60–76	12–16	Gordon *et al.* (1970)
S. sanguineus	>25.5	-12.8	–	100	17.5–18.6	Marusic *et al.* (1981)
Pherallodiscus funebris	93	-46.4*	Humid	–	–	Eger (1971)
Tomicodon humeralis	33	-47.0*	Humid	–	–	Eger (1971)
Tomicodon böehlkei	19	-39.8*	Humid	–	–	Eger (1971)
Gobiesox pinniger	24	-41.6*	Humid	–	–	Eger (1971)
Rockskippers						
Mnierpes macrocepahlus	1.5–3	-21.4	–	60–70	30–32	Graham (1973)
Entomacrodus nigricans	0.125	–	Sun	–	31–37	Graham *et al.* (1985)
E. nigricans	2.8	–	Shade	–	27–29.5	Graham *et al.* (1985)
E. nigricans	2.25	–	–	95	30	Graham *et al.* (1985)
E. nigricans	0.75	–	–	80–90	30	Graham *et al.* (1985)

Table 15.3 Continued

Rocky shore fish						
Blennius pholis	–	-22.0	–	32–93	5–20	Daniel (1971)
B. pholis	96–144	–	–	>93	–	Daniel (1971)
Cebidichthys violaceus	24†	-23.8	–	95.5	15	Horn and Riegle (1981)
Xiphister mucosus	19.9†	-19.8	–	95.5	15	Horn and Riegle (1981)
Xiphister atropurpureus	14.4†	-16.6	–	95.5	15	Horn and Riegle (1981)
Anoplarchus purpurescens	24.0†	-17.5	–	95.5	15	Horn and Riegle (1981)
Xererpes fucorum	14.4†	-20.5	–	95.5	15	Horn and Riegle (1981)
Amphibians						
Rana tigerina	3.0	-23.0	Still, shade	40–55	23–24	Gordon *et al.* (1978)
Rana clamitans	–	-36.0	–	–	–	Horn and Gibson (1988)
Scaphiophus couchi	–	-47.0	–	–	–	Horn and Gibson (1988)

* Calculated assuming water content is 78% of wet weight.
† Calculated for a 12.5 cm fish.
‡ Weight loss which the fish can survive without dying.
Moving or still indicates the wind movement.

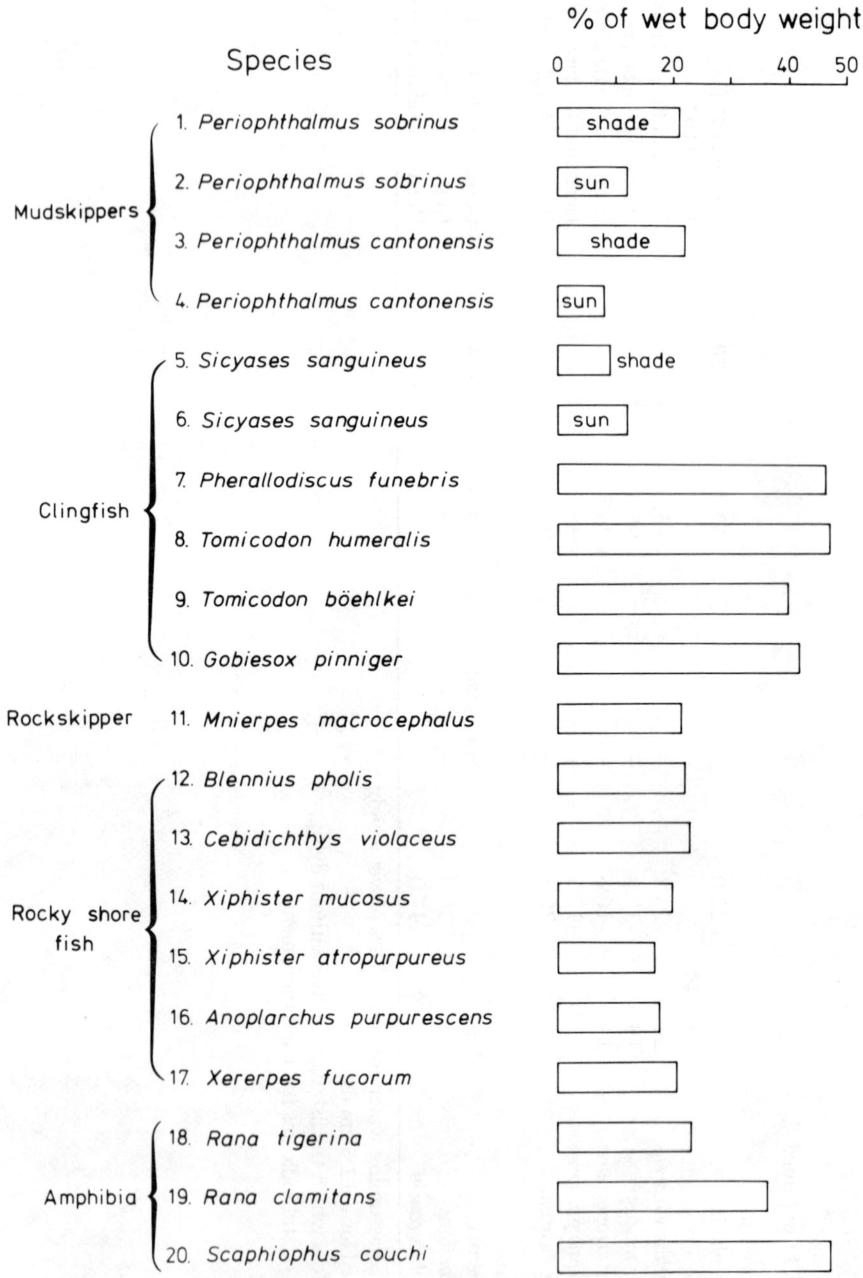

Fig. 15.3 Tolerance to desiccation of intertidal fish in comparison with some amphibians, expressed as the percentage of normal body wet weight to which each species could be reduced without killing the animal.

fucorum slower rates were observed in live fish, indicating active regulation. Rates of evaporative water loss in intertidal fish are low in comparison with amphibians under the same conditions (Gordon *et al.*, 1978), although intertidal fish tend to have a larger surface area to volume ratio. They can survive weight losses equivalent to or larger than those seen in amphibians (Table 15.3; Fig. 15.3; Horn and Riegle, 1981).

Effects of temperature on oxygen consumption

Since temperature has an important effect on many physiological processes, its influence on metabolism may also be a determining factor in survival. In general the mean Q_{10} value of oxygen consumption for intertidal fish is no different from that of other fish, with an overall value of around 2.3 (Bridges, 1988). Some differences are found though, and these shed light on possible adaptations. Morris (1961) demonstrated both seasonal and geographical dependence of Q_{10} in Pacific gobiids and correlated this with their vertical distribution on the shore. Graham (1973) found that the high-shore species was more adapted to warmer water. In *Blennius pholis*, the response to cold appears to be more important, with a low Q_{10} between 2 and 5 °C and a higher value between 5 and 10 °C when fish were acclimated at 10 °C (Campbell and Spencer-Davies, 1975). Acclimation at higher temperatures led to an increase in the Q_{10}. Low-temperature-acclimated animals do not need to respond to temperature decreases, whereas between 5 and 10 °C and in warm-acclimated animals a large decrease in metabolism may be adaptive against the cold. In tropical forms this process appears to be similar. Tamura *et al.* (1976) found that metabolism was relatively temperature independent in mudskippers between 5 and 10 °C, but between 10 and 35 °C Q_{10} increased. Gordon *et al.* (1978) obtained similar findings for the mudskipper and suggest that this difference in Q_{10} at high and low temperatures fits in well with natural behaviour patterns. In the winter, temperatures may fall to 3–4 °C and fish remain in their burrows. A reduction in metabolic rate would be beneficial under these conditions. As environmental temperatures rise, then a rapid response in metabolism at higher temperatures would also enable them to become active more rapidly and to use the available time for foraging.

15.4 OSMOREGULATION

Salinity tolerance or preference

Salinity tolerance has been extensively reviewed by Gibson (1969, 1982) and nearly all intertidal fish are euryhaline. Many intertidal fish exhibit a sensitivity to salinity changes as small as 0.06 to 0.45‰ (Bull, 1938). Most workers have either referred to the finding of fish in extreme salinities or have used an experimental approach. In general, *in situ* values for salinity in coastal lagoons, where fish can survive, rise as high as 142.4‰ (Gunter, 1967). Few authors comment on the finding of intertidal fish at low salinity. Experimentally, salinities (67‰) which are higher than environmental values are withstood by cottid fish (Morris, 1960), with decreasing salinity best withstood by the more freshwater species (Foster, 1969). In the laboratory, *Pomatoschistus microps* acclimated to 6–8‰ could also survive in distilled water for 2 weeks (Kunz, 1969), and an abrupt change in salinity (27.4 to 35.6‰ over 14 h) had no deleterious effects.

Brown *et al.* (1991) showed that rockskippers could survive between 0 and 60‰ salinity, but salinities above 70‰ were avoided. A slight preference was shown for salinities between 30 and 40‰, which is near to the natural range.

Osmoregulatory adaptations

Many intertidal fish are able to regulate the osmolarity of their plasma or serum extremely well against changing salinity (Fig. 15.4 (a, b)). There are potentially three main sites of water and ion transfer – the skin, the gills and the gut – in euryhaline teleosts, and osmoregulatory mechanisms in intertidal fish have been studied in some detail due to their euryhaline status. Diffusional and osmotic permeabilities to water must be modified to minimize osmotic water movements in these species (Evans, 1984). *Blennius pholis* is able to maintain full seawater-acclimated osmolarity down to around 14‰ by maintaining plasma sodium chloride concentrations almost constant (House, 1963). At values below 14‰ a rapid physiological response occurs, with a change from active chloride excretion over the gills in hyperosmotic conditions, to active sodium and chloride uptake in hypo-osmotic seawater. This response is bidirectional and occurs within minutes of the fish being transferred from one salinity to another (House, 1963). Passive loss of salts on transfer to fresh water is therefore offset by active salt uptake, thus slowing the change, and in simulated tidal salinity changes, plasma osmolarity did not change in a 6 h exposure period, indicating the buffering effect of the blood on osmoconcentration (Davenport and Vahl, 1979). Adult fish are much more tolerant of low salinities than immature or juvenile fish. This is associated with the greater uptake of water by younger fish, which have a greater gill area per unit body weight, and the more efficient kidneys of the adults with a larger total glomerular area (Milton, 1971).

In detailed studies, Evans (1967a, b, c) found that *Xiphister atropurpureus* was an excellent regulator of Na^+ and Cl^- concentrations in the plasma and that the intra- and extracellular water space remained relatively unchanged when the fish was transferred to dilute seawater. The low rates of urine flow in dilute seawater, and a low drinking rate in 100% seawater indicate low water permeability. Chloride permeability could not be controlled, but sodium

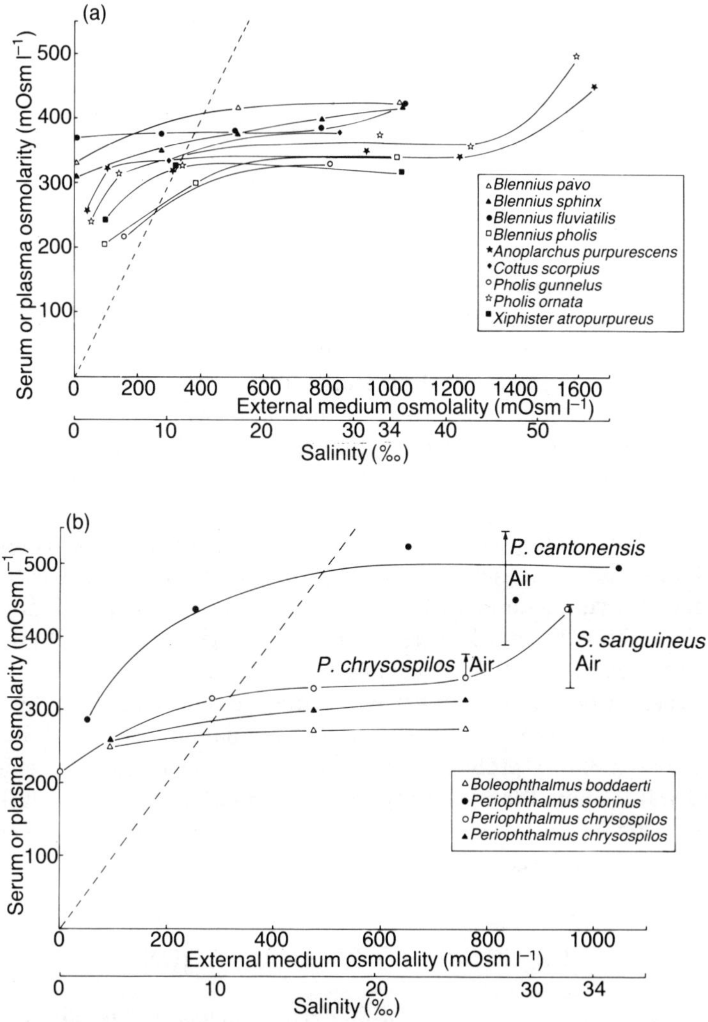

Fig. 15.4 (a and b) Regulation of internal osmolarity ($mOsml^{-1}$) in intertidal fish exposed to osmotic stress shown by the osmolarity in the external medium plotted against serum or plasma osmolarity. The values are taken or calculated from the following authors: (a) *B. pavo*, *B. sphinx* and *B. fluviatilis*, Müller *et al.* (1973); *B. pholis*, House (1963); *A. purpurescens* and *P. ornata*, Barton (1979); *C. scorpius*, Foster (1969); *P. gunnelus*, Evans (1969); *X. atropurpureus*, Evans (1967a); (b) for mudskippers, *B. boddaerti* and *P. chrysospilos*, Chew and Ip (1990); *P. sobrinus*, Gordon *et al.* (1965); *P. chrysospilos*, Lee *et al.* (1987). The values shown by the vertical arrows are changes in osmolarity observed when the different species are exposed to air for 24 h: *P. cantonensis*, Gordon *et al.* (1978); *P. chrysospilos*, Lee *et al.* (1987); *S. sanguineus*, Marusic *et al.* (1981).

permeability declined by 50% on exposure to 10% seawater. The kidney and the gut played only a small part in sodium, chloride and water balance (Evans, 1967c). Foster (1969), in a study on cottids form three different habitats – fresh water, rockpools and open sea – found that the two seawater species were similar in their flux rates of salts and water but not in their salinity tolerance, and that the freshwater form was different in almost all respects, having a lower sodium loss and a decreasing rate of water loss with increasing salinity compared with a constant rate in the other two species.

Mudskippers are euryhaline fish (Fig. 15.4(b)) with the ability to survive between 20 and 100% seawater, even tolerating fresh water after acclimation (Gordon, *et al.*, 1965). Body weight increases by 10–18% on transfer to 20% seawater or fresh water (Gordon *et al.*, 1965; Lee *et al.*, 1987). Regulation of plasma sodium and chloride over a wide range of salinities is seen. After transfer of *Periophthalmus chrysospilos* to deionized water (Lee *et al.*, 1987), the initial regulation of plasma sodium and potassium concentrations was short termed and sodium regulation was by modification of membrane permeability. A high rate of Na^+ efflux at higher salinities indicated the involvement of an active mechanism, and Na^+–K^+ ATPase activity increased. Long-term acclimation to high salinities caused an increase in plasma Na^+ and Cl^- concentrations. Corresponding adjustments were made in intracellular iso-osmotic regulation through an increase in ninhydrin positive substances (NPS). The glutamate dehydrogenase (GDH) levels for amination or deamination also respond to salinity in mudskippers, increasing at higher salinities. In the more brackish-water species *Boleophthalmus boddaerti*, no change in NPS or GDH was observed (Chew and Ip, 1990), which may be beneficial, since this species is naturally exposed to salinity changes twice a day. Intertidal fish may also suffer osmoregulatory problems due to water loss associated with emersion. Figure 15.4(b) shows the osmolarity changes in mudskippers and rockskippers exposed to air. The osmotic changes may be of the same magnitude as those experienced over a large salinity range.

15.5 RESPIRATORY ADAPTATIONS

The respiratory adaptations of intertidal fish have been dealt with in several reviews (Graham, 1976; Gibson, 1982; Bridges, 1988). This section will briefly summarize some of the main points and consider new data where appropriate.

Response to hyperoxia

Metabolism appears not to be influenced by hyperoxia in intertidal fish, as exemplified by the study of Berschick *et al.* (1987) on *Gobius cobitis.* Oxygen consumption remained constant over the P_{O_2} range from 150 to 450 Torr at

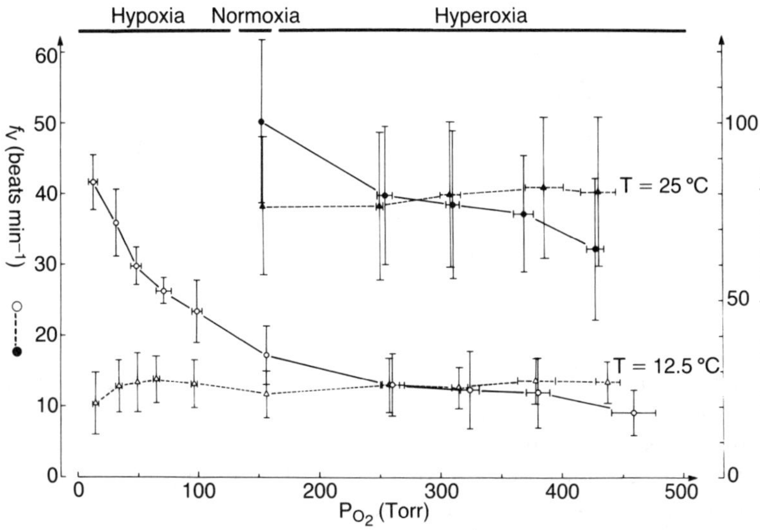

Fig. 15.5 Effect of hyperoxia at 12.5 and 25 °C and hypoxia at 12.5 °C on the ventilatory frequency (f_V, ○ - ●) and heart frequency (f_H, Δ—▲) of *Gobius cobitis*. Values are means; bars denote ± 1 SD. Modified from Berschick *et al.* (1987).

both 12.5 and 25 °C. This constancy in oxygen consumption during hyperoxia also applies to freshwater fish (Dejours, 1973; Wilkes *et al.*, 1981). Hyperoxia per se causes hypoventilation, reducing both ventilation rate and ventilation volume. Dejours *et al.* (1977) reported a decrease in the ventilatory requirement by a factor of two- to fivefold in a number of marine species under hyperoxia while oxygen extraction efficiency remained the same. In *B. pholis*, hyperoxia decreases ventilation by about 20% both in the laboratory and in the field (Bridges *et al.*, 1984; Bridges, 1988). Experiments on *G. cobitis* show the same trend, as reflected by a decrease in ventilatory frequency from 17.3 beats min $^{-1}$ to 9.2 beats min^{-1} at 460 Torr and 12.5 °C (Berschick *et al.*, 1987; Fig. 15.5). A change of similar magnitude occurred at 25 °C. The cardiac responses to hyperoxia have been studied in only a few intertidal fish. However, the available data indicate that neither heart rate (Berschick *et al.*, 1987; Fig. 15.5) nor ventral aortic blood flow and pressure (Eickelmann, 1991) changes during hyperoxia. This lack of response in intertidal fish appears to be different from that of freshwater fish, as Wilkes *et al.* (1981) report a decrease in ventral aortic pressure and heart frequency in *Catostomus commersoni*. Bridges (1988) has examined the respiratory function of haemoglobin in *B. pholis* during hyperoxia. In this species a large Bohr effect may be adaptive during daytime hyperoxia. The hyperoxia-induced hypoventilation increases blood P_{CO_2} and decreases blood pH, as confirmed by recent catheter experiments on *B. pholis*. This respiratory acidosis was more marked at 25 °C than at 12.5 °C and causes a right shift of the oxygen dissociation curve (via the large Bohr effect) that favours O_2 unloading in the tissues. Full saturation at the gills is assured by the high environmental PO_2. In this way the increased oxygen demand due to an elevation in temperature can be sustained without increasing cardiac output. Since acid–base balance can be regulated by physiological mechanisms, the influence of the large Bohr effect can be controlled. Body temperature, however, cannot be regulated in intertidal fish, therefore the low temperature sensitivity of the haemoglobin found in *B. pholis* assures that oxygen affinity remains independent of direct temperature changes (Bridges *et al.*, 1984; Bridges, 1988).

Response to hypoxia

During hypoxia, oxygen consumption in intertidal fish is maintained constant as PO_2 values decline to the critical oxygen tension (*Pc*) where oxygen consumption becomes dependent on oxygen tension. Values for the *Pc* in rocky shore fish vary among species, from 16 to 40 Torr (Congleton, 1980; Innes and Wells, 1985; Pelster, 1985; Berschick *et al.*, 1987). In those species colonizing hypoxic burrows of ghost shrimps lower values are found (9–25 Torr; Congleton, 1974). Values for the *Pc* are low in intertidal fish in comparison with other teleosts and represent an adaptation to the hypoxic conditions sometimes found within the intertidal zone.

Anaerobic metabolism appears to play a greater role after the *Pc* is reached in intertidal fish. Pelster (1985) found that the energy charge remained constant down to a PO_2 of 40 Torr in *B. pholis* and that lactic acid accumulated in both blood and muscle only at PO_2 values below the *Pc*. In mudskippers, lactate accumulated in the muscle and brain tissue after a 15 minute exposure to aerial anoxia or exercise, with a three- to fourfold increase in muscle lactate above resting levels (Bandurski *et al.*, 1968). Similar findings were made by Chew *et al.* (1990) and Ip *et al.* (1991) for mudskippers, where blood lactate increased under hypoxia (25 Torr) and muscle tissue energy charge and lactate concentration remained constant, although functional hypoxia caused a massive rise in tissue lactate levels and a dramatic fall in energy charge. Ip *et al.* (1991) suggest that mudskippers may reduce energy consumption in muscle during hypoxia, an idea which is supported by the small oxygen debt repayment observed in this species during normoxic recovery (Chew *et al.*, 1990).

Ventilatory responses to hypoxia are normally larger than those seen during hyperoxia (Fig. 15.5, *G. cobitis*). Hyperventilation occurs as the PO_2 decreases below normoxia and ventilatory frequency rises by nearly 150% (Berschick *et al.*, 1987). In *B. pholis*, on exposure to hypoxia in the laboratory a 14–19% increase in ventilatory frequency occurred, compared with 100% in field experiments (Bridges, 1988). Laming *et al.* (1982) and Pelster (1985) report similar changes in the same species. Innes and Wells (1985) found a fourfold change in the ventilatory rate of *Helcogramma medium* on exposure to hypoxia. The magnitude of the frequency response is dependent upon changes in stroke volume, which appears not to be altered in intertidal fish (Berschick *et al.*, 1987).

Cardiac responses to hypoxia normally involve a bradycardia, and Bandurski *et al.* (1968) found a bradycardia in the mudskipper exposed to anoxia. In *B. pholis*, heart rate decreased by 50% with a corresponding decrease in ventral aortic blood flow and blood pressure of 25% and 10% respectively (Eickelmann, 1991). In *G. cobitis* no change in heart rate occurs with hypoxia because heart rate is already low (Fig. 15.5). This may be adaptive in that *G. cobitis*, living in high-shore rockpools (Gibson, 1972), is often exposed to hypoxia, and maintaining a constant heart rate may be energetically more efficient than increasing heart rate or stroke volume.

Hyperventilation may cause an increase in blood pH under hypoxia. In *B. pholis* exposed to hypoxia, pHv increases. As shown earlier (Bridges, 1988), a left shift of the oxygen dissociation curve through the Bohr effect could be advantageous in helping oxygen loading and maintaining the venous reserve. Hence $P\mathrm{v}O_2$ values are kept high at around 10 Torr despite hypoxic (30 Torr) water values. However, hypoxia occurs relatively seldom alone and is accompanied by hypercapnia (Fig. 15.1; Tables 15.1 and 15.2). Under these conditions, blood pH remains constant or decreases, thereby initiating a right shift of the curve. Temperature and organic phosphate level changes may then become more important for any compensatory left shift of the oxygen dissociation curve.

Response to aerial exposure

Sayer and Davenport (1991) showed that a large number of factors associated with degradation of the aquatic habitat or biotic factors within it can cause fish to leave the water. In closed systems such as rockpools, physico-chemical factors may play a more dominant role. Intertidal fish differ from air-breathing freshwater fish in that oxygen consumption rates in air are normally lower than in water (Bridges, 1988). This may be due to decreased activity in air or the lack of accessory air-breathing organs as in freshwater air-breathing fish. In facultative air breathers such as *B. pholis*, a value of around unity has been observed for the aerial/aquatic oxygen consumption ratio (Fig. 15.6(a) – cf. Pelster *et al.*, 1988 and Steeger, 1991). Similar results have been obtained for various mudskippers (Fig. 15.6(b) – cf. Gordon *et al.*, 1969, 1978 and Steeger, 1991). In more aquatic species, where terrestrial excursions are shorter, values much lower than unity have been observed (Bridges, 1988; Edwards and Cech, 1990; Brown *et al.*, 1991; Martin, 1991). In most air-breathing freshwater species the respiratory exchange ratio (RER = CO_2 release/O_2 uptake) remains between 0.2 and 0.5 (Randall *et al.*, 1981). This may be due to the inefficiency of the accessory gas exchange organs to excrete carbon dioxide in these species. In most intertidal species so far studied, the RER remains high, between 0.7 and 0.9 in air, and intertidal fish are well adapted

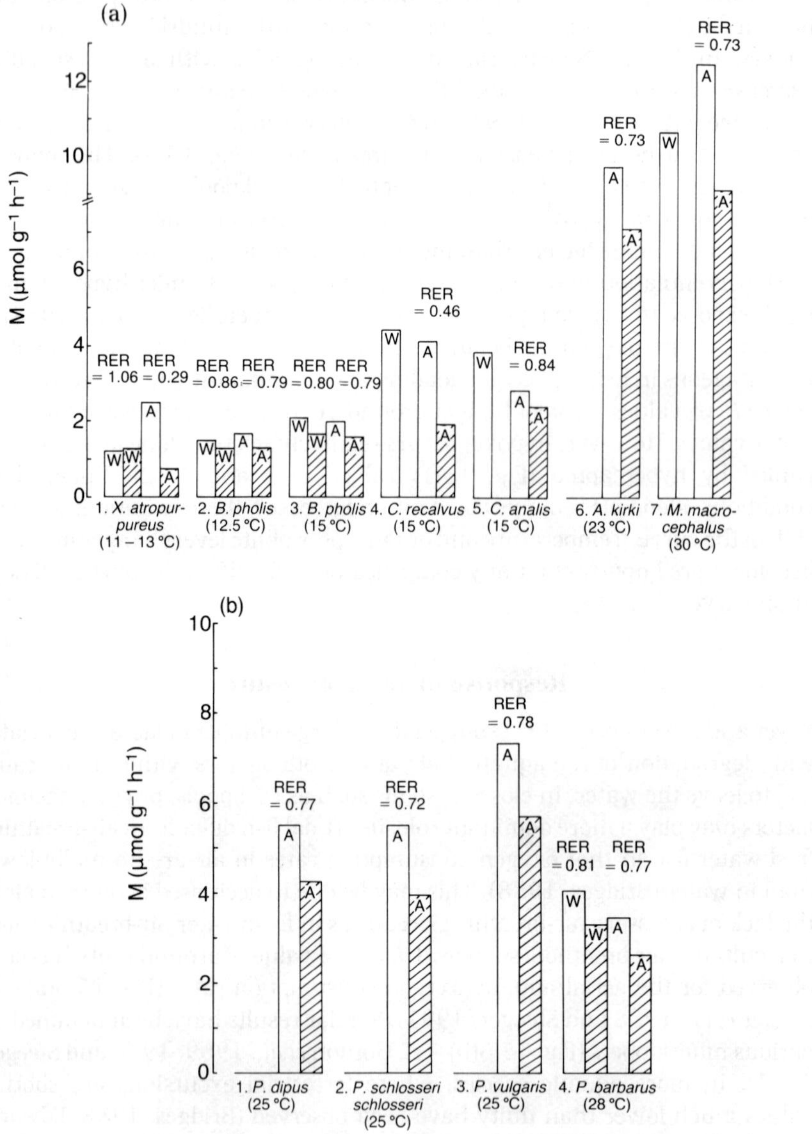

Fig. 15.6 Aquatic versus aerial gas exchange, M μmol g^{-1} h^{-1} in intertidal fish. (a) Rocky shore fish values taken from (1) Daxboeck and Heming (1982), (2) Pelster *et al.* (1988), (3) Steeger (1991), (4) Wright and Raymond (1978), (5) Martin (1991), (6) Martin and Lighton (1989), (7) Graham (1973). (b) Mudskippers: (1–3) Schöttle (1932), (4) Steeger (1991). Open bars, M_{O_2}; hatched bars, M_{CO_2}; A, air; W, water. For explanation of RER (respiratory exchange ratio), see text.

to continue both oxygen uptake and carbon dioxide excretion in air (Fig. 15.6).

No evidence has been found of lactate acid accumulation in blood or muscle of species aerially exposed (Gordon *et al.*, 1969, 1970, 1978; Riegle, 1976; Pelster *et al.*, 1988; Edwards and Cech, 1990; Martin, 1991).

Cutaneous gas exchange plays an important role in air-breathing intertidal fish (Graham, 1973; Tamura *et al.*, 1976), accounting for 40–52% of the total respiration in air in some species. Pelster *et al.* (1988) found that aerial gas exchange was equally divided between head/gills and the tail region in *B. pholis* with similar RER values. Martin (1991) found that in *Clinocottus analis*, which is not well adapted for long periods of aerial respiration, 71–78% of oxygen consumption or carbon dioxide excretion took place through the gills (RER = 0.92) and only 29% or 22% of oxygen and carbon dioxide exchange across the skin (RER = 0.66). Nonnotte and Kirsch (1978) have demonstrated that cutaneous respiration increased in *B. pholis* exposed to air, and Tamura *et al.* (1976) report a 30% increase in cutaneous oxygen uptake in *P. cantonensis* during aerial exposure, indicating some form of control in these species. Graham (1976) has reviewed cardiac and ventilatory responses to aerial exposure in marine fish. A typical response is 'gulping' and the retention of air in the buccopharyngeal region (Laming *et al.*, 1982; Graham *et al.*, 1985; Pelster *et al.*, 1988). Laming *et al.* (1982) also suggested that *B. pholis* my use the oesophagus as a site of gas exchange. Although Pelster *et al.* (1988) were unable to find increased carbonic anhydrase activity for CO_2 exchange in this region of the oesophagus, Jermann (1987), in a morphological study, has shown that the surface area of the oesophagus has been increased by folding in the caudal part. A high capillary density and separate circulation, together with a special sphincter for closing off the oesophagus from the rest of the digestive tract, also suggest a respiratory function.

Cardiac responses to aerial exposure are well documented. Clingfish show a 'diving syndrome' with a bradycardia on air exposure (Gordon *et al.*, 1970). *B. pholis* shows a transient bradycardia on exposure to air, with a rapid recovery to aquatic levels (Daniel, 1971; Laming *et al.*, 1982; Pelster *et al.*, 1988). Recently Eickelmann (1991) confirmed that dorsal aortic blood pressure and blood flow did not change significantly from aquatic levels on aerial exposure in *B. pholis*. Microsphere infusion studies on the gills indicated no difference in gill perfusion during aerial exposure, but the perfusion of the skin increased. *C. violaceus* does not show a bradycardia on exposure to air, but an initial tachycardia is observed with a gradual decrease over the emersion period (Riegle, 1976). Heart rate in the mudskipper *P. sobrinus* was unaffected by shifts between air and water (Gordon *et al.*, 1969), but Garey (1962) reported a bradycardia in *Periophthalmadon australis* exposed to water and much lower rates when the fish was in its burrow.

In conclusion, it is clear that intertidal fish show many adaptations in their thermoregulatory osmoregulatory and respiratory physiology. Many features, however, remain to be studied in more detail, including evaporative water loss and the full mechanisms of osmoregulation in terms of permeability control, active ion transport and perhaps morphological adaptations. Respiratory regulation of cutaneous gas exchange must also be elucidated. This task is made more difficult by the small size of intertidal fish, but they will continue to provide many interesting insights into adaptations to an extreme environment.

ACKNOWLEDGEMENTS

I should like to thank the many colleagues who provided reprints and pre-prints of their work for this review. Ms. A. Köster provided technical assistance and the Marine Biological Station, Roscoff provided support services for our own work. Financial support was provided by the Deutsche Forschungsgemeinschaft. The editors are to be thanked for their helpful advice and patience in organizing the symposium and publication.

REFERENCES

Bandurski, R.S., Bradstreet, E. and Scholander P.F. (1968) Short communication: metabolic changes in the mud-skipper during asphyxia or exercise. *Comp. Biochem. Physiol.*, **24**, 271–4.

Barton, M. (1979) Serum osmoregulation in two species of estuarine blennioid fish, *Anoplarchus purpurescens* and *Pholis ornata. Comp. Biochem. Physiol.*, **64A**, 305–7.

Berschick, P., Bridges, C.R. and Grieshaber, M. K. (1987) The influence of hyperoxia, hypoxia and temperature on the respiratory physiology of the intertidal rockpool fish *Gobius cobitis* Pallas. *J. exp. Biol.*, **130**, 369–87.

Bridges, C.R. (1988) Respiratory adaptations in intertidal fish. *Am. Zool.*, **28**, 79–96.

Bridges, C.R. Taylor, A.C., Morris, S.J. and Grieshaber, M.K. (1984) Ecophysiological adaptations in *Blennius pholis* (L.) blood to intertidal rockpool environments. *J. exp. mar. Biol. Ecol.*, **77**, 151–67.

Brown, C.R., Gordon, M.S. and Chin, H.G. (1991) Field and laboratory observations on microhabitat selection in the amphibious Red Sea rockskipper fish *Alticus kirki* (Family Blenniidae). *Mar. Behav. Physiol.*, **19**, 1–13.

Bull, H.O. (1936) Studies on conditioned responses in fishes. Temperature perception in teleosts. *J. mar. biol. Ass. U.K.*, **21**, 1–27.

Bull, H.O. (1938) Studies on conditioned responses in fish. Part VIII. Discrimination of salinity changes by marine teleosts. *Rep. Dov. mar. Lab.*, **3(5)**, 19–35.

Campbell, C.M. and Spencer-Davies, P. (1975) Thermal acclimation in the teleost, *Blennius pholis* (L.). *Comp. Biochem. Physiol.*, **52A**, 147–51.

Chew, S.F. and Ip, Y.K. (1990) Differences in the responses of two mudskippers, *Boleophthalmus boddaerti* and *Periophthalmus chrysospilos* to changes in salinity. *J. exp. Zool.*, **256**, 227–31.

Chew, S.F., Lim A.L.L., Low, W.P., Lee, C.G.L., Chan, K.M. and Ip, Y.W. (1990) Can the mudskipper, *Periophthalmus chrysospilos*, tolerate acute environment hypoxic exposure? *Fish Physiol. Biochem.*, **8**(3), 221–7.

Congleton, J.L. (1974) The respiratory response to asphyxia of *Typhlogobius californiensis* (Teleostei: Gobiidae) and some related gobies. *Biol. Bull. mar. biol. Lab., Woods Hole*, **146**, 186–205.

Congleton, J.L. (1980) Observations on the responses of some southern California tidepool fishes to nocturnal hypoxic stress. *Comp. Biochem. Physiol.*, **66A**, 719–22.

Daniel, M.J. (1971) Aspects of the physiology of the intertidal teleost, *Blennius pholis* (L.). Ph D thesis, University of London., (227 pp.)

Daniel, M.J. and Boyden, C.R. (1975) Diurnal variations in physico-chemical conditions within intertidal rockpools. *Field Stud.*, **4**, 161–76.

Davenport, J. and Vahl, O. (1979) Responses of the fish *Blennius pholis* to fluctuating salinities. *Mar. Ecol. Prog. Ser.*, **1**, 101–7.

Daxboeck, C. and Heming, T.A. (1982) Bimodal respiration in the intertidal fish, *Xiphister atropurpureus* (Kittlitz). *Mar. Behav. Physiol.*, **9**, 23–33.

Dejours, P. (1973) Problems of control of breathing in fishes, in *Comparative Physiology: Locomotion Respiration Transport and Blood* (eds L. Bolis, W. Schmidt-Nielsen and S.H.P. Maddrett), North/Holland Elsevier, Amsterdam, pp. 117–33.

Dejours, P., Toulmond, A. and Truchot, J.-P. (1977) The effect of hyperoxia on the breathing of some marine fishes. *Comp. Biochem. Physiol.*, **58A**, 409–11.

Doudoroff, P. (1938) Reactions of marine fishes to temperature gradients. *Biol. Bull. mar. biol. Lab., Woods Hole*, **75**, 494–509.

Doudoroff, P. (1942) The resistance and acclimatization of marine fishes to temperature changes I. Experiments with *Girella nigricans* (Ayres). *Biol. Bull. mar. biol. Lab., Woods Hole*, **83**, 219–44.

Duman, J.G. and DeVries, A.L. (1974) The effects of temperature and photoperiod on antifreeze production in cold water fishes. *J. Exp. Zool.*, **190**, 89–98.

Edwards, D.G. and Cech, J. J., jun. (1990) Aquatic and aerial metabolism of juvenile monkeyface prickleback, *Cebidichthys violaceus*, an intertidal fish of California. *Comp. Biochem. Physiol.*, **96A**, 61–5.

Eger, W.H. (1971) Ecological and physiological adaptations of intertidal clingfishes (Teleostei: Bobiesocidae) in the northern Gulf of California. Ph D thesis, University of Arizona. (210 pp.)

Eickelmann, P. (1991) Zur Kreislaufphysiologie des Felstümpelfisches *Blennius pholis* unter extremen Bedingungen. *Diplomarbeit*, University of Düsseldorf. (77 pp.)

El-Ziady, S., Eissa, S.M. and Al-Naqi, Z. (1979) Ecological studies on the mudskipper fish, *Periophthalmus chrysospilos* (Bleeker) in Kuwait. *Bull. Fac. Sci. Cairo Univ.*, **48**, 113–33.

Evans, D.H. (1967a) Sodium, chloride and water balance of the intertidal teleost, *Xiphister atropurpureus* I. Regulation of plasma concentration and body water content. *J. exp. Biol.*, **47**, 513–17.

Evans, D.H. (1967b) Sodium, chloride and water balance of the intertidal teleost, *Xiphister atropurpureus* II. The role of the kidney and the gut. *J. exp. Biol.*, **47**, 519–24.

Evans, D.H. (1967c) Sodium, chloride and water balance of the intertidal teleost, *Xiphister atropurpureus* III. The roles of simple diffusion, exchange diffusion, osmosis and active transport. *J. exp. Biol.*, **47**, 525–34.

Evans, D.H. (1969) Sodium, chloride and water balance of the intertidal teleost, *Pholis gunnellus*. *J. exp. Biol.*, **50**, 179–90.

Evans, D.H. (1984) The roles of gill permeability and transport mechanisms in euryhalinity, in *Fish Physiology*, Vol. X (eds W.S. Hoar and D.J. Randall), Academic Press, London, pp. 239–83.

Foster, M.A. (1969) Ionic and osmotic regulation in three species of *Cottus* (Cottidae, Teleost). *Comp. Biochem. Physiol.*, **30**, 751–9.

Ganning, B. (1971) Studies on chemical, physical and biological conditions in Swedish rockpool ecosystems. *Ophelia*, **9**, 51–105.

Garey, W.F. (1962) Cardiac responses of fishes in asphyxic environments. *Biol. Bull. mar. biol. Lab., Woods Hole*, **122**, 362–8.

Gibson, R.N. (1969) The biology and behaviour of littoral fish. *Oceanogr. Mar. Biol. Ann. Rev.*, **7**, 367–410.

Gibson, R.N. (1982) Recent studies on the biology of intertidal fishes. *Oceanogr. Mar. Biol. Ann. Rev.*, **20**, 363–414.

Gordon, M.S., Boëtius, I., Evans, D.H., McCarthy, R. and Oglesby, L.C. (1969) Aspects of the physiology of terrestrial life in amphibious fishes I. The mudskipper, *Periophthalmus sobrinus. J. exp. Biol.*, **50**, 141–9.

Gordon, M.S., Boëtius, J., Boëtius, I., Evans, D.H., McCarthy, R. and Oglesby, L.C. (1965) Salinity adaptation in the mudskipper fish *Periophthalmus sobrinus. Hvalradets Skr.*, **48**, 85–93.

Gordon, M.S., Fischer, S. and Tarifenos, E. (1970) Aspects of the physiology of terrestrial life in amphibious fishes II. The Chilean clingfish, *Sicyases sanguineus. J. exp. Biol.*, **53**, 559–72.

Gordon, M.S., Gabaldon, D.J. and Yip, A.Y. (1985) Exploratory observations on microhabitat selection within the intertidal zone by the Chinese mudskipper fish *Periophthalmus cantonensis. Mar. Biol.*, **85**, 209–15.

Gordon, M.S. Ng, W.-S. and Yip, A.Y.-W. (1978) Aspects of the physiology of terrestrial life in amphibious fishes. III. The Chinese mudskipper *Periophthalmus cantonensis. J. exp. Biol.*, **72**, 57–77.

Graham, J.B. (1970) Temperature sensitivity of two species of intertidal fishes. *Copeia*, **1970**, 49–56.

Graham, J.B. (1973) Terrestrial life of the amphibious fish *Mnierpes macrocephalus. Mar. Biol.*, **23**, 83–91.

Graham, J.B. (1976) Respiratory adaptations of marine air-breathing fishes, in *Respiration of Amphibious Vertebrates* (ed. G.M. Hughes), Academic Press, New York, pp. 165–87.

Graham, J.B., Jones, C.B. and Rubinoff, I. (1985) Behavioural, physiological, and ecological aspects of the amphibious life of the pearl blenny *Entomacrodus nigricans* Gill. *J. Exp. Mar. Biol. Ecol.*, **98**, 225–68.

Gunter, G. (1967) Vertebrates in hypersaline waters. *Contr. mar. Sci. Univ. Tex.*, **12**, 230–41.

Horn, M.H. and Gibson, R.N. (1988) Intertidal fishes. *Scient. Am.* **258**, 54–60.

Horn, M.H. and Riegle, K.C. (1981) Evaporative water loss and intertidal vertical distribution in relation to body size and morphology of stichaeoid fishes from California. *J. Exp. mar. Biol. Ecol.*, **50**, 273–88.

House, C.R. (1963) Osmotic regulation in the brackish water teleost, *Blennius pholis. J. exp. Biol.*, **40**, 87–104.

Innes, A.J. and Wells, R.M.G. (1985) Respiration and oxygen transport functions of the blood from an intertidal fish, *Helcogramma medium* (Tripterygiidae). *Env. Biol. Fishes*, **14**, 213–26.

Ip, Y.K., Chew, S.F., and Low, W.P. (1991) Effects of hypoxia on the mudskipper, *Periophthalmus chrysospilos* (Bleeker, 1853). *J. Fish Biol.*, **38**, 621–3.

Jermann, T. (1987) Zur amphibischen Lebensweise von (*Blennius (Lipophrys) pholis* (Linne, 1758). Diplomarbeit, University of Basel. (84 pp.)

Klugh, A.B. (1924) Factors controlling the biota of tide-pools. *Ecology*, **5**, 192–6.

Kunz, Y. (1969) Salinity preference and salinity tolerance of the katadromous teleost *Pomatoschistus (Gobius) Microps* Kroyer. *Proc. Roy. Irish Acad.*, **68**, 77–88.

Laming, P.R., Funston, C.W., Roberts, D. and Armstrong, M.J. (1982) Behavioural, physiological and morphological adaptations of the shanny (*Blennius pholis*) to the intertidal habitat. *J. mar. biol. Ass. U.K.*, **62**, 329–38.

Lee, C.G.L., Low, W.P. and Ip, Y.K. (1987) Na^+, K^+ and volume regulation in the mudskipper, *Periophthalmus chrysospilos*. *Comp. Biochem. Physiol.*, **87A**, 439–48.
Martin, K.L.M. (1991) Facultative aerial respiration in an intertidal sculpin, *Clinocottus analis* (Scorpaeniformes: Cottidae). *Physiol. Zool.*, **64**, 1341–55.
Martin, K.L.M. and Lighton, J.R.B. (1989) Aerial CO_2 and O_2 exchange during terrestrial activity in an amphibious fish, *Alticus kirki* (Blenniidae). *Copeia*, **1989**(3), 723–7.
Marusic, E.T., Balbontin, F., Galli-Gallardo, S.M., Garreton, M., Pang, P.K.T. and Griffith, R.W. (1981) Osmotic adaptations of the Chilean clingfish, *Sicyases sanguineus*, during emersion. *Comp. Biochem. Physiol.*, **68A**, 123–6.
Milton, P. (1971) Oxygen consumption and osmoregulation in the shanny, *Blennius pholis*. *J. Mar. biol. Ass. U.K.*, **51**, 247–65.
Morris, R.W. (1960) Temperature, salinity, and southern limits of three species of Pacific cottid fishes. *Limnol. Oceanogr.*, **5**, 175–9.
Morris, R.W. (1961) Distribution and temperature sensitivity of some eastern Pacific cottid fishes. *Physiol. Zool.*, **34**, 217–27.
Morris, S. and Taylor, A.C. (1983) Diurnal and seasonal variation in physico-chemical conditions within intertidal rock pools. *Est. Coast. Shelf Sci.*, **17**, 339–55.
Müller, R., Böke, K., Martin-Neuhaus, U. and Hanke, W. (1973) Salzwasseradaption und Stoffwechsel bei Fischen. *Fortschr. Zool.*, **32**, 456–67.
Nakamura, R. (1976) Temperature and the vertical distribution of two tidepool fishes (*Oligocottus maculosus, O. snyderi*). *Copeia* **1976**, 143–52.
Nonnotte, G. and Kirsch, R. (1978) Cutaneous respiration in seven seawater teleosts. *Respir. Physiol.*, **35**, 111–18.
Norris, K.S. (1963) The functions of temperature in the ecology of the percoid fish *Girella nigricans* (Ayres). *Ecol. Monogr.*, **33**, 23–62.
Pelster, B. (1985) Mechanismen der Anpassung an das Leben in extremen Biotopen: Vergleichende Studien zur Atmungsphysiologie bei *Lumpenus lampretaeformis* und *Blennius pholis*. Ph D thesis, University of Düsseldorf. (168 pp.)
Pelster, B., Bridges, C.R. and Grieshaber, M.K. (1988) Physiological adaptations of the intertidal rockpool teleost *Blennius pholis* L., to aerial exposure. *Respir. Physiol.*, **71**, 355–74.
Pyefinch, K.A. (1943) The intertidal ecology of Bardsey Island, North Wales, with special reference to the recolonization of rock surfaces and the rockpool environment. *J. Anim. Ecol.*, **12**, 82–108.
Ramachandran Nair, P.V., Luther, G. and Adolph, C. (1965) An ecological study of some pools near Mandapam (South India) formed as a result of the cyclone and tidal wave of 1964. *J. mar. biol. Ass. India*, **7**, 420–39.
Randall, D.J., Burggren, W.W., Farrell, A.P. and Haswell, M.S. (1981) *The Evolution of Air Breathing in Vertebrates*, Cambridge Univ. Press, London, pp. 1–133.
Riegle, K.C. (1976) Oxygen consumption, heart rates, whole body lactate levels, and evaporative water loss in the monkeyface eel *Cebidichthys violaceus* (Family: Stichaeidae) an amphibious marine fish from California. MA thesis, California State University. (72 pp.)
Sayer, M.D.J. and Davenport, J. (1991) Amphibious fish: why do they leave water? *Rev. Fish Biol. Fish.*, **1**, 159–81.
Schöttle, E. (1932) Morphologie und Physiologie der Atmung bei wasser-, schlamm- und landlebenden Gobiiformen. *Z. wiss. Zool.*, **140**, 1–114.
Steeger, H.-U. (1991) Untersuchungen zum Gasaustausch bei Litoralfischen. Diplomarbeit, Univ. of Düsseldorf. (49 pp.)
Stephenson, T.A., Zoond, A. and Eyre, J. (1934) The liberation and utilisation of oxygen by the population of rock-pools. *J. exp. Biol.*, **11**, 162–72.

Tamura, S.O., Morii, H. and Yuzuriha, M. (1976) Respiration of the amphibious fishes *Periophthalmus chinensis* in water and on land. *J. exp. Biol.*, **65**, 97–107.

Tolksdorf, W. (1978) Der Einfluβ von Salzgehalt und Temperatur auf Wachstum, Stoffwechsel und Letaltemperatur der Strandgrundel *Pomastoschistus microps* K. *Meeresforschung*, **26**, 15–29.

Truchot, J.P. (1988) Problems of acid–base balance in rapidly changing intertidal environments. *Am. Zool.*, **28**, 55–64.

Truchot, J.P. and Duhamel-Jouve, A. (1980) Oxygen and carbon dioxide in the marine intertidal environment: diurnal and tidal changes in rockpools. *Respir. Physiol.*, **39**, 241–54.

Tytler, P. and Vaughan, T. (1983) Thermal ecology of the mudskippers, *Periophthalmus koelreuteri* (Pallas) and *Boleophthalmus boddarti* (Pallas) of Kuwait Bay. *J. Fish Biol.*, **23**, 327–37.

Wilkes, P.R.H., Walker, R.L., McDonald, D.G. and Wood, C.M. (1981) Respiratory, ventilatory, acid–base and ionoregulatory physiology of the white sucker *Catostomus commersoni*: the influence of hyperoxia. *J. exp. Biol.*, **91**, 239–54.

Wright, W.G. and Raymond, J.A. (1978) Air-breathing in a California sculpin. *J. Exp. Zool.*, **203**, 171–6.

Zander, C.D. (1972) Beiträge zur Ökologie und Biologie von Blenniidae (Pisces) des Mittelmeeres. *Helgoläuder. wiss. Meeresunters.*, **23**, 193–231.

Author index

Note to Nordic readers: this index is arranged according to the English alphabet. Å, Ä and Æ are included under A; Ø and Ö under O

Subject index